AF656428

RECENT ADVANCES IN DEVELOPMENTAL BIOLOGY

Recent Advances in Developmental Biology

Vol. 1
Principles of Developmental Biology

A.P. DIWAN
N.K. DHAKAD

ANMOL PUBLICATIONS PVT LTD
New Delhi-110 002

ANMOL PUBLICATIONS PVT LTD
4374/4B, Ansari Road
Daryaganj, New Delhi-110 002

Recent Advances in Developmental Biology
© Reserved
First Edition 1995
ISBN 81-7488-039-9

PRINTED IN INDIA

Published by J.L. Kumar for Anmol Publications Pvt. Ltd. and
Printed at Mehra Offset Press, Delhi.

Preface

The fundamentals of Developmental Biology are presented within the framework of scientific discovery. Researches in Embryology have made almost incredible strides in the past few decades. Consequently, the existing concepts of the developmental biology have expanded. There has been a revolution indeed in this direction.

The text integrates the descriptive, experimental and biochemical approches into a conceptual framework for the analysis of development. All important points are illustrated diagrammatically. The title is not intended to be comprehensive, nor could it be at this length, but concentrates as putting across the basic principles of the subject as briefly and lucidly as possible. It does this with the aid of careful selected examples, some recent and other classic of the field and with numerous illustrations. The aim is to enthuse the reader with this active and exciting area of research and to lay a solid foundation on which further study of its various facts may be based.

Authors

Contents

	Preface	*v*
1.	Cleavage and Blastulation	1
2.	Gastrulation	67
3.	Basic Embryonic Processes	112
4.	Fate of Germ Layers	158
5.	Mesodermal Derivatives in Vertebrates	307
6.	Endodermal Derivatives in Vertebrates	414
7.	Metamorphosis	456
8.	Differentiation of Sexes	491
9.	Placentation in Mammals	508

1

Cleavage and Blastulation

The development of a normal, adult animal from a zygote involves a series of distinct steps in which specific tasks are accomplished, *cleavage,* the mitotic division of the egg, is the first step after fertilization and is a universally occurring phenomenon among metazoan animals. A series of rapid cell divisions enables a fertilized egg to develop into a complex multicellular animal, perfect in every detail. Penetration of the plasma membrane by the sperm normally activates the egg and stimulates it to start dividing. Fertilization removes the block to cellular division that existed in the egg prior to fertilization, and initiates a type of mitosis.

Cleavage is characterized by the following events :

(i) The unicellular fertilized egg is transformed by consecutive divisions into a heap of cells called *morula,* which by further divisions and arrangement of cells, ultimately transformed into *blastula,* having single layered *blastoderm* around a cavity called the *blastocoel.*

(ii) No growth of the embryo occurs because there is an increasing number of cells of progressively decreasing size.

(iii) There is chemical conversion of reserve food (yolk, glycogen etc.) into active cytoplasm and the active cytoplasmic substances into the molecules of nuclear substances like DNA, RNA and nucleoproteins. However, the qualitative changes in the chemical composition of the egg are limited.

(iv) The general shape of the embryo does not change.

(v) The constituent parts of ooplasm remain unaltered during cleavage i.e., organization of ooplasm of a fertilized egg remains the same in a cleaved egg.

(vi) The ratio of nucleus to cytoplasm is very low at the beginning of cleavage but brought to the level found in ordinary somatic cells at the end of cleavage. This is because the nuclear material (DNA and accompanying proteins) is replicated after each cell division, but the cytoplasm does not increase.

Peculiarities of Cell Division Involved in Cleavage

The cleavage of the fertilized egg is initiated by the division of nucleus *(karyokinesis)* and is followed by the division of cytoplasm *(cytokinesis)*, so that the egg cell divides into two daughter cells, termed the *cleavage cells* or the *blastomeres*. The first two blastomeres divide again and again, thus producing 4, 8, 16, 32 blastomeres and so on. The first few cleavages occur simultaneously in all the blastomeres but sooner or later the synchronization is lost and blastomeres divide at different times, independent of one another.

Though, the division of blastomeres is essentially a typical mitosis as it occurs in adult animals. But the cell divisions in cleavage and in later stages of development as well in the adult organism differ drastically in the following aspects :

(i) In the later embryonic stages and in the adult, after each division the daughter cell grow and when they are approximately doubled in size, they divide again. The cells thus maintaining an average size in every type of tissue. During cleavage, on the other hand, a blastomere does not increase in size before the next division begins. Thus, with each division the resulting blastomeres are only half the original size. It concludes that cleavage begins with one very large cell and ends with a great number of cells which are usually even smaller than most cells of the adult tissue.

(ii) The nuclei of early cleavage cells are considerably larger than they are in ordinary somatic cells of the same animal due to the presence of larger amounts of nuclear sap.

(iii) The duration of interphase between two successive divisions remains very short in mitosis involved in cleavage.

Chemical Changes During Cleavage

Although there is no growth during the period of cleavage, chemical transformations do go on and at least some are markedly intensified, as compared with the conditions in the unfertilized egg. The most obvious change observed during cleavage is a steady increase of nuclear material at the expense of cytoplasm. The number of nuclei is of course doubled with every new division of the blastomeres, and this increase is accompanied by an increase of nuclear substance, which involves an increase of deoxyribonucleic acid — the amount per nucleus of the latter remaining constant.

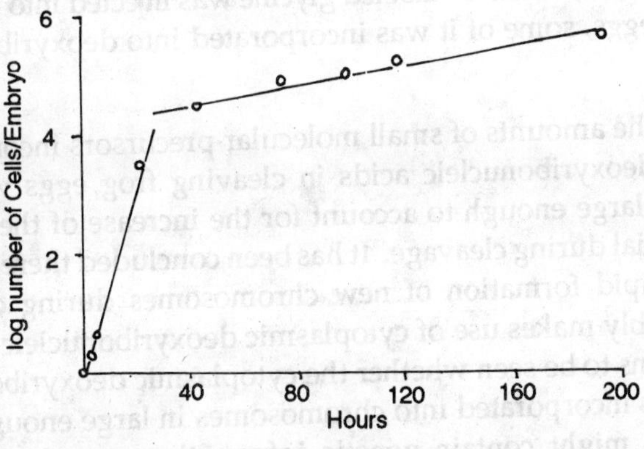

Fig. 1.1. Increase of number of cells during early development of the frog's egg.

The increase of the chromosomal deoxyribonucleic acid, at least during the earlier phases of development, must be at the expense of materials contained in the egg. There are several possible sources of such materials. In the first place, one should mention the nucleic acids present in the cytoplasm of the eggs. In the sea urchin eggs there is a large amount of ribonucleic acid in the egg cytoplasm, and this gradually disappears later in development. Of particular interest is the presence of deoxyribonucleic acid in the egg cytoplasm of many animals. Some of the cytoplasmic DNA is contained in the mitochondria but some is in the yolk platelets and would be liberated during the breakdown of the platelets, and become available for the building up of the chromosomes. DNA may, however, be synthesized in the cleaving egg directly from low molecular weight precursors. This has been proved by supplying such precursors, labeled with radioactive atoms, to cleaving eggs of sea urchins and amphibians. When cleaving sea urchin eggs were kept in sea water containing C^{14} labeled glycine (which may be used for the synthesis of purine groups in the nucleic acid molecule), it was found that the radioactive carbon atoms were incorporated in large amounts into the deoxyribonucleic acid, bypassing the cytoplasmic ribonucleic acid. Also, when C^{14} labeled glycine was injected into fertilized frog eggs, some of it was incorporated into deoxyribonucleic acid.

The amounts of small molecular precursors incorporated into deoxyribonucleic acids in cleaving frog eggs were not quite large enough to account for the increase of the nuclear material during cleavage. It has been concluded therefore that the rapid formation of new chromosomes during cleavage probably makes use of cytoplasmic deoxyribonucleic acid. It remains to be seen whether the cytoplasmic deoxyribonucleic acid is incorporated into chromosomes in large enough units which might contain genetic information, or whether it is broken down into mononucleotides before incorporation. So far there is no satisfactory evidence that the cytoplasmic deoxyribonucleic acid carries genetic information which is incorporated into the chromosomes.

The second important aspect of metabolism during cleavage is the synthesis of ribonucleic acids. This is believed to be very limited, although not absent altogether. In frogs ribosomal

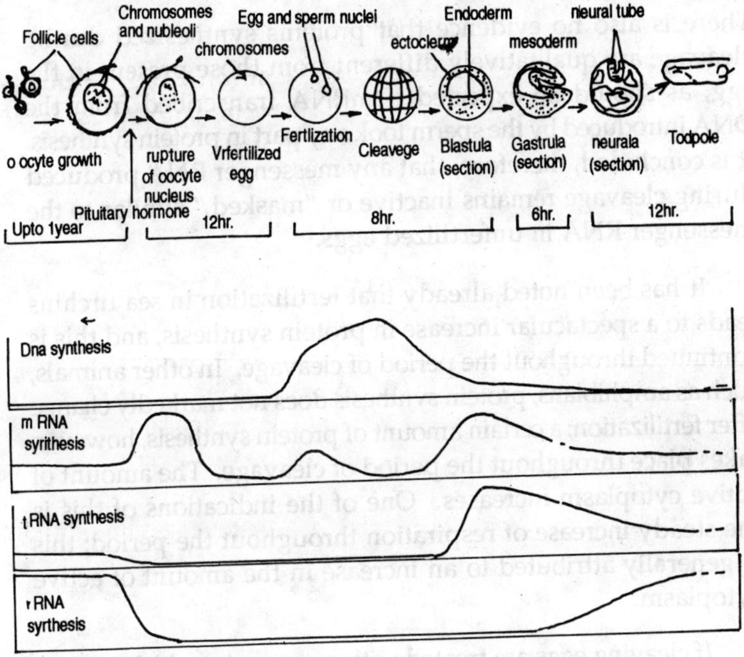

Fig. 1.2. **Changes in the synthesis of various classes of nucleic acids in oogenesis, at fertilization, and during early development in the frog embryo.**

RNA apparently is not produced at all until after completion of cleavage. As the nucleolus is the site of synthesis of rRNA, this organoid is completely lacking in these animals during cleavage. It reappears in the nuclei at the onset of gastrulation simultaneously with the resumption of ribosomal RNA synthesis. In the sea urchin there is only very little ribosomal RNA produced during cleavage, but in both the amphibians and sea urchins synthesis increases drastically at the onset of gastrulation. Messenger RNA and transfer RNA on the other hand are synthesized during cleavage, or at least in the later stages of cleavage. Synthesis of RNA, however, does not seem to be necessary for cleavage, since eggs which are treated with actinomycin D and in which presumably DNA dependent RNA synthesis is suppressed, continue cleaving normally.

There is also no evidence that proteins synthesized during cleavage are qualitatively different from those present in the egg, as should be expected if mRNA transcribed from the DNA introduced by the sperm took any part in protein synthesis. It is concluded, therefore, that any messenger RNA produced during cleavage remains inactive or "masked," similar to the messenger RNA in unfertilized eggs.

It has been noted already that fertilization in sea urchins leads to a spectacular increase in protein synthesis, and this is continued throughout the period of cleavage. In other animals, such as amphibians, protein synthesis does not markedly change after fertilization; a certain amount of protein synthesis, however, takes place throughout the period of cleavage. The amount of active cytoplasm increases. One of the indications of this is the steady increase of respiration throughout the period; this is generally attributed to an increase in the amount of active cytoplasm.

If cleaving eggs are treated with puromycin, which inhibits RNA dependent protein synthesis, cleavage stops immediately, thus showing that protein synthesis is indispensable for cleavage to take place. This is in marked contrast to cleavage being able to proceed in the presence of actinomycin D which effectively prevents the production of new RNA and in particular of new messenger RNA.

It is not altogether clear what purpose is served by the newly synthesized protein. Possible uses for protein might be:

1. Formation of cytasters during cleavage; these, however, may be formed even in the presence of puromycin, but do not lead to cell division.

2. Formation of new cell membrane or cortical cytoplasm. The total amount of cell surface increases enormously while the single fertilized egg is subdivided into many small cells. It has not been estimated in what degree the total amount of

cortical cytoplasm increases in this process but it is fairly obvious that the amount of increase is far in excess of the amount of newly synthesized protein. The new cell membranes and underlying surface cytoplasm must thus be derived from preexisting materials in the egg.

3. It has been suggested that the newly synthesized proteins during cleavage are the enzymes DNA polymerases. Although only small quantities of these enzymes are needed, in their absence, duplication of chromosomal DNA cannot proceed, making cleavage impossible.

Inasmuch as any newly synthesized messenger RNA is not active during cleavage, any protein synthesis taking place is under the control of the messenger RNA produced in the egg (oocyte) before fertilization and thus presents characteristics of the maternal parent only.

Patterns of Cleavage

The way in which the egg is subdivided into the daughter blastomeres is usually very regular. The plane of the first division is as a rule vertical; it passes through the main axis of the egg. The plane of the second division is also vertical and passes through the main axis, but it is at right angles to the first plane of cleavage. The result is that the first four blastomeres all lie side by side. The plane of the third division is at right angles to the first two planes and to the main axis of the egg. It is therefore horizontal or parallel to the equator of the egg. Of the eight blastomeres four lie on top of the other four.

Planes of Cleavage

Cleavage is initiated by appearance of a groove or constrictions called *cleavage furrow*. The furrow appears first at one point of the egg. For example, in *Amphioxus* the first furrow appears at the animal pole. The furrow then deepens and extends downward on both sides. The two ends meet at the vegetal pole. The furrow then extends inwards radially, finally constricting the egg into two blastomeres. The cleavage furrows divide the egg at different angles or planes. There are four main planes of cleavage. They are as follows :

1. Meridional Plane

When the cleavage furrow passes through the centre of animal-vegetal axis and bisects the both poles of the egg, then such plane of cleavage is called *meridional plane, e.g., Rana pipiens, Lepidosteus osseus* and first cleavage furrow of chick.

2. Vertical Plane

The vertical plane of cleavage is a furrow which tends to pass in a direction from the animal pole toward the vegetal pole. But like meridional plane, it does not pass through the median axis of the egg, but courses to one side of this axis, e.g., Third cleavage furrows of *Amia calva, Lepidosteus osseus* and chick.

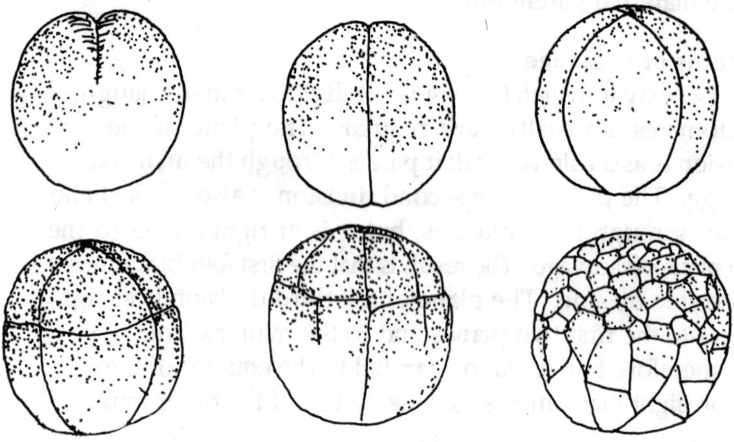

Fig. 1.3. Cleavage in the frog, semidiagrammatic.

3. Equatorial Plane

The equatorial plane of cleavage bisects the egg at right angles to the main axis and half way between the animal and vegetal poles, e.g., Fifth cleavage planes of the eggs of *Ambystoma maculatum,* and first cleavage plane of eggs of higher mammals.

4. Latitudinal Plane

The latitudinal plane of cleavage is similar to the equatorial, but it courses through the cytoplasm on either side of the equatorial plane. It is also called *transverse* or *horizontal plane, e.g.* Third cleavage planes of *Amphioxus* and frog.

PATTERNS OF CLEAVAGE

Most animal eggs incorporate an organizational pattern and this pattern play a far more important role in conditioning cleavage than does whatever amount of yolk may be present. The pattern of cleavage due to organization of egg may be of following types :

1. Radial Cleavage

In radial cleavage the cleavage furrows appear in such that each of blastomeres of the upper tier lies over the corresponding blastomeres of the lower tiers. When viewed from the poles, the blastomeres seem to be arranged on a radial plane. *E.g.*, sea-urchin, sea cucumber etc.

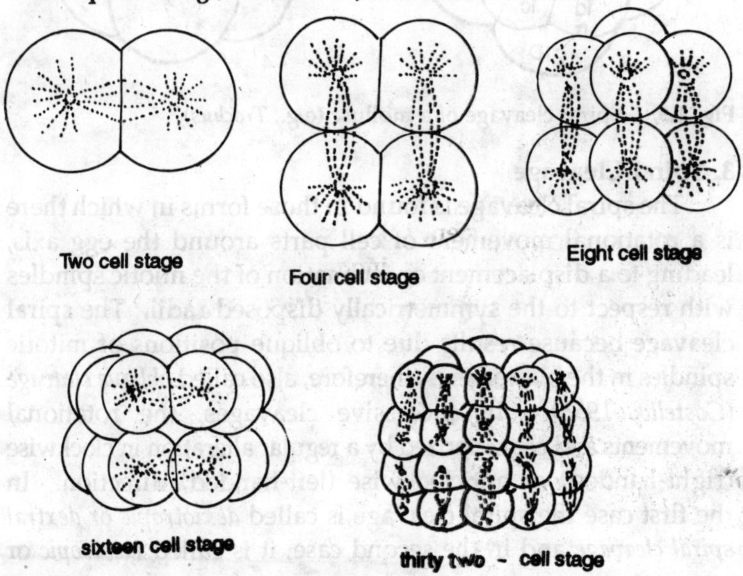

Fig. 1.4. Radial cleavage in sea urchin.

2. Biradial Cleavage

Biradial cleavage is seen only in ctenophora. The first two cleavage planes are moridional. The third cleavage plane is vertical resulting in the formation of a curved plate of cells arranged in two rows of 4 each with the central cells larger than the end ones.

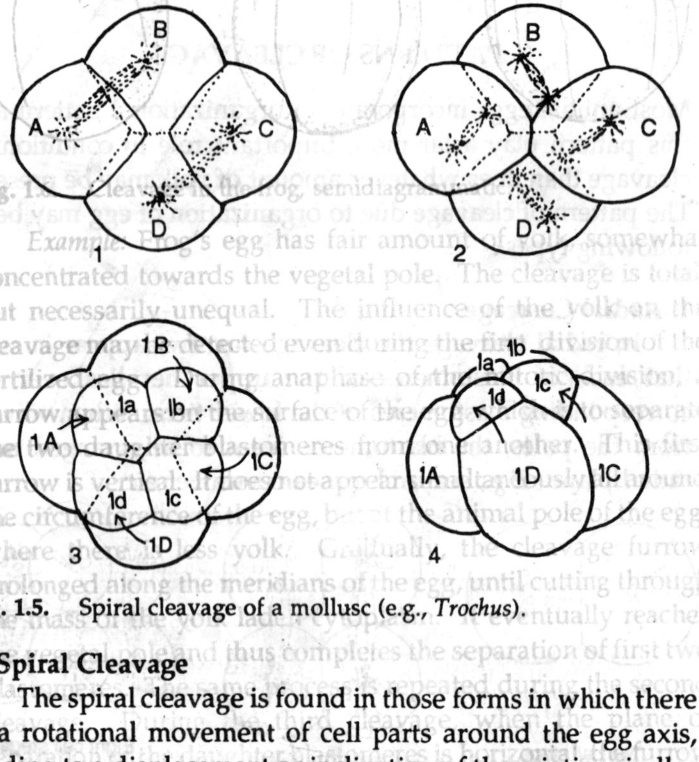

Fig. 1.5. Spiral cleavage of a mollusc (e.g., *Trochus*).

3. Spiral Cleavage

The spiral cleavage is found in those forms in which there is a rotational movement of cell parts around the egg axis, leading to a displacement or inclination of the miotic spindles with respect to the symmetrically disposed radii. The spiral cleavage because results due to oblique positions of mitotic spindles in the blastomeres, therefore, also called *oblique cleavage (Costello, 1955)*. In successive cleavages, the rotational movements and characterized by a regular alteration in clockwise (right-handed) or anticlockwise (left-handed) direction. In the first case the spiral cleavage is called *dexiotropic or dextral spiral cleavage* and in the second case, it is called *laeotropic* or *sinistral spiral cleavage,* e.g., Furbellaria, Rolifers, Armelida, and all molluscs except cephalopods.

4. Bilateral Cleavage

In bilateral cleavage, the mitotic spindles and cleavage planes remain bilaterally arranged with reference to a plane of symmetry which coincides with the medium plane of the embryo, e.g., Ctenophora, Tunicata, *Amphioxus,* Amphibia and higher mammals.

(a) Total or holoblastic cleavage

In holoblastic, total or complete cleavage, the entire egg divides by each cleavage furrow. It may be :

(i) Equal holoblastic cleavage

In microlecithal or isolecithal eggs, it produces blastomeres of approximately equal size e.g., *Amphioxus,* marsupials and placental mammals.

Example : Almost microscopic egg of *Amphioxus,* contains small amount of yolk somewhat concentrated at the vegetal pole. Thus, the cleavage is total and nearly equal. The plane of first cleavage passes through the egg axis from pole to pole. Soon the daughter cells again cleave in a vertical (meridional) plane, but at right angles to the first plane, thus forming four cells. In the third series of divisions the plane of separation is horizontal. As the yolk is comparatively more towards the vegetal pole, the four mitotic spindle lie nearer the animal pole : consequently, in the resulting eight celled stage, the four cells of upper tier are slightly smaller than the lower four. The continuation of practically synchronous cell divisions in alternate planes produce 16, 32, 64 and 128 celled stage. During this period the size of the cells is progressively diminished, while the central cavity filled with a watery jelly, enlarges.

(ii) Unequal holoblastic cleavage

In moderately telolecithal eggs, the yolk is accumulated at the vegetal pole which retards mitosis and a few but larger blastomeres are formed there. Smaller but large number of micromeres are formed at the animal pole, e.g., lower fishes and amphibians.

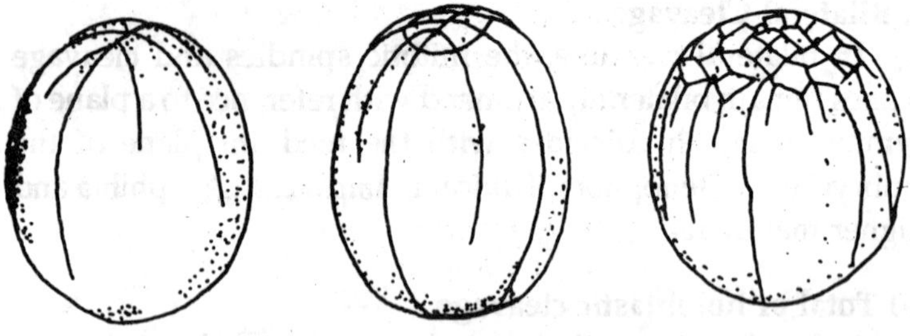

Fig. 1.6. Cleavage in the frog, semidiagrammatic.

Example: Frog's egg has fair amount of yolk, somewhat concentrated towards the vegetal pole. The cleavage is total, but necessarily unequal. The influence of the yolk on the cleavage may be detected even during the first division of the fertilized egg. During anaphase of the mitotic division, a furrow appears on the surface of the egg which is to separate the two daughter blastomeres from one another. This first furrow is vertical. It does not appear simultaneously all around the circumference of the egg, but at the animal pole of the egg, where there is less yolk. Gradually, the cleavage furrow prolonged along the meridians of the egg, until cutting through the mass of the yolk laden cytoplasm. It eventually reaches the vegetal pole and thus completes the separation of first two blastomeres. The same process is repeated during the second cleavage. During the third cleavage, when the plane of separation of the daughter blastomeres is horizontal, the furrow appears simultaneously over the whole circumference of the egg because it meets with an equal resistance from yolk everywhere. Hence, the upper four cells *(micromeres)*, thus cut off, are distinctly smaller than the lower four *(macromeres)*. In further cleavages that follow, the larger yolk-laden cells divide more slowly than the smaller, more protoplasmic ones of the animal hemisphere. After about 32 cells have been formed, tangential divisions begin to occur along with other types already described.

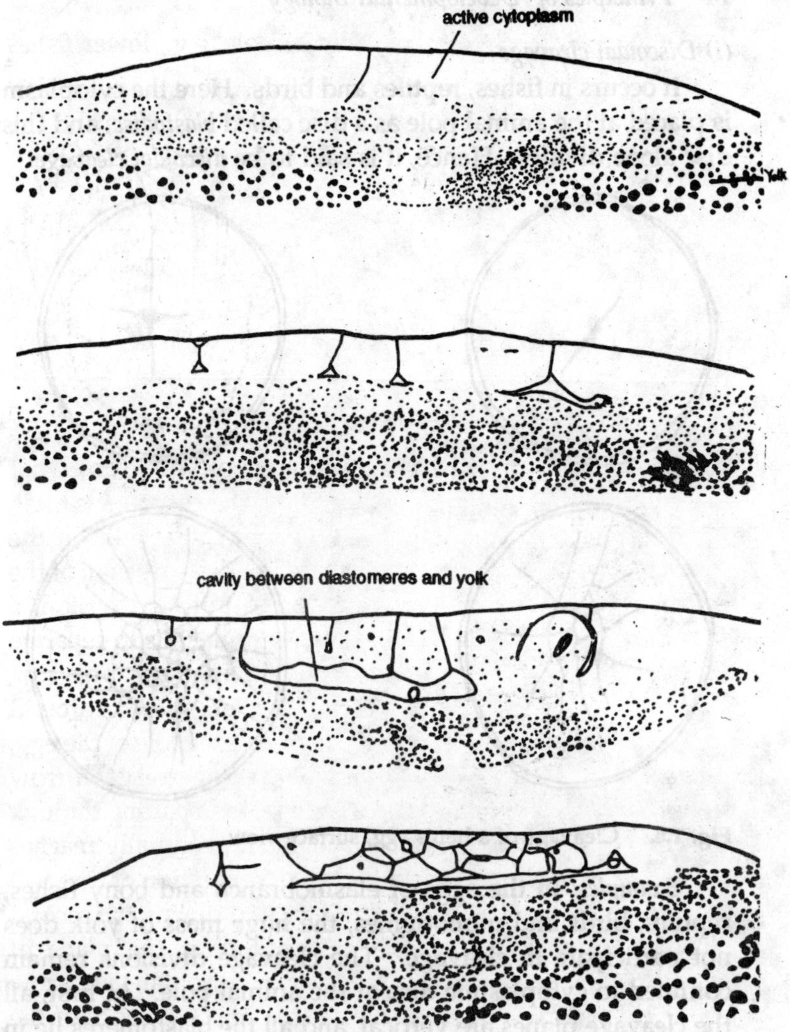

Fig. 1.7. Cleavage of hen's egg (sections).

(b) Meroblastic Cleavage

In meroblastic cleavage, only a portion of the egg divides. It is otherwise called *partial* or *incomplete* cleavage. It is characteristic of telolecithal and megalecithal eggs. The meroblastic cleavage is divided into two types. They are :

(i) Discoidal cleavage

It occurs in fishes, reptiles and birds. Here the cytoplasm is placed at the animal pole as a disc called *blastodisc,* and this disc alone divides. Hence, it is said to be *discoidai cleavage.*

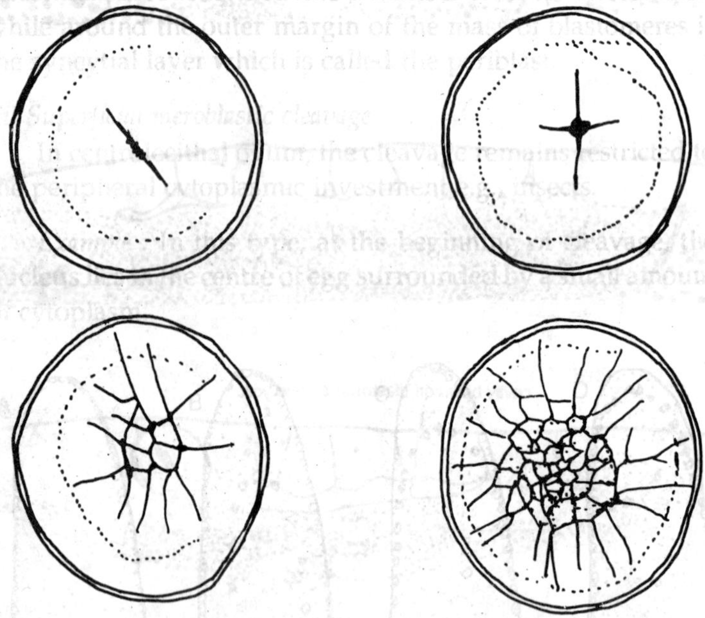

Fig. 1.8. Cleavage of a hen's egg, surface view.

Examples: In the eggs of elasmobranch and bony fishes, reptiles, birds and monotremes, the huge mass of yolk does not participate in cleavage. The cleavage divisions remain confined to cytoplasmic disc at the animal pole. At first, all the cleavage planes are vertical, and all the blastomeres lie in one plane only. The cleavage furrows separate the daughter blastomeres from each other but not from the yolk, so that the central blastomeres are continuous with the yolk at their lower ends, and the blastomeres lying on the circumference are continuous with the uncleaved cytoplasm at their outer edges. As the nuclei at the edges divide, more and more cells become cut off to join the ones lying in the centre but the new blastomeres are also in continuity with the uncleaved yolk underneath.

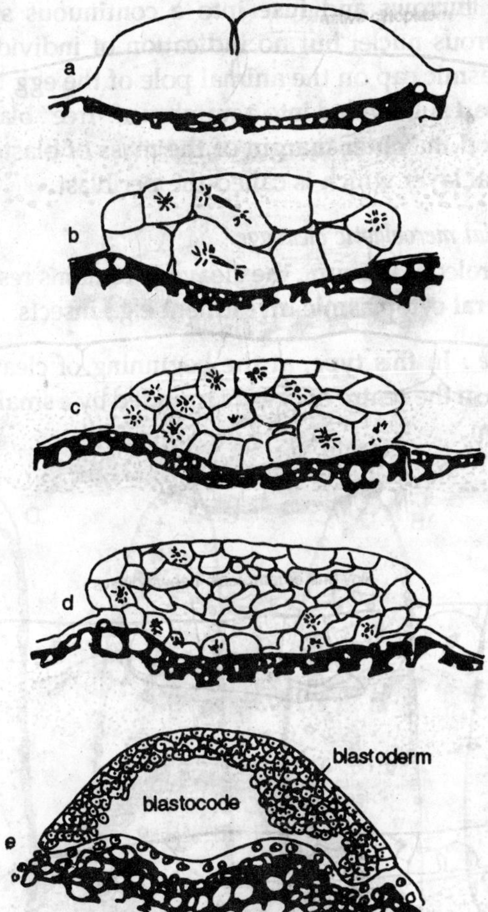

Fig. 1.9. Cleavage and blastula of bony fish. a, b, c, d, Trout e, Muraena.

In the large stage of cleavage the blastomeres of the central area become separated from the underlying yolk in one of the two ways -- either slits may appear beneath the nucleated parts of the cells, or the cell divisions may occur with horizontal *(tangential)* planes of fission, so that one of the daughter cells, the upper one, becomes completely separated from its neighbours, wnile the lower blastomeres retains the connection with the yolk mass. All these blastomeres eventually lose even those furrows and fuse into a continuous syncytium with numerous nuclei but no indication of individual cells. The cytoplasmic cap on the animal pole of the egg is now, in its central part subdivided into a number of "free" blastomeres, while around the outer margin of the mass of blastomeres is the syncytial layer which is called the periblast.

(ii) Superficial meroblastic cleavage

In centrolecithal ovum, the cleavage remains restricted to the peripheral cytoplasmic investment e.g., insects.

Example : In this type, at the beginning of cleavage, the nucleus lies in the centre of egg surrounded by a small amount of cytoplasm.

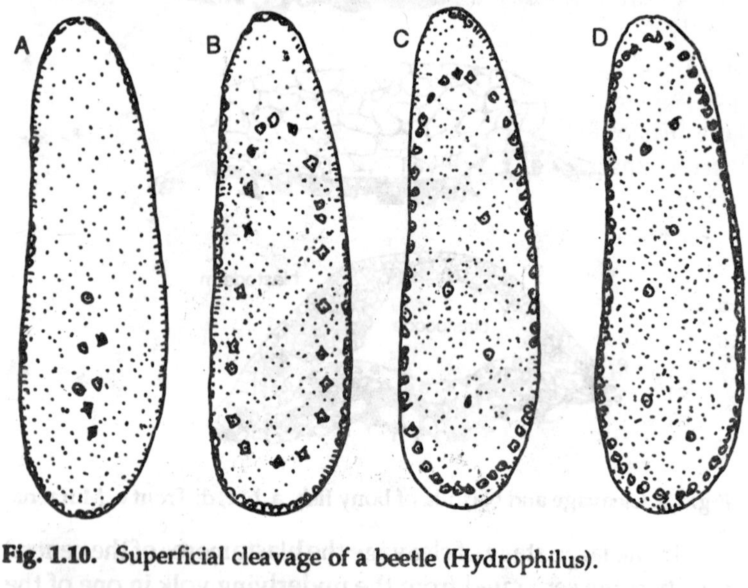

Fig. 1.10. Superficial cleavage of a beetle (Hydrophilus).

The mitotic division of the nucleus starts but it is not followed by the division of the cytoplasm. As a result, a number of nuclei are formed, all embedded in the undivided central mass of cytoplasm. After several divisions have taken place, the nuclei start moving away from the centre towards the surface of the egg. Each nucleus is now surrounded by a small portion of the original central cytoplasm. When the nuclei reach the surface of the egg, the cytoplasm surrounding them fuses with the superficial layer of cytoplasm. The surface layer of the embryo is then a syncytium with numerous nuclei embedded in an undivided layer of cytoplasm. Later, the cytoplasm becomes subdivided by a number of furrows going inward from the surface, into as many sections or the cells as there are nuclei.

Cleavage is further classified into two types on the basis of the *potentiality* of the egg or blastomere. They are : 1. determinate cleavage and 2. indeterminate cleavage.

1. Determinate Cleavage

In some animals cleavage produces blastomeres which are not qualitatively equipotential. When these blastomeres are isolated, they cannot develop into entire embryos. When any blastomere is removed from the embryo, there will be a corresponding defect in the resulting animal. This is because the fate of each blastomere is fixed in the early embryonic stage itself. This type of cleavage is *determinate cleavage* where the fate of the each part of the embryo or the blastomere is already determined. The determinate cleavage is also called *mosaic* development. E.g., annelids, molluscs, ascidians, polyclads (platyhelminthes) and nemerteans.

2. Indeterminate Cleavage

In most of the animals cleavage produces blastomeres which are qualitatively equipotential. When they are isolated, they develop into complete embryos. When a blastomere is removed from the embryo, there will not be any defect in the resulting animal. This is because the fate of each blastomere is not fixed in the early embryonic stage. This type of cleavage is called *indeterminate cleavage* where the fate of each part or blastomere is not already determined. The indeterminate cleavage is also called *regulative development*. *E.g.* vertebrates.

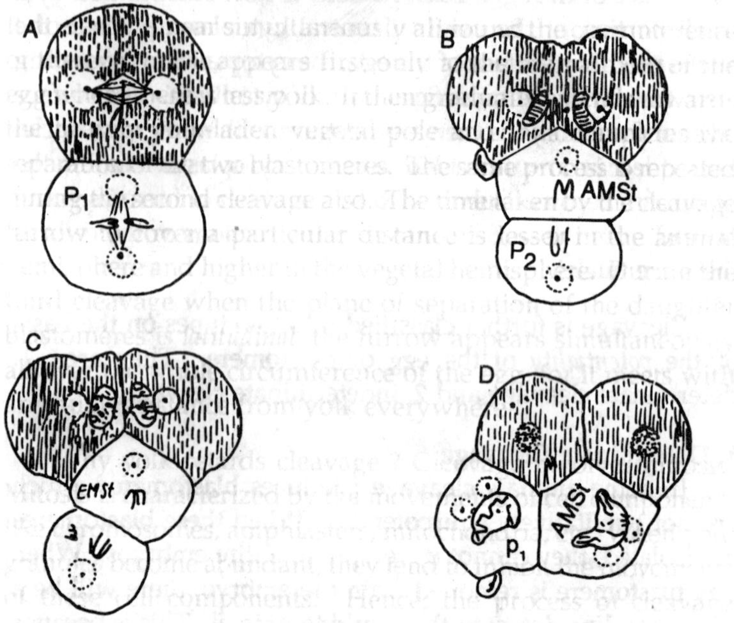

Fig. 1.11. Cleavage of the nematode *Ascaris*, up to four-cell stage.

LAWS OF CLEAVAGE

As already explained that cleavage is an orderly process in which transformation of multicellular body from single celled zygote occurs through a series of mitotic divisions, and the cleavage planes arise according to a fairly definite pattern and in a particular relationship to each other. As a result, certain rules have been formulated to express the simpler geometrical relationship between the blastomeres.

There are four major cleavage laws known by names of those who first emphasized their importance.

1. Pfluger's Law

The spindle elongates in the direction of least resistance.

2. Sach's Law

(a) Cells typically tend to divide into equal parts.

(b) Each new plane of division tends to intersect the previous plane at right angles.

3. Hertwig's Law

(a) The nucleus and its spindle generally are found in the center of the active protoplasm and the axis of any division spindle lies in the longest axis of the protoplasmic mass.

(a) Divisions tend to cut the protoplasmic masses at right angles to their axes.

4. Balfour's Law

The rate of cleavage is inversely proportional to the amount of yolk present in the egg. The yolk tends to impede divisions of both the nucleus and the cytoplasm.

FACTORS AFFECTING CLEAVAGE

The rate and the nature of cleavage are much affected by factors like light, temperature, medium, amount of yolk, cytoplasmic organization, mitosis, etc.

Influence of Yolk on Cleavage

The yolk exerts a very far reaching effect on the process of cleavage. Yolk influences cleavage in two ways :

1. It affects the rate of cleavage and
2. It determines the pattern of cleavage.

When the yolk is abundant, it tends to retard and even inhibit the process of cleavage. As a result, the blastomeres which are rich in yolk tend to divide at a slower rate, and consequently they remain larger than those which have less yolk. The yolk is more concentrated at the vegetal pole. Therefore, at the vegetal pole the cleavage in most retarded, and also where the blastomeres are larger in size.

A good example of the effect of the yolk on cleavage is given by the egg of frog. The first cleavage furrow is *meridional.*

It does not appear simultaneously all round the circumference of the egg, but it appears first only at the animal pole of the egg where there is less yolk. It then gradually extends towards the heavily yolk-laden vegetal pole and thus completes the separation of the two blastomeres. The same process is repeated during the second cleavage also. The time taken by the cleavage furrow to cover a particular distance is lesser in the animal hemisphere and higher in the vegetal hemisphere. During the third cleavage when the plane of separation of the daughter blastomeres is *latitudinal*, the furrow appears simultaneously all over the whole circumference of the egg, for it meets with an equal resistance from yolk everywhere.

Why yolk retards cleavage ? Cleavage involves mitosis. Mitosis is characterized by the movements of cell components like chromosomes, amphiasters, mitochondria, etc. When yolk granules become abundant, they tend to inhibit the movements of these cell components. Hence, the process of cleavage slows down.

The pattern of cleavage is determined considerably by the amount of yolk in the egg. In egg that contain a little yolk and if the yolk is uniformly distributed, the entire egg undergoes segmentation resulting in blastomeres of the same size. This has been exhibited by *Amphioxus* where the cleavage is *holoblastic*. In frog the yolk is concentrated by the vegetal pole and as a result the process of cleavage is rapid at the animal pole and very slow at the vegetal pole. This results in the production of small blastomeres at the animal pole and large blastomeres at the vegetal pole. Hence, the cleavage is unequal. Thus in frog the cleavage pattern is slightly different from the typical holoblastic cleavage. In *fishes* and *birds*, there is maximum yolk. The small amount of cytoplasm with the nucleus is located as a disc at the animal pole. This disc of cytoplasm alone undergoes segmentation resulting in *meroblastic* cleavage.

The Nuclei of Cleavage Cells

In his "germ plasm theory", A. Weismann presented a

hypothesis to explain both heredity and the ontogenetic development of organisms. According to Weismann (1904), every distinct part of an organism (animal or plant) is represented in the sex cell by a separate particle: a determinant. The sum total of determinants would represent the parts of the adult organisms with all their peculiarities. The complete set of determinants would be handed down from generation to generation, which would account for hereditary transmission of characters. The determinants, according to Weismann, are localized in the chromosomes of the nucleus, just as are the genes of modern genetics. However, there is this difference, that the genes are not supported to represent *parts* of the organism but rather properties which may sometimes be discernible in all the parts of the body.

During the cleavage of the egg, the various determinants, according to Weismann, become segregated into different cells. The blastomeres would get only some of the determinants; namely, those which correspond to the fate of each blastomere. The successive segregation of the determinants would eventually result in each cell's having only determinants of one kind, and then nothing would be left to the cell but to differentiate in a specific way in accordance with the determinants present. Only the cells having the sex cells among their descendants, Weismann held, would preserve the complete set of determinants, since these would be necessary for directing the development of the next generation.

Even though Weismann's conception of the properties of determinants is not tenable from the genetic viewpoint, it is still important to know whether the difference in the fate of the blastomeres and parts of the embryo may be attributed to differences in the nuclei of the cleavage cells. Experiments now give convincing information on this point.

Our knowledge of the development of complete embryos from one of the two daughter blastomeres of the egg (as a result of either separating the two first blastomeres or killing one of the first two blastomeres contradicts Weismann's hypothesis about the segregation of determinants during cleavage. The evidence becomes still more conclusive because it has been found that not only are the first two blastomeres capable of developing into complete embryos, but that the blastomeres in later cleavage stages sometimes possess the same ability. In the case of sea urchins one of the first four blastomeres, and even occasionally one of the first eight blastomeres, may develop into a whole embryo with all the normal parts but reduced in size.

However, the method of isolating blastomeres does not permit one to test the properties of nuclei of later generations of cells. In the four-cell stage, the quantity of cytoplasm contained in one blastomere may already be too small for development to take place in an approximately normal fashion. This is true in increasing degree as cleavage proceeds and the individual blastomeres become smaller and smaller.

To push the investigation further, a different method has been devised. A most elegant experiment in this field was carried out by Spemann (1928). Spemann constricted fertilized eggs of the newt *Triturus (Triton)* into two halves with a fine hair, just as they were about to begin to cleave. The constriction was not carried out completely, so that the two halves were still connected to each other by a narrow bridge of cytoplasm. The nucleus of the fertilized egg lay in one half, and the other half consisted of cytoplasm only. When the egg nucleus began to divide, the cleavage was at first restricted to that half of the egg which contained the nucleus. This half divided into two, four, eight cells, and so on, while the non-nucleated half remained uncleaved. At about the stage of 16 blastomeres, one of the daughter nuclei, now much smaller than at the beginning of cleavage, passed through the cytoplasmic bridge into the half of the egg which had hitherto no nucleus. Forthwith this half also began to cleave. After both halves of the egg were thus supplied with nuclei, Spemann drew the hair loop tighter and separated the two halves of the egg completely from each other. They were now allowed to develop into embryos. In a number of cases two completely normal embryos developed from the two halves of the egg as a result of this experiment.

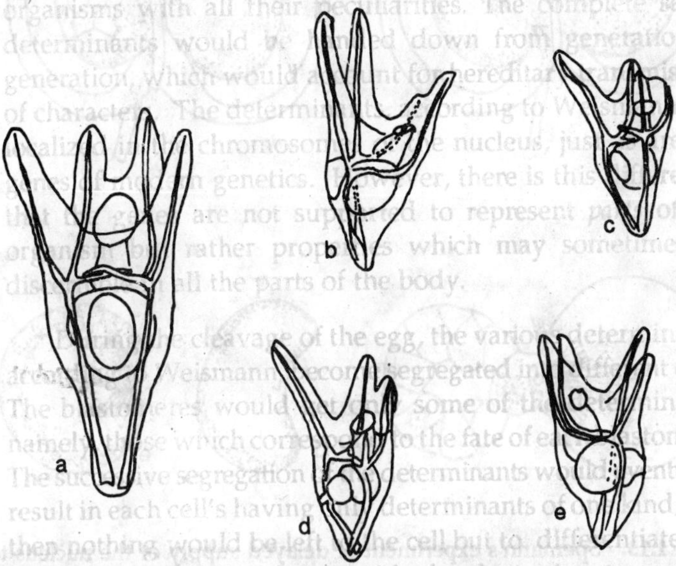

Fig. 1.12. Results of separation of the first four blastomeres of a sea urchin's egg. a, Normal pluteus; b, c, d and e, plutei of normal structure but diminished size, each developed from one of the blastomeres of the four-cell stage.

Of the two embryos, each stated by having one half of the egg cytoplasm, but as to the nucleus they were in a very different position. While one of the embryos possessed 15/16 or even 31/32 of all the nuclear material of the egg, the other got only 1/16 or 1/32 of the nuclear material. This proves conclusively that even in the 32-cell stage every nucleus has a complete set of hereditary factors necessary for the achievement of normal development. All the nuclei in this stage are completely equivalent to one another and to the nucleus of the fertilized egg. The hypothesis of an unequal division of the hereditary substance of the nucleus, of the segregation of

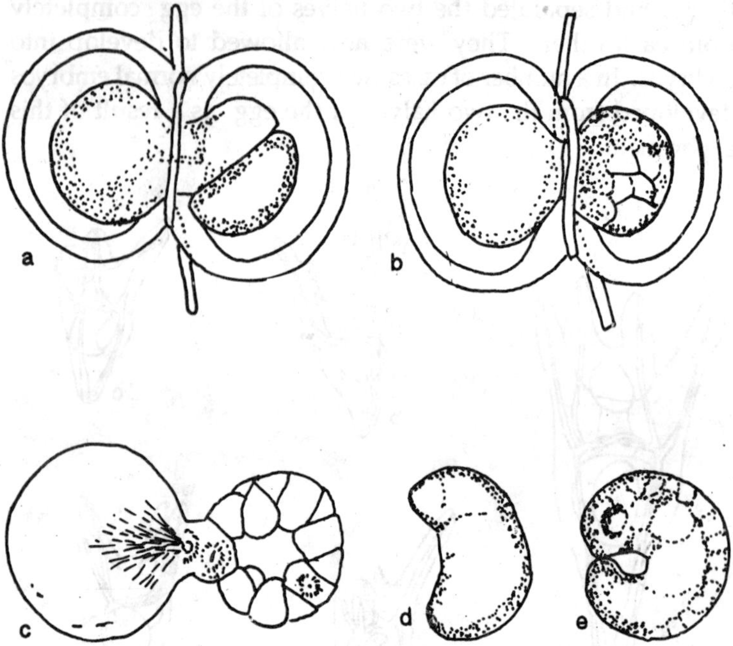

Fig. 1.13. Spemann's experiment of delayed supply of the nucleus to one half of a newt's egg. a, Beginning of cleavage in nucleated half. b, One of the nuclei has penetrated into the uncleaved half, and a cell boundary appears between this and the half which originally contained the zygote nucleus, *c*, Same stage in section, *d*, Embryo from the half with delayed nuclear supply. e, Embryo from the half which contained the zygote nucleus.

determinants or genes to the different cleavage cells, is thus disproved. It is now assumed that every cell of the metazoan body has a complete set of nuclear factors necessary for development (a complete set of genes, in the terminology of modern genetics).

The experiments on the delayed nuclear supply to one half of an amphibian egg have been corroborated by experiments on the eggs of other animals. It will be useful to relate here a corresponding experiment carried out on a very different kind of animal: the dragonfly *Platycnemis pennipes*. In the dragonfly the cleavage is incomplete, and only the nucleus divides at first, the cytoplasm remaining uncleaved. The egg is elongated, and after the first division the daughter nuclei move, one into the anterior and the other into the posterior half of the egg. When eight nuclei are available, these are spaced along the length of the egg. By further divisions nuclei are provided for all cells in the respective regions. In the stage when two nuclei are present, either of them may be killed by a short exposure to a narrow beam of ultraviolet light, which does not damage the cytoplasm to any great extent. The remaining nucleus continues to divide, and its daughter nuclei are distributed to all parts of the egg instead of supplying only one half of it. Completely normal embryos develop, no matter which of the two nuclei is allowed to survive. The two nuclei prove to be completely equivalent for development, although normally they would have supplied very different parts of the embryo.

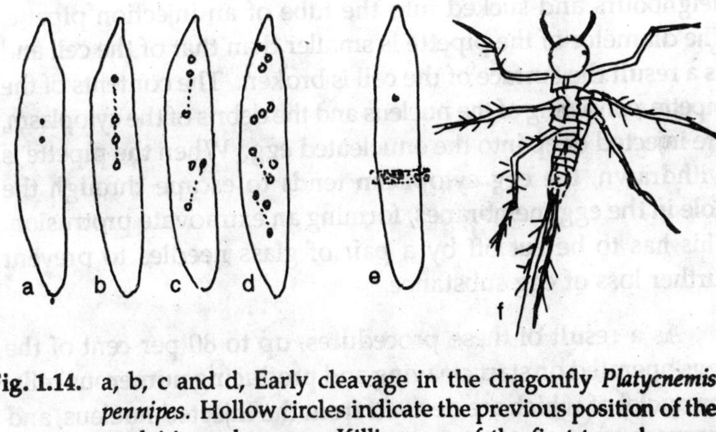

Fig. 1.14. a, b, c and d, Early cleavage in the dragonfly *Platycnemis pennipes*. Hollow circles indicate the previous position of the nuclei in each case. e, Killing one of the first two cleavage nuclei by ultraviolet irradiation. Irradiated area is stippled. f, Larva which developed from an egg treated as in e.

The methods used in the above experiments in testing the properties of the cleavage nuclei are of necessity confined to the earlier stages of cleavage. At present a more universal method is available which allows the investigation to be extended to nuclei of cells of much more advanced embryos, and possibly it may ultimately be applied even to fully differentiated cells of an adult organism. This is the method used for transplantation of nuclei. The transplantation of nuclei from one cell to another was first carried out successfully on *Amoeba,* and the method was then applied to test the properties of nuclei in developing frog embryos. The method, as applied to the frog embryo, consists essentially in taking the nucleus of any cell from a developing embryo and injecting it into an enucleated uncleaved egg.

The egg receiving the nucleus must be specially prepared. The ripe eggs are removed from the oviducts and activated by pricking with a glass needle. The egg nucleus then approaches the surface of the cytoplasm in preparation for the second maturation division and is removed by a second prick with a glass needle at the exact spot where the nucleus is located. Next, a cell of an advanced embryo is separated from its neighbours and sucked into the tube of an injection pipette. The diameter of the pipette is smaller than that of the cell and as a result the surface of the cell is broken. The contents of the pipette, consisting of the nucleus and the debris of the cytoplasm, are injected deep into the enucleated egg. When the pipette is withdrawn, the egg cytoplasm tends to escape through the hole in the egg membranes, forming an extraovate protrusion. This has to be cut off by a pair of glass needles to prevent further loss of egg substance.

As a result of these procedures, up to 80 per cent of the eggs operated on start cleaving and producing numerous cells, the nuclei of which are derived from the injected nucleus, and the cytoplasm of which is from the enucleated egg (the small amount of cytoplasm injected with the nucleus, comprising less than 1/40,000 of the volume of the egg cytoplasm, may be ignored). Not all eggs which start cleaving develop normally later, but at least a small proportion do so and may go through all the stages of embryonic and postembryonic development

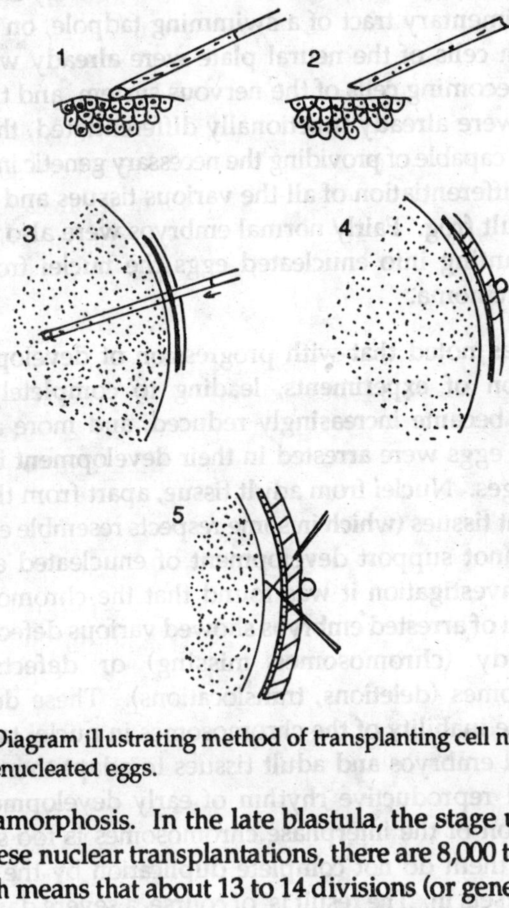

Fig. 1.15. Diagram illustrating method of transplanting cell nuclei into enucleated eggs.

up to metamorphosis. In the late blastula, the stage used for some of these nuclear transplantations, there are 8,000 to 16,000 cells, which means that about 13 to 14 divisions (or generations) of the original nucleus of the fertilized egg have been performed without diminishing the power of the nucleus to support every type of differentiation provided for by the specific genotype. Nuclei of cells of an early gastrula show the same properties.

Further experiments have tested the potentialities of nuclei of cells, which are even more advanced in the process of differentiation by transplanting them into enucleated eggs. Normal tadpoles developing up to and through metamorphosis were obtained by using nuclei from the neural plate of a frog embryo on the one hand, and nuclei from already ciliated cells of the alimentary tract of a swimming tadpole, on the other. Although cells of the neural plate were already well on the way to becoming cells of the nervous system, and the cells of the gut were already functionally differentiated, their nuclei were still capable of providing the necessary genetic information for the differentiation of all the various tissues and cell types of an adult frog. Fairly normal embryos were also produced by implanting into enucleated eggs the nuclei from a frog adenocarcinoma.

It was noted that with progression of development the proportion of experiments, leading to completely normal tadpoles became increasingly reduced, and more and more operated eggs were arrested in their development in various early stages. Nuclei from adult tissue, apart from the cells of malignant tissues (which in some respects resemble embryonic cells) cannot support development of enucleated eggs. On careful investigation it was found that the chromosomes in the nuclei of arrested embryos showed various defects such as aneuploidy (chromosomes missing) or defects within chromosomes (deletions, translocations). These defects are due to the inability of the chromosomes in nuclei taken from advanced embryos and adult tissues to adapt themselves to the rapid reproductive rhythm of early development. The duplication of the interphase chromosomes is too slow, and many of them do not complete duplication by the time cell divisions sets in. The result is, of course, a severe damage and incompletness of the chromosome set. Whatever, the explanation for the defective development, it remains that at least a proportion of nuclei from cells well on the way to differentiation and from those that have become malignant retain their full potentialities for controlling and directing normal development.

The preceding discussion does not mean that nuclei of differentiating cells are in every respect similar to the nuclei of fertilized eggs. It experiments with frogs, nuclei transplanted into the egg cytoplasm undergo a change which makes them resemble early embryonic nuclei. The nuclei of newly fertilized eggs, although small in proportion to the egg cytoplasm, are much larger than the nuclei at later stages. Accordingly, transplanted nuclei increase up to 30-fold in volume during the first 40 minutes after transplantation. The functioning of the nuclei also changes drastically. Nuclei of advanced embryos do not divide rapidly and accordingly the synthesis of DNA, necessary for the replication of chromosomes, is slow. On the other hand, they synthesize large quantities of ribosomal RNA and as this kind of RNA is produced in the nucleolus they have prominent nucleoli. Nuclei of cleavage cells synthesize DNA rapidly but do not synthesize any ribosomal RNA and therefore do not show any nucleoli. Transplanted nuclei cease to synthesize ribosomal RNA and their nucleoli disappear. Instead, they start to synthesize DNA rapidly as do normal cleavage nuclei. By the time the embryo reaches the stage of gastrulation, DNA synthesis in the descendants of the transplanted nucleus slow down, and ribosomal RNA synthesis is renewed; nucleoli reappear. Thus a transplanted nucleus reverts in every respect to the condition in which nuclei would have been after normal fertilization.

Transplantation of nuclei into enucleated eggs has also been performed in insects. In the fruitfly, *Drosophila,* nuclei from the late cleavage embryos (blastoderm with 200 to 2000 nuclei) allow the embryo to start developing and reach the hatching larva stage. There appear to be no reasons why in further experiments more advanced development could not be achieved in insects as well as in frogs.

Distribution of Cytoplasmic Substances of the Egg During Cleavage

The fact that the nuclei of the developing embryo are fully equivalent to each other does not allow us to ascribe to the nucleus the origin of differences in the fate of the cells during development of the embryo. What then is the cause of the observed differences in the developmental behaviour of cells ? By exclusion, we may suppose that the cause of the differences lies in the cytoplasm, and that the cells produced during cleavage do not receive the same cytoplasmic substances. There is direct experimental support in favour of this supposition. Curiously enough, a further elaboration of Spemann's experiment of delayed nuclear supply to one half of the *Triturus* egg can be brought to bear on this problem.

It has already been mentioned that pigment is unequally distributed over the surface of amphibian eggs (sections 4-3 and 6-6). The animal hemisphere of the egg is more or less heavily pigmented; the region around the vegetal pole is poor in pigment or completely white; the equatorial, or rather sub-equatorial, marginal zone is more lightly pigmented than the animal hemisphere but more darkly pigmented than the vegetal region. On one side of the egg the marginal zone is broader than on the other, and this region is known as the gray crescent (from the colour of this region in the common European frogs). The gray crescent indicates the future position of the dorsal lip of the blastopore and also corresponds to the dorsal region of the embryo.

In the egg of the newt the pigment is not very abundant, but the gray crescent is readily distinguishable in the uncleaved egg. It is therefore possible to determine the position of the ligature in respect to the gray crescent. If the ligature subdivides the gray crescent cytoplasm equally between the two halves of the egg, two more or less normally developed embryos are found to develop. If the ligature comes to lie so that all the gray crescent cytoplasm is contained in one half of the egg, whereas the other half does not get any of it, then a complete embryo is developed only from the first half. The second half remains highly abnormal; it does not develop a nervous system, a notochord or segmented muscle; the only parts that can be discerned are epidermis, unsegmented (lateral plate) mesoderm and yolk endoderm. All these are parts of the ventral region of the body, thus justifying the name "belly" (Bauchstuck) given by Spemann to the defective embryos.

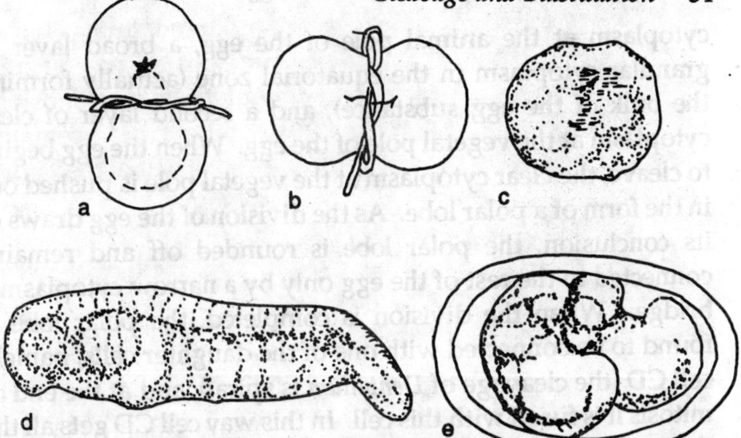

Fig. 1.16. **Constriction of a newt's egg in the frontal *(a)* and medial *(b)* plane. After frontal constriction the ventral half developed into a "belly" *(c)*, the dorsal half into a complete embryo *(d)*, After medial constriction *(b)* both halves developed into complete embryos *(e)*.**

Now it is found that the fate of the egg half, whether it develops into a complete embryo or into a "belly", does not depend on which of the two halves had retained the egg nucleus originally and which was delayed in receiving a nucleus. A half with a delayed nuclear supply will develop into a complete embryo if it contained the gray crescent cytoplasm. On the other hand, the egg half without gray crescent cytoplasm will become a "belly" even if it contained the egg nucleus right from the beginning. The difference in the time of nuclear supply and the amount of nuclear material received is completely overridden by differences in the cytoplasmic composition of the two egg halves. The importance of cytoplasmic substances in the egg for the differentiation of parts of the embryo is further proved by a number of experiments carried out on the eggs of various animals. The following will be chosen as examples:

In the egg of the mollusc *Dentalium* there may be distinguished three layers of cytoplasm : a layer of clear cytoplasm at the animal pole of the egg, a broad layer of granular cytoplasm in the equatorial zone (actually forming the bulk of the egg substance), and a second layer of clear cytoplasm at the vegetal pole of the egg. When the egg begins to cleave, the clear cytoplasm at the vegetal pole is pushed out in the form of a polar lobe. As the division of the egg draws to its conclusion, the polar lobe is rounded off and remains connected to the rest of the egg only by a narrow cytoplasmic bridge. When the division is completed, the polar lobe is found to be connected with one of the daughter cells, namely cell CD (the cleavage of *Dentalium* is spiral), and at the end of mitosis it is fused with this cell. In this way cell CD gets all the three types of cytoplasm, while the other daughter cell, cell AB, gets a half of the animal pole's clear cytoplasm, a half of the granular cytoplasm, but no vegetal cytoplasm. Cell CD is consequently slightly larger than cell AB.

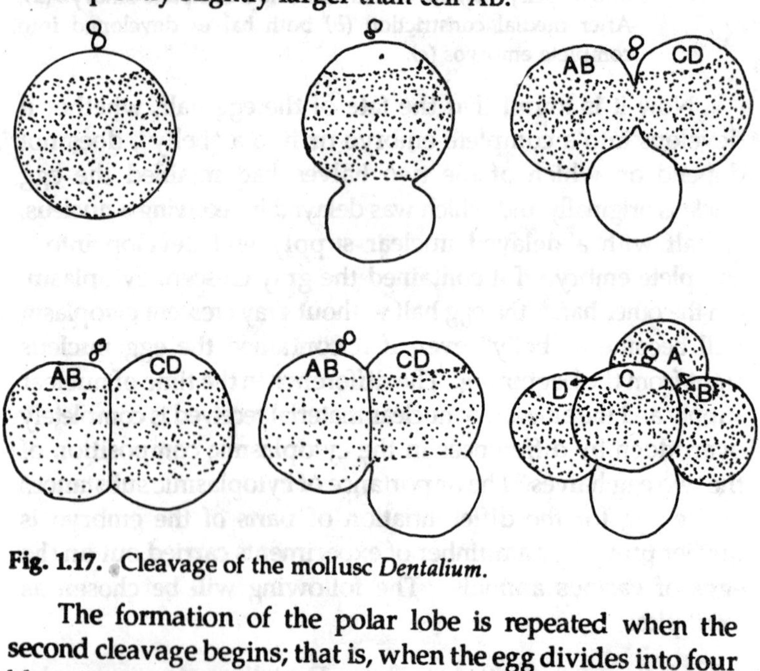

Fig. 1.17. Cleavage of the mollusc *Dentalium*.

The formation of the polar lobe is repeated when the second cleavage begins; that is, when the egg divides into four blastomeres. Of the two daughter cells of blastomere CD, the polar lobe passes to blastomere D, while blastomere C receives only the clear cytoplasm of the animal pole and the granular cytoplasm.

The two blastomeres of the two-cell stage and the four blastomeres of the four-cell stage are therefore not equivalent as to their cytoplasmic composition, only one blastomere (CD or D, respectively) containing the whole of the vegetal polar lobe material. The behaviour of the blastomeres is also found to be different. If the first two blastomeres are separated and allowed to develop, blastomere CD is found to produce a complete larva (a trochophore in this case), though of diminished size. Blastomere AB also develops into a larva, but the larva is defective in some respects, the most significant defect being that it completely lacks the rudiment of the mesoderm normally contained in the posterior or "post-trochal" part of the larva. If the blastomeres are isolated in the four-cell stage, they are again found to differ in their development. Only blastomere D, which possesses the clear cytoplasm of the vegetal pole, develops into a complete, though smaller, trochophore larva; the other three blastomeres develop into defective larvae lacking the mesoderm rudiment. It has been concluded, therefore, that the clear cytoplasm of the vegetal pole is necessary for the development of the mesoderm.

This conclusion is further supported by the following experiment :

At the time of the first or second cleavage the polar lobe containing the clear cytoplasm of the vegetal pole can be nipped off. The remainder then contains all the clear cytoplasm of the animal pole and the granular cytoplasm. It also contains all the nuclear material but no vegetal polar lobe material. The result is that the larva development from such an egg is defective and lacks the mesoderm rudiment. This is a very clear proof that it is the cytoplasm and not the nucleus that makes blastomere D different from the three others and able to produce a certain differentiation (the mesoderm rudiment) which the other blastomeres are incapable of producing.

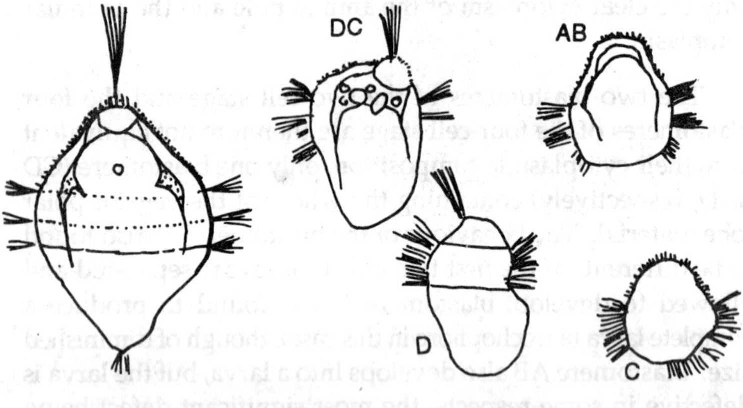

Fig. 1.18. Larvae of *Dentalium* developing from a whole egg (left) and from separated blastomeres (right). The letters indicate from which blastomere each larva has developed.

In the egg of the ascidian *Cynthia partita,* no less than four cytoplasmic substances may be distinguished. because of differences in the colour of these substances. In the egg at the beginning of cleavage, the animal hemisphere consists of clear transparent cytoplasm. The vegetal pole of the egg is taken up by a slaty gray cytoplasm, rich in yolk. In between, slightly below the equator, may be discerned two crescentic areas on opposite sides of the egg. One crescentic area consists of light gray cytoplasm, and the other consists of yellow cytoplasm, the yellow colour due to the presence of yellow granules.

During cleavage the four kinds of cytoplasm are distributed to different cells. Later the cells containing clear cytoplasm give rise to ectoderm; the cells containing slaty gray cytoplasm become endoderm; the cells containing yellow cytoplasm develop into mesoderm; and the cells containing light gray cytoplasm become neural system and notochord. The assumption may be made, therefore, that the development of the parts of the embryo enumerated is caused by the kind of cytoplasm that is contained in the cells. This can actually be proved by centrifuging the eggs before they have begun cleaving.

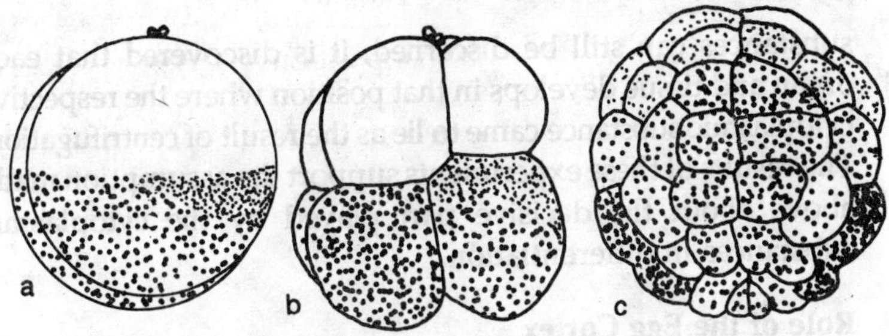

Fig. 1.19. Three cleavage stages of the ascidian Cynthia partita. *a*, Two-cell stage, *b*. Eight cell stage. *c*, 64-cell stage.

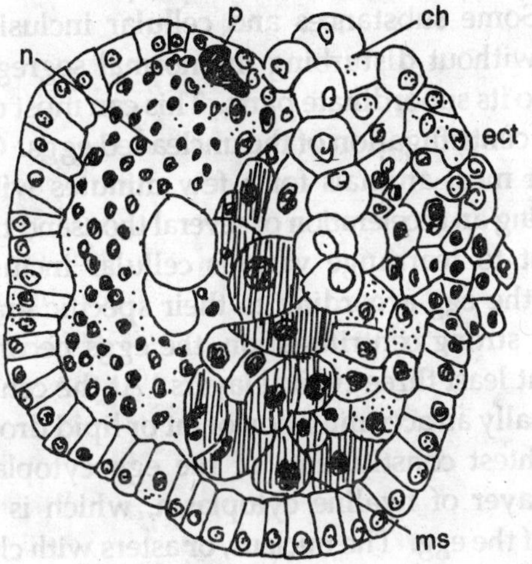

Fig. 1.20. Displacement of organ rudiments in an embryo of *Cynthia partita* as a result of the centrifugation of the egg. *n*, Neural system; *p*, eye pigment; *ch*, notochord; *ect*, ectoderm; *ms*, muscle cells.

In this way the cytoplasmic substances within the egg may be displaced, and when the eggs begin to cleave, the various kinds of cytoplasm are held in abnormal positions when the substance of the egg is subdivided into cells. An embryo may develop from a centrifuged egg, possessing recognizable tissues and rudiments of organs, but these are arranged in a chaotic way. As the distinctions between the various cytoplasmic substances can still be discerned, it is discovered that each organ and tissue develops in that position where the respective cytoplasmic substance came to lie as the result of centrifugation. The results of these experiments support the assumption made above about the decisive part played by the cytoplasmic substances in differentiation.

Role of the Egg Cortex

During cleavage in the egg and in the embryo the distribution of visible materials is, however, not always of crucial importance for the localization of parts in the developing embryo. Some substances and cellular inclusions may be displaced without disturbing the normal segregation of the embryo into its subordinate parts. This can most conveniently be done by centrifugation of the uncleaved eggs. Centrifuging the eggs of most animals for a few minutes with moderate speeds, giving an acceleration of several thousand times gravity, is sufficient to rearrange various cellular inclusions in the interior of the egg according to their specific gravity. After sufficiently strong centrifugation, the eggs become-stratified and show at least three typical layers. At the centripetal pole there is usually an accumulation of fat or lipid droplets, which are the lightest constitutents of the egg cytoplasm. There follows a layer of hyaline cytoplasm, which is the ground substance of the egg. The nucleus, or asters with chromosomes (if the centrifuged egg was in meiosis or mitosis), are also found in the hyaline layer. The yolk, as the most dense and heavy constituent of the egg, accumulates at the centrifugal pole.

In most of the cases the vegetal pole is so much heavier, owing to the presence of yolk, that it becomes oriented centrifugally during centrifugation. In this case the yolk is not displaced from its normal site at the vegetal pole but is only more concentrated. To displace the yolk in these cases the eggs have to be fixed in a desired position, so that they cannot freely rotate. This can be done sometimes by sucking them into narrow capillaries or by embedding the eggs in gelatin

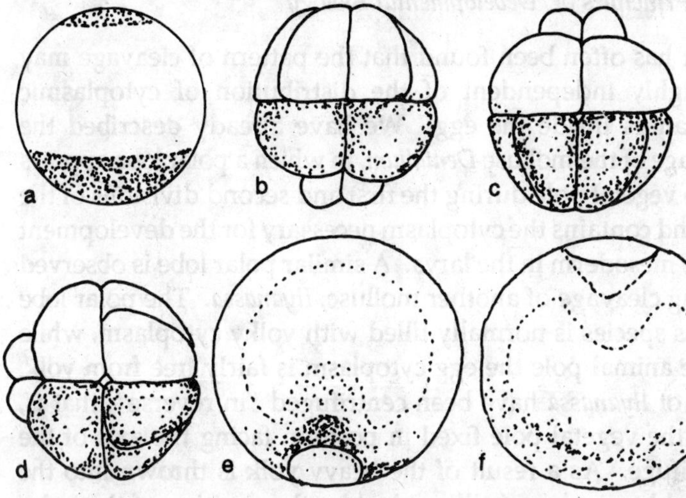

Fig. 1.21. Centrifugation of the egg of the sea urchin *Arbacia*. *a*, Egg immediately after centrifugation red pigment (stippling) thrown to centrifugal pole, oil droplets assembled at the centripetal pole. Drawings show position of red pigment in respect to the main axis during cleavage (b, c, d) and at the beginning of gastrulation (e, f).

prior to centrifugation. If the main axis of the embryo lies at random to the centrifugal force, as often happens, cellular inclusions will be dislocated to different parts in individual eggs. Figure shows the results of centrifuging eggs of the sea urchin *Arbacia*. The red pigment granules present in these eggs are concentrated at the centrifugal end of the egg. Some scattering of the granules occurs after the centrifuging is stopped and before cleavage progresses sufficiently to prevent further redistribution of the granules by subdividing the egg into blastomeres. It now becomes evident that the granules are concentrated in different positions in respect to the egg axis : near the vegetal pole, near the animal pole, or toward one side. Independently of the position of the granules, the invagination of the blastopore occurs at the vegetal pole, so that the region of the blastopore will contain the pigment granules in some embryos but not in others.

It has often been found that the pattern of cleavage may be highly independent of the distribution of cytoplasmic substances inside the egg. We have already described the cleavage of the mollusc *Dentalium,* in which a polar lobe appears at the vegetal pole during the first and second divisions of the egg and contains the cytoplasm necessary for the development of the mesoderm in the larva. A similar polar lobe is observed during cleavage of another mollusc, *Ilyanassa.* The polar lobe in this species is normally filled with yolky cytoplasm, while at the animal pole the egg cytoplasm is fairly free from yolk. Eggs of *Ilyanassa* have been centrifuged "in reverse"; that is, with the vegetal pole fixed in position facing the axis of the centrifuge. As a result of the heavy yolk is thrown into the animal hemisphere (still marked by the position of the polar bodies), and the hyaline cytoplasm and lipid droplets are concentrated at the vegetal pole. Nevertheless, the polar lobe appears at the vegetal pole when the egg starts cleaving, although the lobe now contains mainly hyaline cytoplasm and lipid instead of the yolk granules. It is obvious that the formation of the polar lobe is not dependent on the yolk normally located at the vegetal pole, but on something that is not displaced by the centrifugal force.

What can this something be ? There are two possibilities. The first is that there exists in the cytoplasm some fixed network with sufficiently broad meshes to allow for the free movement of yolk granules and other inclusions without itself being torn or distorted. The existence of such a network cannot, however, be supported by any observation on the physical state or ultramicroscopic structure of cytoplasm. There remains the other alternative that the fixed system which is not displaced by centrifugation is the cortical layer of cytoplasm, or the cortex of the egg. This is in conformity with direct observation; the cortical granules are not displaced by centrifugation. The layer of cytoplasm in which they are embedded is therefore sufficiently viscous to resist the forces usually generated in centrifugation experiments.

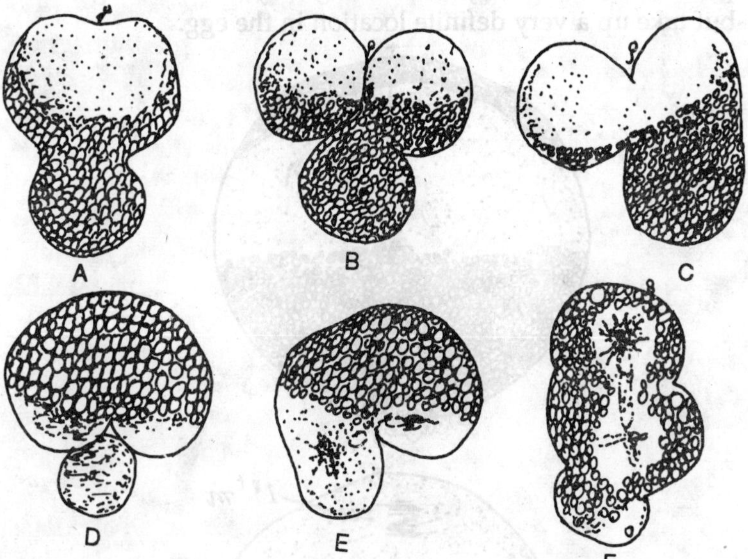

Fig. 1.22. *A, B, C,* Normal Cleavage of the mollusc *Ilyanassa, D, E, F,* Cleavage stages after centrifugation "in the reverse", with the heavy yolk displaced toward the animal pole.

Since the immovable cortex of the egg appears to determine the point at which invagination begins in centrifuged *Arbacia eggs,* as well as to determine the position of the vegetal polar lobe in *Ilyanassa,* the further suggestion may be made that the cortex is the actual carrier of the polarity of the egg, or that the polarity of the egg is ingrained in its cortex. If this were correct, the distribution of the substances in the anterior of the egg might be expected to be controlled or determined by the egg cortex. This suggestion finds support in some further results of centrifugation experiments; namely, the fact that cell constituents tend to return to their normal positions after the cessation of centrifuging. A scattering or mixing up of the strata into which the egg contents had been arranged by centrifugation could be a result of random movement of particles. But this explanation does not apply to cases in which after centrifugation certain particles not only move out from the position to which they were brought by the centrifugal force but take up a very definite location in the egg.

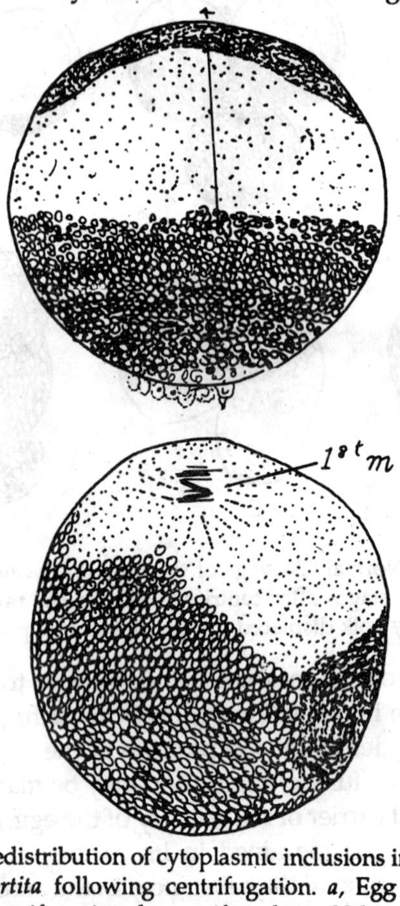

Fig. 1.23. Redistribution of cytoplasmic inclusions in the egg of *Cynthia partita* following centrifugation. *a*, Egg immediately after centrifugation; *b*, centrifuged egg 22 hours later (the yellow granules, shown black, have taken up a subequatorial position on one side of the egg). Ist *m*, Spindle of first meiotic division.

We have seen that in the egg of the ascidian *Cynthia* different kinds of egg cytoplasm may be displaced by centrifugation. Immediately after centrifugation the eggs show a very clear stratification. The yellow granules, which go into the formation of mesoderm, are displaced to the centripetal pole; the yolk is displaced to the centrifugal pole; and the hyaline cytoplasm remains as a layer in between. When the eggs are left to themselves after centrifugation, the egg substances start following and rearranging themselves in the interior of the egg. If the cell divisions set in sufficiently soon, this rearrangement is stopped by partitions appearing between the cleavage cells, and the result is the formation of the abnormal embryos dealt with earlier. If, however, the eggs are centrifuged well in advance of the beginning of cleavage, or if the first divisions of the egg are delayed, the redistribution of the egg contents may proceed so far that normal conditions are attained. Figure shows the end result of such a redistribution: the yolk is at the vegetal pole, the animal hemisphere is filled with hyaline cytoplasm, and the yellow granules take up a subequatorial area on one side of the egg, corresponding to the mesodermal "yellow crescent" of normal development.

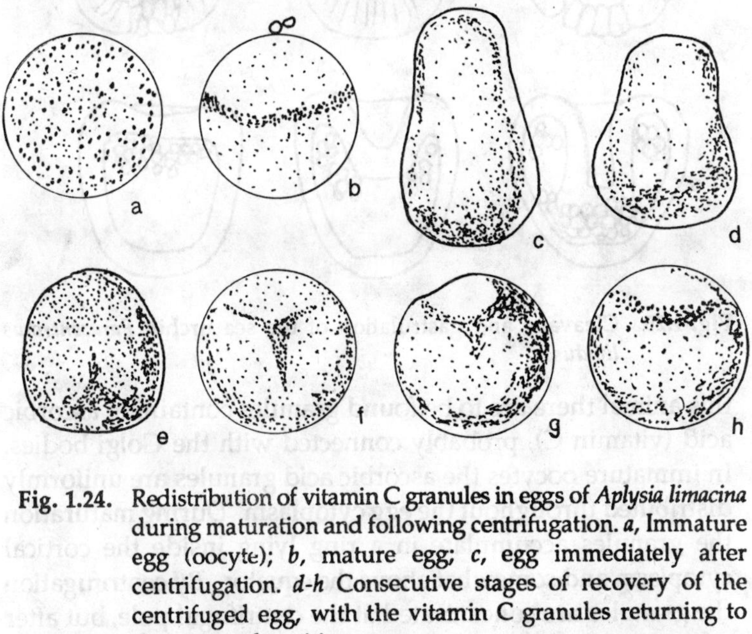

Fig. 1.24. Redistribution of vitamin C granules in eggs of *Aplysia limacina* during maturation and following centrifugation. *a,* Immature egg (oocyte); *b,* mature egg; *c,* egg immediately after centrifugation. *d-h* Consecutive stages of recovery of the centrifuged egg, with the vitamin C granules returning to their normal position.

Another and perhaps even more impressive example is that of the egg of the mollusc *Aplysia limacina*. In the eggs of

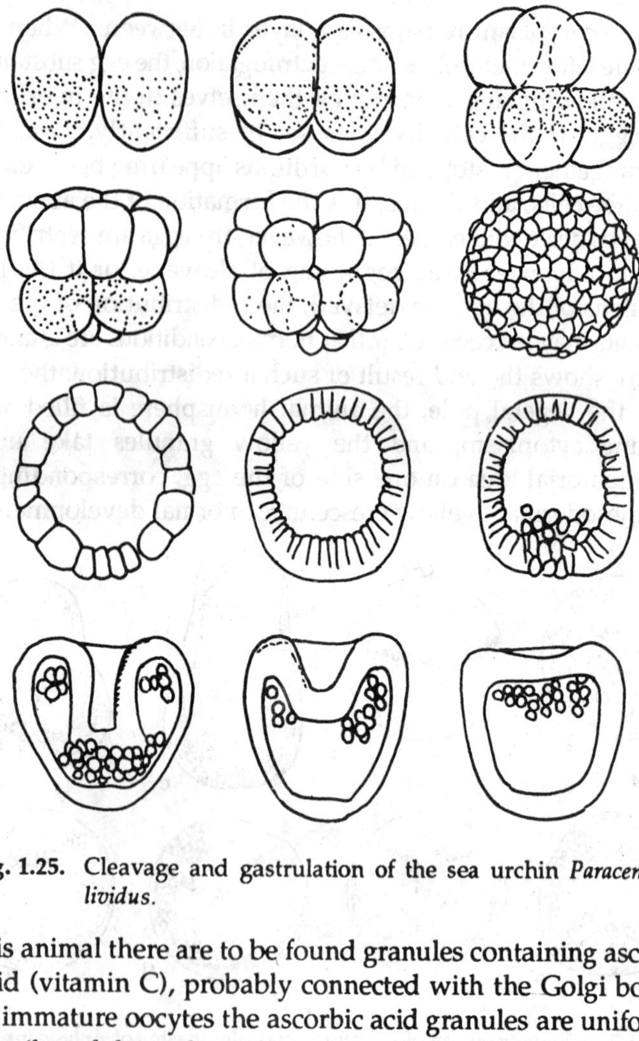

Fig. 1.25. Cleavage and gastrulation of the sea urchin *Paracentrotus lividus*.

this animal there are to be found granules containing ascorbic acid (vitamin C), probably connected with the Golgi bodies. In immature oocytes the ascorbic acid granules are uniformly distributed throughout the egg cytoplasm. During maturation the granules accumulate in a ring lying inside the cortical cytoplasm and somewhat above the equator. By centrifugation the granules are concentrated at the centrifugal pole, but after cessation of centrifugation the ascorbic acid granules soon start moving and again take up their normal position as a supra-equatorial ring.

The most plausible explanation of the last two experiments is that the displaced cytoplasmic particles tend to return to the proximity of certain regions of the egg cortex which had remained in their respective positions all the time the egg was being centrifuged. It would follow that in normal development the cytoplasmic substances in the egg are distributed in relation to local differences in the egg cortex, and that it is the egg cortex, therefore, that foreshadows in some way the pattern of the future development of the embryo.

The Morphogenetic Gradients in the Egg Cytoplasm

While we have learned that the cytoplasm of the egg, and in particular its cortex, starts the chain of reactions which eventually leads to the differentiation of parts of the embryo, it should not be imagined that the structure of the future embryo is rigidly determined by local differences in the cytoplasm of the egg. The local peculiarities of the egg cytoplasm are only some of the factors necessary for the formation of organ rudiments. That this is so can be shown by examining the development of the sea urchin, for instance.

In the 16-cell stage of the sea urchin, the blastomeres are of three different size. The animal hemisphere consists of eight blastomeres of medium size, the mesomeres, which are destined to produce most of the ectoderm of the larva. The vegetal hemisphere consists of four very large blastomeres, the macromeres, and of four very small blastomeres, the micromeres, which lie at the vegetal pole of the egg. The macromeres contain some more material for the ectoderm and all the material for the endoderm. In a subsequent cleavage the future ectoderm and endoderm are segregated from each other into an upper tier of macromeres and a lower tier of macromeres. The upper tier of macromeres lying immediately under the equator of the embryo contains the ectodermal material; the lower tier of macromeres, lying nearer to the vegetal pole, contains the endodermal material. The micromeres develop into mesenchyme, which later produces the larval skeleton consisting of calcareous spicules. In the sea urchin

Paracentrotus lividus, the cytoplasm of the macromeres possesses a surface layer of red pigment granules, making the macromeres easily distinguishable from the other cleavage cells. The red pigment is already present in the egg at the beginning of cleavage as a broad subequatorial zone.

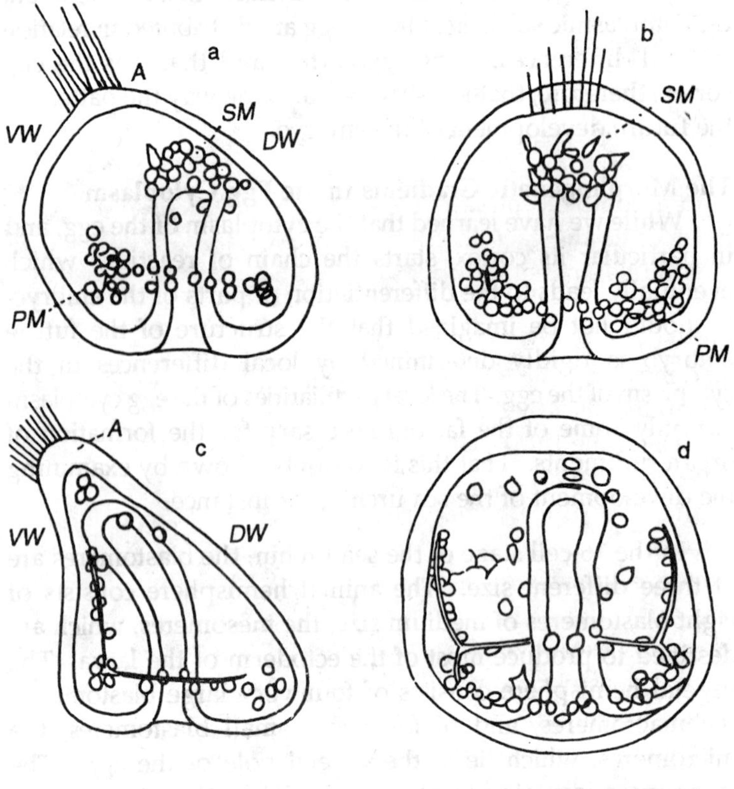

Fig. 1.26. Early larvae of the sea urchin *Parechinus* showing development of skeleton. *pM*, primary mesenchyme, *sM*, secondary mesenchyme; *A*, apical tuft of ciliae; *VW* and *DW*, ventral and dorsal body walls.

In the blastula stage, the cytoplasmic substances are found in the same arrangement as at the beginning of cleavage. Subsequently, the descendants of the micromeres migrate into the blastocoele where they develop the skeletal spicules; the descendants of the lower tier of macromeres invaginate, forming a pocketlike cavity — the gut, and the ectoderm produces a ciliary branch serving for locomotion, and a tuft of rigid cilia at the former animal pole. The ectoderm also sinks inward to produce a stomodeum, coming into communication with the endodermal gut at its anterior end, while the original opening of the pocketlike invagination the blastopore) becomes the anal opening.

The embryo of the sea urchin develops into a larva called a pluteus. If the blastomeres of the sea urchin are separated in the two-cell state or in the four-cell stage, each of them develops into a complete pluteus of diminished size. This may be related to the fact that the first two cleavage planes are meridional, passing through the main axis of the egg. All of the first four blastomeres therefore get equal portions of the three cytoplasmic regions of the egg (ectodermal, endodermal and mesenchymal). A different result is observed if the egg is separated into the animal and vegetal halves after the third cleavage, the third cleavage plane being equatorial. In this case both halves produce, as a rule, defective embryos. The animal half differentiates as an ectodermal vesicle; it does not produce a gut. Even purely ectodermal structures are abnormally differentiated (i.e., the ciliary band is not developed; the tuft of long cilia on the animal pole, on the other hand, develops excessively, growing over a much greater surface than it does normally. The vegetal half is differentiated into an ovoid embryo with a disproportionately large endodermal gut but without a month. There may be a few irregular skeletal spicules but no arms and no ciliary band.

In later cleavage stages, it is possible to cut the morula transversely below the equator. In such cases the vegetal part develops still more abnormally: it produces a large endodermal gut and a small ectodermal vesicle. The gut does not lie inside the ectodermal vesicle but is evaginated to the exterior (turned inside out), due to an interference with the normal processes of gastrulation. This phenomenon is known as exogastrulation. The mesenchyme cells migrate into the interior, but they usually produce no skeletal spicules or only very small and abnormal ones.

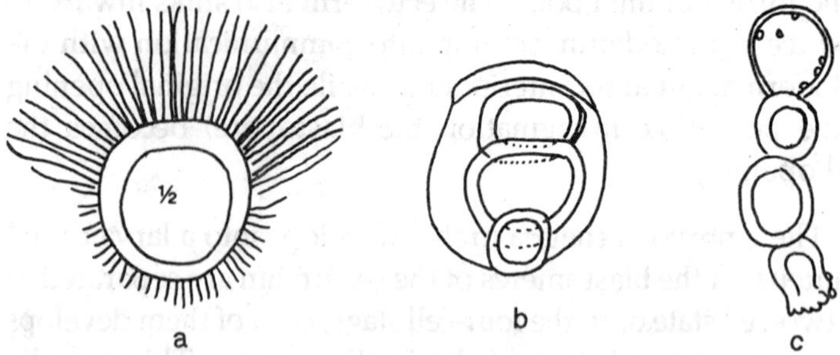

Fig. 1.27. Animalized *(a)* and vegetalized *(b, c)* larvae of sea urchin.

The result of this experiment is obviously due to the fact that each of the two halves lacks some parts contained in the other half. However, each half does not simply produce what would normally have been the fate of the respective part. Instead, the differentiation of each half seems to be "exaggerated" as compared with its normal fate. This is especially clear in the case of the increase animal tuft of cilia.

The same "exaggeration" of the ectodermal or of the endodermal differentiations can be achieved by exposing the developing eggs to certain chemical substances, even without removing any parts of the egg or embryo. If the fertilized sea urchin egg is exposed to sea water containing some lithium salts in solution, the embryo develops just as if it were only the isolated vegetal half. The gut is increased and tends to exogastrulate instead of invaginating toward the interior. The skeleton is absent or abnormal; the ectoderm is represented by an epithelial vesicle and fails to differentiate further. The increase in the size of the gut occurs at the expense of the ectoderm, and in extreme cases most of the embryo differentiates as an enormous everted gut, with the ectodermal vesicle reduced to a tiny appendage. The opposite effect is achieved if, before fertilization, the egg is exposed to artificial sea water lacking calcium ions but with sodium thiocyanate (NaSCN) added to it. In this case the gut is diminished or completely absent, the ciliary bands in the ectoderm fail to develop, and the tuft of stiff cilia at the animal pole is increased in size.

It appears that all these phenomena may be accounted for by assuming that there exist in the sea urchin's egg two factors or principles which are mutually antagonistic yet interact with each other at the same time, and that the normal development is dependent on a certain equilibrium between the two principles. Each has its center of activity at one of the poles of the egg. The activity diminishes away from the center, producing a gradient of activity. The two gradients of activity decline in opposite directions : the one from the animal pole, the other from the vegetal pole. Taking into account this type of distribution of activities, the factors or principles have themselves been called gradients. The two gradients are therefore the animal gradient, with a center of activity at the animal pole, and the vegetal gradient, with a center of activity at the vegetal pole.

According to this theory normal development depends on the presence of both gradients and an equilibrium between them. If the animal gradient is weakened or suppressed, the vegetal gradient becomes preponderant and the embryo is vegetalized, i.e., it develops to excess parts pertaining to the vegetal gradient, such as the gut. If the vegetal gradient is weakened or suppressed, the animal gradient becomes preponderant and the embryo is animalized, i.e., it develops to excess parts pertaining to the animal gradient, such as the tuft of stiff cilia at the animal pole. Other structure of the embryo, such as the ciliary bands and the skeleton, can only develop if both gradients are active, and the development is the more nearly normal the more the two gradients approach a correct equilibrium.

The equilibrium between the two gradients may be upset in various ways. The animal gradient may be weakened, with concomitant vegetalization of the embryo, by removing its center of activity (the blastomeres at the animal pole of the egg) or by the action of lithium salts. The vegetal gradient may be weakened by removing its center of activity (the vegetal part of the egg) or by the action of sodium thiocyanate.

The gradient concept aptly covers the result of various experiments on sea urchin's eggs. In addition we shall consider the following experiments. It is possible to separate from each other the three groups of blastomeres in the 16-cell stage-the mesomeres, the macromeres and the micromeres — and then recombine them at will. In isolation the three groups develop as follows :

Isolated mesomeres develop into a vesicle with a tuft of cilia (as mentioned before), due to a preponderance of the animal gradient and animalization.

Isolated macromeres plus micromeres develop into an extremely vegetalized embryo with evaginated gut, the result of a preponderance of the vegetal gradient.

Mesomeres plus macromeres develop into a practically normal embryo; the macromeres bear a sufficiently strong vegetal gradient to counterbalance the animal gradient. Mesenchyme and skeleton develop in such embryos in spite of the absence of micromeres.

Micromeres alone are not capable of any development, since they do not keep together but fall apart.

Mesomeres (ectoderm) plus micromeres (mesenchyme) develop into a complete and more or less normal pluteus. This combination is especially illuminating since the gut of such embryos develops in spite of the absence of the macromeres which should normally have supplied the material for the gut. However, the embryo possesses the two gradients, the animal gradient borne by the mesomeres and the vegetal gradient borne by the micromeres, and the possession of the two gradients seems to create the necessary conditions for normal development.

What has been said is sufficient to show that the presence of different cytoplasmic substances in the egg does not necessarily mean that there is a direct relation between the substances in the cytoplasm of the egg and certain specific organs, in the sense that the cells containing the particular kind of cytoplasm develop directly into the corresponding organ. In the last experiment quoted, the absence of the cytoplasm with red pigment did not prevent the development of the gut because the necessary conditions for the gut development were provided for in another way.

Physicochemical Nature of the Animal-Vegetal Gradient System in Sea Urchin Eggs. The existence of animal and vegetal gradients is proved not only by the patterns of morphogenetic processes occurring after certain interventions in the normal development of the egg, but also by the direct demonstration of the gradients as peculiar physicochemical states of the cells of the developing embryo. One way of proving their existence is to study the reduction of dyes by the embryo under conditions of anaerobiosis. Sea urchin embryos (or embryos of other animals for that matter), slightly stained with the vital dye Janus green (diethylsafraninzaodimethylaniline), are placed in a small chamber sealed off with petroleum jelly. After some time all the free oxygen contained in the chamber is used up as result of the respiration of the embryos. After that the embryos respire using Janus green as an acceptor of hydrogen, which is first reduced to a red dye, diethylsafranin, and further to a colourless substance, leucosafranin. The light grayish blue color of Janus green first changes into red, and then, as a second step in the reduction of the dye, the colour disappears completely.

The reduction of Janus green, shown by the colour change, does not occur simultaneously in the whole embryo but follows a very characteristic sequence. In the late blastula and early gastrula of the sea urchin, the change in colour is first noticeable at the vegetal pole, as the point where the primary mesenchyme is given off. Then the colour change spreads out involving the whole vegetal hemisphere, and reaches the equator. At this stage a spot of changed colour appears at the animal pole and also gradually increases, so that the bluish colour remains only in the form of a ring lying well above the equator in the animal hemisphere. Even this ring eventually disappears. The red colour now begins to fade in the same sequence, starting first from the vegetal pole and then from the animal pole.

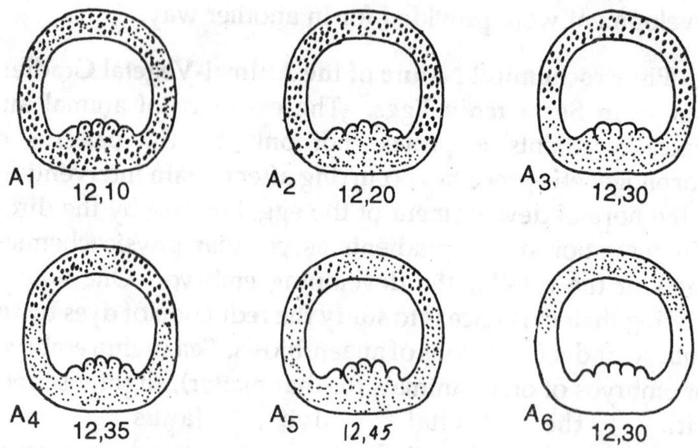

Fig. 1.28. Reduction of Janus green in a late blastula of a sea urchin. Large dots represent blue colour, small dots, red; no dots show a complete fading of the stain. Figures indicate time of observation.

The order in which the dye is reduced in the embryo shows such a nice correspondence to the postulated animal and vegetal gradients that this alone would justify our mentioning these experiments. However, the connection goes much further. Where either the animal or vegetal gradient is suppressed, the corresponding gradient of reduction also disappears. In isolated animal halves of sea urchins' eggs the animal tendencies of development are predominant and it is found that such animalized embryos start reducing Janus green at the animal pole only; there is no center of reduction at the vegetal pole. In isolated vegetal halves the animal tendencies of development are suppressed and the vegetal tendencies are supreme. Correspondingly, the only center of reduction is at the vegetal pole; no center of reduction appears at the animal pole. Embryos animalized or vegetalized chemically show reduction gradients similar to isolated animal and vegetal halves of the egg.

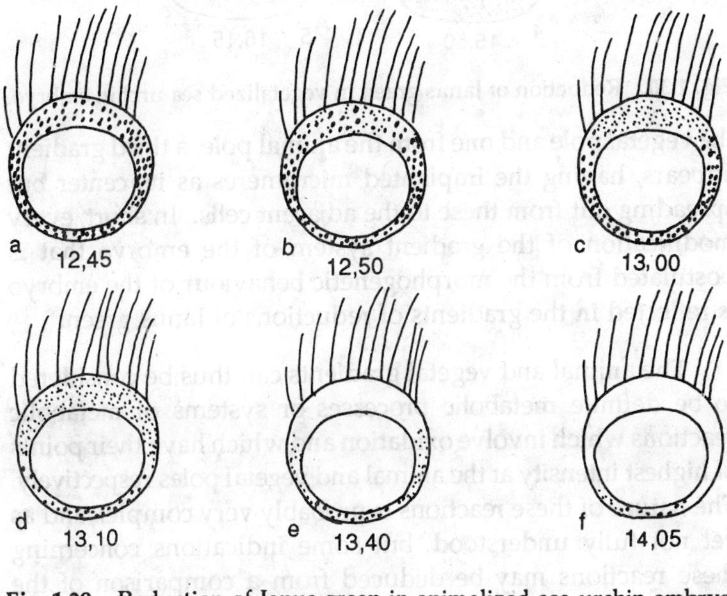

Fig. 1.29. Reduction of Janus green in animalized sea urchin embryo (isolated animal half).

We have seen that the micromeres are carriers of the highest point of the vegetal gradient, and they preserve this property after transplantation. Accordingly, the micromeres can also serve as a center of a reduction gradient.

If a group of four micromeres is implanted laterally, and the embryo is tested for reduction of Janus green, it can be seen that in addition to the two normal gradients, one from

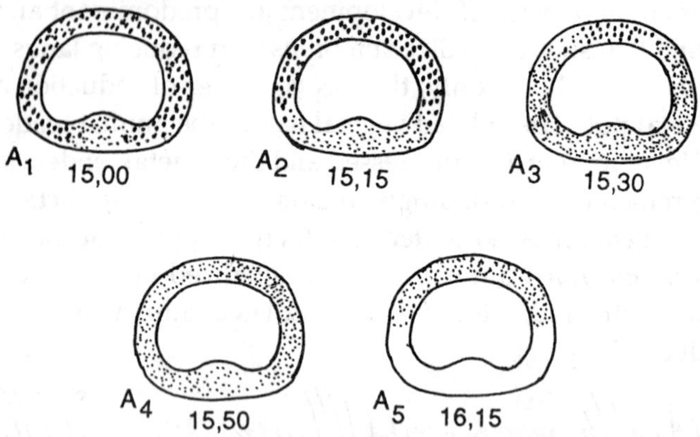

Fig. 1.30. Reduction of Janus green in vegetalized sea urchin embryo.

the vegetal pole and one from the animal pole, a third gradient appears, having the implanted micromeres as its center but spreading out from these to the adjacent cells. In short, every modification of the gradient system of the embryo that is postulated from the morphogenetic behaviour of the embryo is reflected in the gradients of reductions of Janus green.

The animal and vegetal gradients can thus be considered to be definite metabolic processes or systems of metabolic reactions which involve oxidation and which have their points of highest intensity at the animal and vegetal poles respectively. The nature of these reactions is probably very complex and as yet not fully understood, but some indications concerning these reactions may be deduced from a comparison of the chemical agents which may cause vegetalization or animalization.

Apart from the use lithium ions, vegetalization may be caused by sodium azide and dinitrophenol. Both of these substances belong to a group of enzyme poisons; the azide is known to inactivate the cytochrome oxidase system, and the dinitrophenol disturbs respiration by preventing oxidative phosphorylation, that is, the formation of energy-rich bonds between phosphoric acid and adenosine diphosphate (with the formation of adenosine triphosphate). This might mean that vegetalization is based essentially on a disturbance of oxidative processes in the embryo and, perhaps even more specifically, a disturbance of the processes of phosphorylation.

That the action of lithium ions is along the same lines may be deduced from a number of observations of which we shall single out the following. Lithium salt treatment suppresses the rise of oxygen consumption which occurs normally at the beginning of gastrulation, that is, at the same time as the morphological effects of vegetalization begin to be apparent. Furthermore, inorganic phosphate accumulates in the sea water during development of lithium ion-treated embryos — again in a hint that these embryos are incapable of utilizing the energy of oxidation for phosphorylation and synthesis of adenosine triphosphate.

The whole sequence of reactions involved in vegetalization must, of course, be far more complicated. Thus, it has been found that the relative amounts of various amino acids change as a result of vegetalization in lithium ion-treated embryos. Furthermore, if a modification of the respiratory system of the embryo is the essential feature of vegetalization, it remains to be discovered why the processes of morphogenesis at the animal pole of the embryo (essentially the ectodermal organs) and more severely affected than those of endodermal parts developing at the vegetal pole.

As to animalizing agents, the first of those mentioned above, sodium thiocyanate, has been found to be rather unreliable. Some batches of eggs treated by this chemical do not reach at all. In other batches individual embryos become animalized to greatly varying degrees. Subsequently, many other substances were found whose animalizing action is much more predictable. These belong to several distinct groups :

1. *Some metals* : zinc, mercury.

2. *Some proteolytic enzymes* : trypsin, chymotrypsin.

3. *Many acidic dyes* : in particular some possessing sulfonic (HSO_3) groups in their molecules, such as Evans blue, chlorazol sky blue, trypan blue, Congo red, and others possessing carboxyl groups (COOH), such as uranin and rose bengal.

4. Some other sulfonated organic compounds: germanin and others.

5. Some anionic (acid) detergents.

Animalization may be achieved by extreme dilutions of a chemical; for instance, 1:50,000 in the case of Evans blue.

It seems that the common property of most if not all the above agents is their ability to attack proteins or to form compounds with proteins, especially with basic proteins. In the case of the very large number of sulfonated and carboxylated dyestuffs, it is fairly certain that their acidic groups become attached to the functional side groups of protein molecules. The two metals mentioned above, zinc and mercury, also become easily bonded to side groups of protein molecules. The case of mercury is very remarkable : mercuric chloride, a powerful poison used therefore as a fixative for proteins, curic chloride, a powerful poison used therefore as a fixative for proteins, causes animalization when applied for a very short time in great dilutions (1 : 90,000). It is probable that by blocking the active side groups of the protein molecules, the agents referred to "immobilize" these molecules, preventing them from interacting normally with other cell constituents. Even the steric configurations of protein molecules may thus be altered, since these are based on lateral bonding between molecules and their parts. A far-reaching change of the properties of proteins may thus be brought about.

In the case of animalizing agents, we are in the fortunate position that their point of attack can be clearly demonstrated. The dyes which have been used for this purpose stain the cells of the embryo, and it can be seen that the first cells to be stained are the primary mesenchyme cells at the vegetal pole.

With higher concentrations of the dye or with a longer duration of the treatment, the adjoining presumptive endodermal cells show the staining too. The zinc taken up into the cells of embryo may be made visible because it gives a pink colouration with dithizone. In embryos first treated with zinc and later immersed in a dithizone solution, the pink colour is detectable in the same position as in the case of animalizing dyes. It is thus quite clear that animalization is due to damage to the center of the vegetal gradient.

Considering what has been said about the mechanism of action of the vegetalizing and animalizing agents, we may conclude that neither animalization nor vegetalization is ever achieved by a stimulation of one of the gradients, but always by a depression of the opposite gradient (or of the metabolic system which is responsible for the gradient). The increased development of any of the structure is then brought about by the alternative morphogenetic system expanding when the one is reduced in vigor. The competitive relation between the two gradients is thus demonstrated in a most convincing way.

Manifestation of Maternal Genes During the Early Stages of Development

In the previous sections evidence has been presented that: (1) The differences between the various parts of the embryo are not due to differences in their nuclei but to differences of the cytoplasm. From this it may be inferred that (2) the nuclei of the cleavage cells and the genes contained therein do not control the earliest stages of development of the animal egg.

This does not mean, however, that genetic factors have nothing to do with the early stages of development. The genetic factors, the genes, do play a part in this period but in a very special way. It is not the genes contained in the nuclei of the blastomeres but the genes in the cells of the material body that determine the peculiarities of the egg and its early development. The best-known example of this type of gene action is concerned with the inheritance of coiling in gastropod molluscs.

In the snails, as a rule, the shell is coiled spirally, and the internal organization of the animal shows a corresponding dislocation, some of the organs (heart, kidney and gills) being twisted around through an angle of nearly 180°. The direction of coiling of the shell is clockwise, if viewed from the apex of the shell. This type of coiling is called dextral. As an exception, the coiling may be counterclockwise, or sinistral. Individuals having a sinistral coiling of the shell have the internal organs dislocated in the opposite direction from that in dextral individuals. In short, dextral and sinistral individuals comport themselves in every detail of organization as mirror images of each other.

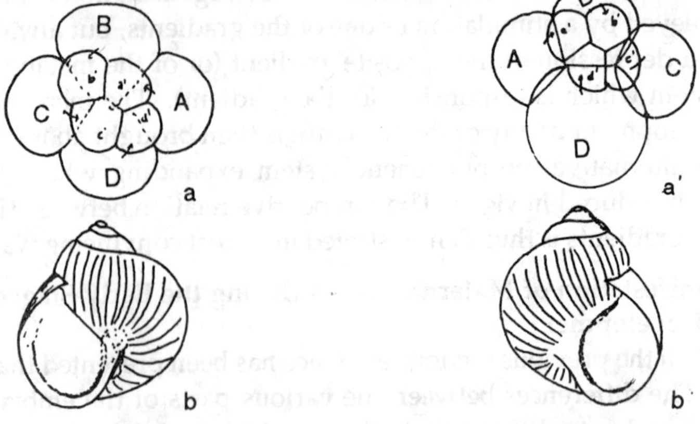

Fig. 1.31. Correlation of sinistral *(a)* and dextral *(a')* cleavage with sinistral *(b)* and dextral *(b')* coiling of the shell in gastropods.

The eggs of gastropod molluscs show a spiral type of cleavage, and it has been noted that the type of cleavage of the egg has a definite relation to the type of coiling of the adult. If the cleavage is dextral—that is, if the cleavage spindles show a clockwise spiraling — the adult snail also has a dextrally coiled shell. If the cleavage is sinistral, the coiling of the shell is likewise sinistral. The connection between the type of cleavage and the coiling of the shell is established through the position of blastomere *D* in respect to the other blastomeres. Since it is from blastomere *D* that the mesoderm develops, its situation is reflected in the position of the internal organs, and the coiling of the shell is one of the secondary expressions of this asymmetry.

There are some species of gastropods in which all the individuals are sinistral, but the main interest attaches to a species in which sinistral individuals occur as a mutation among a population of normal dextral animals. Such a mutant was discovered in the fresh-water snail *Limnaea peregra*. Breeding and cross breeding of dextral and sinistral snails showed that the difference between the two forms is dependent on a pair of allelomorphic genes, the gene for sinistrality being recessive (1), and the gene for the normal dextral coiling being dominant (L). The two genes are inherited according to mendelian laws, but the action of any genetic combination is manifested only in the next generation after the one in which a given genotype is found.

Thus, if the eggs of a homozygous sinistral individual are fertilized by the sperm of a dextral individual, the eggs cleave sinistrally, and all the snails of this F_1 generation show a sinistral coiling of the shell. The genes of the sperm do not manifest themselves, although the genotype of the F_1 generation is L_1.

If a second generation (F_2) is bred from such sinistral individuals, it is all dextral, instead of showing segregation as would be expected in normal mendelian inheritance. In fact, segregation does take place in the F_2 generation so far as the genes are concerned, but the new genic combinations fail to manifest themselves, since the coiling is determined by the genotype of the mother. The genotype being L1, the gene for dextrality dominates and is responsible for the exclusively dextral coiling of the second generation. Only in the third generation (F_3) does segregation in the proportion of 3 to 1 become apparent, and then not as segregation between individuals in each brood, but as a segregation of broods.

Each brood — that is, the offspring of an individual of the F_2 generation — shows a coiling that is determined by the maternal genotype. Since the individuals of the F_2 generation had the genotypes 1 LL, 2 L1, 1 ll, three quarters of them, on the average, produce eggs developing into dextral snails, and one quarter produce eggs developing into sinistral individuals. Of the dextral broods, one third breed true, giving only dextral offspring, and two thirds show further segregation among the broods of the next generation.

It is easy to understand that the results of a reciprocal cross -- that is, of the fertilization of the eggs of a homozygous dextral individual by the sperm of a sinistral individual — will lead to a somewhat different type of pedigree: the F_1 generation will be all dextral (with genotype L1) and the F_2 generation again all dextral (with genotypes LL, L1, 1L and ll). The F_2 generation will show segregation among broods, just as in the cross examined first.

The whole case becomes clear if it is realized that the type of cleavage depends on the organization of the egg which is established before the maturation divisions of the oocyte nucleus. The type of cleavage is therefore under the influence of the genotype of the parent (the mother). The haploid state of the egg nucleus continues for only a very short time and cannot materially affect the organization of the egg. The sperm enters the egg after this organization is already established. Similarly, we should expect that the elaboration of cytoplasmic organ-forming substances is under the control of maternal genes, even though this cannot be proved because no mutants are known which produce a difference in these substances.

It is conceivable that maternal genes might produce conditions in the egg leading to morphogenetic processes, which take place at a later stage of development, without visibly modifying the cleavage pattern. Several cases of such an influence are actually known.

In the axolotl *(Ambystoma mexicanum)* a lethal gene "o"

has been discovered which has a maternal effect. The gene is recessive, so that heterozygote individuals (the + o individuals) are completely normal. Homozygote animals (oo) produced by a cross between two heterozygote parents develop normally in the early stages but in later life shown a slight retardation of growth. Their regeneration capacity is severely reduced so that amputated legs are not restored as would occur in normal axolotls. The homozygote males are sterile : their testes are underdeveloped and spermatogenesis does not go beyond the spermatogonial stages. The homozygote females produce eggs capable of fertilization and normal cleavage, but at the onset of gastrulation the development is retarded. The embryos start gastrulation but usually do not go beyond the crescentic blastopore stage and then die. Occasionally gastrulation is completed but the affected embryos never enter the stage of organogenesis. This abnormal course of development is in no way affected by the genotype of the spermatozoon : the spermatozoon may carry the normal + gene or a mutant o gene (if the male is a heterozygote). The arrest in development is in both cases exactly the same : this shows that the particular type of abnormal early development does not depend on the genes present in the cells of the developing embryo, but exclusively on the genes of the mother (maternal effect).

The effect of the maternal genotype oo may be neutralized by injecting into freshly fertilized eggs or into eggs at the beginning of cleavage (two blastomeres) the cytoplasm from a normal, ripe egg. As little as 1 to 5 per cent of normal cytoplasm improves development and causes most of the embryos to complete gastrulation, enter organogenesis and sometimes produce swimming larvae. This experiment proves that the lethal embryos lack something which is present in the cytoplasm of the normal embryo. This "something" which has been termed the "corrective factor" must be the result of the activity, in the mother, of the unmutated + gene. In the homozygous female the o gene is unable to produce the corrective factor; the maternal lethal effect is thus due to a specific deficiency.

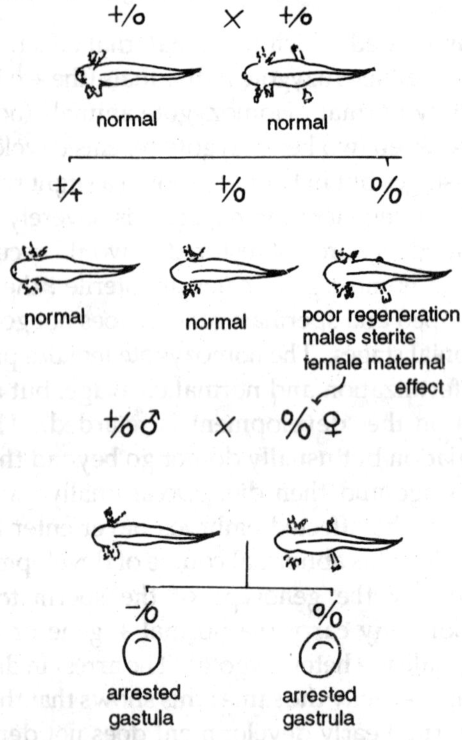

Fig. 1.32. Maternal inheritance and manifestation of the o gene in the axolotl.

Further experiments showed that in the oocyte a high concentration of the corrective factor is contained in the nuclear sap (0.2 to 0.5 per cent is sufficient to improve development) but that the cytoplasm of the oocyte when injected into a lethal egg has hardly any effect. It is only when the nuclear membrane of the oocyte ruptures in the process of maturation and the nuclear sap mixes with the cytoplasm that the cytoplasm acquires the ability to correct development of the lethal embryos. Centrifugation of the cytoplasm of normal eggs showed that the corrective agent remains in the supernatant after sedimentation of all cellular inclusions (yolk, pigment, mitochondria and ribosomes) but that it sediments after further prolonged high speed centrifugation. It is thus a macromolecular substance which is not bound to any cellular organoids. The substance is inactivated by heating to 50-55° C, or by treatment with trypsin, and it may be sedimented by ammonium sulphate. Thus it shows characteristics of a protein.

From this work it emerges that the normal allele of the gene o is responsible for the accumulation in the nuclear sap of immature oocytes of a protein-like substance, which, upon the nuclear membrane's breaking up during maturation, enters the cytoplasm and is indispensable if the embryo is to go through gastrulation and enter the phase of organogenesis.

The corrective substance is of course not the only substance produced under the influence of the maternal genes while the egg is developing in the ovary. Very many, if not all, of the cytoplasmic substances in the egg mentioned in section 4-3 must have a similar origin. The cytoplasmic substances determining the properties of blastomeres during cleavage may thus be an after-effect of the action of genes at a previous stage of the reproductive cycle.

BLASTULATION

The blastomeres in the early cleavage stages tend to assume a spherical shape like that of the egg before cleavage. Their mutual pressure flattens the surfaces of the blastomeres in contact with each other, but the free surfaces of each blastomere remain spherical, unless these outer surfaces are also compressed by the vitelline membrane. The whole embryo acquires, in this stage, a characteristic appearance reminiscent of a mulberry. Because of this superficial similarity, the embryo in this stage is called a morula (Latin for mulberry).

The arrangement of the blastomeres in the morula stage may vary in the different groups of the animal kingdom. In coelenterates it is often a massive structure, blastomeres filling all the space that had been occupied by the uncleaved egg. Some of the blastomeres then lie externally, others in the interior.

(Some embryologists apply the name morula to this type of embryo only). More often, as the egg undergoes cleavage, the blastomeres become arranged in one layer, so that all the blastomeres participate in the external surface of the embryo. In this case a cavity soon appears which at first may be represented just by narrow crevices between the blastomeres, but which gradually increases as the cleavage goes on. This cavity is called the blastocoele.

As the cleavage proceeds, the adhesion of the blastomeres to each other increases, and they arrange themselves into a true epithelium. In cases where a cavity has been forming in the interior of the embryo, the epithelial layer completely encloses this cavity, and the embryo becomes a hollow sphere, the walls of which consist of an epithelial layer of cells. Such an embryo is called a blastula. The layer of cells is called the blastoderm, and the cavity is the blastocoele already mentioned.

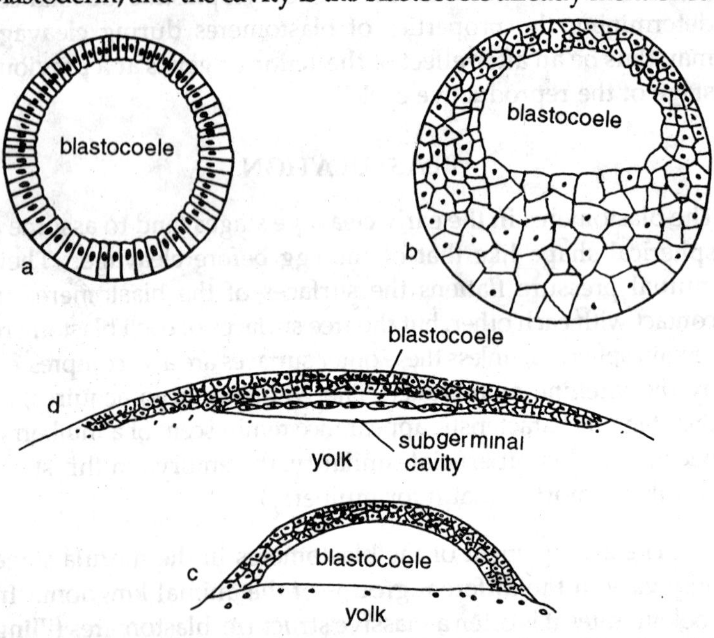

Fig. 1.33. Diagrammatic comparison of blastulae of an echinoderm *(a)*, a frog *(b)*, a bony fish *(c)* and a bird *(d)*.

The types of blastulae greatly varies among different animals depending upon various factors such as the egg size, the amount and distribution of yolk, and the rate and pattern of cleavage. The various types of blastulae can be classified into four types : (i) stereoblastula, (ii) coeloblastula, (iii) periblastula, and (iv) discoblastula.

(1) Holoblastically Formed Blastulae

In holoblastic embryos, the blastula is either solid (stereoblastula or hollow (coeloblastula).

(i) *Stereoblastula.* This type of blastula is composed of densely-packed but larger sized and relatively small number of cells. Blastocoelic space in the centre is very small or virtually absent. The blastomeres in a stereoblastula may reach from the surface to the centre ; or separate blastomeres may occupy the interior. Examples : Insects, some worms (e.g., *Nereis)* and some mollusks (e.g., *Crepidula).*

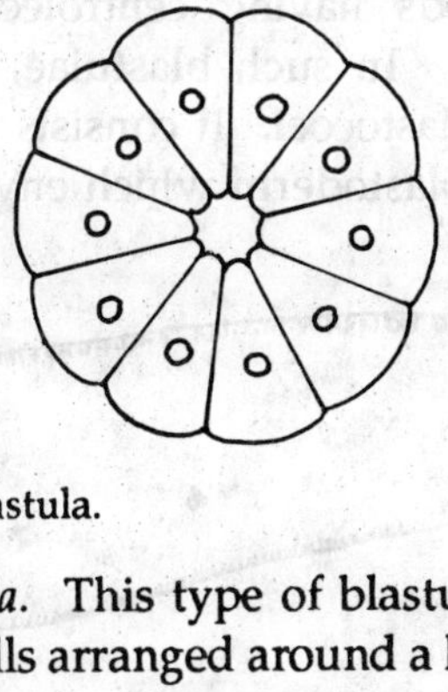

Fig. 1.34. A stereoblastula.

(ii) *Coeloblastula.* This type of blastula consist of one or several layers of cells arranged around a large blastocoel. It is of two types :

(a) *Adequal.* The cells of the blastoderm are of about the same size throughout. The blastocoel is centrally placed, e.g., echinoderms.

(b) *Inequal.* The wall of the vegetal half is thicker than that of the animal half. The blastocoel is reduced in volume and distinctly *eccentric, i.e.,* displaced towards the animal pole, e.g., *Amphioxus,* amphibians.'

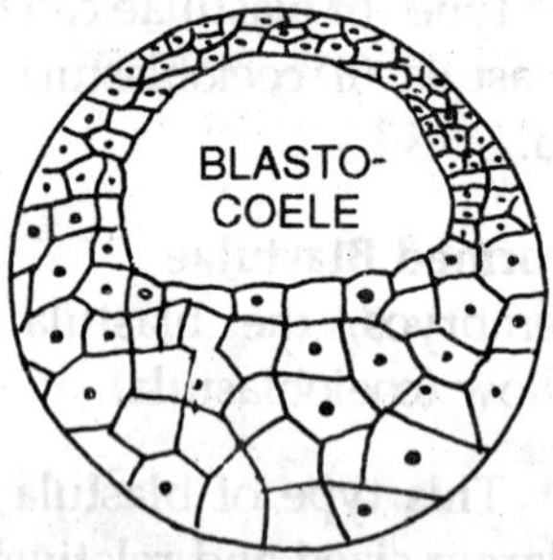

Fig. 1.35. An inequal coeloblastula.

(2) Meroblastically Formed Blastulae

The blastula stage consists of a layer of cells over yolk.

(i) *Periblastula.* This type of blastula is found in insects and some arthropods having centrolecithal eggs with a superficial cleavage. In such blastulae, there is no cavity comparable to the blastocoel. It consists of a single layer of epithelial cells, the blastoderm which envelopes the mass of nucleaved yolk.

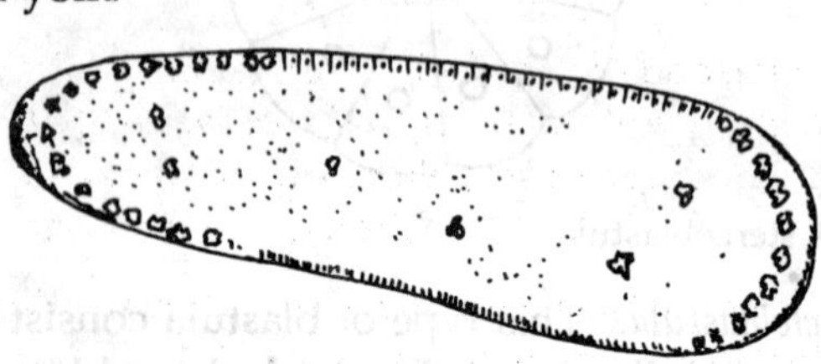

Fig. 1.36. A Periblastula of an insect.

(ii) *Discoblastula.* It is found in sharks, bony fishes, reptiles, birds and prototherian mammals. In these animals, eggs are highly telolecithal. They undergo discoidal cleavage which results in the formation of a *blastodisc* on top of the undivided yolk. The disc consists of several layers of cells.

In a *bony fish* or a *shark,* cleavage is restricted to the cap of

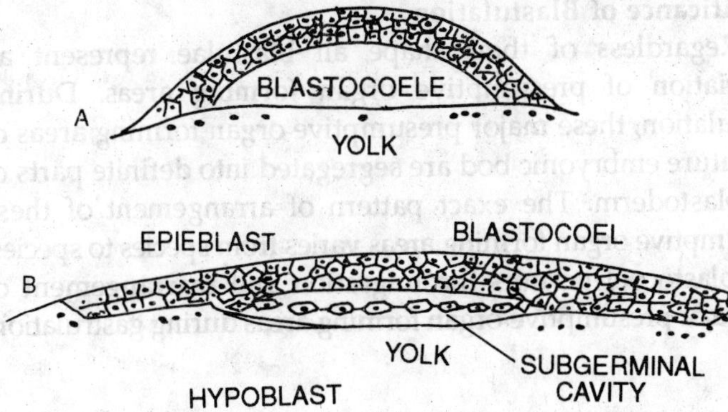

Fig. 1.37. A. Blastula of a bony fish. B. Blastula of a bird.

cytoplasm on the animal pole of the egg where formation of blastoderm takes place. Therefore, the blastoderm assumes the shape of a disc called *blastodisc* lying on the animal pole. The blastodisc is more less convex. It encloses, between itself and the uncleaved residue of the egg, a cavity representing the *blastocoel.*

In *reptiles* and *birds* the blastocoel forms by a lifting away of some of the cells of blastoderm. The cells are incorporated into a superficial layer, called the *epiblast*. A small number of cells remain underneath the superficial layer, forming a thin flat epithelium, called the *hypoblast.* The cavity between epiblast and hypoblast in birds and reptiles corresponds to the *blastocoel* of amphibians and fishes. The space between the hypoblast and the yolk is designated as the *subgerminal cavity.*

The cavity in birds and reptiles is formed only under the central part of the blastoderm. No cavity is formed under the layer of blastomeres in the periphery of the blastodisc. The disc in thus subdivided into two parts : *(i)* the more transparent central part called the *area pellucida* and, *(ii)* the opaque peripheral part, the *area opaca*. Area pellucida is transparent because of the presence of the cavity under the layer of blastomeres. The peripheral area looks opaque as the blastomeres rest directly on the yolk and there is no cavity underneath.

Significance of Blastulation

Regardless of their shape all blastulae represent an association of presumptive organ forming areas. During blastulation, these major presumptive organ forming areas of the future embryonic bod are segregated into definite parts of the blastoderm. The exact pattern of arrangement of these presumptive organ forming areas varies from species to species. The blastocoel permits the migration and rearrangement of the major presumptive organ forming areas during gastrulation.

2

Gastrulation

The events which transform a blastula into a gastrula are collectively called *gastrulation.* According to *McEwen (1923)* gastrulation refers to the formation of the primordial gastric or gut cavity called the *archenteron.* In addition to the formation of gastrular cavity, the process also involves the development of two of the three primordial *germ layers,* referred to as the *ectoderm* (epiblast) and *endoderm* (hypoblast). The third layer called *mesoderm* (mesoblast) is derived from one or the other or both of the two layers ; the endoderm is the inner layer ; and the mesoderm is the middle layer.

Lankester (1875) and *Hubrecht (1906)* were of the opinion that the gastrulation is a process during which the single layered blastula is converted into a two-layered (e.g., Amphioxus) or *three-layered* (e.g., most vertebrates) embryonic stage called *gastrula.*

Gastrulation begins as soon as blastulation is accomplished. With the appearance of this new process we think of the embryo as progressing into the next phase of its development, commonly called the "gastrula stage" (diminutive from Greek word *gaster,* stomach). In forms with little yolk, such as *Amphioxus,* gastrulation is essentially an inpocketing of the blastula. A double-layered cup is formed from a single-layered hollow sphere much as one might push in a hollow rubber ball with the thumb. The new cavity in the double walled cup is termed the *gastrocoele* or *archenteron.* The opening from the outside into the gastrocoele has traditionally been called the *blastopore,* although it would be far more logical to call it the *gastropore.* Thus in gastrulation the single cell layer of the blastula has been differentiated to form two layers. The outer cell layer is known as the *ectoderm,* and the inner layer as the *entoderm.* These germ layers, as they are called, differ from each other in their positional relationship to the embryo and to the surrounding environment. Each has different potentialities, and each will in the course of development give rise to quite different types of tissues and organs.

The intriguing thing about these primary germ layers is not so much what we can see in them when they first arise, as the unseen developmental potentialities we know lie hidden within their cells. For we know that cells which look all alike to us when they first emerge as part of the ectoderm are destined to produce descendants as different as the horny outer cells of our skin and the elaborate photosensitive rods and cones of our retina. Cells of the endoderm will become as unlike as the iodine greedy cells of the thyroid which produce the hormone thyroxin, and the ciliated epithelial cells moving the cleansing film of mucus over the lining of our respiratory passages. It is a realization of the divergent developmental routes to be followed by their cells that makes the emergence of the germ layers of so much interest. For the same reasons we now think of gastrulation more as the process which begins the segregation of the germ layers than as the mechanism of forming the gastrula cavity.

THE PROMINENT FEATURES OF GASTRULATION

The process of gastrulation is basically characterized by following events:

1. A rearrangement of the cells of the embryo takes places by means of *formative* or *morphogenetic movements.*

2. The pace of *cellular division* is slowed down.

3. *Growth,* if any, is trifling.

4. The rate of oxidation is intensified and the type of *metabolism changes.*

5. The *nuclei* become more active in governing the activities of the embryonic cells.

6. The influence of the *pateraal chromosomes* becomes clear during gastrulation.

7. New kinds of *proteins*, that were formerly not present in the egg, begin to be synthesized.

FATE MAPS

A correct interpretation of gastrulation is only possible when we have a perfect knowledge of the position of presumptive germinal layers which occupy in the blastula. A *chart, showing the fate of each part of an early embryo, in particular a blastula, is called a fate map.* The fate maps of different groups of animals may be constructed by following methods:

Construction of Fate Maps by Natural Marking

The cytoplasm of the fertilized eggs of some animals such as some ascidians (e.g., *Styela, Ciona, Cynthia partita)*, has natural colour differences in its various regions. For example, on the basis of pigmentation differentiation, four general regions may be distinguished in the egg-cytoplasm of *Styela*: an upper hemisphere of light protoplasm a yellow crescent postero-ventrally, a gray crescent antero dorsally, and a vegetative area of dark gray yolky substance. Because, of the natural differences in pigmentation, the developmental fate of these regions has been worked out with considerable precision. Knowing what these developmental prospects are, we may then view the areas as the components of a "map" of prospective organ regions. Accordingly, the upper clear plasm represents prospective epidermal ectoderm, the yellow crescent is prospective mesoderm, dark gray yolky plasm is prospective endoderm, and gray crescent is prospective neutral ectoderm and notochord.

By vital staining marking methods, the prospective organ region maps or fate maps have been made for a number of verte-brate species. The fate maps of an ascidian, *Amphioxus,* frog and chick will be described in sufficient details in coming chapters.

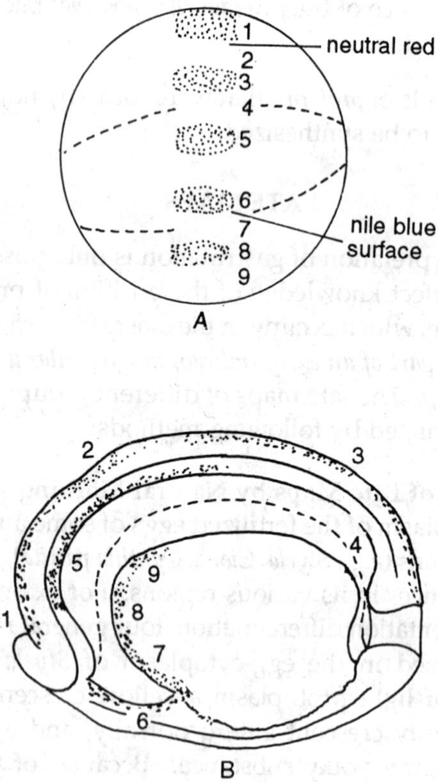

Fig. 2.1. The development of nine areas marked by vital dyes during gastrulation of a urodele.

Carbon particle marking. Independently of vital staining, another method has also been advised by *Spratt* (1946) for marking cells of a developing embryo. This method consists in applying tiny particles of carbon to the surface of the embryo. Carbon particles stick to the surface of the cells and can thus be used as markers enabling the investigator to follow the movements of cells and draw up fate maps. Recently, radioactive labelling of blasto-meres to ascertain their prospective destiny, has been profitably employed.

GENERAL PROCESSES INVOLVED IN GASTRULATION

Gastrulation is an integrated, dynamic process, controlled largely by intrinsic forces bound up in the specific physico-chemical conditions of the various presumptive organ forming areas of the late blastula and early gastrula. These forces, in turn, are correlated with external conditions. In most vertebrates, it includes different kinds of morphogenetic movements.

Morphogenetic Movements

The movement of cells in the embryo from one place to another, to establish a particular form or structural arrangement called the *morphogenetic movements.* The cell movements are not only involved in the formation of the blastula where these movements are slow, or in gastrulation where the cell migrations are dynamic and rapid but also in the later developments or even in adult animals. However, there is fundamental difference between one occurs during embryonic stages and the adult stage. The morphogenetic movements of an adult animal are of a *reversible* nature, while that of embryonic stages are of *irreversible* nature.

During gastrulation some specific type of morphogenetic movements of blastomeres occur. For the sake of convenience, these movements will be described separately. However, it should be noted that in the majority of cases two or often more of them take place together. At the moment, these morphogenetic movements of cells are grouped under two heads :

(A) Epibolic morphogenetic movements of epiboly.

(B) Embolic morphogenetic movements of emboly.

Both these words, namely *epiboly* and *emboly*, have Greek origin and they mean two different kinds of motion. The word emboly means to throw in or thrust in, while the word epiboly, on the other hand, denotes a throwing on or extending upon. Thus, these words, though have opposite meaning, but occur simultaneously during gastrulation.

(A) Epibolic Morphogenetic Movements

The process occurs only in the prospective ectodermal blastomeres (epidermal and neural areas) which have an inherent property of flattening, multiplication and rearrangement of cells. The ectoderm expands and extends in the antero-posterior direction and eventually envelops the inwardly migrating prospective mesodermal and endodermal blastomeres. In *holoblastic* eggs of *Amphioxus and frog* with rounded blastulae, the prospective ectodermal blastomeres actually engulf and surround the inwardly moving presumptive notochordal, mesodermal and endodermal blastomeres, while in *meroblastic* eggs of teleost fishes, reptiles and birds where the blastulae are flattened epibolic movements of cells are concerned largely with anterio posterior extensions, associated with peripheral migration and expansion of the ectodermal blastomeres.

Embolic Morphogenetic Movements

The embolic morphogenetic movements are concerned with inward migration of prospective chorda—mesodermal and endodermal blastomeres from the external surface of the blastula and their extension along antero-posterior axis of the developing embryo. This inward movement of the cells is due to innate forces within various groups, some apparently are autonomous i.e., they arise from forces within a particular cell group while others are dependent upon the movement of other cell groups.

The inward morphogenetic movements may occur in the following ways:

(i) Invagination

The simplest method of gastrulation is by invagination. The movements of *"infolding" or "inward bending"* of the endoderm and mesoderm are known as *invagination.* In this process, the vegetal half of the blastula is pushed in or invaginated until it almost touches the animal half opposite to it. Thus, a simple spherical body becomes converted into a double-walled cup, as if one side of the wall of an elastic hollow ball had been pushed in by an external force. The original blastocoel becomes virtually obliterated and the cavity arising through the invagination is called the *primary gut* or *archenteron.* The opening leading from this cavity to the exterior is called the *blastopore.* The blastopore, being the opening leading into the primary gut, has been likened to the mouth. Its rims, therefore, are usually referred to as the lips of the blastopore. As such, we may distinguish the *dorsal lip, the ventral lip, and lateral lips* of the blastopore, respectively.

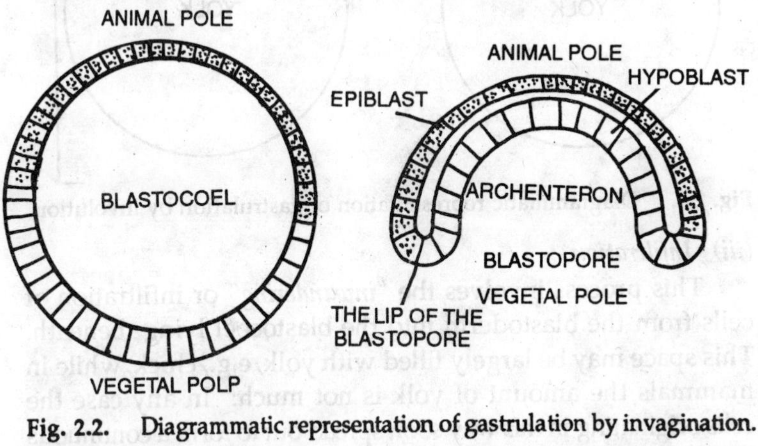

Fig. 2.2. Diagrammatic representation of gastrulation by invagination.

The archenteron is lined by the original vegetal cells which give rise to endoderm and mesoderm. The part of the gastrula that remains on the surface becomes ectoderm.

(ii) Involution or inflection

The term involution means a *"turning in" or "rolling under"* of chorda-mesodermal cells along the blastoporal lip. It is very common among the vertebrates and is often accompanied with invagination. The blastocoel is greatly reduced in eggs that gastrulate by involution. At some point on the edge of the blastoderm the dividing cells begin to be turned over the blastoporal rim i.e., *involuted* in the blastocoel. As a result, the cells located on the external margin of the blastoporal lip, roll-over the lip and are brought to the inside edge of the lip just beneath the ectodermal blastomeres. These inturned cells then constitute the hypoblast (chorda-mesoderm) while those which remain with out are epiblast (ectoderm). The involution is most active on the dorsal lip of the blastopore.

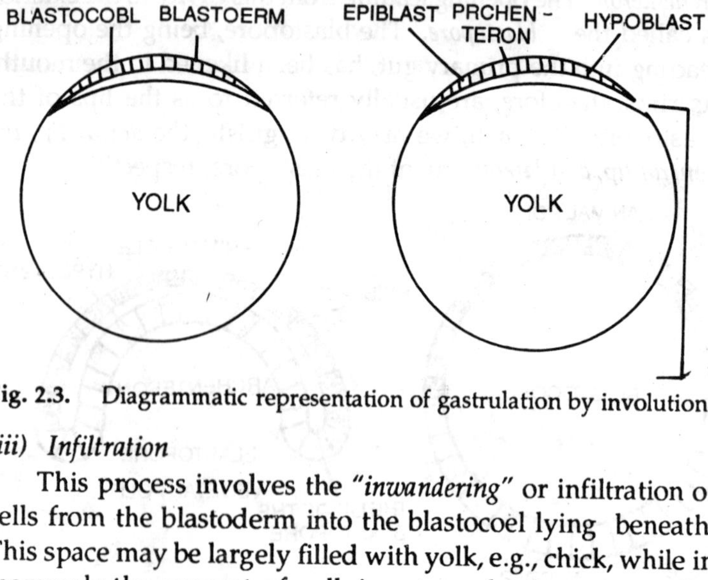

Fig. 2.3. Diagrammatic representation of gastrulation by involution.

(iii) Infiltration

This process involves the "*inwandering*" or infiltration of cells from the blastoderm into the blastocoel lying beneath. This space may be largely filled with yolk, e.g., chick, while in mammals the amount of yolk is not much. In any case the cells originating in this way soon spread out to form a continuous layer called *hypoblast*. The cavity which formerly represented blastocoel, now becomes the *archenteron*.

(iv) Concrescence

The word denotes the movements of masses of cells towards each other, particularly in the region of the blastopore, and implies the idea of fusion of cell groups from two bilaterally situated areas. This term is used in older description of gastrulation.

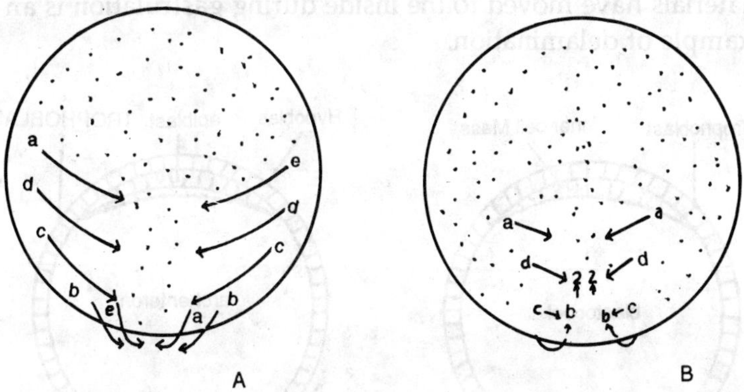

Fig. 2.4. Diagrammatic representation of concrescene and convergence. A. Surface view of the blastoderm at the beginning of the gastrulation. B. Surface view of the same near the completion of gastrulation. Most of the marginai material indicated by alphabets moves toward the median line and the dorsal blastoporal lip of the future embryo. Some of the marginal material represented by alphabets *a* and *b* "turns in" over the dorsal blastoporal lip. The material, thus involved, becomes "invisible" as shown by these alphabets in dots.

(v) Ingression

The word ingression is suitable for use in cases where a cell or small groups of cells separate from other layers and migrate into the segmentation cavity or into spaces developed within the developing body. In the primitive streak area of reptiels, birds, and mammals, for example, mesodermal cells detach themselves from the primitive streak and migrate into the space between the *epiblast* and *hypoblast*. In the formation of two layered embryo in *Echidna,* the inward migration of small entodermal cells to form the hypoblast may be regarded as cellular ingression.

(vi) Delamination

This denotes a mass separation of groups of cells from

other cell groups. The separation of notochordal, mesochordal, and endodermal tissue from each other to form discrete cellular masses in such forms as in teleost fish or the frog, after these materials have moved to the inside during gastrulation is an example of delamination.

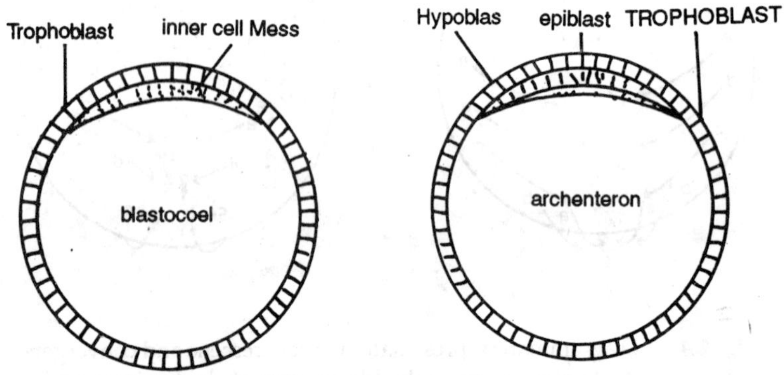

Fig. 2.5. Gastrulation by delamination.

(vii) Extension

The elongation of the presumptive neural and epidermal areas externally and of the notochordal, mesodermal, and endodermal materials after they have moved inward beneath the neural plate and epidermal material are examples of extension. The extension of cellular masses is a prominent factor in gastrulation in all chordates starting from *Amphioxus* to mammal. As a result of this tendency the entire gastrula in many cases begins to elongate in the antero-posterior axis as gastrulation proceeds. The presumptive noto-chordal material possesses great autonomous powers for extension, and hence, during gastrulation it becomes extended into an elongated band of cells.

(viii) Divergence

This movement particularly in true to the lateral plate and ventral mesoderm in the frog. It also happens in the lateral plate and extra embryonic mesoderm in reptile, bird or mammal. In this, the cells have involuted over the blastoporal lips during gastrulation, they migrate and diverge to their future position within the forming gastrula.

(ix) Constriction

This term is used with reference to the convergence of cell areas which results in the gradual closure of blastapore. It occurs in frog's egg of gastrulation and is due to a narrowing of the marginal zone and a pull of the dorsal lip.

Significance of Gastrulation

1. Gastrulation brings the presumptive structures from a positions on the external surface of the blastula into their definitive positions in the embryo.

2. It is necessary for the morphogenesis of nervous system.

GROWTH

Growth is the developmental increase in mass. It results due to synthesis of new protoplasm which includes both the cytoplasm and nucleus of cells. Besides, protoplasmic products such as the fibres of connective tissue and the matrix of bone and cartilage are also responsible for growth. Increase in mass is usually accompanied by cell division. Cell multiplication continues throughout the development and to a variable degree, throughout the life of the individual. Thus, cell multiplication is a distinguishing characteristic of growth.

DIFFERENTIATION

The increasing diversification of form and function in an organism is called *differentiation.* It refers to the events by which cells or other parts of an organism become different from one another and also different from their previous state. The differentiation may be of three types : (i) morphological, (ii) behavioural, and (iii) chemical differentiation.

1. Morphological Differentiation

During multiplication, individual cells and groups of cells

differentiate structurally from other cells. For example, nerve cells and *epidermal cells* which start from a common point in *generalized ectoderm*, develop diagnostic features of size, shape and internal organization.

2. Behavioural Differentiation

Although all cells process common basic characteristics, such as irritability and metabolism, yet special functional responsibilities are laid upon these general properties. For example, nerve cells conduct neural impulses, muscles function to contract, epithelium serves to protect, skeletal elements provide support, and so on.

3. Chemical Differentiation

It based on enzymes. Enzymes direct the synthesis of the organic compounds that provide the cell its uniqueness. Cell constituents such as carbohydrates, lipids, and organic acids are synthesized under the influence of enzymes. Therefore, a given battery of enzymes will operate the pattern of these organic compounds. For example, as the various vertebrates eggs undergo cleavage and progressively become converted to blastula, gastrula, and definitive embryo; individual cells, and areas of cells in each case become biochemically different from the other. It results in the formation of a different and distinct adult from a particular egg.

CHEMICAL CHANGES DURING GASTRULATION

The gastrulation is the most active phase of embryonic development during which extensive or profound morphogenetic activities of different types of blastomeres occur and rate of metabolism of gastrula as a whole become rapidly increased. Its metabolism can be studied under following headings.

(a) Catabolism

The morphogenetic movements during gastrulation cause an increased expenditure of energy rich ATP molecules and consequently, an increase oxidation. It has been found that the oxygen consumption during the gastrulation of frog and sea urchin shows a further increase as compared with the blastula stage. During gastrulation the food reserves (e.g., glycogen and yolk) is oxidized for the manufacturing of ATP molecules.

(i) Oxidation of glycogen

One of the substances particularly involved in the supply of energy during gastrulation in amphibians and other animals, is glycogen. It has been shown in frogs that amount of glycogen becomes considerably diminished in the invaginating cells of the dorsal lip of the blastopore. Rapid breakdown of the glycogen in the dorsal lip suggests a particularly active respiration in this area.

(ii)Breakdown of yolk

Besides glycogen, the yolk of the blastomeres of broken down for energy metabolism or assimilation. The breakdown of yolk granules has been investigated in amphibian embryos both electron microscopically and biochemically. With the electron microscopy, it can be seen that in amphibian embryos, the first change in the yolk platelets consists in the disappearance of the amorphous or granular peripheral layer which contains, besides proteins, considerable quantities of ribonucleic acids. The disappearing material goes into solution in the cytoplasm and becomes available for synthetic processes. The solubilization of the peripheral layer occurs in the invaginating chorda-mesoderm during gastrulation, in the neural plate during late gastrulation and early neurulation and still later in the epidermis. The solubilization of the crystalline core ("main body") of the yolk platelets occurs considerably later, and in endoderm, it is delayed tili just preceding the stage when the larvae start feeding.

(b) Anabolism

Throughout gastrulation, the volume of the embryo does not change appreciably. Every expansion in one directions occurs at the expense of a contraction in another direction. Division of cells by mitosis continues, throughout gastrulation and this mean, that there is an increase of nuclear material at the expense of the cytoplasmic substances. The anabolic chemical activities of gastrulation includes following important functions :

(i)Nucleic acid synthesis

During gastrulation, synthesis of different kinds of nuclei acid molecules occurs. The replication of DAN is needed for the duplication of chromosomes during each mitosis of gastrulation.

The onset of gastrulation, actually recognised at the molecular level by the start of transcription of DNA dependent rRNA molecules. During gastrulation , the rate of transcription of new mRNA and tRNA molecules become greatly increased which indicates towards the possibility of increased protein synthesis in the clytoplásm.

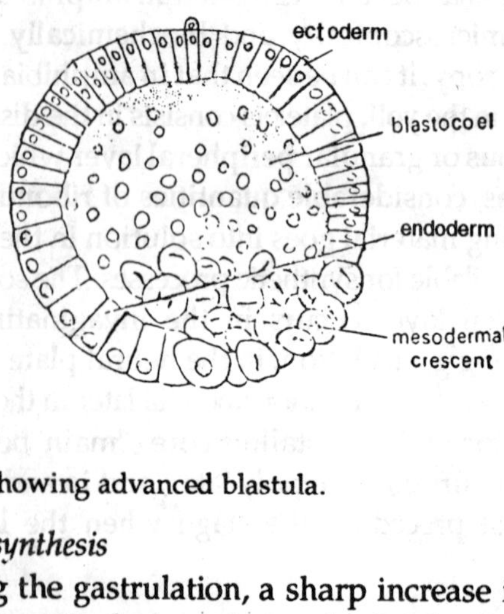

Fig. 2.6. Showing advanced blastula.

(ii)Protein synthesis

During the gastrulation, a sharp increase in protein turn over and in particular protein synthesis take place. The source of materials for the protein synthesis is mainly the protein yolk, contained in the eggs of most animals. The protein turn over after the beginning of gastrulation seems, to be essentially different from what had been going on during cleavage. The new proteins are qualitatively different from those present in the egg. These different kinds of proteins are manufactured in gartrula, only due to active participation of paternal genes in the production of new mRNA molecules during this phase.

In sea urchins and amphibians, it has been found that their gastrulae contain *antigen proteins*, capable of causing the formation of *antibodies*, which were not present before.Clayton, 1951).

Gastrulation in Amphioxus

Gastrulation in vertebrate animals (like blastulation) is easiest understood when studied in *Amphioxus*. We have already studied the formation of the blastula and identified the presumptive major organ-forming areas, geographically isolated on its surface. The arrangement of the cells into a cup-shaped gastrula is the next step in development.

As gastrulation begins, the cells in the region of the dorsal crescent (presumptive notochord and neural plate cells) and ventral crescent (presumptive mesodermal cells) increase their mitotic activity. The prospective endodermal cells at the vegetal pole do not divide, but form a flat plate that gradually folds inward, invaginating into the blastocoel. This endodermal plate leads the way and eventually becomes apposed to the presumptive ectoderm forming a cuplike structure with an outer and inner wall. The endodermal cells that push inward will eventually give rise to the primitive gut. The cavity or the gut is known as the *archenteron*, or *gastrocoel*, and communicates to the outside by an opening, the *blastopore*. Since the blastopore marks the entrance of the cells into the gut, it is often referred to as a mouth. The rim of the blastopore, therefore, becomes the lips of the blastopore, distinguishable as dorsal, lateral, and ventral lips.

The notochord cells, which originally make up a part of the rim of the cup, roll inside (involute) along with the endodermal cells, and come to lie in the middorsal area of the developing embryo. The presumptive mesoderm cells of the blastula form a continuous layer with the endodermal cells. As the endodermal plate continues to invaginate, the presumptive mesoderm cells converge dorsally toward the roof and take up a position on either side of the notochord cells. As a result of the archenteron formation, the blastocoel becomes obliterated.

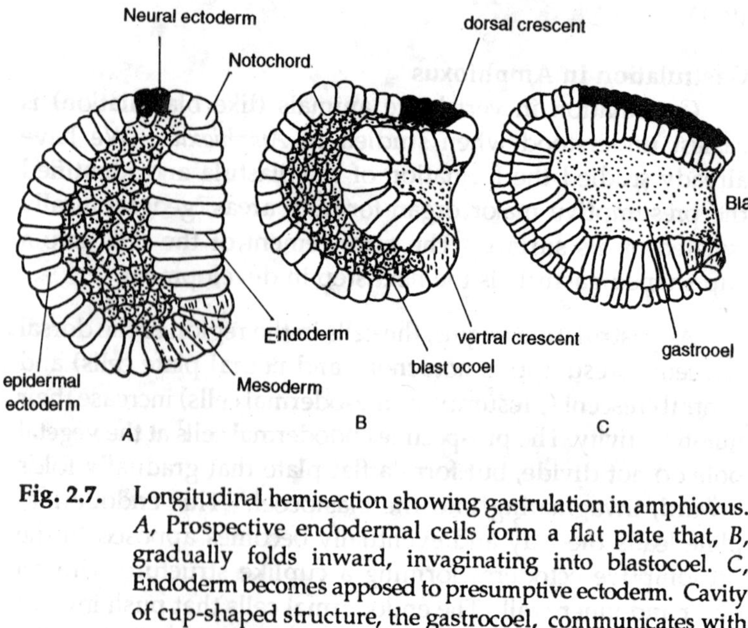

Fig. 2.7. Longitudinal hemisection showing gastrulation in amphioxus. *A*, Prospective endodermal cells form a flat plate that, *B*, gradually folds inward, invaginating into blastocoel. *C*, Endoderm becomes apposed to presumptive ectoderm. Cavity of cup-shaped structure, the gastrocoel, communicates with the outside by an opening, the blastopore.

As the mesodermal and notochordal cells move inside the gastrula, the lips of the blastopore contract and the opening grows smaller.

The embryo elongates in an anteroposterior direction as the prospective epidermal and neural plate cells actively divide. The antero-posterior extension of the embryo causes the presumptive neural ectoderm, originally crescent-shaped in the blastula, to be pulled into an elongated strip on the middorsal surface. At the end of gastrulation, the outer shell of the embryo consists of the neural ectoderm cells, covering the dorsal surface of the gastrula, and epidermal ectoderm, composing the remaining portion of the shell. The presumptive notochordal cells lie in the middorsal region under the neural ectoderm cells. Lateral to the notochord are the presumptive mesodermal cells, completing the roof of the archenteron. The presumptive endodermal cells compose the sides and floor of the archenteron and make up most of the gut. This is best seen in cross section of the gut. This is best seen in a cross section of the gastrula. The only entrance into the interior of the gastrula is by way of the blastopore.

The third germ layer forms in amphioxus when the notochord, mesoderm, and endoderm separate from one another. Crevices appear at their junctions, separating presumptive notochord from presumptive mesoderm and mesoderm from presumptive endoderm. They become three distinct entities. The presumptive notochordal cells round up and become a discrete rodlike structure running the length of the embryo in the middorsal region. It is considered a notochord at this stage, although differentiation must occur before it is identical to the notochord found in the adult. The presumptive mesoderm, which runs the length of the animal on either side of the notochord, pushes outward toward the ectoderm, forming pouchlike structures. The anterior portion contains a narrow cavity continuous with the archenteron. No pocket forms in the more posterior part of the mesoderm material.

After the mesoderm separates from the notochord and endoderm cells, transverse crevices form along the bilateral strips of mesoderm. These crevices divide the strips into a longitudinal series of blocks or somites located on each side of the notochord. In the anterior region, each block of mesodermal cells still contains a cavity. The solid blocks of mesoderm in the posterior end eventually acquire cavities. These cavities will contribute the body cavity of the adult. After the separation of the notochord and mesoderm from the rest of the cells, the freed dorsal edges of the presumptive endodermal layer (originally lateral and ventrally located) move up and fuse beneath the notochordal rod. The endoderm thus completes the roof of the primitive gut.

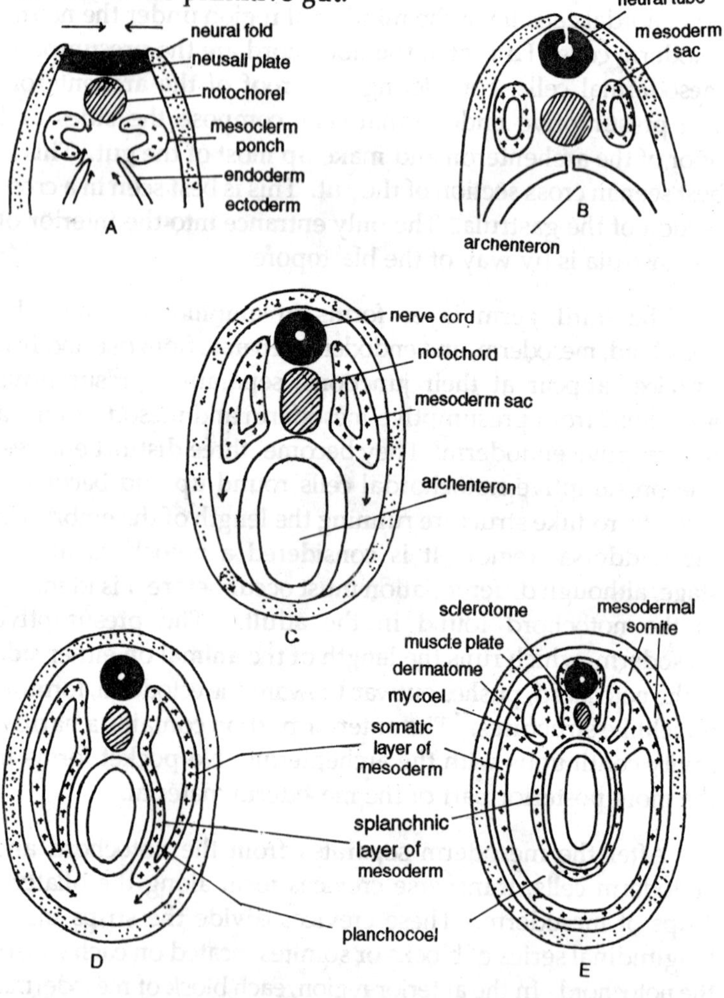

Fig. 2.8. *Amphioxus*—Successive stages in neurogenesis, notogenesis, mesogenesis and the formation of coelom (Diagrammatic).

Significance of Amphioxus Development

The study of embryology, along with paleontology, offers an insight into the history of animals. Animals now living have become so modified that it is difficult to follow their descent from common ancestral types. It is especially difficult to trace the evolution of protochordates, since their bodies, were soft and rarely fossilized. Paleontology therefore, tells us little about the phylogeny of the chordates, a story that man in his egocentric way finds most intriguing. On the other hand, the study of the embryology of living organisms reveals basic patterns of development. Those animals that follow similar path-ways are believed to be related, since these similarities suggest development from common ancestoral types.

The formation of the blastopore and its position in the developing embryo are characteristics that divide bilateral animals into two main evolutionary groups. In one group the blastopore marks the area of the mouth. These animals are referred to as *Protostomia* and include the flat-worms, annelids, mollusks, arthropods, and other smaller related groups. In the other group, the blastopore identifies the posterior region of the animal. The anus arises from or near the blastopore, and a new mouth forms at the anterior end by the invaginating ectoderm. This branch of animals, known as the *Deuterostomia* (second opening), embraces the echinoderms, the chordates, and several smaller phyla. Both groups evolved from a common coelenterate stage in which neither anterior nor posterior ends were determined. However, the development of the animals in each of these two main branches differs considerably and indicates two main evolutionary lines.

A second main difference between groups of animals is seen in the way they form mesoderm. Triploblastic animals, that is, those that have a third layer located between the "skin" and "gut", use two different embryonic pathways to form this third layer. In one group, the mesoderm arises as solid masses of cells derived from the posterior regions of the embryo. Cavities form in these solid masses and give rise to

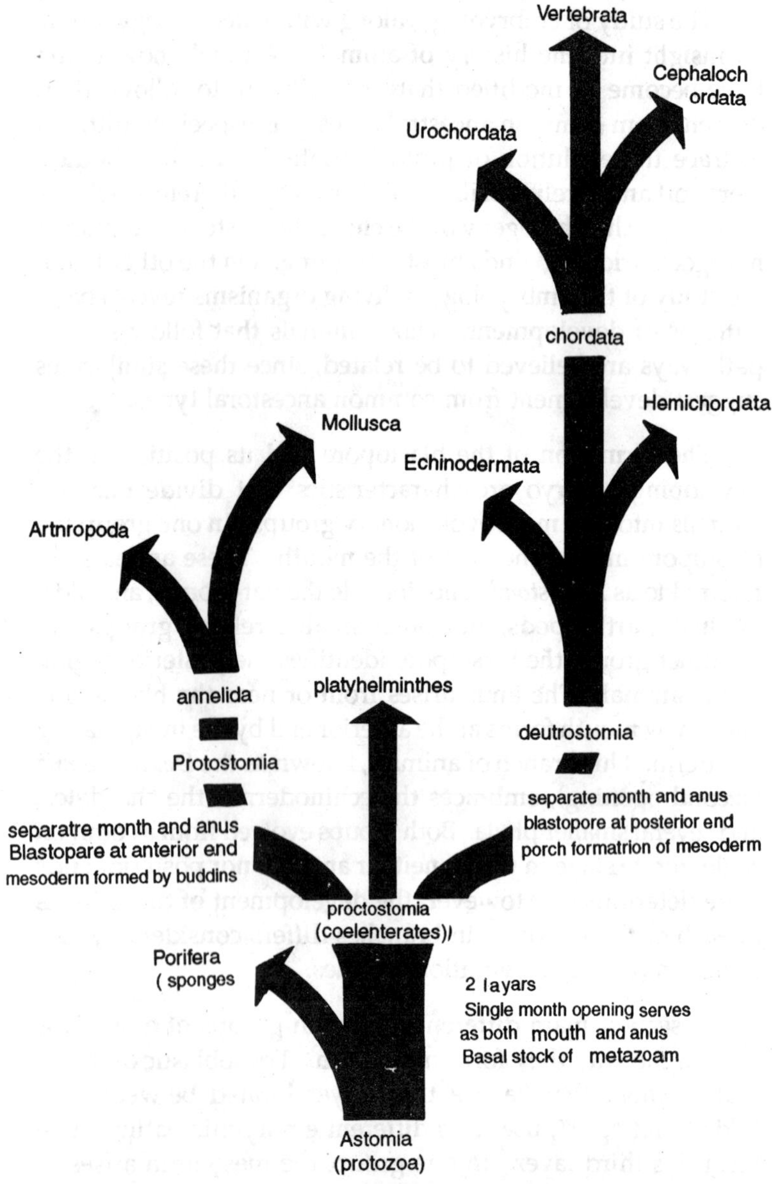

Fig. 2.9. Phylogeny of the animal kingdom suggesting the possible lines of evolution of the vertebrates.

the body cavity. In the second type the mesoderm arises in the form of pouches. These pouch cavities become the body cavities of the adult. The first method of mesoderm formation is found in the annelid worms, mollusks and arthropods. The second is found in the echinoderms and amphioxus.

Although the pouchlike formation of mesoderm never occurs in the vertebrates, mesoderm formation is comparable in the Hemichordata and in the invertebrate phylum Echinodermata. This similarity of mesoderm formation has led many biologists to relate the origin of the chordates to the echinoderms. Mesoderm forms in the vertebrate by a modification of the echinoderm and amphioxus pattern. The mesoderm of the vertebrates pushes out as a sheet between ectoderm and endoderm and only later shows segmentation and cavitation.

Based on the location of the blastopore and manner of mesoderm formation, as well as certain other factors such as larval types, muscle chemistry, and similarity of sera proteins, most biologists agree that, above the coelenterates, two stocks of invertebrates arose forming a Y-shaped "tree". The ancestors of present-day echinoderms are located at the base of one branch with most of the invertebrates on the other. The best clues relate the chordates to the branch originating with the echinoderms.

Gastrulation in Amphibia

The sparsely scattered yolk in the egg of amphioxus enables those cells that give rise to muscle, notochord, and gut to invaginate and migrate anteriorly. Invagination appears to be the primary process responsible for gastrulation in these embryos. In contrast, the yolk-laden, vegetal cells of the frog fill much of the interior of the blastula and react passively. It is mechanically difficult for them to move about as an amphioxus. Consequently, the process of gastrulation is modified in those animals possessing moderately telolecithal eggs, for example, the amphibians.

However, the end result of gastrulation remains the same. Areas on the surface of the blastula destined to become internal organs move to the interior where they arrange themselves in a proper relationship to each other and to the embryo as a whole. Involution plays a larger pole in the gastrulation of amphibians than that of amphioxus. Cells enter the interior of the blastula by involution and circumvent the yolk-filled endodermal cells, gaining their proper position in the gastrula.

Presumptive Endoderm

Those endodermal cells of the marginal zone, destined to form the foregut, are among the first cells to move inside the frog blastula. They lie below the gray crescent, in the center of the area, and on the axis of bilateral symmetry. The cells elongate and actually push inward (invaginate), forming a dark cleftlike depression or groove about halfway between the equator and the vegetal pole. This inward movement of cells toward the blastocoelic space can best be seen in sagittal section. As gastrulation continues, the groove enlarge and forms a cavity leading into the interior. This new cavity, surrounded by the invaginated cells, is the *archenteron*, or *gastrocoel*, of the amphibian gastrula. As the cells continue to migrate toward the cephalic end of the embryo, the archenteron increases in size and crowds the blastocoel to one side, showing the eventual obliteration of it. The rim of the groove marks the beginning of the dorsal lip of the blastopore stage or early gastrula.

These first invaginating endodermal cells enter the blastula by actually changing their shape. The surface cells located near the prospective blastopore become bottle-shaped. One end of a cell remains attached to the surface of the blastula while the cell itself becomes extended inward and the bulk of the cell mass shifts into the deep endoderm. As a result of this inward pull, the blastula becomes indented where the cells cohere at one end. This indentation pulls other endoderm cells inside and is the beginning of the blastopore. If blastoporal endodermal cells are dissected away and placed on an explant of blastocoelic endoderm, they immediately cohere at the surface of the explant and sink into the deeper endodermal cells. According to Trinkaus (1965) the bottle cells apparently initiate gastrulation by invaginating at the region of the presumptive blastopore.

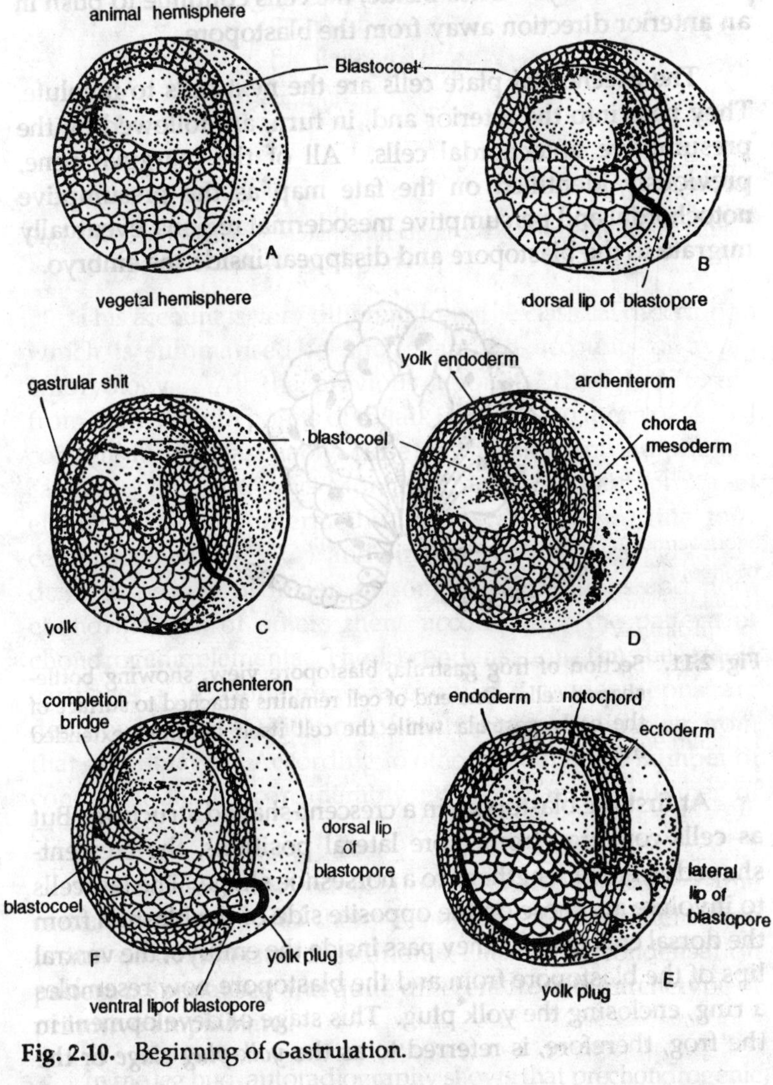

Fig. 2.10. Beginning of Gastrulation.

The shape of the blastopore changes as gastrulation proceeds. If the cells of the blastula are marked and their migration is followed, one can see that the stained areas stretch toward the small crescent-shaped groove; the cells migrate to the rim of the blastopore, roll over its surface, and disappear inside the embryo. Once inside, the cells continue to push in an anterior direction away from the blastopore.

The prechordal plate cells are the next cells to involute. They pass into the interior and, in turn, are followed by the presumptive notochordal cells. All of the marginal zone, previously identified on the fate map as the presumptive notochordal and persumptive mesodermal regions, eventually migrate to the blastopore and disappear inside the embryo.

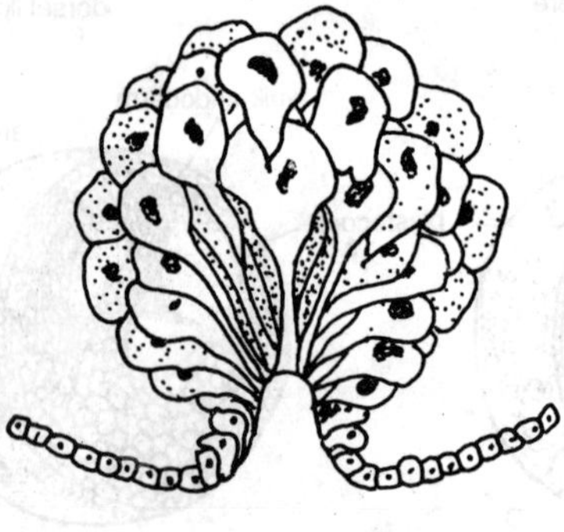

Fig. 2.11. Section of frog gastrula, blastopore view, showing bottle-shaped cells. One end of cell remains attached to surface of the early gastrula while the cell itself becomes extended inward.

At first the blastopore in a crescent-shaped structure. But as cells converge from more lateral positions, the crescent-shaped opening lengthens to a horseshoe shape. The last cells to involute are those on the opposite side of the blastula from the dorsal crescent. As they pass inside the embryo, the ventral lips of the blastopore from and the blastopore now resembles a ring, enclosing the yolk plug. This stage of development in the frog, therefore, is referred to as the *yolk-plug stage* or the late gastrula. More material passes over the dorsal lips than over the lateral ones, and more over the lateral lips than the ventral. As these presumptive layers move to the inside, the expanding ectodermal cells replace them on the surface. The position of the blastopore is not fixed, but as the cells migrate to it and then involute, the blastopore moves away from the animal pole toward the vegetative pole.

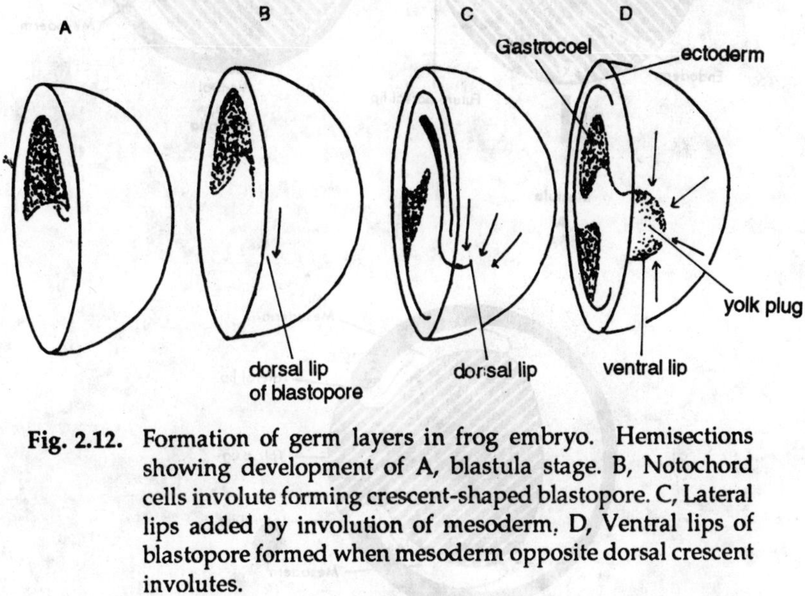

Fig. 2.12. Formation of germ layers in frog embryo. Hemisections showing development of A, blastula stage. B, Notochord cells involute forming crescent-shaped blastopore. C, Lateral lips added by involution of mesoderm. D, Ventral lips of blastopore formed when mesoderm opposite dorsal crescent involutes.

By the end of gastrulation the vegetal region of the blastula is drawn inside the embryo. In its new location it is in a position to give rise to the gut and its derivatives. As a result of the formation of the archenteron and the obliteration of the blastocoel, the force of gravity causes the embryo to rotate and the endodermal cells take up a ventral position. Those cells that were originally located at the vegetal pole (marked X on the diagram) are now located on the surface of the archenteron floor in the midgut region. Those endodermal cells, destined to form the foregut and the first to invaginate, are now located properly at the anterior end of the gut. As gastrulation continues,

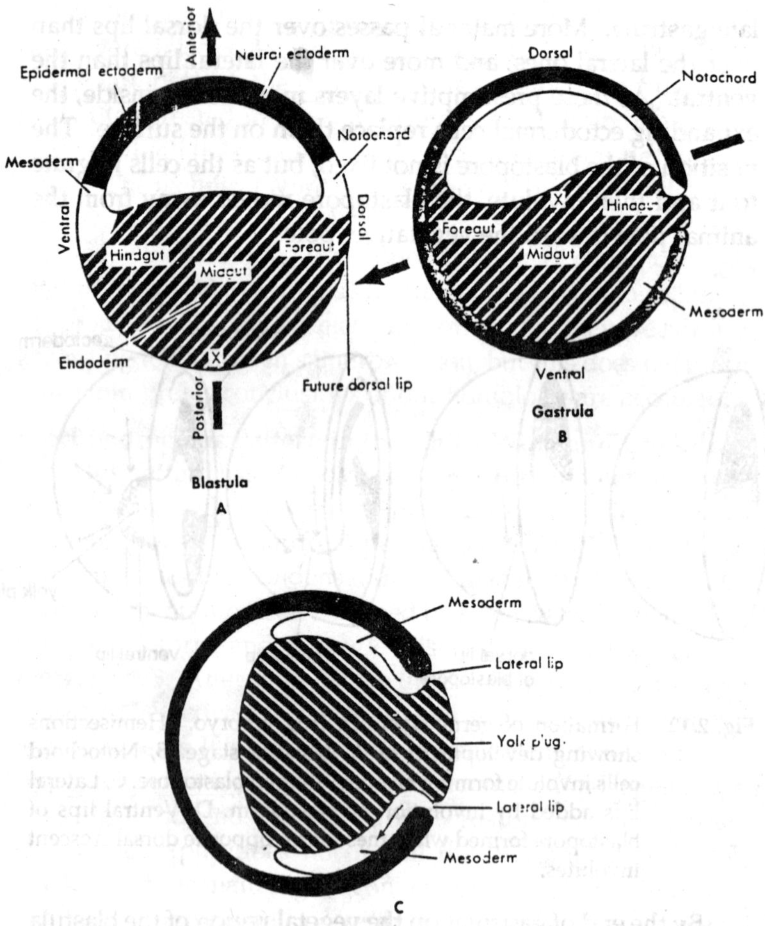

Fig. 2.13. Location of organ-forming areas in frog blastula. A, and gastrula, B, As a result of gastrulation the foregut cells are located at anterior end of gut and notochord is lined up under neural plate ectoderm. X indicates vegetal pole cells of original blastula. C, Horizontal section of gastrula through the yolk plug showing involution of mesoderm between endoderm and ectoderm and the formation of later lips.

the embryo loses its spherical form and under the influence of the archenteron elongates, establishing the anteroposterior and dorsoventral axes of the embryo. The formation of the gray cresent determined bilateral symmetry at the time of fertilization.

Presumptive Notochord

The endodermal cells, followed by prechordal plate cells, continue to push forward in the middorsal line toward the animal pole, along the inner surface of the presumptive ectoderm. They are succeeded by the presumptive notochordal cells, which do not become bottle shaped as did the earlier endodermal cells but, rather, orient the larger mass of their cells toward the exterior of the embryo surface.

The migration of the presumptive notochord cells can be compared to movement of a ribbon-like structure. The anterior end of the ribbon leads the way inside the embryo and the rest of the ribbon follows. As some cells closest to the forming blastopore involute and follow the prechordal plate cells inside the embryo, the next cells move to the blastopore on the outer surface layer and then roll over the rim to the inside. There is a point in gastrulation at which some notochordal cells are inside the embryo travelling toward the animal pole region, some are involuting at the dorsal lip, and some remain on the surface of the embryo. Eventually all of the presumptive notochordal cells will be "pulled" inside and none will be left on the surface of the gastrula. These cells take up a position in the midline below the neural plate, forming a middorsal strip in the roof of the archenteron. Later these cells will give rise to the notochord, a transient embryonic structure. As the notochordal cells disappear inside the embryo, the neural ectoderm cells are pulled in an anteroposterior direction. They form an elongated plate, the medullary plate, on the surface of the embryo.

Presumptive Mesoderm

The mesodermal cells located in the marginal zone of the blastula also converge toward the blastopore, involute over its edges, and disappear within the embryo. These movements are synchronized with those of the presumptive endoderm,

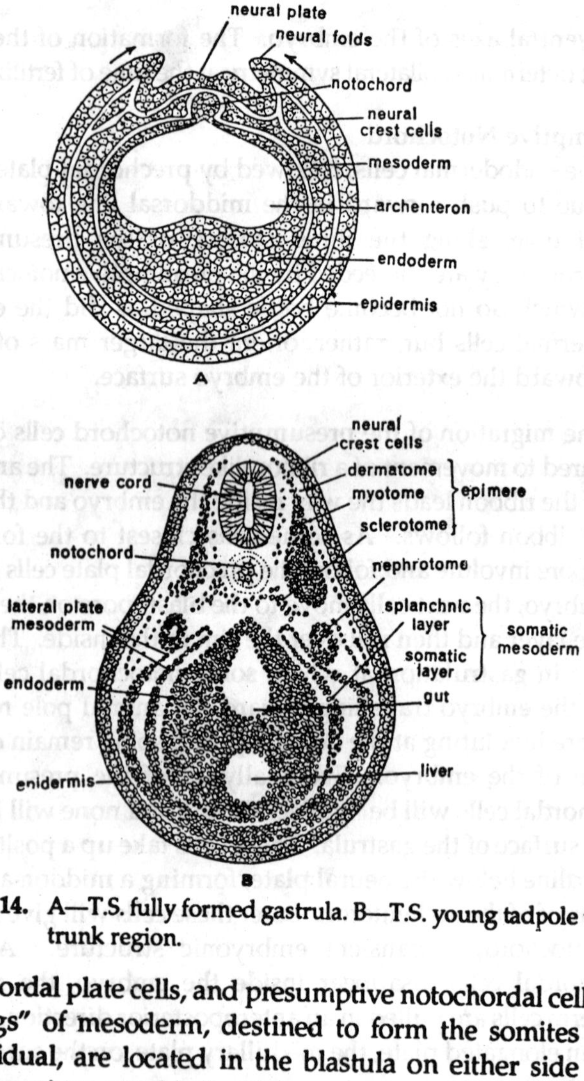

Fig. 2.14. A—T.S. fully formed gastrula. B—T.S. young tadpole through trunk region.

prechordal plate cells, and presumptive notochordal cells. The "wings" of mesoderm, destined to form the somites of the individual, are located in the blastula on either side of the presumptive notochord. They migrate in a lateral direction toward the blastopore and involute over the lateral lips. That part of the somitic mesoderm located closest to the blastopore in the early gastrula in the first to reach the blastopore and roll over its rim. These mesodermal cells then migrate in an anterior and dorsal direction where they give rise to the most anterior blocks of mesoderm (somites).

More laterally placed somitic mesoderm involutes later and forms the somites in the posterior region of the embryo. In this way the consecutively arranged, presumptive somites involute, on after the other, and take up their position along the notochord in an anteroposterior orientation. The mesoderm located laterally to the somitic mesoderm converges and passes interiorly through the lateral lips; the most ventral presumptive mesoderm, located in the opposite side of the embryo from the gray crescent, involutes over the ventral blastoporal lips.

Once inside the embryo, the mesoderm inserts itself between the outer shell of presumptive ectoderm and the inner layer of endoderm, and moves away from the blastopore in an anterior direction. It does not remain in a position comparable to its site in the blastula, that is, in the marginal zone, but migrates dorsally up the sides of the gastrula, eventually establishing a close association with the notochord. The combined sheet of notochordal and mesodermal cells is known as the *chorda-mesodermal mantle.* Because of this dorsal migration, the mesodermal layer is much thicker in the dorsal region of the embryo where it forms the roof of the archenteron. This relationship can be seen in a cross section of the late gastrula. Laterally, the mesodermal layer is less thick toward the more ventral areas of the embryo. Although the mesodermal sheet migrates anteriorly, it does not reach the most anterior regions of the embryo. Later, the mouth will form in this area.

Presumptive Ectoderm

As the cells on the surface of the early gastrula migrate to the blastopore and roll over the lips to the inside, they are replaced on the outside by the spreading presumptive epidermal cells. The stretching and thinning of the epidermal and nervous layers compensate for this increase in surface area. Several layers of cells compose the presumptive ectoderm in early gastrulation but only two layers of cells remain by late gastrulation.

In the process of covering the surface of the gastrula, the neural crescent becomes elongated in an anteroposterior direction and decreases in width. It remains on the dorsal surface of the embryo and assumes the shape of a shield, with the broad end of the shield directed anteriorly.

As the inward migration of the marginal zone draws to a completion, the epidermal ectoderm approaches the edges of the blastopore until finally, when all mesodermal and notochordal cells disappear inside, ectodermal cells compose the lips of the blastopore. After about 36 hours, only a small area of endodermal cells surrounded by blastopore remains visible. This stage of development, identified as the yolk-plug stage.

Closure of Blastopore

When the material of the marginal zone finally disappears from the surface of the gastrula, the blastopore closes. The yolk plug withdraws inside the gastrula and the lateral lips come together. The closing of the blastopore marks the termination of the gastrula stage. The neurula stage follows as the various regions of the gastrula begin to differentiate.

Gastrulation in Chick

The chick egg is usually studied as the representative type of an extreme telolecithal egg, since it has practical convenience, is easily handled, and available in any season of the year. Its development was studies as far back as Aristotle, Harvey, Malpighi, Wolf, Pander, and von Baer, and the sequence of its events is better known than that of any other vertebrate. The large amount of yolk modifies not only cleavage but also the morphogenetic movements involved in the organization of the germ layers. Various other animals, such as elasmobranchs, bony fishes, and reptiles, also have heavily yolked eggs, but the chick represents the climax of yolk accumulation that started in the chordates with the isolecithal egg of amphioxus. Thus we shall use it as our example and limit our discussion to those processes involved in the formation of the germ layers in the chick.

The same basic processes occur in the flattened blastoderm of the chick as in the rounded gastrula of the frog. Certain cells on the surface of the blastodisk move inside and establish correct spatial relationships. Interactions then take place between the cells; the organ primordia form and begin to differentiate. In the bird there is so much yolk present that the cells, make no attempt to move around it, as they do in the frog. Instead, all cellular activities occur in the small disk of cytoplasm that sits on top of the yolk. It is in this cap that morphogenetic activities take place and the cells undergo convergent and divergent streaming, involution, invagination and epiboly.

At the end of blastulation, the blastodisk consists of two layers, epiblast (prospective ectoderm and mesoderm) and the beginnings of the hypoblast (endoderm) next to the yolk. The flattened space between the layers is considered the blastocoel. Gastrulation begins in the chick soon after the egg is laid. After 3 to 4 hours of incubation, there is the suggestion of a thickening of the blastoderm in the posterior region of the epiblast in the area pellucida. By 8 hours the thickening has lengthened and broadened as a result of the convergence of surface cells to the area. This densely packed area of epiblast cells marks the caudal end of the animal. The epiblast also extends over the area opaca, but it contributes to extra-embryonic membranes and will not be considered in the discussion of the development of the embryo itself.

The cells composing the epiblast destined to become mesoderm, migrate to this posterior region of the area pellucida, pile up, finally involute, and then spread anteriorly and laterally between the epiblast and hypoblast layers. This concentration of cells at the surface (the formation of the thickening described above) occurs because the cells migrate and proliferate faster in that area than they can spread out under the epiblast.

Spratt (1946) grew the developing chick blastoderm on a plasma clot and studied the cellular migrations by placing carbon particles on the epiblast. He followed the path of migration of the carbon-marked cells. It is seen that the particles moved posteriorly and toward the midline as gastrulation proceeded. The elongated thickening produced in the posterior region of the area pellucida by the sixteenth hour of incubation is identified as the *primitive streak.* This stage of development is the *primitive-streak stage.*

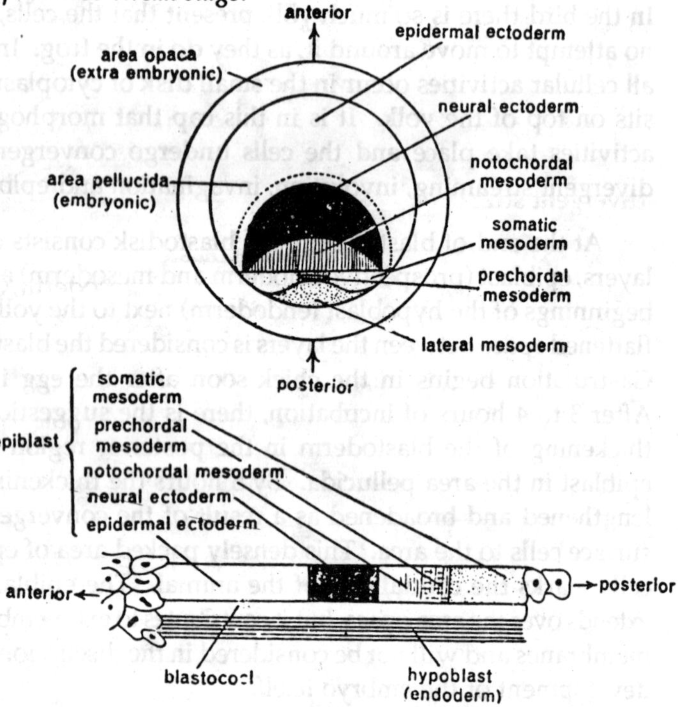

Fig. 2.15. Fate maps of chick blastoderm. A—surface view to show the probable locations of main prospective areas; B—diagrammatic section of discoblastula passing through antero-posterior median axis.

Since the cells move toward the primitive streak and disappear from the surface of the blastodisk at this point, the formation of the streak is a progressive affair. The cells

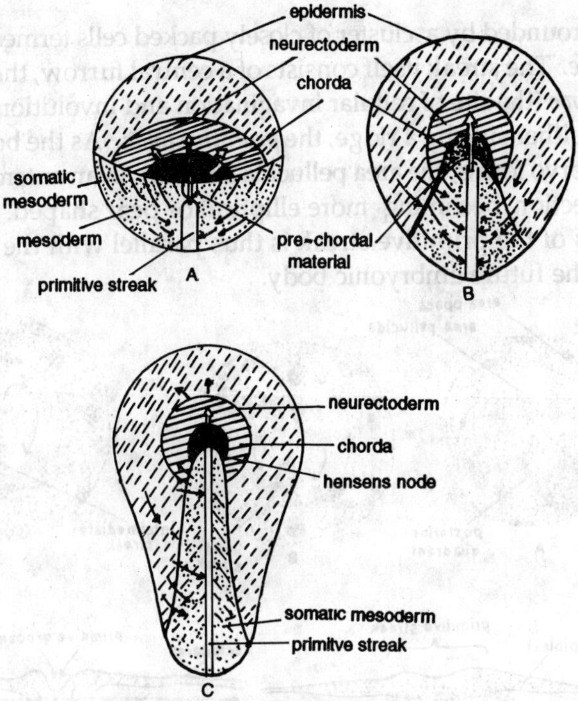

Fig. 2.16. Stages in the gastrulation in chick (dotted arrows indicate movements beneath the surface).

composing it constantly change. Although in general, the cells of the streak contribute to the middle layer of mesoderm, they are also the source of ectoderm and endoderm. In the early stages of gastrulation the streak is a small thickened region in the posterior part of the area pellucida. However, as more cells migrate to the area, it grows anteriorly and posteriorly. Also, according to mitotic counts, cell proliferation goes on at a higher rate in this region than in others of the area pellucida. The primitive streak grows, therefore, because of the proliferation of cells as well as of the addition of cell that migrate to it.

After about 20 hours of incubation, the streak stretches two thirds to three fourths the distance of the area pellucida. Different regions can be identified after 12 hours. The anterior end terminates in the *primitive pit*, a funnel-shaped structure surrounded by a cluster of closely packed cells termed *Hensen's node.* The streak itself consists of a central furrow, the *primitive groove* (the site of cellular invagination and involution), flanked on either side by a ridge, the *primitive folds.* As the body of the embryo forms, the area pellucida stretches in an anteroposterior direction, becoming more elliptical or pear shaped. The long axis of the primitive streak is thus parallel with the long axis of the future embryonic body.

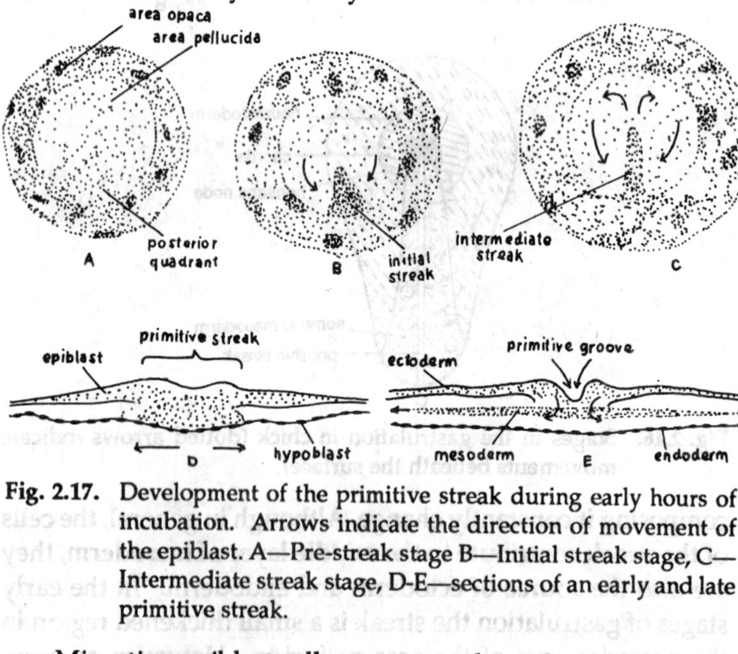

Fig. 2.17. Development of the primitive streak during early hours of incubation. Arrows indicate the direction of movement of the epiblast. A—Pre-streak stage B—Initial streak stage, C—Intermediate streak stage, D-E—sections of an early and late primitive streak.

Migrating epiblast cells converge along the entire length of the primitive streak including the primitive pit and Hensen's node. As in the amphibian, the prechordal plate cells are among the first to travel inside. They migrate over Hensen's node, involute at the funnel-shaped pit, and then move in a direction that is anterior to the streak. They are followed (as in the amphibian gastrula) by the presumptive notochordal cells from the notochordal crescent. Once inside, the presumptive notochordal cells migrate anteriorly and eventually establish the rodlike notochord lying in the area pellucida, anterior to the streak and in the midline of the future embryo.

On the fate map, the presumptive mesodermal cells occupy approximately the posterior half of the area pellucida. They must migrate from the surface of the blastodisk to its interior and insert themselves between the endoderm and ectoderm. They converge from either side of the epiblast to the primitive streak, involute, and then diverge anteriorly and laterally to form the broad middle layer. An examination of cross sections of the area anterior to Hensen's node shows the notochord flanked on either side by the mesoderm.

As a result of gastrulation the entire posterior area of the blastoderm, the one identified as presumptive mesoderm, disappears from the surface. The remaining cells of the epiblast (presumptive ectoderm) follow the migrating presumptive mesoderm as it moves toward the primitive streak, but do not themselves involute. Eventually when all the cells destined to form the notochord and mesoderm have made their way inside the embryo, there are no more cells on the surface to replace them. Then the streak starts to regress.

According to Spratt, the carbon-marked cells of Hensen's node migrate posteriorly as more and more notochordal cells move inside. As the streak regress, cells destined to give rise to the embryo are left behind. They increase in mass and give rise to that part of the embryo located anterior to the streak. For example, the marking of cells shows that as the anterior part of the streak regress, cells are added to the notochord. The notochord does not grow forward but, rather, grows posteriorly as the streak shortens. The primitive streak continues to shrink in size until, by 48 hours or incubation, the posterior end is incorporated into the tail bud of the developing chick. The embryo is said to be in the *head-process stage* when the notochord takes up its position inside the embryo, anterior to the streak, and the streak, itself regresses.

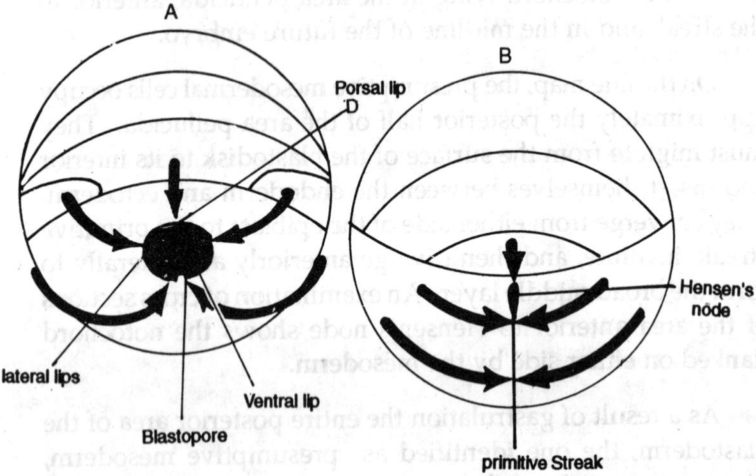

Fig. 2.18. **Comparison of frog blastopore with chick primitive streak. Notochord cells involute over the dorsal lip of the frog. A, and Hensen's node in the chick. B, Mesoderm involutes over lateral and ventral lips in the frog and through primitive streak in the chick. In both examples, surface cells undergo convergent streaming, disappear at blastòpore or streak, and take up a position inside the embryo between the epiblast and hypoblast, establishing the third germ layer.**

Comparison of Blastopore and Primitive Streak

Many biologists consider the primitive streak of the chick and the blastopore of the frog to be homologous structures with comparable functions. The movement of cells toward the blastopore in the frog resembles the migration of chick cells toward the primitive streak. The cells of the frog involute through the lips of the blastopore; the cells of the chick move into the interior through the primitive groove. In both cases, surface cells undergo convergent streaming, disappear at the streak or blastopore and then take up a position inside the embryo between the epiblast and hypoblast, establishing the third germ layer.

The large amount of yolk in the chick egg is responsible for the main difference between the two structures, that is, the blastopore and the primitive streak. The primitive streak is never open as in the frog, but is comparable to the fused lips of the blastopore. Unlike the frog blastopore, it never communicates with the archenteron. The dorsal lip of the amphibian gastrula and Hensen's node appear to correspond to one another, since it is in these respective areas that the notochordal material makes its way inside the embryo. The presumptive mesoderm of the frog involutes over the lateral lips of the blastopore; the presumptive mesoderm of the chick converges to the primitive streak.

The regression of the primitive streak can be compared with the closure of the amphibian blastopore. When the amphibian blastopore closes, the less active regions of morphogenetic activity, that is, lateral and ventral lips, close first, whereas the dorsal lips, the most active site of cellular activity, close last. A comparable situation occurs in the chick as the primitive streak regresses. Regression occurs faster in the posterior portions of the streak. Hensen's node remains functional until the end.

Gastrulation in Mammals

since there is a lack of yolk in the eggs of viviparous mammals, food must be supplied to the embryo. It depends on the mother for food, oxygen, and the elimination of wastes. An early differentiation of tissues in the developing ovum of mammals (including man) gives rise to fetal membranes. These will function in intimate contact with the maternal tissue and carry out the vital functions of the embryo. By the time implantation occurs two of these fetal membranes, the amnion and the yolk sac, have started to develop. But gastrulation lags behind. There is a precocious appearance of extraembryonic mesoderm in the human embryo before intraembryonic mesoderm forms; in fact, even before the appearance of the primitive-streak stage. Essentially, the developing egg first establishes the mechanics for maintaining the embryo before it constructs its body form.

Instead of following a chronological description of mammalian development, we shall continue, as in the frog and the chick, with a discussion of the development of the embryo proper, specifically the differentiation of the germ layers. The membranes that form early in development are associated with viviparity and are adaptive structures. They do not contribute to the definitive body form. The description of these membranes is delayed until a later chapter.

Differentiation of Inner Cell Mass

The blastocyst stage of the mammal is complete when the cells divide into the inner cell mass and trophoblast cells. This sorting of cells according to their future function occurs during the 3 days it takes the egg to travel down the fallopian, tube. The blastocyst is an adaptive structure, it makes possible intrauterine development in the mammal. However, during the first few days, it is still enclosed by the zona pellucida. In mice, the blastocyst expands, thinning and rupturing the zona. Also, a lytic agent, probably emanating from the uterus, lyses the membrane. Loss of the zona pellucida allows the mouse blastocyst to "hatch" and exposes the trophoblast cells to the uterine wall. They are free to attach to the uterine mucosa.

Gastrulation begins in the mammal when cells on the undersurface of the inner cell mass apparently segregate and arrange themselves in a layer of cuboidal cells, the endoderm. The formation of the endoderm in the mammal is reminiscent of its formation in the bird; in both examples the endoderm migrates or delaminates from the epiblast layer. No one has been able to mark the cells of the human embryos; therefore, the details of early development in these forms are tentative.

Around 7 days the human ovum starts to implant into the uterine wall. At this time there develops a cavity that separates the upper surface of the inner mass cells from the outer wall of trophoblast cells. This cavity is destined to become the amniotic cavity. It fills with fluid and bathes the embryo as it develops. The trophoblast cells, usually considered to be of ectodermal origin, form the chorion, one of the embryonic membranes, and contribute to the development of the placenta.

The endoderm cells not only cover the lower surface of the inner cell mass, they also extend beyond the margins of the disk. Here they become continuous with cells believed to be of trophoblast origin, the cells of Heuser's membrane. As a result, endoderm cells and the cells of Heuser's membrane line the large cavity of the blastocyst. This cavity is known as the *primary yolk sac*, although no yolk is present. Once the endoderm differentiates and the amniotic cavity forms, the inner cell mass flattens and forms a bilaminar disk. This disk consists of columnar epiblast cells closely applied to endodermal cells. This two-layered disk is circular in outline and gives rise to all the tissues of the embryo.

The endodermal cells form the roof of the yolk sac. Some of the cuboidal endodermal cells in the anterior regions become more columnar and mark the site of the future *prechordal plate*. It, in turn, determines the cephalocaudal orientation of the embryo. As the cephalic end of the embryo continues to develop, the prechordal plate marks the anterior limits of the foregut and eventually contributes to the formation of the oral plate.

Increasing amounts of precociously developed mesoderm appear between the trophoblast cells and the endoderm that lines the primary yolk sac. It is referred to as extraembryonic mesoderm. Appearing in this mesoderm are spaces that merge and give rise to the extraembryonic coelom. The coelom expands, crowding up around the embryo and leaving only a small stalk of cells effectively suspending the embryo from the chorion (mesoderm plus trophoblast). This remaining mesodermal condensation will become the body stalk of the embryo, connecting its caudal end to the chorion. The mesoderm next to the trophoblast cells is considered somatic (parietal) mesoderm and that covering the yolk sac is splanchnic (visceral) mesoderm. This extraembryonic mesoderm does not contribute to the formation of the embryo, but to the membrane composing the chorion and yolk sac. It is only after the inner cell mass and trophoblast cells have developed to this point that the primitive streak forms in the bilaminar disk and the body of the embryo begins its differentiation.

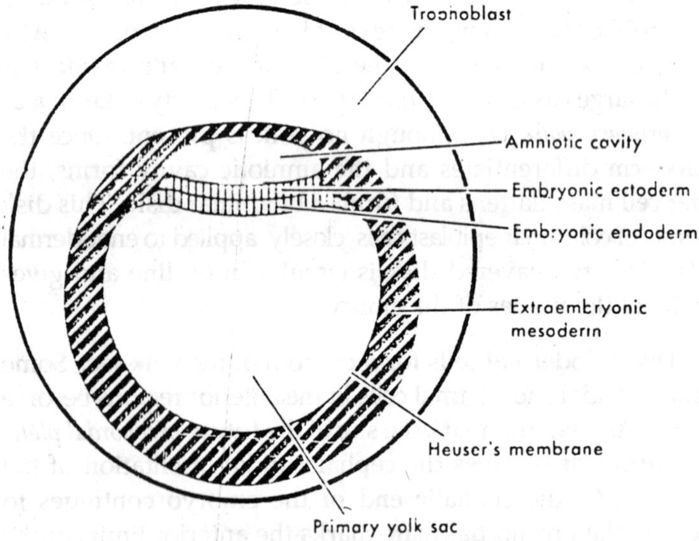

Fig. 2.19. Drawing showing formation of amnion and yolk sac in the human embryo.

Differentiation of Germ Layers

During the third week of human development, cells segregate into germ layers in a manner similar to that of lower animals. By the fifteenth day cells accumulate in the posterior region of the disk. This phenomenon distinguishes the caudal end of the embryo and initiates the beginning of the primitive streak. The primitive streak of the mammal, like that of the chick consists of a primitive groove, primitive folds, Hensen's node, and the primitive pit. Cells from the surface of the disk migrate to the primitive streak, move inside, and take up a position between the endoderm and ectoderm. A transverse section through the 16-day-old human embryo shows the mesoderm as a definite layer between the ectoderm and endoderm. This mesoderm contributes to the formation of the embryo proper and is referred to as the intraembryonic mesoderm, to distinguish it from the extraembryonic that will form the membranes. The intraembryonic mesoderm, once inserted as a third layer, spreads anteriorly and laterally until it becomes continuous with the extraembryonic mesoderm of the amnion and yolk sac. It also spread caudally and contributes mesoderm to the connecting stalk. The formation of the intraembryonic mesoderm transforms the embryo to a trilaminar structure. It now contains three germ layers.

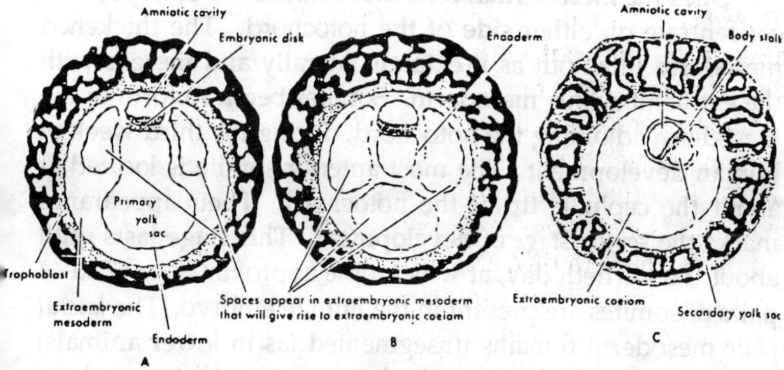

Fig. 2.20. Drawing illustrating the formation of extraembryonic coelom and body stalk. A and B, Primary yolk sac is partially filled in with mesoderm. Spaces appear in the mesoderm that merge to form the extraembryonic coelom. C, Extraembryonic coelom expands, crowding up around the embryo, until only a small stalk of cells is left suspending the embryo from the chorion. A secondary yolk sac is established.

As the intraembryonic mesoderm forms, Hensen's note appears as a thickening in the cephalic end of the primitive streak. The primitive pit, as in the chick, probably represents the dorsal lip of the blastopore described for the lower animals. The presumptive notochordal cells pass through the pit. They then migrate anteriorly to a position intermediate between the ectoderm and endoderm to form the head process of the embryo.

As the presumptive mesodermal cells move inside, the ectodermal cells immediately over the notochord and to each side of it, arrange themselves in an elongated ectodermal band. This band of cells extending in an anteroposterior direction is the neural plate. Immediately surrounding the plate, separating it from the rest of the ectoderm, is an area of special cells, the *neural crest cells*. The remaining ectoderm gives rise to epidermal ectoderm and provides the external surface of the embryo, as it did in the lower forms previously described.

Once the mesodermal cells move inside the embryo, they concentrate on either side of the notochord. The thickened mesoderm thins out as it extends laterally and merges with the extraembryonic mesoderm. Somites begin to form in the mesoderm adjoining the notochord, during the third week of human development. The most anterior somite is located at about the cephalic tip of the notochord. Their appearance marks the *somite stage* of development. This stage lasts until about the thirtieth day, at which time approximately 42 to 44 pairs of somites are present in the human embryo. The *lateral plate* mesoderm remains unsegmented (as in lower animals) and is attached to each somite by the *intermediate mesoderm*. The lateral plate mesoderm splits and gives rise to the *intraembryonic coelom*.

When the endodermal cells separate from the rest of the inner cell mass, they form the embryonic gut. Cuboidal endodermal cells compose the roof of the primary yolk sac and Heuser's membrane composes the walls. Mesoderm, proliferated by the inner surface of the trophoblast, surrounds the original primary yolk sac, the spaces within the mass of embryonic coelom are described above. The primary yolk sac decreased in size as a result of the mesodermal proliferation and the extraembryonic coelom expansion. The endodermal cells rearrange themselves and line the much smaller secondary yolk sac, which still adjoins the embryonic disk.

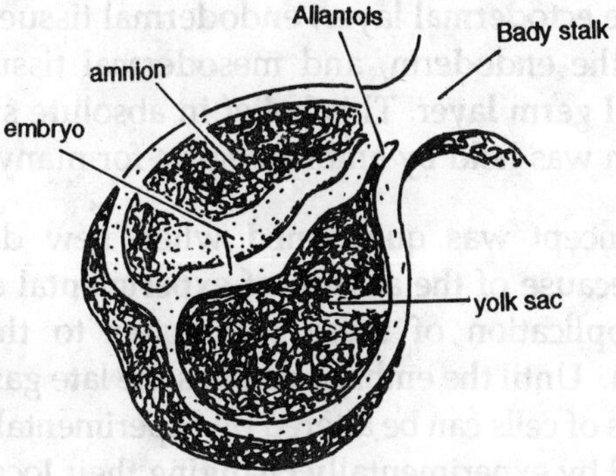

Fig. 2.21. Small diverticulum, the allantois, grows out from yolk sac in to body stalk.

At this stage the extraembryonic coelom of the human embryo surrounds the embryonic disk and amnion except at one point, the body stalk, which attaches the caudal end of the disk to the trophoblast. A small diverticulum (pocket) from the yolk sac grows into the connecting stalk. This small extension of the yolk sac is the *allantois.* It never expands in the mammalian embryo as it does in the chick and reptile but will contribute to the formation of the urinary bladder of the adult. Its blood vessels, on the other hand, will be important when the fetal tissues make contact with the uterine wall. It is these fetal allantoic arteries and veins that will carry the substances to and from the maternal vascular system.

GERM LAYER CONCEPT AND ITS LIMITATIONS

Construction of the organs of the body from the germ layers is an old concept. Von Baer, often referred to as the father of comparative embryology, spoke of the "layers" of the embryo. He believed that there were four instead of three, since he considered the middle layer to be divided into two parts. As originally conceived it was a rigid concept and specified that certain cells of the blastula were destined to form specific (germ) layers. A germ layer, in turn, could contribute only to certain tissues. Epidermal or nervous tissues were limited in origin to the ectodermal layer; endodermal tissue could arise only from the endoderm, and mesodermal tissue from the mesodermal germ layer. This belief in absolute specificity of tissue origin was held by embryologists for many decades.

The concept was questioned when new data became available because of the advent of experimental embryology and the application of these techniques to the study of gastrulation. Until the embryo reaches the late gastrula stage, the destinies of cells can be altered by experimentally changing their altered by experimentally changing their location within the embryo. Each cell has the potential of forming more than one specific layer. The ultimate fates of the cells were not as *predetermined* as previously assumed. The presumptive epidermal ectoderm cell of the early blastula is destined to give rise to the epidermis of the individual. However, if it is transplanted at the early gastrula stage to another location near the dorsal lip, it will involute and be carried inside the gastrula where it can give rise to kidney, muscle, or skeletal structures usually associated with other germ layers.

The germ layer concept can be used in generalizing the structure of the embryo, but we now know that similar structures do not always arise from corresponding germ layers. Cartilage, for example, usually forms from mesoderm, but in some animals, neural crest cells, considered to be ectoderm, contribute to the cartilage of the visceral arches. Teeth have both endodermal and ectodermal origins. The thymus may also develop from both of these layers, adding further evidence to the idea that the potencies of the germ layers are not as fixed as supposed by the early embryologists.

The important factor in determining the fate of a cell may

be the type of cell involved and the sphere of influence in which it becomes located. The interaction between a cell and its environment is believed to determine the direction of its differentiation. The idea that cell potency can be influenced by the cell's microenvironment is generally accepted by today's embryologists. The accumulation of such information shattered the faith of some biologists in the germ layer concept and they discarded it as meaningless. Others, placed it in a new, more modern perspective, since under normal developmental conditions cells of the blastula do follow established pathways of migration, take up regional positions with in the gastrula, undergo mitosis, and then differentiate into predictable tissues and organs.

Modern embryologists consider that the germ layer theory says nothing about the potentialities of cells within the embryo or about their reactions to different environments. It is purely a morphological concept. However, it certainly remains a useful means of categorizing development and is of value when one discusses morphology. The concept is used in this manner in these chapters, and the origins of structures of the body are traced to one or more of the three germ layers.

3

Basic Embryonic Processes

Ontogenetical changes continue throughout the life of the individual, but the prenatal changes that are carried out before hatching or birth occur faster and involve greater architectural molding than those that continue into the posthatching or postnatal period. These early embryonic stages can be further divided into *prefunctional* and *functional periods*. The prefunctional state implies a lack of cellular or organ structure necessary to carry out the specialized physiological activities. The initiation of function must await the completion of basic embryonic structure, which for many organs is established about the end of the neurula stage. All organs do not begin to function at the same time. For example, the heart of the chick starts pumping blood after 40 hours of incubation have passed, and the blood circulates in the human embryo by the fourth week of gestation, but the development and differentiation of skeletal muscle and the initiation of its contraction begin later. During the earliest embryonic stages (prefunctional) the greater structural changes occur and result from differentiation and growth of the embryo as the fertilized egg develops into the multicellular and multilayered organism. These changes depend on morphogenetic movements of cells or layers of cells. The formation of the germ layers and the establishment of the basic body form (with all its organ primordia) limit further drastic changes in structure. In the human embryo the major organs and systems form by the end of the eighth week. Once established, further development primarily involves continued differentiation and growth.

Structural development occurs in an orderly stepwise manner under the control of genes. Gene activity covers the spectrum. They act singly to produce a variety of effects *(pleiotropism)*, or they co-operate with other genes to produce a singe effect. Most of the genes of the multicellular animal are in a repressed state until gastrulation. After gastrulation further development of the embryo results as the genes become sequentially derepressed.

Development is an extremely complex process, however, and can be better described and understood if broken down into some of its components. For convenience of discussion we shall divide the processes responsible for establishing body form into the following three areas : (1) growth, (2) differentiation, and (3) morphogenesis.

Growth

Growth is an important process in development and concerns the permanent enlargement of parts of the organism or of the whole. Although it is a difficult word to define, it refers to the increase in cell size and number, or total mass. In general, an increase in protoplasm produces growth, but the non-living materials manufactured by the cells and deposited in the form of fibers, ground substances, or jelly also contribute to the mass. Growth is more than cell division. For example, the oocyte of *Rana pipiens* over a period of 3 years increases in size by a factor of 27,000. In fact, cleavage is usually not considered as growth since the mass of the fertilized egg is merely subdivided into smaller and smaller units.

Growth is a regulated process; it does not continue unchecked. Many years ago Mendel crossed dwarf and giant varieties of pea plants and demonstrated the importance of inherited factors in determining their size. These genetic factors control the metabolic processes of the cell, which in turn affect cell division and synthesis of new materials. In general, giantism and dwarfism result from an increase or decrease in cell numbers rather than cell size, since the size of cells in plants and animals is limited. Growth of cells, organs, or organisms also reflect the conditions of their surrounding environment. The available space, the proper mixture of nutrients, including oxygen, and the accumulation of cell products are important factors. For example, maternal malnutrition limits the size of the fetus. Hormones also play a role in regulating growth. Hormones released by the pituitary, thyroid, adrenal glands, and gonads stimulate or inhibit cell proliferation in developing organs.

There are several ways of measuring growth. Cell division is a part of growth and is demonstrated by merely counting the increase in cell numbers of a measured sample. Growth is also associated with the synthesis of new protoplasm. Protein growth can be calculated by measuring the increase in protein nitrogen per unit of DNA. A third way of measuring growth is by weighing the organism at certain regular intervals over a period of time. The shape of the growth curve plotted from such data is usually a sigmoid curve, i.e., it resembles the letters. Growth starts out very slowly, increases rapidly in the midperiod, slows down again, and then essentially stops. When cell death equals cell growth, no further increase in overall growth occurs, stabilizing the weight of the organism.

The growth of the organism is an expression of the total growth of the individual parts. Individual organs grow by increasing their cell numbers. Although growth curves are similar for all organs, each organ does not necessarily grow at the same rate or at the same time. The change in position and shape of the various organs during development reflects this differentiated growth rate. For example, the proportions of an organism constantly change. As the human body develops from embryo to adult the alternations in the form can easily be observed as shown in the figure.

Growth and differentiation may or may not go on at the same rate at the same time. A certain antagonism exists between cell division and cell differentiation in some cells. For example, after myoblasts fuse to form the myotube, they do not normally

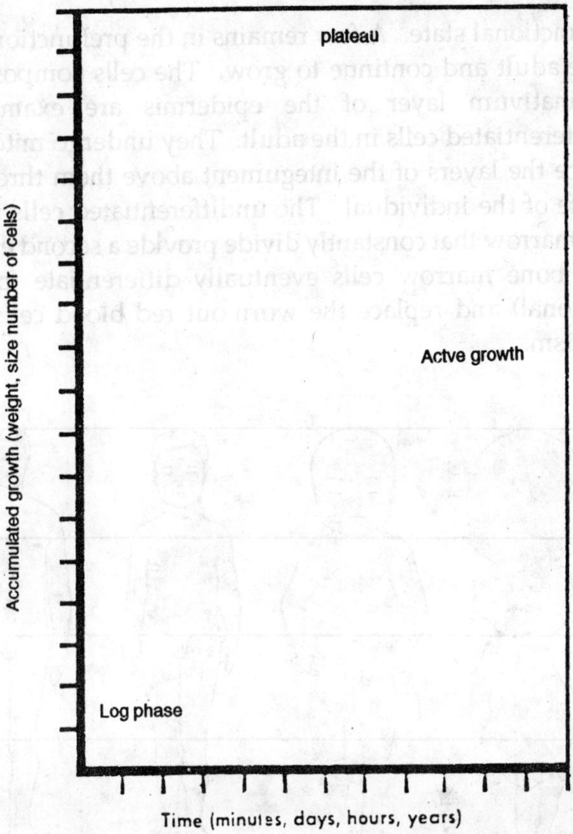

Fig. 3.1. Growth curve of cells, organisms, or populations is S shaped. The abscissa is an expression of time, and the ordinate is some measurement of growth. Growth starts out slowly, enters a period of rapid increase, and then reaches a plateau.

divide. Further, a highly differentiated cell, such as a nerve cell cannot undergo cell division, yet it increases in size. It is capable of limited growth as evidenced by the regeneration of a cut nerve fiber. On the other hand, the cloning and subcloning of retinal pigment cells and cartilage cells, both considered differentiated phenotypes, provide evidence that these specialized cells, after dividing several times, still retain their specialized characteristics. In a few cases, only some of the cells composing an organ undergo differentiation and enter the functional state. A few remains in the prefunctional state in the adult and continue to grow. The cells composing the germinativum layer of the epidermis are examples of undifferentiated cells in the adult. They undergo mitosis and replace the layers of the integument above them throughout the life of the individual. The undifferentiated cells in adult bone marrow that constantly divide provide a second example; some bone marrow cells eventually differentiate (becomes functional) and replace the worn-out red blood cells of the organism.

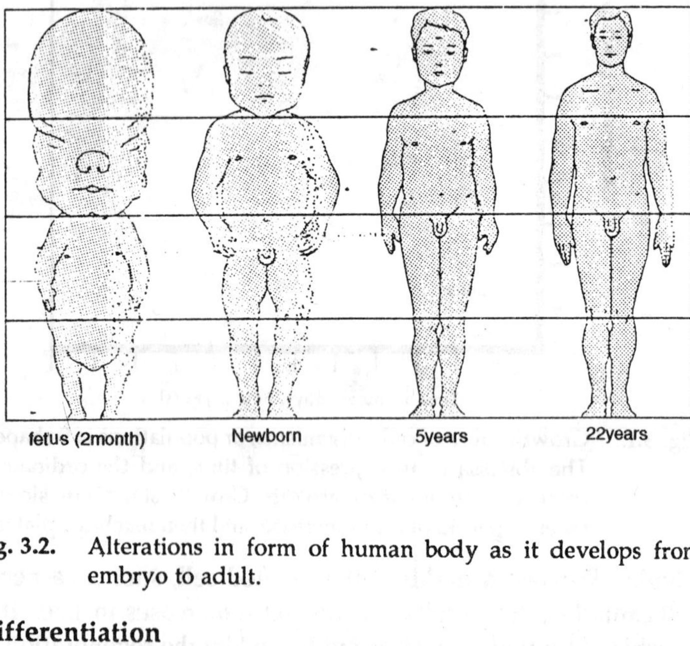

Fig. 3.2. Alterations in form of human body as it develops from embryo to adult.

Differentiation

The cells of the early embryo look much alike although they differ in size, amount of yolk present, and amount of pigment granules. According to electron micrographs, embryonic cells contain few organelles; the various membrane systems of the endoplasmic reticulum are often absent in the embryonic or undifferentiated cell. As development continues, the cells become *differentiated*. They become different from one another and from the state in which they previously existed.

The establishment of the germ layers by the end of gastrulation gives rise to layers of cells that are differentiated but remain prefunctional. For example, the cells that make up the neural plate at this time are different from the surrounding epidermal ectoderm and from their previous state of presumptive nerve ectoderm. They are still unable to function, however. They cannot carry out physiological activities until they undergo a second more special type of differentiation, *histological differentiation*. As a result of this second level of differentiation, new proteins appear in the cell and a specialized cell results. These special proteins allow certain cells to perform one function and not another. Muscle cells contract, nerve cells conduct, and gland cells secrete.

Induction

By the end of gastrulation, cells have moved to new locations within the embryo and established new relationships. No cell exists in isolation but is in intimate association with other cells. These new relationships apposed new cell surfaces and allowed new exchanges between cells, new stimulations, and new inhibitions. This interrelationship plays a major role in development as groups of cells become more restricted and give rise to tissues and organs.

During normal development, the chorda-mesoderm cells, those that form the dorsal lip of the blastopore of the early animal, line up beneath the neural plate and influence the formation of the brain in the anterior region. The surgical removal of the dorsal lip cells so that they cannot occupy this position beneath the neural plate prevents the formation of the brain. Normally, the presumptive neural ectoderm cells become determined and give rise to nerve cells; under the conditions described, when not in contact with the chorda-mesoderm cells, they remain in an undifferentiated state. The term *induction* is used to describe the influence of a group of cells on the morphogenetic development of an adjacent group of cells. It is the primary event in embryonic differentiation. The response of the neural plate to the chorda-mesoderm is considered a classical example of embryonic induction. It is through the inductive process that cells lose their lability and become determined.

The specific influence exerted by the roof of the archenteron (chorda-mesoderm cells) on the neural plate to form brain vesicles is called *primary induction.* It occurs not only in the frog, but in amphioxus, fishes, and other amphibians. In amniotes, the anterior part of the primitive streak, considered homologous to the dorsal lip, produces similar responses. Induction is a part of a continuum that releases certain reactions inherent in a cell. It results in the induced cell's ability to synthesize specific proteins which, in the case of nerve cells, would be nerve-specific proteins.

Not only must the chorda-mesoderm cells have the ability to induce changes in the neural ectoderm, but the neural ectoderm must be able to respond to the inductive stimuli. This ability to react is referred to as *competence* and the condition exists for only a certain period of time. Transplantation experiments show that the competence of neural ectoderm decreases from the blastula to neurula stage. At the late gastrula stage, ectoderm that comes in contact with the inductor for the first time lacks competence to form brain structures. By the late neurula stage, the ectoderm is unable to form neural tissue at all, and once the ectoderm has become epidermis, it remains epidermis.

The cells of the dorsal lip of the blastopore responsible for bringing about embryonic induction in the amphibia are often referred to as the *organizer.* This term, coined by Spemann, refers to the ability of these cells to organize a regionalized central nervous system. Spemann and his student, Hilde Mangold (1924), transplanted the dorsal lip of one salamander embryo to the lateral surface of another and induced a second embryo to form at the site of the transplant. The second embryo had axial organs that were more or less complete and that were formed partly from grafted tissue and partly from host. It was the first clear demonstration of a primary induction, and Hans Spemann was awarded the Nobel Prize in 1935 for developing his concept of the organizing effects of the dorsal lip.

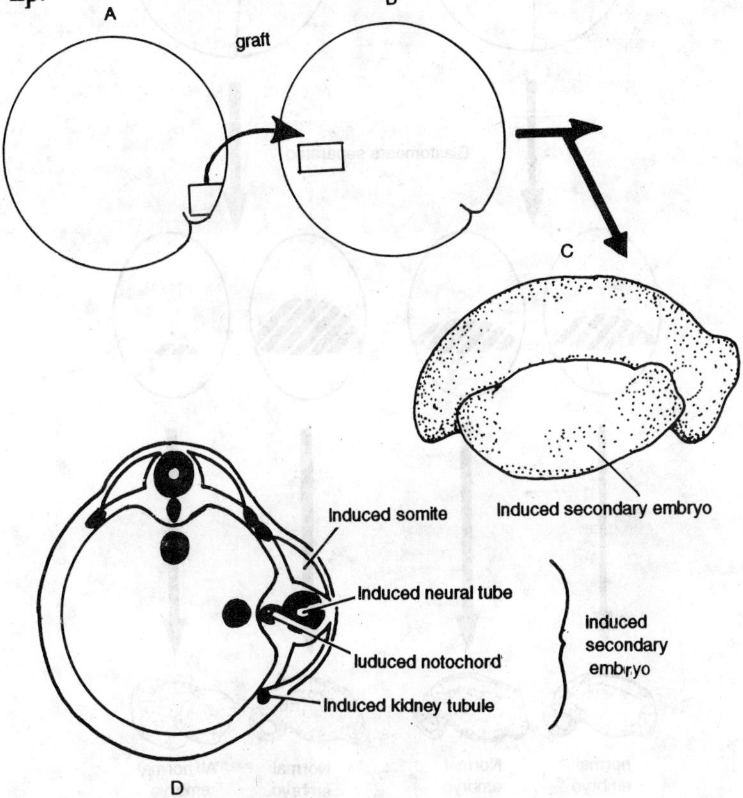

Fig. 3.3. Diagram showing the results of Spemann's and Mangold's experiment. The dorsal lip of embryo *A* was transplanted to a second embryo, *B*. In *C*, the transplanted dorsal lip induced formation of a second embryo on the lateral surface of *B*. Cross section through the resulting embryo *D* shows normal development of host and the development of the induced secondary embryo.

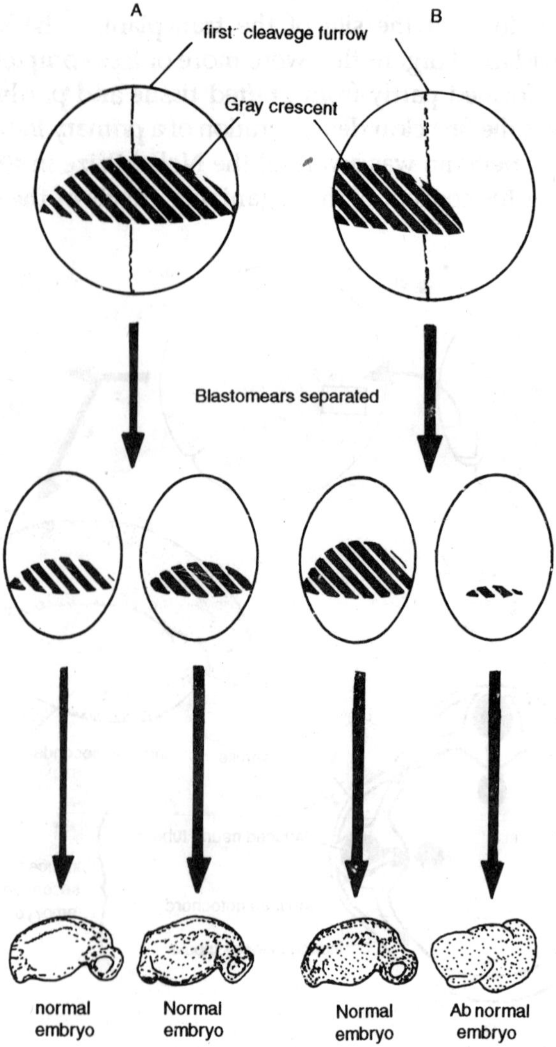

Fig. 3.4. A, If the first cleavage furrow bisects the gray crescent and the resulting blastomeres are separated, two small norms individuals will develop. B, If the first furrow divides the egg so that only one blastomere receives most of the material of the gray crescent, only this blastomere will give rise to a normal individual.

The cells that make up the dorsal lip of the frog can be traced back to the gray crescent. If the two blastomeres formed at the end of the first cleavage are separated, each will develop into a smaller but complete individual, provided that the first cleavage bisected the gray crescent. When this happens, the inductive material is present in sufficient amount in each of the separated blastomeres to bring about normal development. If the first cleavage furrow divides the egg so that only one blastomere receives the gray crescent material then only this blastomere has the potential of developing into a whole individual. The blastomere that lacks a significant portion of gray crescent substance is unable to complete its development.

The chemistry and mechanisms of embryonic induction have been the subject of much research since the time of Spemann's experiments. A great deal of frustration resulted from these studies, which often yielded contradictory results. We still do not completely understand the exact nature of inductors, and the role of the tissue responding to the induction is even more of an nigma. Only a few systems have been defined and their timing and spatial arrangements described. Investigators agree that the influence occurs over a period time and in a stepwise manner. Jacobson (1966) describes a succession of inductive systems (at last three) that express their influence over a period of time and bring about the determination of the lens in *Taricha torosa*, a West Coast newt. In the early gastrula stage, the endodermal wall of the future pharynx underlying the head ectoderm acts as the first inductor of the lens. Removal of this endodermal wall prevents lens formation. Later in gastrulation after the mesoderm involutes and migrates anteriorly in the embryo, its anterior edge acts as the second inductor of the lens. Its removal also inhibits lens formation. Finally after neurulation, the optic vesicles bulge out from the brain and the future retina contacts the future lens. This retinal contact results in the third induction, which continues until the adult stage.

In order for induction to occur, actual physical contact is not necessary, but the cells must be in the same microenvironment. Induction involves some kind of morphological or chemical communication between cells. There is evidence that induction is attributable to the release of some chemical by one group of cells that causes another group to differentiate, but the molecule involved may not be the same for all inductions. The effective substance can pass through filters. Since its activity is destroyed by such protein-destroying enzymes as trypsin, pepsin, and chymotrypsin, it is likely that some type of protein is the responsible agent. Despite the fact that the mechanism of induction is not understood at the molecular level, induction should be considered a basic mechanism in the development of the embryo. It is specific and sequential, since more than primary induction is involved. A hierarchy of organizers are presumed to exist, leading to the development of the final embryonic form.

Restriction of Cell Potency

The restriction of cell potency *(determination)* is associated with the gastrulative process and leads to cell differentiation. It limits the pathways of differentiation open to a cell or a population of cells. The older concept of differentiation was based on the idea that development occurred as a result of a series of restrictions on cell capacity. These restrictions led the cell down a narrowing pathway that eventually allowed it to form only a specific structure. The more modern concept of differentiation rests on a molecular basis. It recognizes the appearance of new macromolecules (nucleic acids, proteins, etc.) in the cytoplasm of the cell. It is the interaction of these different macromolecules with similar hereditary machinery that causes the cells to become progressively determined, eventually bringing about a fully differentiated cell.

When cells become determined, areas of the embryo differ visibly from other areas. As a result, organ rudiments or organ primordia form. Certain chemical changes occur within the cell, however, before any external difference is apparent.

The basis of differentiation is chemical (for example, the appearance of fetal hemoglobin or myosin), but the ability to perform special functions depends on the presence of cytoplasmic mechanisms. The nerve cell develops long processes. Myofibrils appear in muscle cells. The cells in the middorsal areas of an early gastrula give rise to neural ectoderm. The neural ectoderm differentiates (in the prefunctional sense) into neural plate cells. As development proceeds, the neural plate cells differentiate histologically into nervous tissue. The nervous tissue in the anterior region of the embryo gives rise to the brain and in the more posterior region, to the spinal cord. This is the normal destiny of neural ectoderm cells brought about by a series of determinative steps. Once the highly specialized nerve cell forms, it can no longer express the complete potential contained within the nucleus of the cell. That is, although presumably endowed with identical genomes of other cells, its developmental pathway is restricted and it can only form nervous tissue. It is determined. Once an embryonic cell become determined, under normal circumstances the process is irreversible.

By carrying out a simple transplantation, experiment we can show that the determination of neural ectoderm occurs over a brief period of time. A group of the neural ectoderm cells are surgically removed from an early unpigmented gastrula and reinserted into a new, more ventral location on the surface of a second pigmented gastrula of the same stage. The donor cells can be distinguished from the host cells on the basis of their pigmentation and their activities followed. The embryo of the newt is quite often used in such experiments since there are pigmented *(Triturus taeniatus)* and non-pigmented *(Triturus cristatus)* species available for study. If non-pigmented donor cells, which normally give rise to neural ectoderm, are transplanted to this new location in a pigmented host, the pigmented cells become epidermal ectoderm. The transplanted cells conform and develop into whatever the surrounding cells become. Although the normal destiny of these cells is to form neural ectoderm, they can be influenced by their surrounding environment to give rise to something else. As a matter of fact, if the presumptive neural ectoderm cells are placed close to the dorsal lips of the early gastrula, they involute with the other cells and give rise to chordamesoderm.

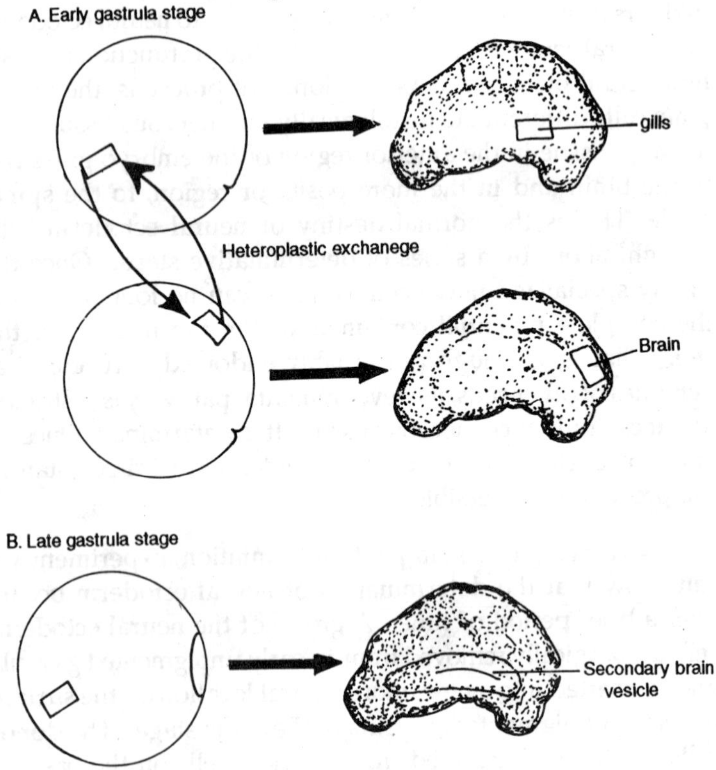

Fig. 3.5. A, in early gastrula stage, if presumptive neural ectoderm (unpigmented) is exchanged with presumptive epidermis (pigmented), the unpigmented neural ectoderm develops into epidermis and the presumptive epidermal ectoderm becomes a part of the brain. B, If the experiment is carried out at late gastrula stage and the non-pigmented neural ectoderm is transplanted to epidermal layer of pigmented host, the neural ectoderm continues to develop into brain vesicles. It has become determined.

At this stage of development, that is, early gastrula, the cells composing the neural ectoderm have the ability to develop into other types of cells, which, in the above examples, are epidermal ectoderm and chorda-mesoderm. This labile characteristic is not limited to neural ectoderm but includes other cells as well. If the transplantation experiment is carried out at the *end* of gastrulation, the results differ. A change has occurred. If some neural plate cells of the late gastrula or early neurula (the next stage of development) are inserted into the host of the same stage, as in the previous experiment, the cells of the neural plate follow their own destiny and form only brainlike structures. They can no longer give rise to epidermal ectoderm. The cells are said to be determined; they have started down a development pathway that is irrevocably fixed.

GENE ACTIVATION

Protein Synthesis

A cell differs from its neighbour, that is, it is differentiated, because of the presence of its particular proteins. These proteins either give rise to the structure of the body or form enzymes that direct the synthetic activities of the cell. Some of the enzymes of a liver cell are the same as the enzymes of, for example, a pancreas cell, but some are different. It is the appearance of these unique substances in the developing cell that indicates the beginning of its histological differentiation. Since the genetic composition of a liver cell and a pancreas cell in the same animal is identical the synthesis of different proteins is believed to occur when certain parts of the chromosome are turned, "on" or "turned off." An understanding of protein synthesis, which eventually is expressed morphologically, involves an understanding of gene regulation.

The control of gene action in the synthesis of proteins has been explored primarily in bacteria and viruses. Beadle and Tatum were the first to show that genes control protein synthesis. They suggested the one gene-one enzyme hypothesis. They worked with the relatively simple bread mold *Neurospora*, which is able to synthesize all of its purines, pyrimidines, and vitamins. They irradiated the spores of Neurospora and were able to demonstrate a mutant that synthesized a defective enzyme. As a result of the mutation, the mold was enable to synthesize vitamin B_6, pyridoxine. Only recently have investigations been expanded to explore protein synthesis in the more complex embryo of a multicellular animal. However, some of the information obtained from bacterial studies is believed to be basic to all types of cells.

Influence of Male Genes

During gastrulation the nuclear genes of the diploid cell begin to express themselves. If the egg of one species is fertilized by the sperm of another species cleavage often goes on but the interspecies hybrid cannot carry out the gastrulative process. This has been interpreted to indicate that the mRNA stored in the egg was capable of synthesizing the proteins needed for cleavage, but could not satisfy the needs for differentiation. Apparently newly synthesized proteins are essential for gastrulation because there is an increase in mRNA synthesis by the end of blastulation, especially in the presumptive endoderm and mesodermal cells. This lack of synthetic ability in the hybrid is caused by some nuclear genetic incompatibility of the foreign genes of the sperm that prevents their expression at blastulation.

Sometimes, the appearance of new proteins in the embryo can be traced directly to the influence of the paternal genes. If sea urchin eggs of *Paracentrotus lividus* are fertilized with sperm of *Psammechinus microtuberculatus*, there result true hybrids that contain the maternal cytoplasm, a haploid set of maternal genes, as well as a haploid set of paternal genes. According to immunological techniques, proteins (antigens) unique for the male species appear in the hybrid sea urchin embryos about the time of gastrulation. These proteins were not present in the egg and, therefore, must have been synthesized under the influence of the sperm genes. The formation of *andromerogons* is another way of demonstrating the influence of the male genes. In this case the female pronucleus is removed immediately after fertilization before it has time to fuse with the male pronucleus. If the sperm is from a species different from that of the egg, the resulting zygote is said to be a hybrid andromerogon. As a result of this technique the zygote formed has the maternal cytoplasm and paternal nucleus. No maternal nuclear material is present. Thus, these individuals are usually haploid. Any new protein synthesized must be the result of the co-operation between the cytoplasm and the nucleus. In those viable embryos (the hybrid andromerogons are extremely fragile) cleavage patterns are always from the maternal side, but the characteristics of the larva, pigmentation, for example, are intermediate between the two species, indicating that these factors are under the control of the male genes.

There is much more evidence available that demonstrates the importance of gastrulation as a time in which new structures are formed. The above two examples, however, are sufficient to show that at gastrulation the two sets of chromosomes (maternal and paternal) brought together at the time of fertilization, start to co-operate with one another in the production of the characteristics associated with the next generation.

Constancy of Genetic Information

For many years differentiation was considered at tributable to one of two processes. Either each cell received a different genetic endowment (genome) from its parent, which accounted for its differentiation, of the genetic endowment remained the same in each cell and the environment with which it interacted changed. Around 1900, A. Weismann suggested that when a fertilized egg cleaved, each daughter cell did not receive identical genetic information. He called the hereditary material determiners. According to him, the egg contained all the determinants (or determiners) necessary to make a complete embryo. With each cleavage they were parceled out in such a manner that different cells received different determinants. As a result of this sorting out of the hereditary material, different cells differentiated along different lines and developed into distinct cell types, depending on the types of determinants they received. This hypothesis was tested by separating cells at the various cleavage stages.

These experiments showed that at the two-cell stage each blastomere (of certain animals) was capable of developing into a complete individual: each cell had received all the necessary instructions or determinants. Weismann's hypothesis, therefore, was not upheld by experimental evidence.

As pointed out earlier, the third cleavage divides the cytoplasm of a telolecithal egg in such a way that the blastomeres of the animal pole contain much less yolk than those of the vegetal pole. Although the genetic endowment of each blastomere remains the same, the environment provided by each blastomere is different. The synthetic process that goes on because of the interaction between the hereditary material and substrate provides different metabolites, and these, in turn, are believed to result in different morphological expressions.

The composition of the chromosomes has been studied and noted that they consisted of a double spiral of linearly placed nucleotide units. Deoxyribose sugar, phosphoric acid, and nitrogen bases compose these units. The specificity of the hereditary information contained in the chromosome depends on the sequence of the nitrogenous bases, since the "backbone" of the molecule, the deoxyribose sugar and phosphoric acid, is the same for all animals and plants. The sequence of bases in the DNA molecule codes for the of amino acids in proteins. Three bases (a triplet), referred to as a *codon,* are always necessary to code for one amino acid. A gene responsible for protein synthesis must consist of a sequence of codons (triplets).

During cell division the chromosomes replicate. As a result, an exact copy of the code is distributed to each daughter cell. However, when protein synthesis occurs, instead of replicating as in mitosis, the DNA molecule containing the necessary code separates into two strands and serves as a template for the synthesis of another molecule, RNA. This process is referred to as *transcription.* The fact has been shown experimentally that only one of the strands of DNA codes for the amino acid sequence. The other strand lacks the code.

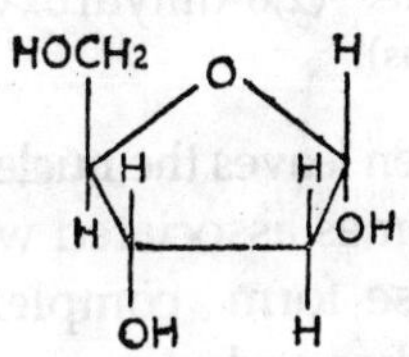

Fig. 3.6. Chemical formula of deoxyribose.

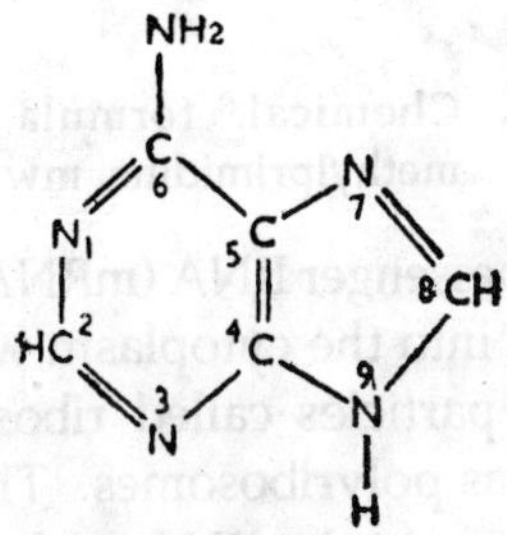

Fig. 3.7 Chemical formula of adenine (6-aminopurine, mw = 135.13 daltons)

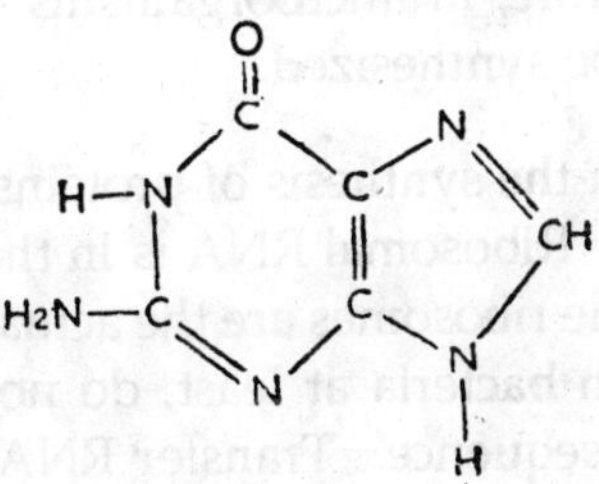

Fig. 3.8. Chemical formula of guanine (2-amino-6-hydroxypurine, mw=151.13 daltons).

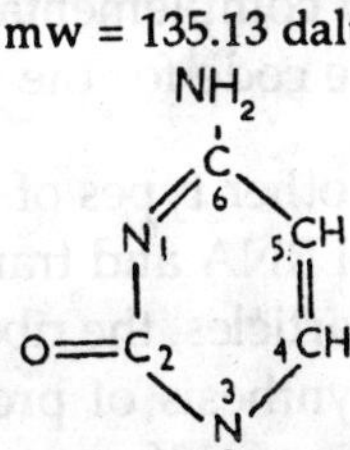

Fig. 3.9 Chemical formula of cytosine (2-hydroxy-6-aminopyrimidine, mw-111.10 daltons).

As a result of transcription, long chains that lack double spiral make up the RNA molecule. It is similar in composition to DNA, with the exception that the sugar molecule involved is ribose instead of deoxyribose. The nitrogenous bases remain the same as those in DNA with one exception. Instead of thymine, uracil is found in its place. The new nucleic acid is

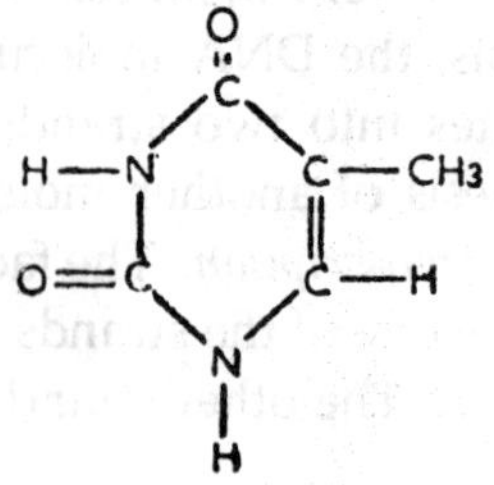

Fig. 3.10. Chemical formula of thymine (2,6-dihydroxy-5-methylprimidine, mw=126.12 daltons).

called messenger RNA (mRNA), since it then leaves the nucleus, passing into the cytoplasm where it becomes associated with several particles called ribosomes. These form complexes known as polyribosomes. The total number and sequence of the codons in the RNA molecule corresponds to those on the DNA molecule, and its base triplets are in the same order (although complementary). Therefore, in microorganisms it carries the code for the protein to be synthesized.

Two other types of RNA aid in the synthesis of proteins: ribosomal RNA and transfer RNA. Ribosomal RNA is in the form of particles, the ribosomes. The ribosomes are the actual sites of synthesis of protein but, in bacteria at least, do not themselves specify an amino acid sequence. Transfer RNA, also called soluble RNA (sRNA), acts as an adapter molecule. It is able to pick up specific amino acids, with the help of enzymes and link them to their codons. These small molecules of transfer RNA are specific for each of the 20 amino acids. In a key location on the molecule, the transfer RNA for a particular amino acid contains a sequence of bases complementary to a specific triplet on the mRNA. Each molecule of transfer RNA, therefore, attaches to an amino acid and its codon in messenger RNA. It is the "pairing" attraction between the triplets of tRNA and mRNA that is responsible for holding specific amino acid-tRNA complexes in specific places long enough for the amino acids to react and hook together. Of course, these reactions are speeded up by enzymes. When the bond between tRNA and amino acid is broken, the energy released is used to form peptide bonds between the amino acids. The sequence of amino acids is ultimately dictated by the base sequence of the original DNA molecule in the nucleus of the cell.

Protein Synthesis in Higher Animals

The synthesis of proteins in metazoan animals is less clear, since the specificity of RNAs has not been defined. Although we are certain that RNA acts as a template for protein synthesis, we have no evidence that mRNA actually carries a message. For example, mRNA is synthesized at the puffs of dipteran giant chromosomes. However, it is not possible to prove that the mRNA contains a message that can be identified with the final product. We know little or nothing about the RNA of higher organisms, but the biochemical mechanism of mRNA formation is believed to differ radically from that of bacteria. RNAs are identified primarily on the basis of their size and base composition. We are not even certain of the source of RNA; not only nuclear DNA but cytoplasmic DNA or cytoplasmic RNA have been suggested as possibilities.

The problems of determining the source of RNA in the cells of higher animals lies in a lack of a nuclear precursor for a specific mRNA functioning at the ribosomal level. It has been suggested that the RNA present in the cytoplasm may be a part of a high molecular weight nuclear RNA molecule, designated heterogeneous, nuclear RNA (HnRNA). Darnell. Jelinek. and Molloy, 1973). Biochemists are in the process of trying to match the existence of sequences of nucleotides in HnRNA with a specific sequence in a specific RNA. One such sequence discovered many years ago is now known to be shared by both HnRNA and mRNA. This sequence, consisting of about 200 nucleotides containing adenylic residues, is termed "poly (A)." Its presence in both HnRNA and mRNA (but not in rRNA and tRNA) strengthens the idea that mRNA in eukaryotic cells is derived from a modification of a larger RNA-precursor molecule.

An alternative hypothesis suggests that an informational transfer occurs between the nucleus and cytoplasm. He suggests that small segments of nuclear DNA gene copies (I-DNA) pass out of the nucleus into the cytoplasm in the form of I-somes of different sizes. There is evidence, for example, that DNA constantly turns over in eukaryotic cells, even in nondividing cells. Also, DNA has been identified in the cytoplasm. According to the I-some theory of information transfer, the I-somes in the cytoplasm are transcribed to yield an RNA template for protein synthesis. During differentiation copes of genes necessary for each phase of development are excised from the nuclear genome and pass out into the cytoplasm where they carry out their specific tasks. To prove this hypothesis correct, the next step obviously is to isolate these genes in the cytoplasm of the cell.

Gene Regulation

The next important problem concerns the selectivity of protein synthesis. How are the activities of the various genes co-ordinated ? What determines what gene will be activated and consequently what protein will be synthesized ? All cells possess common enzymes for catalyzing common metabolic pathways, but different cells in higher organisms possess also unique enzymes. There is general agreement that the differentiation of a cell depends on its ability to synthesize these unique proteins. The synthesis of a specific protein depends in turn on the gene transcribing its code to mRNA, which carries the message out into the cytoplasm and directs the synthesis. Obviously all genes do not synthesize all proteins in all cells at the time. For example, probably all the cells in the body contain the genes for making myosin, but only the genes of a muscle cell are turned on and direct its synthesis. There must be some kind of a programmed mechanism to co-ordinate and regulate the thousands of chemical reactions that go on in the body. The kind and amount of protein must be programmed, as well as the time and sequence of its appearance.

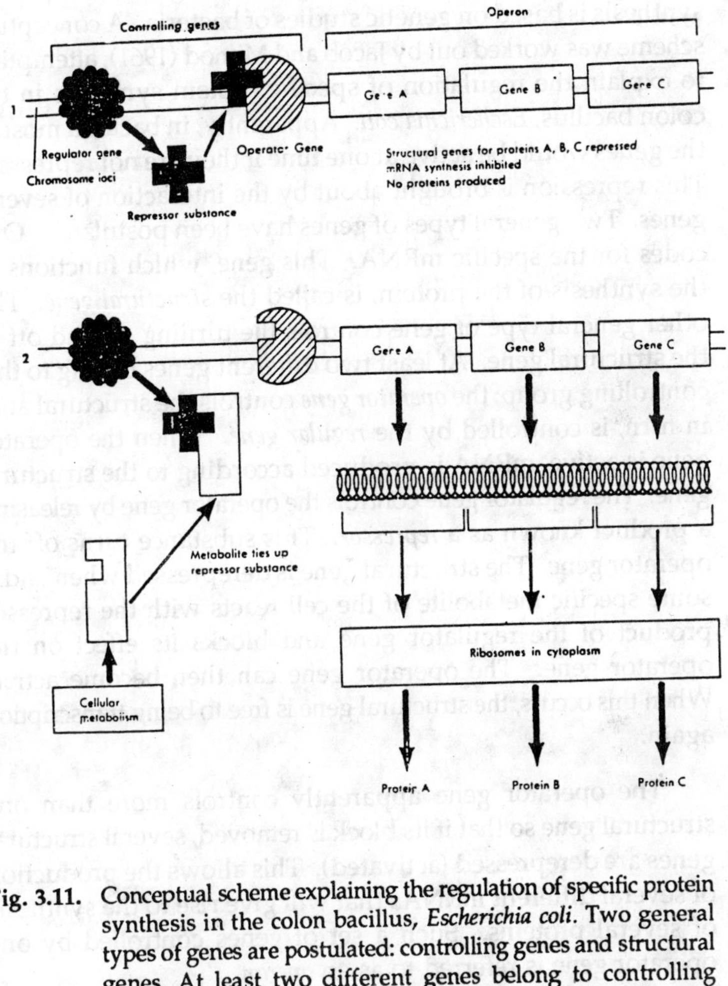

Fig. 3.11. Conceptual scheme explaining the regulation of specific protein synthesis in the colon bacillus, *Escherichia coli*. Two general types of genes are postulated: controlling genes and structural genes. At least two different genes belong to controlling group: regulator gene and operator gene. 1, Regulator gene releases a repressor substance that reacts with the operator gene "turning off" its activity. As a result the structural genes for proteins A, B and C are repressed and mRNA synthesis is inhibited. 2, When a cellular metabolite reacts with the repressor substance, the operator gene becomes active. As a result the structural genes are stimulated to produce mRNAs that will give rise in the cytoplasm to synthesis of specific A, B and C proteins.

Much of our present knowledge of the regulation of protein synthesis is based on genetic studies of bacteria. A conceptual scheme was worked out by Jacob and Monod (1961), attempting to explain the regulation of specific protein synthesis in the colon bacillus, *Escherichia coli.* Apparently, in bacteria most of the genes would be active at one time if they are not repressed. This repression is brought about by the interaction of several genes. Two general types of genes have been postulated. One codes for the specific mRNA. This gene, which functions in the synthesis of the protein, is called the *structural gene.* The other general type of gene controls the turning on and off of the structural gene. At least two different genes belong to this controlling group; the *operator gene* controls the structural and, in turn, is controlled by the *regular gene.* When the operator gene is active, mRNA is produced according to the structural gene. The regulator gene controls the operator gene by releasing a product known as a *repressor.* This substance turns off the operator gene. The structural gene is derepressed when and if some specific metabolite of the cell reacts with the repressor product of the regulator gene and blocks its effect on the operator gene. The operator gene can then become active. When this occurs, the structural gene is free to being transcription again.

The operator gene apparently controls more than one structural gene so that if its block is removed, several structural genes are derepressed (activated). This allows the production of several different mRNAs that will give rise to the synthesis of several proteins. Such a set of genes controlled by one operator gene is referred to as an *operon.*

At the present time, we do not know if the control of gene activity in higher organisms resembles that in microorganisms. There is scattered evidence that induction and repression of enzymes that occur in bacteria also take place in vertebrates, for the most part in the liver. One may logically assume this function for the liver, since it is the organ subjected to a varying supply of nutrients. It helps to maintain a constant environment for such organs as muscles, brain and heart. Undoubtedly, gene control would be more complex in a metazoan animal and must be modified. For example, the genome of multicellular animals is much larger than in microorganisms and differentiation requires the co-operation of non-adjacent genes. Also a large amount of repetitive nucleotide sequences (redundancy) occurs scattered through the genome. Only a small portion is active at any one time.

A modification of Jacob and Monod's scheme for gene regulation was suggested by Britten and Davidson (1969). According to their model, sensor and integrator genes substitute for regulator genes of bacteria. They stress the multiplicity of changes in gene activity that can result from a single initiatory event. Their model includes a highly integrative system in which certain genes are sensitive to the products of other genes. A single event may activate several controlling genes, which, in turn, activate a vast number of structural (producer) genes. They speak of a magnitude of batteries of genes that may be necessary for the maintenance of the cell state, and they point out that there must be precise co-ordination of these batteries of structural genes if the differentiated state is reached.

Probably the most universal control of gene regulation occurs at the level of transcription. However, we do not understand how it occurs. Certain proteins associated with DNA may "mask" it in some way and inhibit it from transcription. Histones, proteins manufactured in the cytoplasm, are closely associated with the chromosomes and have been implicated in the control of gene activity. For example, addition of histones in vitro to nuclear preparations represses gene activity. Removal of histones caused depression. There is also evidence that non-histone proteins may move from the cytoplasm into the nulceus and also act to repress gene activity.

Certain hormones increase the rate of transcription and therefore the amount of RNA synthesized. Administration of estrone and estradiol to ovariectomized rats induces in the uterus a considerable increase in protein synthesis, which is blocked by administration of actino-mycin D. Some biologists suggest that the hormone reacts with a protein receptor and that the complex in some way releases a repressed state of an operon. Hormones may also exert their influence by changing cell permeabilities. The influx of various ions in some way may regulate the activity of the gene. Gene regulation, however, occurs at many more levels than the operon. In some instances differential gene replication is correlated with certain stages in the development of the embryo. For example, in *Xenopus*, 28S and 18S ribosomal RNA's (designated according to their size) are synthesized during the early stages of the developing egg (oogenesis). Their synthesis ceases when the egg matures and does not begin again until the initiation of development. The synthesis of rRNA apparently occurs in the nucleoli and during oogenesis extra chromosomal nucleoli form. The result of molecular mapping demonstrates that the genes for rRNA are redundant; that is, there are repeated sequences. The oocyte has many more genes for rRNA than does the nucleus of a somatic cell — some 1000 times more. During oogenesis the oocyte of *Xenopus* makes extra copies of genes for rRNA and stores them in the nucleoli. Presumably they are used as templates to support the increased rate of synthesis of rRNA that occurs during oogenesis, since they become non-functional at later stages of development. According to Brown and Dawid (1968) the immature oocyte of X. *laevis* synthesizes rRNA at a rate comparable to the equal weight of liver tissue, comprising some 2000 cells.

There is biochemical evidence that the type of chromatin varies among cells. Chromatin is the term given to the complex of DNA with protein and RNA. It exists in a condensed state during most of the mitotic cycle and in the extended state during the interphase. It is during the extended state that transcription occur. Different cells transcribe different amounts of the genome, and the transcribable DNA may differ from cell to cell. For example, the transcribable DNA from liver cells is organ specific and differs from that extracted from bone marrow or kidney.

The control of gene activity in multicellular organisms remains largely speculative. Most investigators believe that transcription and translation of genetic information follow pathways similar to those of microorganisms. However, we still do not understand the genetic mechanisms that ensures the progressive, sequential programmatic reading of the genetic code over defined time intervals.

Evolutionary Significance

One can see from this brief and necessarily superficial review that the transmission of hereditary material from parent to offspring implies the constancy of the double-stranded DNA molecule in the nucleus of the cells. The correct duplication of this material, its passage to the next generation in the egg or sperm, the new gene combinations occurring at fertilization, and the activity of the various types of RNA result in the appearance of parental characteristics in the offspring. The mechanisms of DNA transcription and RNA activity are remarkably similar not only in all vertebrates and invertebrates, but in all living things. The DNA always consists of the same four nucleotides. Multitudes of different proteins are assembled from the same 20 amino acids because of the activity of RNA. It has even been shown that under circumstances the purified RNA of one organism (polio virus) can direct the protein-synthesizing system of another organism *(E. coli)* to synthesize protein. However, diversity is obvious when the living organisms of the world are considered. Diversity appears among organisms and in the parts of the organisms. It is this diversity or variation among organisms at the anatomical, physiological, and behavioural levels that allows them to adapt to different environments. Charles Darwin realized that animals produce many more offspring than could survive. While some animals flourish, others are eliminated. Those that could best adapt to their environment continued to produce offspring whereas those that could not were "weeded out." As stated in Chapter 1, this process is referred to as *natural selection* and depends on the great diversity among animals.

The deviation of animals from this basic biological uniformity can be explained by turning once again to the DNA molecule. It is here that any alteration must express itself if it is to be passed to the next generation. If a change occurs in the sequence or composition of the base triplets of a cell's DNA, then the code is changed and results in a different amino acid being incorporated into a particular spot in the assembled protein. This change is synthesized protein expresses itself in the function or structure of the cell. Alteration of the code on the DNA molecule is referred to as a *mutation.* Mutation is the basis for evolution, because it is the only means of producing new genes. If the alteration is great, such as if the affects some vital process in the cell, then the mutation is usually lethal to the organism. Obviously only slight changes can be tolerated by an organism at any one time. Spontaneous mutations occur very rarely, but over a period of geological time mutant genes accumulate within a natural population and result in gradually altered from (phenotype). The process of evolution selects those genotypes that can live most successfully under the conditions prevalent at the time. The early developmental process of all vertebrates are conservative and follow known pathways that produce successful results. Over the ages, mutations must have occurred during the early stages of development but proved lethal. For example, a mutation at the gastrula stage could place the foregut in positions that would not allow further development. On the other hand, at later stages of development, there probably occurred mutations that caused the gut of one embryo to develop into a crop and a gizzard, and the gut of another to develop into the mammalian stomach. These modifications introduced at a late stage of development gave rise to organisms that have different survival values. The crop and gizzard of the grain-eating bird enable it to cope with one type of environment much better than if its gut were of the mammalian type.

The similarity of embryonic form was observed by the embryologists over a century ago. They noted that all embryos start out as an egg, undergo cleavage, form a blastula, and then form a gastrula in which the germ layers are established, before developing further to the neurula and transforming to the adult form. Many years before the theory of evolution became popular and before knowledge of the gene and its activity was accepted, von Baer noted that the more general characteristics of the embryo are established before the more specific features become apparent. The acceptance of Darwin's theory of evolution in the mideighteenth century focused attention on the embryo from an evolutionist's viewpoint. It seemed obvious to the biologists of the day that the basic similarity in vertebrates embryonic form was attributable to the fact that all were related in some way. They had evolved one from another. Those characteristics held by a large group of animals must have been held by a common ancestor.

Ernst Haeckel, a prominent German morphologist of this period, was convinced that the development of the individual (ontogeny) passed through the various stages of adult ancestors (phylogeny) and suggested what has become the aphorism, the "biogenetic law." The law is usually expressed as "ontogeny recapitulates phylogeny," or in more general terms, the embryo repeats in its development the evolutionary stages of its adult ancestors. It climbs its own family tree. Demonstrates the obvious similarity of comparable embryonic stages of vertebrates. The embryos closely resemble one another in stage I when the pharyngeal arches form. It is difficult to tell the examples of one class from another. By the time the embryos reach stage III the more specialized characteristics that allow them to be categorized into the various classes appear. Haeckel believed that evolution occurred by the addition of new stages at the end of certain life histories. Therefore, the developmental history of any animal included all the previous stages of development.

Time has proved Haeckel's theory to rest on false premises. Gill pouches, produced by the outpocketing of the pharynx, are found in all chordate embryos from fish to man. The pouches either remain intact or they become perforated to form gill slits. In the aquatic animals, vascular lamellae develop on the sides of the pouches and function in the exchange of gases in respiration. These become the gills of the aquatic animal. Human adults, of course, have no gills and neither do human embryos. In the embryo the pharyngeal pouches do appear but are transitory structures. Similarities are present, therefore, only in the embryonic stages. The embryo of a mammal passes through stages that resemble the embryos of the lower forms, but it never passes through stages that resemble the adults of these forms.

Basic Embryonic Plan

At the end of gastrulation epidermal ectoderm covers the cylindrically shaped embryo. The neural plate lies in the middorsal region with the prechordal plate cells lined up beneath its anterior regions. By the end of the gastrulation stage the mesoderm takes up a position between the ectoderm and endoderm. At first the mesoderm, closely associated with the notochord, is known as the *chorda-mesodermal mantle*. In the next developmental stop the notochord segregates itself, forming a rod-shaped structure that extends as an axis down the length of the embryo in the middorsal region under the neural tube.

When the notochord material separates from the mesoderm, the mesoderm disassociates itself from the endoderm. Only in amphioxus does the mesoderm originate as a pouchlike structure, in all vertebrates it is a solid band, which later delaminates or splits to form a lumen.

The gut of the embryo, composed of endodermal cells, extends from the head to the tail regions. Once the blastopore fuses in the amphibians, the gut is a closed cavity. In meroblastic animals the ventral area of the gut is open and rests on the

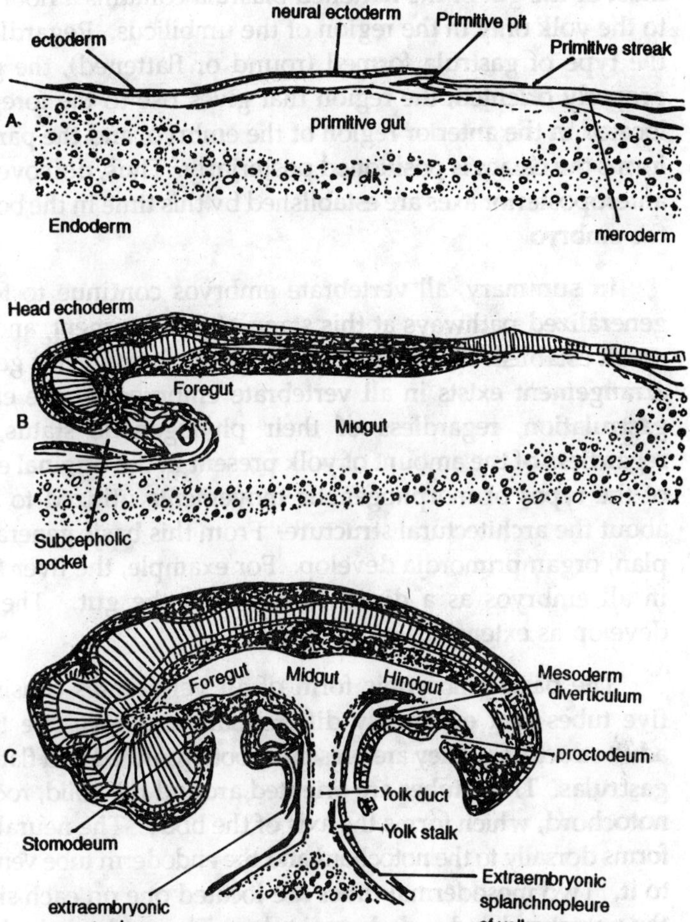

Fig. 3.12. Schematic drawing (sagittal view) showing stages in the formation of the body of chick embryo. A, Chick body develops anterior to the area of primitive streak. B, Folds roll under head process, separating it from underlying yolk by a subcephalic pocket. Floor to foregut is established. C, As a result of subcephalic, lateral, and subcaudal folds, the body of the embryo becomes defined and separated, from extraembryonic regions.

yolk. Subcephalic and subcaudal folds progressively undercut the anterior and posterior ends of the body until eventually most of the gut of the flattened blastula contains a floor open to the yolk only in the region of the umbilicus. Regardless of the type of gastrula formed (round or flattened), the gut is properly oriented; the region that gives rise to the foregut is located in the anterior region of the embryo, and the part that forms the cloaca is positioned posteriorly. Thus, dorsoventral, anteroposterior axes are established by this time in the body of the embryo.

In summary, all vertebrate embryos continue to follow generalized pathways at this stage of development, and as a result establish similar, basic, embryonic forms. This general arrangement exists in all vertebrate embryos by the end of gastrulation, regardless of their phylogenetic status, and regardless of the amount of yolk present in the original egg or of the types of morphogenetic movements utilized to bring about the architectural structure. From this basic generalized plan, organ primordia develop. For example, the liver forms in all embryos as a diverticulum from the gut. The eyes develop as extensions of the brain.

The basic embryonic form of all vertebrates consists of five tubes that eventually differentiate and give rise to the adult structures. They are present in both rounded and flattened gastrulas. These tubes are oriented around the solid, rod-like notochord, which forms the axis of the body. The neural tube forms dorsally to the notochord and the endoderm tube ventrally to it. Two mesodermal tubes are located one on each side of the notochord and endodermal tubes. The epidermal tube encases the entire embryo.

As the late gastrula stages develop into the neurula, the body of the embryo flattens laterally and extends in an anteroposterior direction. The dorsal region of the body arches, pushing the neural tube, notochord, and somites (mesoderm) more dorsally in the body. The anterior extension of the embryo is known as the *cephalic* or *head region;* a similar projection in the posterior region becomes the *caudal* or *tail* outgrowth. The remainder of the embryo is designated as the truck area.

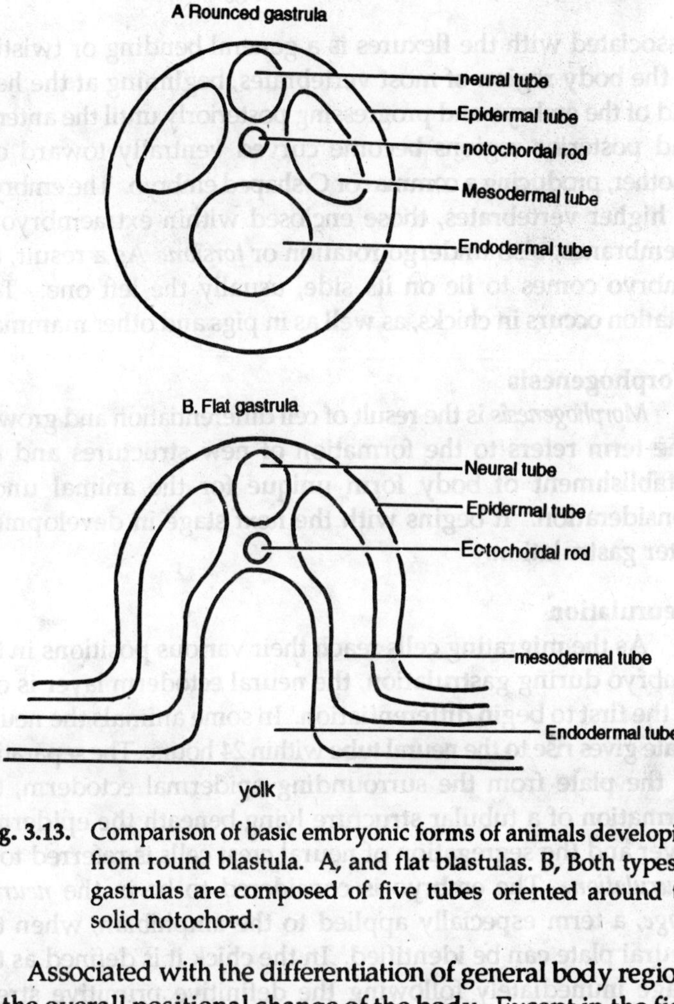

Fig. 3.13. Comparison of basic embryonic forms of animals developing from round blastula. A, and flat blastulas. B, Both types of gastrulas are composed of five tubes oriented around the solid notochord.

Associated with the differentiation of general body regions is the overall positional change of the body. Except in the fish, the anterior region of the neural tube becomes bent or flexed toward a ventral position. This flexure involves the midbrain and results in its growth perpendicular to the axis of the notochord. The midbrain bends toward the cardiac region in what is known as the *cephalic* or *cranial flexure*, readily observable in the 48-hour chick embryo. The cervical flexure appears in the region of the hindbrain at a later stage of development.

Associated with the flexures is a general bending or twisting of the body region of most vertebrates, beginning at the head end of the embryo and progressing posteriorly until the anterior and posterior regions become curved ventrally toward one another, producing a comma- or C-shaped embryo. The embryos of higher vertebrates, those enclosed within extraembryonic membranes, also undergo rotation or *torsion.* As a result, the embryo comes to lie on its side, usually the left one. This rotation occurs in chicks, as well as in pigs and other mammals.

Morphogenesis

Morphogenesis is the result of cell differentiation and growth. The term refers to the formation of new structures and the establishment of body form unique for the animal under consideration. It begins with the next stage in development after gastrulation.

Neurulation

As the migrating cells reach their various positions in the embryo during gastrulation, the neural ectoderm layer is one of the first to begin differentiation. In some animals the neural plate gives rise to the neural tube within 24 hours. The separation of the plate from the surrounding epidermal ectoderm, the formation of a tubular structure lying beneath the epidermal layer and the segregation of neural crest cells is referred to as *neurulation.* The embryo is considered to be in the *neurula stage,* a term especially applied to the amphibian, when the neural plate can be identified. In the chick it is defined as the stage immediately following the definitive primitive streak stages. The neural tube forms on the dorsal surface and extends the length of all chordate embryos. The formation of a single, dorsal, hollow neural cord separates the chordates from the invertebrate animals, whose neural cord is solid, ventrally located, and sometimes double.

As gastrulation ends, the cells of the oval area of neural ectoderm elongate because of the presence of microtubules and form a flat neural or medullary plate. The neural plate divides into several presumptive areas that develop into certain defined structures, provided that the necessary induction takes place. The anterolateral regions of the neural plate form the olfactory area. Posteriorly, the plate divides into the presumptive forebrain and optic area, midbrain, and hindbrain, respectively. The epidermal ectoderm over the hindbrain eventually forms auditory vesicles, and this region of the plate develops into the auditory area.

The cells of the neural plate become flask-shaped and differentiate from the more squamous (flat) cells of the epidermal epithelium. The plate extends from the dorsal lip of the blastopore, in those animals with round gastrulas, and from Hensen's node anteriorly in those embryos with a disk-shaped blastoderm. There are two general procedures by which vertebrates bring about tubulation of the neural tissue; the neural-fold method and the thickened-keel method. Although we can describe neurulation, we do not understand all the forces responsible for the formation of the neural tube.

In the majority of vertebrates the basic process of neurulation is one of folding. This folding is caused by the apical constriction of the neural plate cells, probably brought about by a micro-filament that encircles the apex of the neural cell result, in purse string fashion. As a result, in the middorsal region of the medullary plate a longitudinal depression forms, flanked by lateral neural folds. The folds are the result of the upward movement of the epidermal ectoderm at the margins of the plate. As the folds push upward and toward the midline, the depression deepens and a neural groove forms. The neural folds appear in the amphibian soon after closure of the blastopore and are first evident in the cephalic end of the chick around the three-somite stage (approximately 21 to 23 hours of incubation). Neural folds are definitely identified in the human embryo at the seven-to-eight-somite stage (18 to 20 days after fertilization).

Differential growth causes the lateral folds to move toward

one another where a double fusion takes place. The margin of the neural ectoderm detaches from the epidermal ectoderm and fuses to form the neural tube, first in the midbody region, and then continues in either direction zipper-fashion. At the same time the lateral epidermal ectoderms also fuse with one another, filling the spot once occupied by the nerve ectoderm. As a result, the neural tube comes to rest under the external surface of the embryo. For a time the neural tube is open to the outside at its anterior and posterior ends, but these openings eventually close. The neural tube is hollow from the beginning, with its lining consisting of some of the original surface cells of the blastula. The cavity of the neural tube is larger at the anterior end and develops into the vesicles of the brain. The smaller more posterior extension of the hollow tube is destined to form the spinal cord, with its cavity becoming the central canal of the cord.

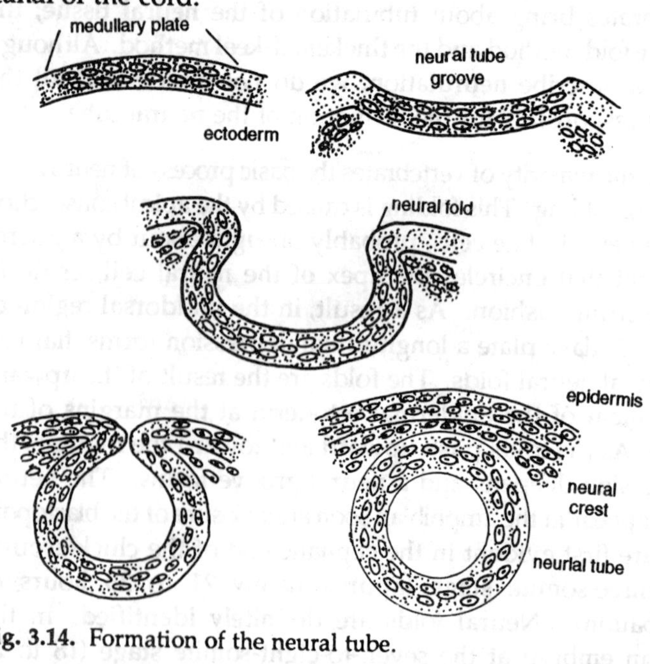

Fig. 3.14. Formation of the neural tube.

The thickened-keel method is the second method by which vertebrates, primarily certain fish, achieve neurulation. The material composing the medullary plate is in the form of a rod or keel running down the length of the animal. This keel detaches itself from the surrounding epidermal ectoderm and sinks down into the dorsal area of the embryo. The epidermal ectoderm covers the surface where it was located. Eventually a cavity forms in the keel, transforming it into a longitudinal nerve tube.

As neurulation continues, the embryo stretches along its anteroposterior axis, with the neural tube participating in the elongation process. The stretching that occurs in the posterior region of the embryo causes the body to extend beyond the blastopore into what is known as the *tail bud*.

Formation of Neural Crest Cells

After the neural tube separates from the epidermal ectoderm, certain other ectoderm cells collect in the midline in the space between the tube and epidermal surface. These are the *neural crest cells,* which are neither part of the epidermal ectoderm nor included in the neural tube. As soon as the crest cells form, they migrate to various parts of the body where they play a definite role in specific organogenesis. If they are surgically removed from the embryo, those structures to which they normally contribute do not form. Such excising techniques reveal that normally they contribute spinal ganglia, sensory nerves, sheath cells, and sympathetic neurons to the nervous system. They may also give rise to some of the colour-bearing cells of the body (chromatophores), the medulla of the adrenal gland, and the muscles and cartilage.

Differentiation of Epidermal Ectoderm

The ectoderm makes a major contribution to body structure by forming the nervous system. In addition, the part of the ectoderm that is neither neural ectoderm nor neural crest cells becomes the epidermis of the skin. In the early embryo it consists of two layers, an outer layer (periderm), which develops into certain sensory organs in the amphibian and is shed by the birds and mammals, and an inner layer, which becomes

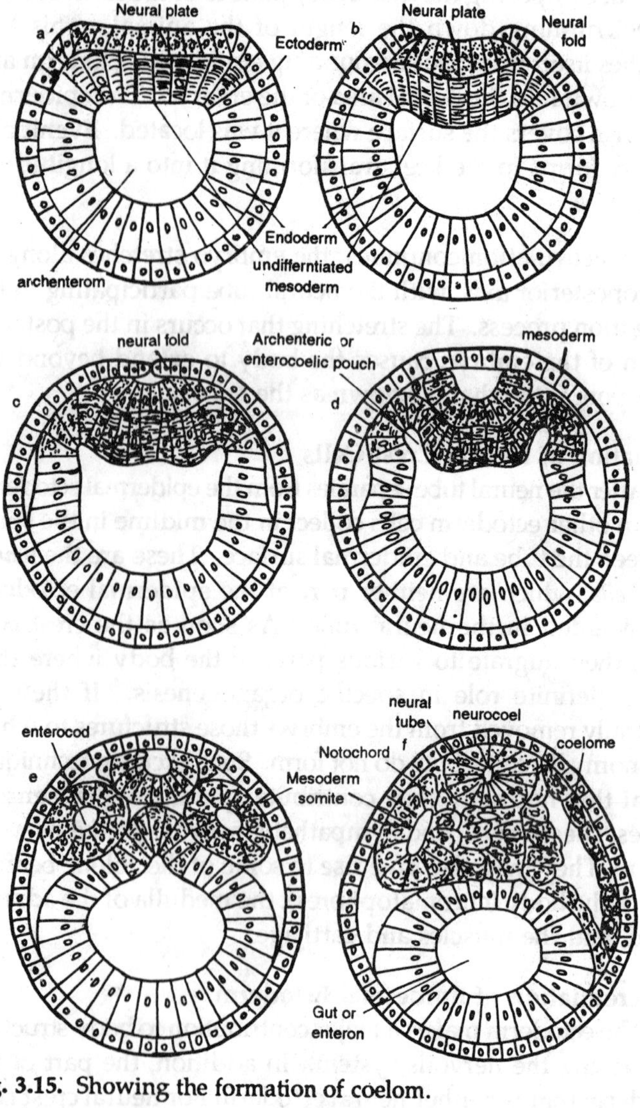

Fig. 3.15. Showing the formation of coelom.

the stratum germinativum of the epidermis. These cuboidal cells retain the ability to divide during the life of the organism and replace those layers of the skin that are sloughed off or worn away. The epidermis in combination with the dermis (mesoderm) composes the integument of the organism. It is from the epidermis and dermis that certain accessory structures (glands, hair, horns, feathers, etc. develop and certain outgrowths, limb buds, fins, and external gills) form.

The ectoderm eventually also gives rise to platelike thickenings of the epidermal epithelium, called *placodes*. They are associated with certain sensory organs and are discussed in greater detail when we describe the histogenesis of specific organs. However, there are five or six placodes that can be mentioned appropriately at this time because they are involved in organo-genesis. In the anterior end of the animal, thickenings form the olfactory placode, which will give rise to the olfactory sacs. The forebrain induces the cells of the lens placode to elongate and form the lens of the eye. The thickenings associated with the hindbrain become auditory placodes, which eventually convert to auditory vesicles. Aquatic animals have lateral-line organs stretching from the anterior to the posterior end of the body. Other thickenings appear in the head region and give rise to certain cranial nerves.

Differentiation of Mesoderm

As neurulation proceeds, the mesoderm is the second area to differentiate actively. It contributes much to the growing embryo. The thickened mass of mesoderm adjacent to the notochord becomes transversely divided into a number of segments, as described for the human embryo in the previous chapter. Each segment is referred to as a *somite* or *epimere* and consists of a block of cells on each side of the notochord, that is, there is always a pair of somites in each segment. *Segmentation or metamerism* is a basic characteristic of the invertebrate phyla, especially of the Annelida and Arthropoda, in which there is obvious serial repetition of body parts. Vertebrates also show segmentation, but in a much more limited fashion. Neither the skin nor gut are segmented, but the mesoderm is. This characteristic is best evident in the embryonic stage. The blocks of mesodermal material give rise to the early segmentation of muscles and vertebrae.

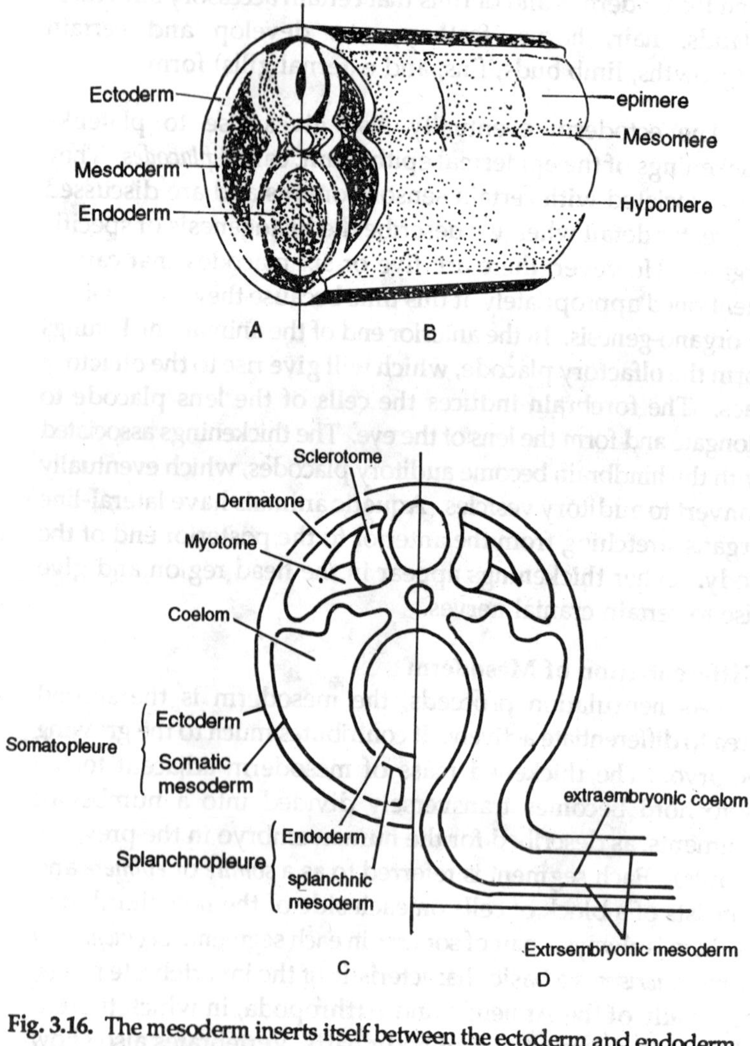

Fig. 3.16. The mesoderm inserts itself between the ectoderm and endoderm in the early gastrula, A. As development continues, the mesoderm divides into three areas: epimere, mesomere, and hypomere, B. These, in turn, undergo further differentiation as shown in C and D. Rounded gastrula, C, and flat gastrula, D, differ since extraembryonic mesoderm is present in the flat gastrula and will split to form the extraembryonic coelom and give rise to the extraembryonic membranes.

Although the dorsal mass of mesoderm divides into blocks of cells, the more lateral extension of the mesoderm remains unsegmented. It spreads laterally and ventrally as *lateral plate mesoderm, the hypomere.* Each somite remains attached to the lateral plate by way of a stalk, the *intermediate mesoderm* or *mesomere,* which will form the urogenital system of the embryo. The lateral plate, or hypomere, separates into two layers, best seen in a cross section of the embryo. The layer closest to the endoderm becomes the *visceral layer,* or *splanchnic mesoderm.* Both the cardiac muscles and the smooth muscles of the gut develop from the splanchnic mesoderm. It is usually closely applied to the endoderm and together with this layer makes up the *splanch-nopleure.* The layer nearest the epidermis gives rise to *parietal* or *somatic mesoderm.* Along with the ectoderm layer it is known as the *somatopleure.* The cavity formed by this mesodermal splitting is the *coelom,* which will become the peritoneal cavity, the body cavity of the adult. In those forms that have the flattened gastrula the lateral mesoderm continues to extend outside the embryo proper where it is known as the *extraembryonic mesoderm.* It will contribute to those structures that are not a part of the embryo proper, the extraembryonic membranes. When this extraembryonic mesoderm splits into two layers, the cavity (the extraembryonic coelom) remains continuous with the coelom inside the embryo for many hours.

During later development each somite or epimere separates from the intermediate mesoderm and differentiates into three areas destined to form specific areas in the adult. The inner portion of mesodermal block gives rise to loose aggregates of migrating cells known as *mesenchyme cells,* which take up positions around the notochord and neural tube. These cells become known as the *sclerotome* and give rise to the vertebral column. The remaining block of cells further differentiates into *dermatome* and *myotome.* The more lateral dermatome cells migrate outward and cover the inner surface of the ectoderm, where they become the dermis of the skin in the dorsal and dorsolateral areas of the body. The myotomes give rise to the voluntary skeletal muscles of the body. In the adult amphioxus the myotomes remain as segmented, V-shaped muscle blocks arranged from the anterior end to the posterior end of the body on each side of the animal. In higher animals a lateral longitudinal septum divides the myotomes into the dorsal *epaxial muscles* and the *ventral hypaxial* group. The longitudinal septum and the two muscle groups are easily seen in the adult shark.

Differentiation of Endoderm

The method by which the endodermal material is converted into a primitive gut tube varies among the different animals and depends on the type of cleavage involved. However, regardless of the mechanism involved, the resulting tube-like structure soon divides into three unequal portions: the foregut, midgut, and hindgut regions. Various evaginations or diverticula form from this endodermal tube and give rise to organ primordia. .The thyroid, trachea, liver, pancreas and urinary bladder are examples of structures originating from the primary endoderm. As development continues, mesoderm usually combines with the endoderm and the splanchnopleure contributes to the histological pattern of the adult structure. Endoderm lines the gut, and the splanchnic mesoderm contributes the connective tissue, smooth muscles, and serosa (peritoneal lining) of the gut.

By the neurula stage, the chordamesodermal mantle of amphibians separates into its respective units and the mesoderm extends ventrally between the ectodermal and endodermal layers to the midventral line. The segregation of the mesoderm leaves the edges of the endoderm free to migrate dorsally, completing the roof to archenteron. As a result, yolk-filled endoderm cells compose the walls, floor, and roof of the archenteron. By the neurula stage these yolky cells move inside the embryo and the yolk plug diminishes in size. Finally, the lateral lips of the blastopore fuse , obliterating the blastopore, and the endodermal cells are no longer visible from the outside.

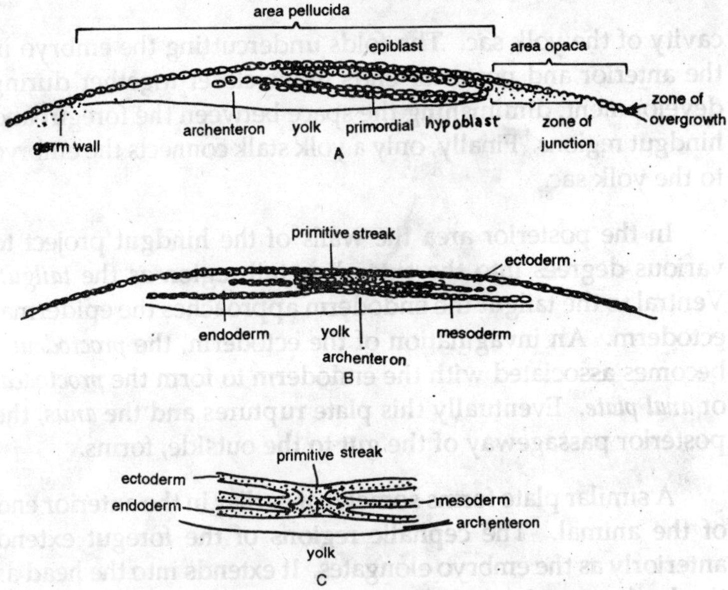

Fig. 3.17. Differentiation of ectoderm, mesoderm and endoderm.

In birds and mammals the gut tube arises in a different manner. The formation of the head process in the anterior end of the embryo elevates the flat sheet of endoderm, resting on the yolk, along with the ectoderm. The formation of the head process is the beginning of the separation of the embryo from the yolk. The development of the head process involves both an upward and anterior projection of the two layers off the blastoderm as well as a caudal movement. The inner surface of the fold that undercuts the head process forms the lining of the gut, and the outer surface contributes to the yolk sac as illustrated in sagittal view. As a consequence of the undercutting, a space forms between the head fold and the blastoderm, the subcephalic pocket, which results in the cephalic region of the embryo projecting free from the blastoderm. Since the head fold is lined with endoderm, an anterior chamber forms the *foregut,* consisting of a roof, sides and a floor. The rest of the endoderm rests on the yolk. Eventually a similar elevation and undercutting in the caudal end of the embryo produce a hindgut and leave only the midgut open to the cavity of the yolk sac. The folds undercutting the embryo in the anterior and posterior ends move closer together during development, diminishing the space between the foregut and hindgut regions. Finally, only a yolk stalk connects the embryo to the yolk sac.

In the posterior area the walls of the hindgut project to various degrees into the extending tail region as the *tailgut*. Ventral to the tailgut the endoderm approaches the epidermal ectoderm. An invagination of the ectoderm, the *proctodeum*, becomes associated with the endoderm to form the *proctodeal* or *anal plate*. Eventually this plate ruptures and the *anus*, the posterior passageway of the gut to the outside, forms.

A similar plate forms somewhat earlier in the anterior end of the animal. The cephalic regions of the foregut extend anteriorly as the embryo elongates. It extends into the head as the *head gut* and expands toward the ectoderm beneath the brain as a large chamber, the *pharynx*. In the anterior end of the embryo, an epidermal invagination, the *stomodeum*, approaches the endodermal cells. As in the posterior region of the embryo, a plate forms, the *stomodeal* or *oral plate*.

Histogenesis

After gastrulation and neurulation the embryo enters the next phase of its development, *organogenesis*. The cells of the embryo's three undifferentiated germ layers undergo mitotic divisions and differentiate into four fundamental tissues of the adult organism. These are usually classified as (1) epithelial, (2) nervous, (3) connective, and (4) muscle tissues. *Histogenesis* is the name given to the process by which the undifferentiated or prefunctional cells of the germ layers differentiate into specialized or functional cells of the body. Histogenesis can be considered as the first step in organogenesis.

Each *tissue* consists of a group of similar cells that carry out comparable functions united by intercellular substances. The cells of the tissues are specialized cells, but epithelial and connective tissues are considered less specialized than nervous and muscle tissues. Their general form and internal structure are modified during development in such a way that each tissue can carry out a particular function. The appearance of myosin, for example, responsible for a cell's ability to contract, is the first discernable difference in the cells of future muscle tissue. Pepsin starts to accumulate in the gastric cells of the digestive tract.

The formation of an intercellular substance, usually secreted by the cell and often referred to as the *matrix* of the tissue separates each cell from a neighbouring cell. Epithelial cells, which are closely associated with one another in lining external and internal surfaces, have little intercellular substances. In some cases, however, it is the matrix that is important in giving form and function to the body. The intercellular substances of bone and cartilage are examples of this type of tissue. Blood is a unique tissue in which the cells, the formed elements of the blood, are suspended in a fluid matrix.

As we begin our study of organogenesis, a brief summary of the contributions made by each germ layer to specific tissues may be of value. The ectoderm gives rise to the nervous tissue and the outer epithelial covering of the body. The endoderm forms the gut epithelium and all its derivatives. The mesoderm contributes to the organs located between the other two layers, that is, muscles, blood, skeleton, and organs of excretion and reproduction.

Following is a more detailed list of the specialized cells derived from the germ layers :

Ectodermal Layer

Epithelium of the skin and its derivatives hair, feathers, nails, scales, sweat glands, sebaceous glands, and mammary glands; primary dental laminae, which give rise to enamel organs of teeth; lens of the eye, part of cornea, tympanic membrane and pituitary gland;

Nervous system, consisting of brain, spinal cord, peripheral nerves, ganglia, rods and cones of eye, various receptors of skin, auditory and taste receptors;

Neural crest cells, which give rise to pigment cells of the hair, feathers, and skin; cartilages in some animals, adrenal gland medulla;

Mesodermal Layer

Skeletal, cardiac, and smooth muscle, components of the circulatory system, connective tissue, lymphatic organs, lining of pleura, pericardium, and peritoneum, adrenal cortex, and epithelium of urogenital system.

Endodermal Layer

Epithelium of digestive tract and its derivatives, the liver, pancreas, and gallbladder; epithelial lining of respiratory tract (larynx, trachea, bronchi, and probably alveoli of lungs); primordial germ cells of some animals; epithelium of tympanum and eustachian tube: thyroid, parathyroid, and thymus glands.

Organ Systems

Groups of specialized cells (tissues) can function independently, but usually the four fundamental tissues combine to form the *organs* of the animal. An organ (kidney, heart, or stomach) is a large functional unit whose combination of tissues enable it to carry out some general function. Certain organs, in turn, co-operate with other related organs to constitute *organ systems*. This level or organization is of interest to the comparative anatomists and comparative physiologists. The organ systems carry out the main functions of the body, such as respiration, excretion, and reproduction.

The organ systems of the adult vertebrate body (which is emphasized in the remaining chapters) may be categorized by the following scheme, reflecting their general germ layer derivation :

1. Ectodermal layer

   a. Integumentary system
   b. Nervous system.

2. Endodermal layer

   c. Digestive system
   d. Respiratory system.

3. Mesodermal layer

   e. Muscular system
   f. Skeletal system
   g. Circulatory system
   h. Reproductive system
   i. Excretory system.

4

Fate of Germ Layers

There are three germinal layers — the ectoderm, the mesoderm, and the endoderm. These layers are formed during the pregastrular process and give rise to all body organs of the related organism. We are taking the various derinatives of these germinal layers are by one.

ECTODERMAL DERIVATIVES

By the time the primitive body form develops, the neural ectoderm segregates from the epidermal ectoderm by a process called *neurulation*. It lies along the dorsal surface of the embryo as a primary organ rudiment, the neural tube, where it undergoes extensive differentiation. Epidermal ectoderm forms a continuous covering over the surface of the cylinder-shaped embryo. This ectodermal layer is only one cell thick in such animals as the shark, chick, pig, opossum, or human, but is two cells thick in the fish and amphibian. The one-called epidermis soon divides into two layers so that the epidermal layer of all vertebrates eventually consists of two layers: an inner one, the *stratum germinativum,* that has the ability to proliferate, and an outer transitory, protective layer referred to as the *periderm.* The periderm first appears in the human embryo at 1-1/2 months of age.

In general, the ectodermal layer gives rise to the skin. The vertebrate skin, including all its accessory structures, is defined as the *integument* or covering of the organism. Its functions are obvious, since it separates the internal environment of the individual from its surrounding environment. It protects the body from physical injury, acts as a barrier to microorganisms, insulates the body, and aids in the maintenance of its temperature. It may also act as an organ of excretion and respiration. It helps the individual to find food, or a mate, and to escape from predators by means of mechanoreceptors, radioreceptors, and chemoreceptors located either over the entire integument or in localized areas. As the members of the various chordate classes adapted to the diverse niches available to them, the adult integument also changed. It varies from a one-cell-thick covering in amphioxus to a stratified layer of several cells in those terrestrial animals in which desiccation presents a threat to survival. The multilayered skin of land animals is more impermeable than that of animals bathed by an aquatic medium. In any case, the surface cells of the vertebrate skin are not exposed to the environment, but are protected by some kind of seal. A cuticle or a layer of mucus covers the body of aquatic animals; keratin, deposited in the cells of land animals, protects the surface of terrestrial forms. The accessory structures of the integument develop mainly from ectoderm, but often a papilla of mesoderm is present. Such accessory structures include unicellular and multicellular glands, scales, nails, beaks, horns, antlers, feathers, and hair and reflect the adaptation of the organism to its environment. This chapter is limited to a discussion of the skin and its accessory structures. Specialized structures such as the lens of the eye, the cornea, and certain cranial ganglia also form from epithelium. Thickenings, called *placodes,* develop into the organs of special senses : the olfactory organ, the ear, and the lateral-line organs.

Epidermal ectoderm also gives rise to the lining of the orifices of the body and to some of the structures located within the cavities. After the formation of primary organ rudiments in the embryo, the mouth and anus form. The ectoderm gives rise to a depression, the *stomodeum,* in the anterior region of the embryo near the cranial tip of the notochordal process. When the stomodeum contacts the endoderm, or oral plate forms and soon ruptures, giving rise to a mouth and an oral cavity. In the human embryo this occurs about the twenty-sixth day. After the rupture of the oral plate, it is difficult to determine the depth of stomodeal ectoderm penetration. Since it is almost impossible to distinguish the exact boundary line between ectoderm and endoderm, the specific embryonic origin of some of the structures located in the oral cavity cannot be described with certainty. In general, most embryologists agree that ectoderm contributes to the lips, cheeks, gums, teeth, tongue, palate, and salivary glands, although the endoderm may also contribute to these same structures. Since these organs are usually discussed in most anatomy books as part of the digestive system, for the sake of convenience and conformance with the generally accepted policy.

In some holoblastic vertebrates a part of the blastopore becomes the anal opening. In others, it forms in a similar fashion to that of the mouth. The skin ectoderm in the posterior region forms a depression, referred to as the *proctodeum.* A proctodeal plate results when the cells forming the depression come in contact with the hindgut endoderm. The further development of this plate in the mammal is intimately related to the division of the cloaca into the dorsal rectum and ventral urogenital sinus. Eventually the plate breaks down, establishing continuity between those structures derived from the cloaca and the outside. It is as difficult to distinguish germ layer boundaries in the formation of the cloacal cavity as it was with the mouth. The ectoderm, however, is acknowledged to contribute to the lower part of the anal canal and to the terminal portions of the genital and urinary tracts. Since the development and evolution of the cloaca is usually considered a part of the urogenital system.

SKIN

Development of Skin

The adult skin is composed of three layers : a superficial

stratified epithelium, the *epidermis;* a dense fibrous *dermis;* and beneath these layers, a loose, fatty *subcutaneous layer*. The skin has a dual origin and any discussion of it must include not only its ectodermal but its mesodermal component as well.

The stratum germinativum forms directly from ectoderm and gives rise to the epidermis, but the dermis differentiates from a layer of mesodermal tissue. The origin of the dermis is not always clear but seems to vary, depending on the species involved. It may arise from head mesenchyme, from dermatome, possibly from neural crest cells, or from mesenchyme derived from the more lateral and ventral somatic mesodermal regions of the body. In general, the vascular dermis supports, cushions, and nourishes the avascular epidermis. Neural crest cells, considered ectodermal in origin, invade the epidermis and dermis and give rise to certain pigment-containing cells referred to as *chromatophores*.

The formation of the integument is an excellent example of tissue interaction during development since the ectoderm depends on the mesoderm for the direction of its differentiation. If the epidermal cells are separated from the underlying mesodermal cells and grown in tissue culture, the germinativum layer loses the ability to divide. If dermis is then added to the culture, cell proliferation begins again. When undifferentiated epidermis is combined experimentally with different types of dermal cells, it will respond according to the type of dermis present. For example, it is possible to induce beaks, feathers, scales, or mucus secretions to form from the epidermis by placing the appropriate dermis adjacent to the maturing epidermal cells. Thus, the dermis determines whether the epidermis clears and forms the cornea of the eye, becomes keratinized and gives rise to the skin, or whatever.

However, the presence of mesoderm is probably not the only answer to epidermal differentiation. Further studies indicate that some kind of substratum also appears necessary for epidermal development. On the other hand, although it needs the contact of a substratum on one of its surfaces, the outer surface of the ectoderm must be free of cellular contact for it to develop normally.

The differentiation of vertebrate skin is a very complex subject and involves many factors, some of which we do not understand at the present time. As is true of most organs, at an early stage of development the ectoderm has greater developmental potentiality than at later stages. It acquires specialized characteristics about the time of the gastrulation and begins to show regional specialization. The fates of the different areas become definitely determined at various times depending on the organ and the species involved. Usually this occurs soon after gastrulation, when new interrelationships established by the tissues play a vital role in their further organization.

As is true for most inductive interactions, we do not understand the exact mechanisms by which the various dermas influence the epidermis. Presumably the dermis acts either by affecting the genes of the ectodermal cells directly or by influencing the cytoplasm of the epithelial cells. In any event, in vitro studies demonstrate that *isolated* epidermis eventually becomes necrotic (dies).

Evolution of Vertebrate Skin

The skin of amphioxus is an example of a chordate covering in its simplest form, probably representing the state of the early chordates living 400 million years ago in the Silurian period. It is similar to the skin of some invertebrates. It differs from the skin of all other chordates since the ectoderm in the adult animal develops into a single layer of ciliated columnar cells. These cells rest on a thin layer of mesodermal tissue, the basement membrane or cutis. A non-cellular cuticle containing many pores and presumably secreted by the epidermal cells covers their outer surface. Amphioxus has only a very thin dermal layer and no epidermal derivatives. This thin barrier, no doubt, allows some exchange of materials between the blood and the aquatic medium.

There is a large gap in our knowledge between the present day protochordates, of which amphioxus is an example, and the earliest known vertebrate fossils, the ostracoderms, since the ostracoderms were heavily armored with bone, a dermal (mesoderm) derivative. One theory postulates that the invertebrates of the period, the eurypterids, preyed upon the tiny ostracoderms, and the presence of a bony armor offered the ostracoderms definite survival value.

A second theory attempts to associate the presence of a bony armor to a seal of the body. The earliest vertebrates left the saltwater of their origins and invaded the freshwater streams and ponds where absorption of water became a problem unless kept out by some kind of a seal. The kidneys of these animals, evolved to handle saltwater, are believed to have been incapable of coping with the hypotonic pond water. The bony armor of the ostracoderms may have helped in keeping the pond water on the outside of the animal.

The dermal portion of the integument gave rise to bony plates of the ostracoderms and placoderms and should not logically by discussed in a chapter devoted to ectoderm. However, it is impossible to gain an insight into the structure of the vertebrate skin without some knowledge of its evolutionary development. This, of course, includes a consideration of the mesodermal component, the dermal contribution to the integument, so important during the early days of vertebrate history.

Many biologists consider the presence of a bone-filled dermis in fossil vertebrates to represent the vertebrate primitive state. The mesodermal component of the dermis of these animals apparently formed bone, a type of connective tissue, which was covered by a thin layer of epidermis. Later in evolutionary history, as the kidneys and an impervious skin evolved along with a biting mouth, the animal changed from a sedentary to a predacious animal. The lack of a cumbersome armor and an increase in motility were a definite advantage.

Mesenchyme cells are capable of differentiating into a variety of tissues. Under normal circumstances they give rise, not only to bone, but also to different types of connective tissue. It is not a great change, therefore, for mesenchyme to stop forming bone and start forming dermis. For example, instead of producing a bony matrix in which salts are deposited, the mesenchyme probably differentiated into a tissue whose matrix lacked a deposition of salts. As a result, the dermis evolved, consisting of fibers of different types intertwined in different degrees of denseness and producing the leathery types of integument associated with modern animals.

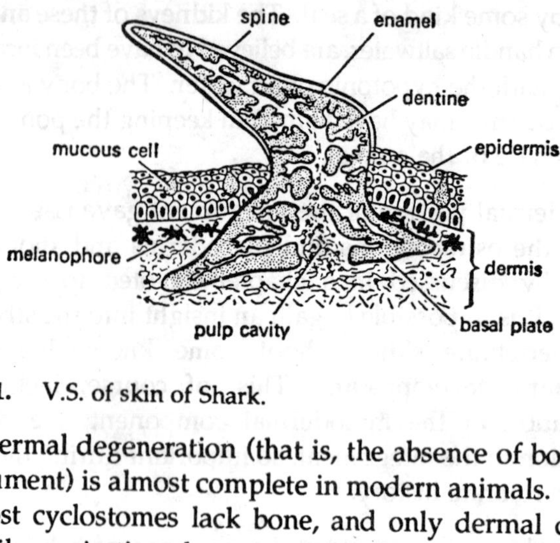

Fig. 4.1. V.S. of skin of Shark.

Dermal degeneration (that is, the absence of bone in the integument) is almost complete in modern animals. The skin of most cyclostomes lack bone, and only dermal denticles, toothlike projections known as placoid scales, remain on the surface of sharklike fishes. These scales projecting through the epidermal layer. Bony fishes retain dermal scales, but they are lost in most land animals, with the exception of the abdominal ribs (gastralia) of lizards, crocodiles, and sphenodon, the bony plates of the turtles, and the flat bones of the vertebıate skull.

In summary, when one considers the integument from an evolutionary standpoint, a decrease in ossification and a decrease in thickness of the skin occurs as anamniotes evolved. On the other hand in land forms the epidermal layer and especially the dermal component increase in thickness. These changes prevent the evaporation of water and help to maintain body temperature.

Basic Integument

The basic integument of vertebrates consists of an epithelium of several layers (a stratified epithelium) resting on a dermal layer. It ranges from the simple, thin skin of the cyclostomes to the thick hides of mammals. In cyclostomes, all cells remain alive covered by a thin, nonliving cuticular layer. On the other hand, in mammals several layers of cells compose the skin, with the outer layer consisting of dead keratinized cells.

In fish the basal layer (stratum germinativum) of the epidermis contains three different cell types. They differentiate into mucus-secreting cells (the major cell type), club cells, and granular cells containing some keratin filaments. However, the stratum germinativum of most vertebrates above the fish appears to be morphologically uniform and to consist of only one type of cell. It gives rise to cuboidal cells that are the only cells in the epidermis to undergo mitosis. Cell division occurs in the stratum germinativum either continuously or periodically in all vertebrate classes. It may go on at a fast or slow rate; in either case, after they divide, the daughter cells follow one of three alternatives. Either they both migrate to the surface of the integument, both remain in the basal layer where they act as reserve cells giving rise to future skin cells, or one daughter cell remains and one moves toward the surface.

During their migration to the surface of the integument, the cells usually become keratinized, a process that involves the stepwise deposition in the cell of a number of fibrous proteins all loosely referred to as keratin. Keratinization occurs when the proteins are cross-linked through disulfide bridges: however, because of its insoluble properties, the characterization and identification of the specific protein composition of keratin is incomplete. The cells of fish lack morphologic keratinization, but keratin is present in amphibians and is most pronounced in land animals : the reptiles, birds and mammals. In the human embryo it occurs by the fourth month of pregnancy. As a result of keratinization, the cuboidal cells flatten, the nucleus degenerates, the cell loses its capacity to divide, and the fibrous keratin replaces the cytoplasm.

X-ray diffraction patterns that reflect macro-molecular organization distinguish between two types of keratin, alpha or beta. For example, keratin that has the beta pattern composes avian feathers, but the cells of mammalian hair, stratum corneum, and nails are filled with keratin of the alpha type.

The fully keratinized cell consists of not much more than a tough cell membrane filled with an abundance of filaments. It acts as a barrier, keeping such materials as water inside the body and preventing other substances from entering. It is resistant to abrasive substances. As a result of keratinization, a layer of dried, horny cells known as the *stratum corneum* covers the surface of the skin. These cells are constantly lost (desquamated).

Replacement of the cells composing the epidermis, lost either through wear and tear or actual shedding, occurs continuously. Either the epidermis sloughs off in the form of small fragments, such as dandruff, or the loss may involve practically the entire outer covering of skin, scales, feathers, or hair. If the loss is extensive. It is usually referred to as a *molt*. The periodic shedding of integument allows the animal to change from a heavier to a lighter coat or from dull to bright plumage, or merely to repair those cells that have worn away.

Sometimes the cells are shed periodically and are associated with seasonal and reproductive cycles. Genetic and neuroendocrine factors interact in the animal's response to environmental stimuli. In some animals, such as the amphibian, the thyroid gland plays an important role in molting. Also, the gonadal hormones may act in a synergistic fashion with other hormones to influence growth of the integument or its derivations. Progesterone, for example, stimulates molting in birds. The integument of the vertebrate represents one of the few areas of the body in which growth and cellular differentiation continue into adulthood.

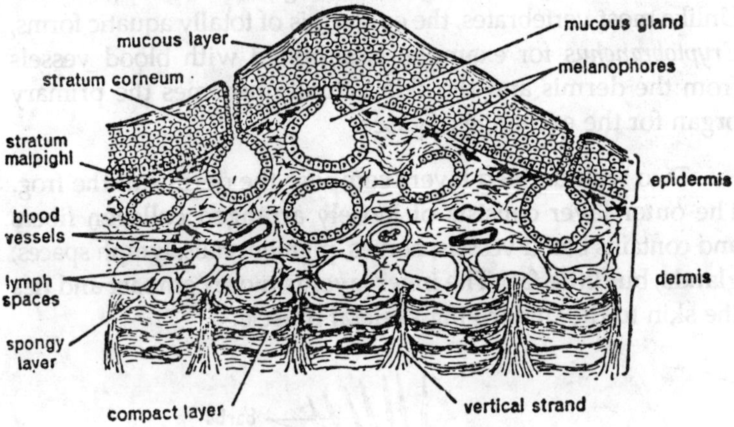

Fig. 4.2. V.S. of skin of frog.

The epidermis of the amphibian, unlike that of the fish, contains only one type of differentiated cell. It produces both mucus and keratin filaments. After cell division as the cells migrate from the basal layer to the surface of the skin, they secrete and release mucus into the intercellular spaces. Along the way to the surface of the animal, however, the mucus-secreting cells transform into horny cells filled with keratin.

Although bony scales covered the body of the earliest amphibians, the labyrinthodonts, most living forms lack scales. Only in the Apoda do remnants of these ancestral scales remain. Instead in modern amphibians, those that are primarily land dwellers, a dead outer *stratum corneum* appears, an adaptation to terrestrial life. The integument of the amphibians varies from the aquatic *Necturus,* in which a cuticle covers the skin, to the frog in which a few keratinized cells appear, to the more terrestrial toad in which the deposition of keratin forms a thick and horny skin. These strata cornea of salamanders, frog, and toads contain alpha keratin in their cells. The secretion of mucus keeps the thin integument of the frog moist and helps it to retain some of its permeable characteristics. The dermis is very vascular and a large proportion of respiratory needs of the frog is carried out through this moist epidermis. Unlike most vertebrates, the epidermis of totally aquatic forms, *Cryptobranchus* for example, is supplied with blood vessels from the dermis and the skin actually becomes the primary organ for the exchange of gases.

Two well-defined layers compose the dermis of the frog. The outer layer consists of loosely arranged collagen fibers and contains blood vessels, nerves, pigment cells, lymph spaces, glands, but little fat. The inner layer is more compact and ties the skin to the underlying skeletal muscle.

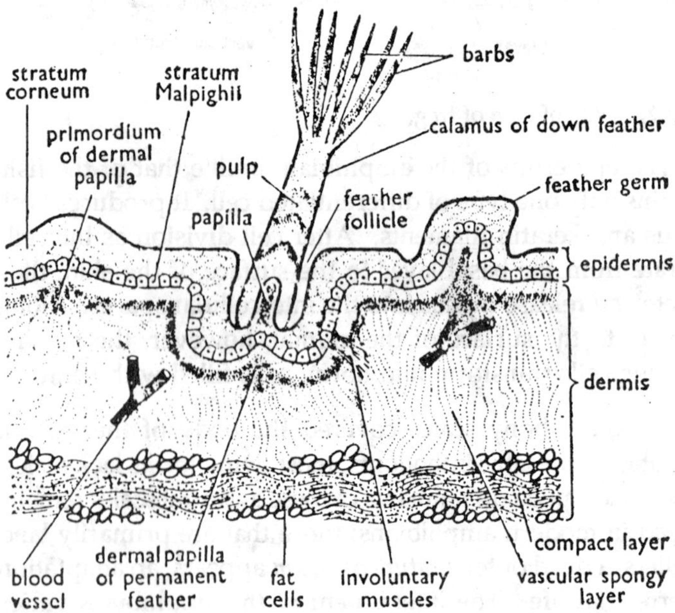

Fig. 4.3. V.S. Skin of a bird.

The nature of the epidermis changed as the reptiles moved on to land. There was a loss of mucus-secreting cells and an increase in keratinization. According to some authors, the development of a flexible, lightweight keratinized skin is one of the factors that allowed aquatic animals to evolve into terrestrial forms and attain the large size of many species. The keratinized scales of the reptilian skin provide a flexible covering that protects the animal against abrasion and desiccation and aids in locomotion. The skin allows for growth, since it is periodically shed and replaced.

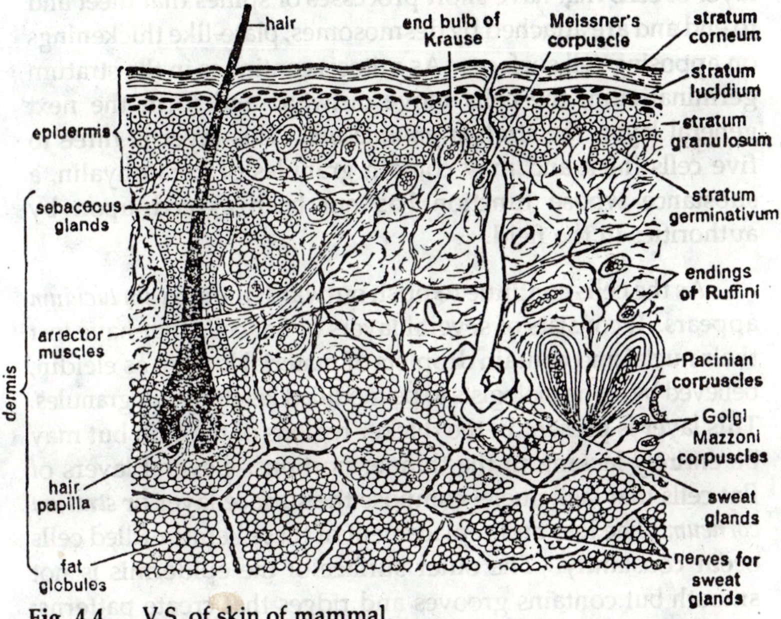

Fig. 4.4. V.S. of skin of mammal.

The main disadvantage of keratinized scales, however, is that they do not control the loss or gain of heat. This problem was not handled until the evolution of feathers and hair, both epidermal structures. The epidermis of reptiles and birds, in general, consist of a living germinative layer, an intermediate zone of cells and an outer covering of flat, dead cells, the *stratum corneum*. The dermis layer, upon which the epidermis rests, varies in thickness, depending on the species. The integument of reptiles and at least one bird, the ostrich, is thick enough that it can be prepared for leather.

The epidermis of mammals may contain as many as four different layers, depending on the location in the animal. In man, the skin of the soles of the feet and palms of the hands is much thicker and exhibits all four layers, as compared to the thinner skin covering the eye lid. A vertical section through the thick part of the epidermis of man shows the first (or deepest) layer to consist of the cuboidal to columnar cells of the *stratum germinativum.* These are associated with another layer of cells that have short processes or spines that meet and to end and are attached by desmosomes, plate-like thickenings on apposing cell surfaces. As mitosis continues in the stratum germinativum, the cells move outward and form the next general layer, the *stratum granulosum.* This layer is three to five cells thick and may contain granules of keratohyalin, a substance whose function has not been agreed upon by authorities in the field.

As the cells continue to push outward, the *stratum lucidum* appears. It consists of several layers of flat cells that have lost their nuclei and contain droplets of material known as eleidin, believed by some biologists to be modified keratohyalin granules. This layer is present in the thicker regions of the skin but may be entirely lacking in the thinner portions. Several layers of flat cells that lack nuclei compose the surface layer or *stratum corneum.* The loss and replacement of these keratin-filled cells occur constantly. The outer surface of the epidermis is not smooth but contains grooves and ridges that create patterns; the most well-known are responsible for the fingerprints. These patterns are so variable that they can be used to identify the individual. The stratum germinativum rests on the dermal (mesoderm) component of the skin. The point of contact between dermis and epidermis is not a smooth one, but rather the interface is away. The projections of the dermis into the epidermis are known as *dermal papillae* and contain many of the receptors described in the following chapter. The periodic downward invasions of epidermis into the dermis (as shown in vertical sections of skin) are referred to as *epidermal ridges.*

The mesodermal component of the mammalian integument is much thicker than the epidermis, it consists of a variety of different-sized collagen and elastic fibers that run in different directions and form network of various degrees of denseness. The epidermis usually is avascular; the dermis, on the other hand, serves as a kind of packing tissue for blood vessels, music cells, fat, nerves, and nerve ending. Certain derivations of the epidermis such as hair follicles, sweat glands, and sebaceous glands extend down into the dermal layer to various levels. The dermis, in turn, rests on a subcutaneous layer, the *hypodermis*, also derived from mesoderm. The layer contains a continuation of some of the dermal fibers along with different amounts of adipose tissue (fat) and muscle. The hypodermis anchors the integument in a loose fashion to the muscles, bones, and tendons of the body.

EPIDERMAL DERIVATIVES

Any discussion of the integument must include not only a consideration of the skin but of its derivatives as well. These structures can be roughly divided into three groups. Some form solely from epidermis although the mesoderm usually plays an inductive role in that some sort of papilla or core is present. Some of the more major epidermal structures are hair, feathers, horns, reptilian scales, scutes, beaks, claws, nails, chromatophores, and glands. A second group of structures involves the participation of both epidermal and dermal components. Examples of this type of co-operation may be seen in dermal denticles, vertebrate teeth, and some scales. The third group contains derivatives formed solely from the dermis (mesoderm) with no epidermal contribution. Such structures as antlers and the dermal bones of vertebrates may be classified in this manner.

The first group, those derived primarily from ectoderm, are considered in this chapter, and those structures derived from a combination of mesoderm and ectoderm or of mesoderm only are discussed later in the appropriate chapters.

Keratinized Epidermal Structures

The epidermis of the reptile folds upon itself about small dermal projections, the papillae, to form *honey scales*. The epidermis on both sides of a papilla becomes keratinized. The living mesodermal component then mostly withdraws leaving the two upper and lower epidermal layers in close apposition. These scales vary from the small projections of the chameleons to the overlapping, thin, keratinized scales of snakes.

Horny scales are a product of the stratum corneum layer of the epidermis and should not be confused with the bony scales of the fish that arise from the dermal layer. In the turtle the scales form flat scutes or plates and cover the bony plates of the carapace and plastron. Muscles attach to the transversely arranged bands of ventral scales in the snake and elevate and depress them. Reptilian scales, therefore, can be used in locomotion.

The remnants of the reptilian scales remain in the two higher groups of vertebrates. They persist on the legs of birds and the tails of certain rodents, insectivores, and marsupials and offer evidence of the reptilian ancestry of these forms.

The stratum corneum also gives rise to *claws* in the reptile and some mammals, which become modified to form *nails* and *hoofs* is most mammals. All three structures cover the digits and resemble epidermal scales as illustrated (Fig. 4.6). They possess a long hard nail plate *(unguis)* and a softer less cornified pad of tissue, the *subunguis*. They differ from one another, depending on the varieties of unguis and subunguis. Hoofs, for example, differ from rails in that they possess more than one shortened nail plate and the subunguis forms a pad within the curve of the hooftip.

The horny plate that makes up the nails covers the *nail bed* and is surrounded laterally and proximally by *nail wall*. A *nail groove* separates the bed from the wall. The nailfold usually covers a whitish crescentic portion, the lunula (or "moon") located near the roof of the nail. This fold has all the layers of the skin. Its underlying region, usually referred to as the *nail matrix,* forms new nail substances from the dead residues of the cornified epithelial cells. Thus the matrix forms the nail. Only the stratum germinativum covers the surface of the nail beds. Most authors agree that it does not contribute to the new nail substances but merely provides a surface for the nail to glide over.

The skin of the jaw in some amniotes becomes modified. The stratum corneum that covers the upper and lower jaw gives rise to a *bill* or *beak* and takes the place of teeth in turtles, tortoises and birds. The bill of the mammalian duckbill platypus is soft and pliable, unlike that of the birds.

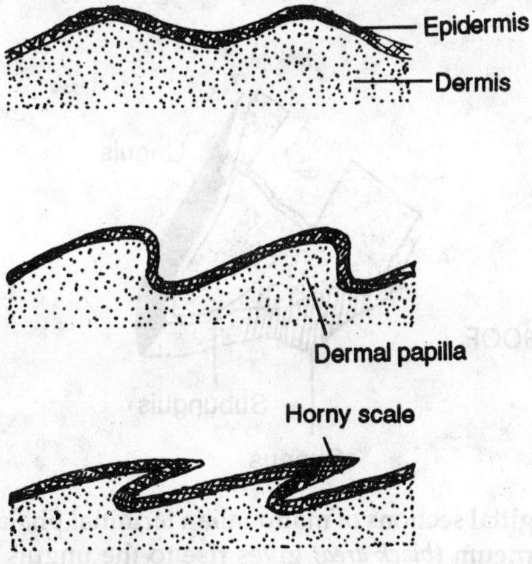

Fig. 4.5. Stages in development of the keratinized (epidermal scales of reptiles).

Horns are found only among mammals. These are non-living structures composed of keratin and have no nerves or blood supply. If horns are removed, they do not regenerate. The horns of sheep, goats, cattle, and antelopes have a core of bone, the *os cornu,* sheathed in keratin. They are not branched like the solid bony antlers of the deer and are never shed. Rhinoceroses have the simplest and most primitive horns

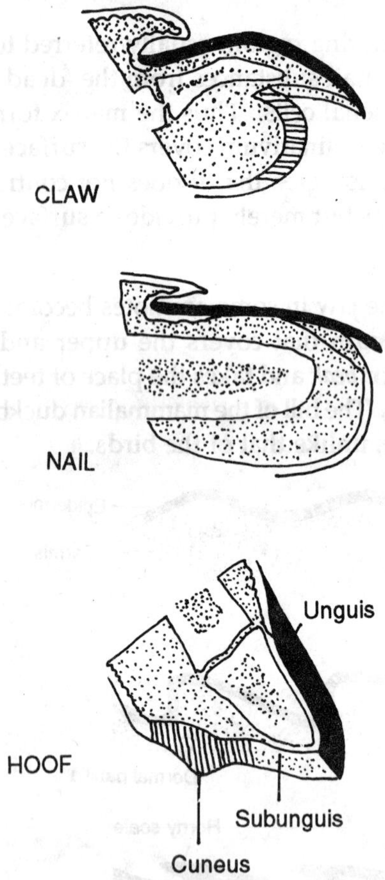

Fig. 4.6. Sagittal sections of mammalian terminal phalanges. Stratum corneum *(black area)* gives rise to the unguis of claws, nails and hoofs.

composed of hardened, solid keratin. The pronghorn antelope *(Antilocapra americana)*, a native of North America has a horny covering over a bony protection. In this type of horn, the covering is periodically shed.

Feathers

Although the reptilian scales persist on the legs and feet of birds and become specialized to form beaks and claws, they

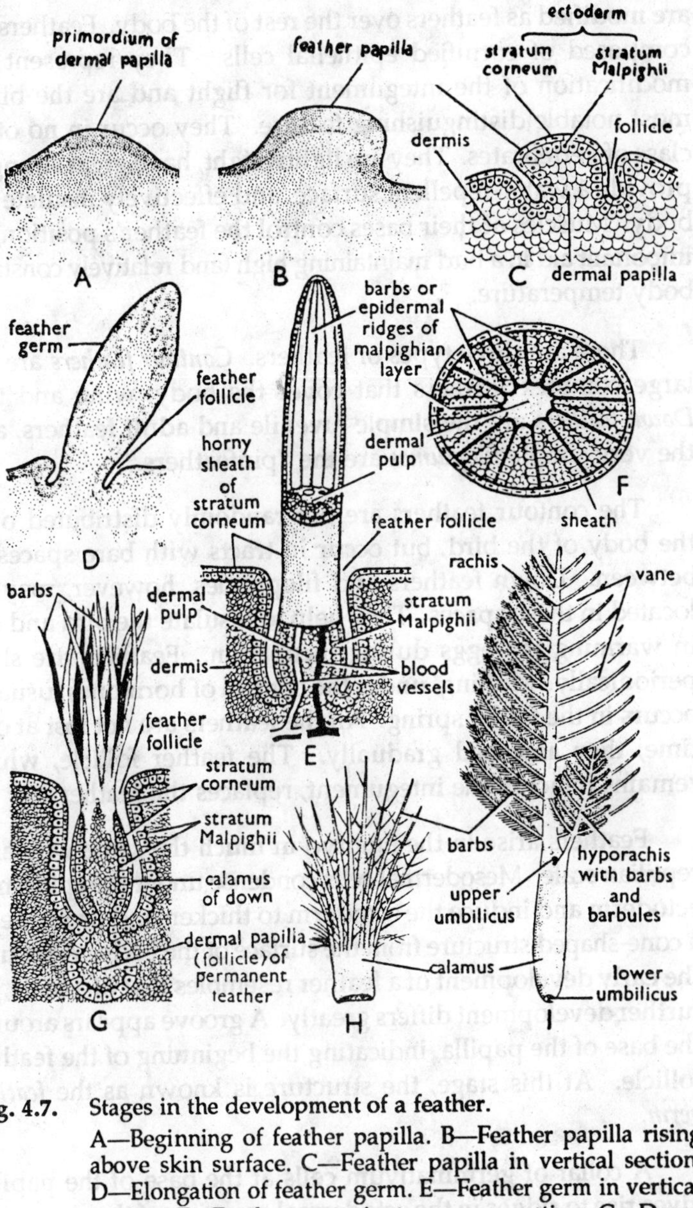

Fig. 4.7. Stages in the development of a feather.

A—Beginning of feather papilla. B—Feather papilla rising above skin surface. C—Feather papilla in vertical section. D—Elongation of feather germ. E—Feather germ in vertical section. F—Feather germ in transverse section. G—Down feather in its follicle. H—Mature down feather of a newly hatched bird. I—Young contour feather.

are modified as feathers over the rest of the body. Feathers are composed of cornified epithelial cells. They represent the modification of the integument for flight and are the bird's most notable distinguishing feature. They occur in no other class of vertebrates. They are lightweight, have great strength, provide a water-repellent surface, and effectively insulate the body. Muscles at their bases control the feather's position, an important aid in a bird maintaining high (and relatively constant) body temperature.

There are three types of feathers. *Contour feathers* are the large complex feathers that cover the body, wing and tail. *Down feathers* are the simple juvenile and adult feathers, and the very simple *filoplumes* are the "pinfeathers."

The contour feathers are not randomly distributed over the body of the bird, but occur in tracts with bare spaces in between. Down feathers and filoplumes, however, may be located in these spaces. They help to insulate the bird and aid in warming the eggs during incubation. Feathers are shed periodically. Molting, under the control of hormones, usually occurs in the fall or spring. All the feathers are not lost at one time; they are shed gradually. The feather follicle, which remains intact in the integument, replaces the feather.

Feathers arise in the embryo in much the same way as a reptilian scale. Mesodermal cells condense under the epidermal ectoderm and induce the ectoderm to thicken and protrude as a cone-shaped structure from the surface of the body. Although the early development of a feather resembles that of a scale, its further development differs greatly. A groove appears around the base of the papilla, indicating the beginning of the feather follicle. At this stage, the structure is known as the *feather germ*.

A collar of germinativum cells at the base of the papilla gives rise to ridges in the ectodermal covering of the elongated papilla. These ridges form the *barbs* of the down feather, which, in turn, give rise to the lateral outgrowths, the *barbules.* Blood vessels located in the mesodermal pulp (the papilla) nourish the developing structure. The base of the papilla becomes a short cylindrical *stalk* or *quill* and contains the radially arranged barbs. The dermal pulp and stratum germinativum gradually retract, leaving dried pithy remnants in the quill.

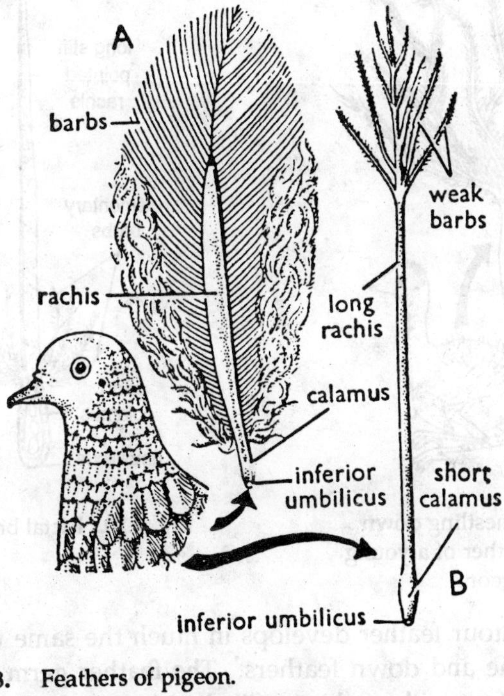

Fig. 4.8. Feathers of pigeon.
A—Contour feather. B—Filoplume.

The down feathers appear externally as a small spray of branches. They cover the young bird and are known as "nestling down." Contour feathers replace these first feathers during the early life of most birds. Filoplumes are even more simple in structure. The quill gives rise to a short slender shaft from which arises a small tuft of barbs. These feathers are hidden by the contour feathers and become visible when feathers are plucked as in preparing a bird for cooking. They are usually singed off.

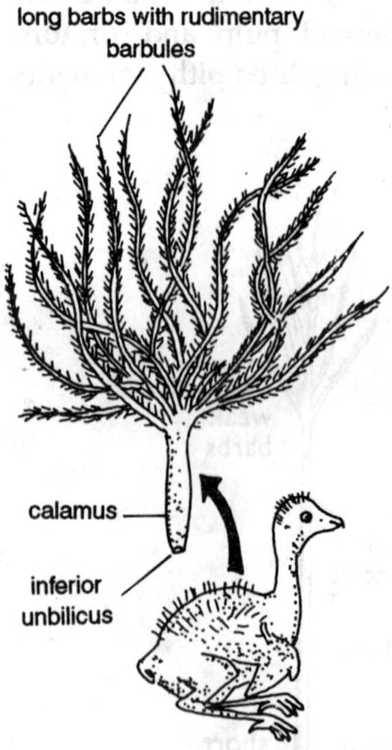

Fig. 4.9. A nestling down feather of a young pigeon.

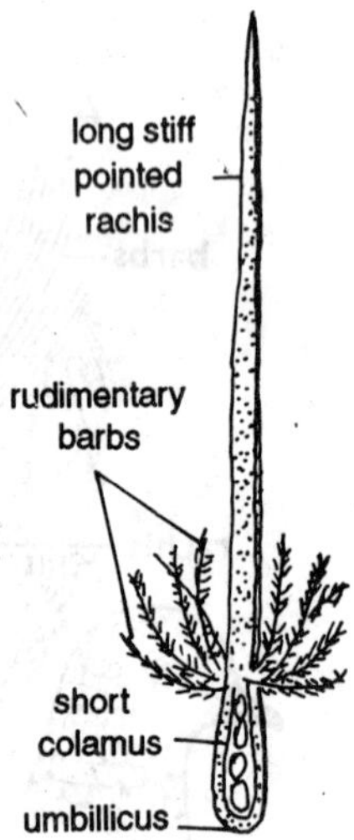

Fig. 4.10. Rictal bristle.

The contour feather develops in much the same way as the filoplume and down feathers. The feather germ forms, consisting of an epidermal cone filled with mesodermal pulp. An epidermal sheath covers the entire structure. As in the example of the down feathers, radially arranged barbs develop from the germinativum collar. Eventually one particularly strong process grows out and gives rise to the shaft of the feather. The barbs migrate onto the shaft. These obliquely located branches (the barbs) form subbranches, the barbules, which interlock with one another. As a result of the interlocking barbules, the feather appears as a continuous sheet.

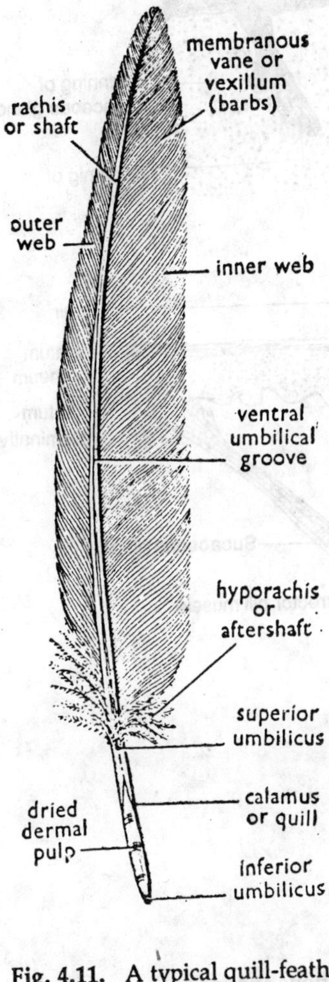

Fig. 4.11. A typical quill-feather (remex) of pigeon in ventral view.

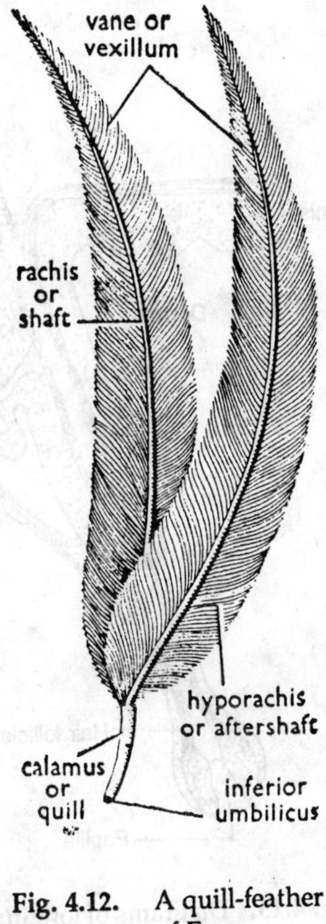

Fig. 4.12. A quill-feather of Emu.

Hair

Hair is a uniquely mammalian characteristic and in most forms covers nearly the entire integument. Like the featner, it is composed entirely of epidermal tissue but unlike the feather it is believed to have evolved from tactile sensory pits rather than from reptilian scales.

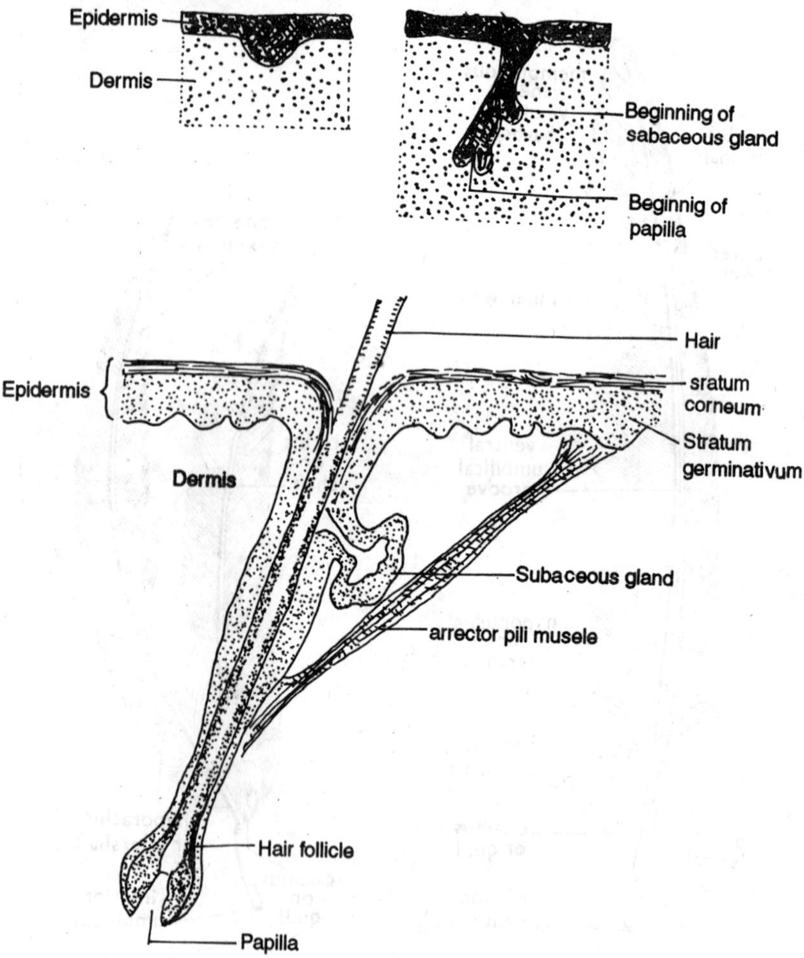

Fig. 4.13. Diagrams of longitudinal sections showing successive stages in the development of mammalian hair.

The bodies of all mammalian embryos at sometime during development are covered with fine hair known as the *lanugo.* It is well-developed in the human embryo by the seventh month but is usually shed by the time of birth.

The shedding and replacement of hair, like feathers, occur either periodically or constantly. If the stratum germinativum composing the hair follicle is lost through injury, no regeneration is possible. The density of hair varies in different mammals as well as over the body of the same animal. Manes, eyelashes, and eyebrows all differ from that covering the rest of the body.

The development of hair is initiated when the epidermal ectoderm grows obliquely down into the dermis. The base of the column becomes bulb-shaped and the dermal mesenchyme pushes into the bulb to form a papilla. Blood vessels form in the papilla and supply nutritive material to bulb cells, the only living part of the hair. The papilla and its covering of epidermal cells make up a hair follicle. A hair consists of its *root* and, the main part, its *shaft*.

As the germinativum cells of the follicle undergo cells division, the new cells are pushed toward the surface through the original epidermal column where they form part of the root. As they continue to move outward, the cells gradually give rise to the snaft of the hair. Dead keratinized cells compose both the root and shaft of the hair.

Two thickenings occur on the side of the follicle. One consists of a cluster of epidermal cells that give rise to a *sebaceous gland* opening into the hair follicle. The other group of cells provides an attachment for the *arrector pili muscle,* which forms from the mesenchyme of the dermal layer and is associated with each hair shaft. Since each hair shaft is at an angle, the contraction of the arrector pili muscle causes the hair to stand erect.

According to cross sections, two or three layers of cells form a hair. A thin cuticle of overlapping, transparent cells covers the hair. Hard keratin fibers with intervening air spaces compose the cells of the *cortex* located under the cuticle. The *medulla* makes up the central area of the fiber and consists of keratin, shrunken cells, and air spaces. The medulla may be absent in thin hair. If the section is made through the wall of the hair follicle, several layers of cells present can be divided into two main layers, the internal and outer root sheath.

Glands

A gland consists of a cell or a group of cells that not only carry on their own metabolic processes, but also manufacture and release certain specific materials. Such substances are either utilized (digestive enzymes) or eliminated from the body (sweat). A unicellular gland is located near the surface where it synthesizes and releases its specific secretion. Several cells (multicellular) formed from invaginations of the germinative layer compose most glands. These invaginations differentiate into a secretory portion and a non-secretory portion, or duct, that carries the secretion to the surface. The term, *exocrine,* refers to this type of gland. *Endocrine glands,* on the other hand, lose their connection with the surface epithelium and are ductless.

Exocrine glands exist in a variety of shapes. If the secretory cells are arranged in a simple tubelike form that opens into an unbranched duct, the gland is said to be *simple.* However, if the duct system is branched, the gland is *compound.* The secretory portions of either simple or compound glands may be arranged as tubules, in which case it is known as a *tubular gland.* If the secretory portion expands to form a sac-like structure, the gland is either a simple or compound *alveolar gland.*

Glands can be classified not only by their shape but also according to the type of material secreted. *Mucous glands* secrete a viscous, slimy material; the secretion of *serous glands* is more watery. Some *mixed* glands produce both types of secretion. The cells composing the glands release their secretory products in different ways. A *merocrine gland* releases its products through the cell membrane with the cell remaining intact. If some of the apical cytoplasm is lost along with the material secreted, the gland is referred to as *apocrine.* In a *holocrine gland,* the entire cell disintegrates, forming a part of the secretion.

Glands develop in the epidermis of all vertebrates. In

general, the glands of fish are unicellular, although some multicellular glands exist.

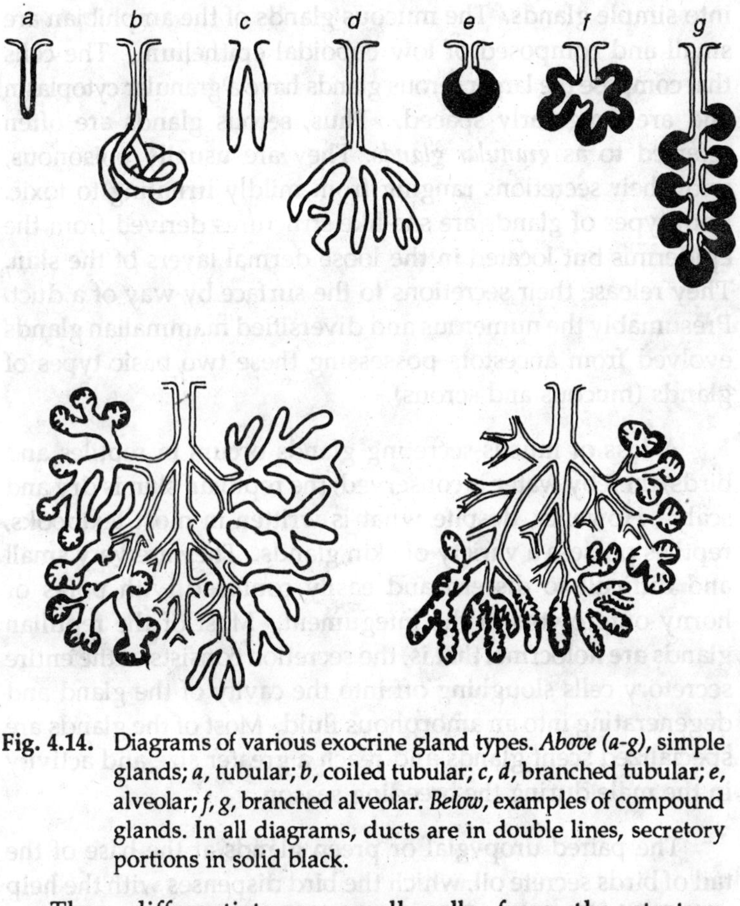

Fig. 4.14. Diagrams of various exocrine gland types. *Above (a-g)*, simple glands; *a*, tubular; *b*, coiled tubular; *c, d*, branched tubular; *e*, alveolar; *f, g*, branched alveolar. *Below*, examples of compound glands. In all diagrams, ducts are in double lines, secretory portions in solid black.

They differentiate as small cells from the stratum germinativum and migrate toward the surface. Most are mucus-secreting glands; some are serious glands. The mucus forms a layer on the surface of the fish and makes it more impermeable to water. In some species, multicellular poison glands are associated with the base of the spine of the fins.

The aquatic amphibians continue to secrete mucus that

aids in the maintenance of a moist skin and protects against loss of water. However, instead of the individual mucus cells of the fish, the cells of the amphibian are grouped together into simple glands. The mucous glands of the amphibian are small and composed of low cuboidal epithelium. The cells that compose the larger serous glands have a granular cytoplasm and are irregularly spaced. Thus, serous glands are often referred to as *granular glands*. They are usually poisonous, with their secretions ranging from mildly irritating to toxic. Both types of glands are sac-like structures derived from the epidermis but located in the loose dermal layers of the skin. They release their secretions to the surface by way of a duct. Presumably the numerous and diversified mammalian glands evolved from ancestors possessing these two basic types of glands (mucous and serous).

A loss of mucus-secreting glands occurs in reptiles and birds, thereby water is conserved; the reptilian skin is dry and scaly. However, despite what is written in most textbooks, reptiles possess a variety of skin glands. They are very small and difficult to discern and easily confused with pores or horny outgrowths of the integument. Most of the reptilian glands are holocrine, that is, the secretion consists of the entire secretory cells sloughing off into the cavity of the gland and degenerating into an amorphous fluid. Most of the glands are specialized scent glands and reach a greater size and activity in the male during the breeding season.

The paired uropygial or preen glands at the base of the tail of birds secrete oil, which the bird dispenses with the help of its beak. This secretion is apparently necessary to maintain the structural integrity and water-repellent characteristic of the feathers, these glands are especially well-developed in aquatic birds. There are also traces of sebaceous-like glands in the ear and trunk regions of some birds.

The mammalian skin contains many glands, but basically they are either modified *sebaceous glands* or *sweat glands*. The sebaceous glands are spherical or ovoid in shape and are simple or branched alveolar glands. They are believed to be homologous with the amphibian alveolar mucous glands. In the mammal the sebaceous gland secretes oil and is associated mainly with the hair follicles. They develop first as thickenings of the outer sheath cells. Their ducts open into the neck of the follicle and their oily secretion appears to be the direct product of cellular disintegration. Several types of sebaceous glands are distributed over the body where hair is absent. They are located, for example, at the corners of the mouth and lips, and on the nipple, the eyelids, and glans penis.

Sweat glands may have evolved from the amphibian serous glands. They are found only in mammals. They appear in the embryo as solid, cylindrical ingrowths of the stratum germinativum. The deeper part of the ingrowth coils on itself and forms the body of the gland. The central cells eventually degenerate and produce a tubular structure that is simple coiled or branched. The coiled portion rests in the sub-cutis region of the skin and a long duct extends through the dermis, entering the epidermis between two papillae. The duct opens to the surface by way of a sweat pore. Some of the surrounding ectodermal cells give rise to myoepithelial cells and, presumably, are able to contract.

The sweat glands are apocrine glands since the distal ends of the cells are discharged as part of the secretion. The sweat glands are important excretory organs. Sweat is a watery secretion containing certain salts, urea, and other wastes; it is a significant method of eliminating metabolic wastes of various kinds. It also helps maintain the body temperature, since the evaporation of sweat on a hot day produces a cooling effect.

Mammary glands, ceruminous (wax) glands of the ear, and certain scent glands also have an apocrine type of secretion and probably evolved from sweat glands. Mammary glands, present only in mammals, are responsible for the name of the class. Mammary glands assume various forms. In the monotremes they consist of two glands in the abdominal wall that secrete a sticky material onto the surface; the secretion is licked off this mammary patch by the young. The glands are located in the pouch of the marsupial. At birth the young crawl into the pouch, and each attaches to a nipple. In ungulates, several ducts empty into a common collecting chamber and a single duct opens to the outside. The primate's mammary gland basically consists of several lobes (15 to 20) each, of which is considered to be a compound gland with a duct opening into the tip of the nipple.

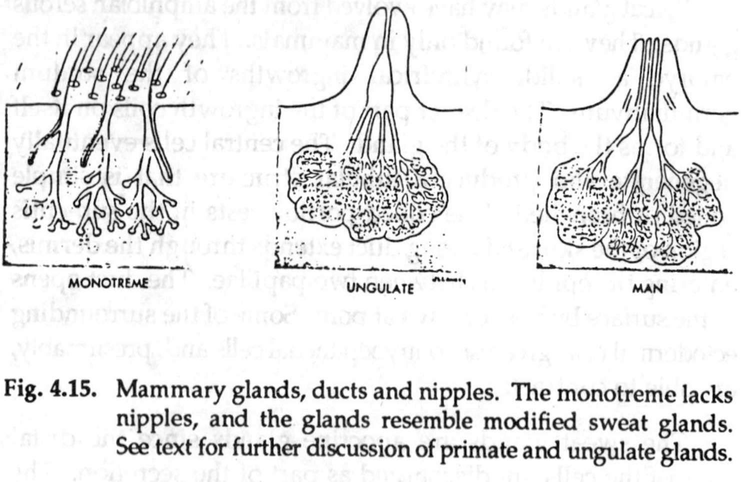

Fig. 4.15. Mammary glands, ducts and nipples. The monotreme lacks nipples, and the glands resemble modified sweat glands. See text for further discussion of primate and ungulate glands.

Mammary glands first form as linear ectodermal thickenings, known as the milk, or mammary, ridges running down the ventrolateral region of the embryo of both sexes. Various areas of this region continue to develop; the precise number of glands usually corresponds with the number of offspring produced. The development of the mammary ridge normally occurs only in the pectoral region in man and the elephant and only in the inguinal region in the sheep cow and horse, but it is extensively developed down the middle of the ridge in such forms as the cat, dog, or pig.

Thickenings of the epidermis represent the beginnings of the nipples. The ectodermal cells undergo proliferation and become bulbous; they grow in toward the mesenchyme layer. Several buds (about 15 to 20 in the human embryo) push out into the dermis from the epidermal bulbous area. These buds are the primordia of the mammary gland ducts. Near the end of embryonic life, they undergo further branching, but then until puberty the mammary glands of both sexes remain in an infantile state. The secretion of estrogen by the female at sexual maturity stimulates duct growth; progesterone stimulates the growth of the alveolar secretory cells. During pregnancy when the estrogen and progesterone levels remain high because of production by both ovaries and placenta, the ducts and alveolae develop extensively, establishing a presecretory state. The lactogenic hormone (luteotrophic hormone, LTH) from the pituitary gland appears necessary in certain species for secretion to take place.

CHROMATOPHORES

The various colour patterns of vertebrate animals function in various ways and are determined by heredity. They are involved in camouflage, protection, sexual attraction, and warning. Colour has definite survival value. The pigment melanin, for example, may function in some cyclostomes as a screen, protecting the central nervous system against the sun. Since melanin is associated with both locomotion and spine response in the invertebrates (sea urchin), some authors suggest that pigment may play a role in regulating the vertebrate central nervous system. Integumental pigmentation has undoubtedly been important in evolution, but we do not understand its specific role. Pigment-containing cells also penetrate more deeply than the integument. They are often found along the blood vessels, along the walls of the coelom, and in the liver, all areas where it is difficult to determine a functional advantage for the pigment.

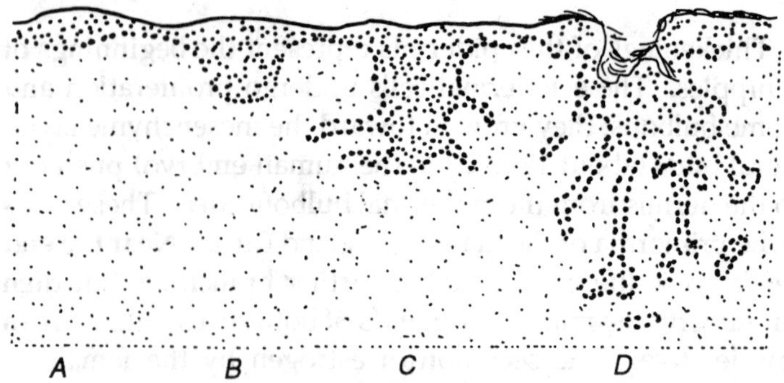

Fig. 4.16. Successive stages of mammary gland development. A, Equivalent to a human embryo at 6 weeks. B, Equivalent to a human embryo at 9 weeks and to mouse embryo at 16 days. C, Intermediate stage. D, At birth. Gray area represents dermis.

Neural crest cells invade the epidermis and dermis and differentiate into colour-forming cells known by the general term *chromatophores*. Since they have been traced to the neural crest cells, they are considered ectodermal in origin. Soon after the completion of neurulation some of the neural crest cells (melanoblasts) migrate out to the various locations they will occupy in the postembryonic body. The specific colour pattern of vertebrates depend on the distribution and activities of these cells. They allow the animal to respond to many environment stimuli by changing colour.

Since the chromatophores are stellate and have long slender processes, they are sometimes referred to as *dendritic cells*. Pigment granules may either be concentrated in the center of the cell in one mass, in which case little colour appears, or else the pigment may be distributed throughout the cell including the long dendritic processes. When dispersal of the pigment occurs, the colour effect is maximum. The colour-bearing cells are also able to shift their relative positions in the body of the animal by ameboid movement. The colour changes displayed by such animals as the flounder and chameleon are attributable to both these mechanisms — the migration of the cells and the amount of dispersal of the granules within the individual cells.

Chromatophores are under hormonal control and are influenced by pituitary thyroid, gonad, pineal, and adrenal gland secretions. For example, the central nervous system influences the intermediate lobe of the pituitary to release its melanotrophic hormone (intermedin) into the bloodstream of fish, amphibians, and some reptiles. Dermal capillaries carry the hormones to the pigment-synthesizing cells. Colour changes controlled by intermedin take minutes or hours to occur.

In poikilotherms (cyclostomes, elasmobranchs teleosts, amphibians, and reptiles), the chromatophores are also effector cells and respond to nervous stimuli. Consequently their response is rapid. Colour pattern in the environment influences colour change in the animal by pathways leading from the eyes to the brain of the organism. If a single cutaneous nerve in a fish is destroyed, it is unable to change colour in the area of the cut. Nerves produce their effect by liberating a chemical a *neurohumor*, at their terminal ends that diffuses to the chromatophore and brings about the change. Epinephrine-like substances released by the nerve terminals cause the melanin to concentrate in the center of the cell and acetylcholine-like materials induce melanin granules to disperse into the dendritic arms.

Birds and mammals apparently lack a pigmentary effector system, since colour changes associated with these classes of animals involve relatively slow alterations in the amount of pigment. The cells respond to seasonal changes, allowing the skin to tan in the summer or the coats of certain animals to change from brown to white. In some cases these colour variations are caused by the amount of pigment present and in other cases to the type of pigment.

There are various kinds of pigments in the integument of all vertebrates. The specific type present depends on the synthetic process of the chromatophores. The chromatophore that synthesizes *melanin* the predominant vertebrate pigment, is referred to as a *melanophore* and is located in both the epidermis and dermis of at least some member of each class of vertebrates. The skin colour of man depends on the presence and amount of melanin deposited in the epidermal layers. Melanin granules vary in colour from yellow through orange and reddish brown to dark brown, depending on their state of oxidation. Melanin also may be present as reduced leukoform, a bleached form. Thus it acts as an oxidation-reduction indicator; in its oxidized form melanin is dark, in the reduced state it is bleached.

Although the *epidermis* contains only melanophores, other types of chromatophores are present in the *dermis*. These cells may originate from a common cell type and differentiate into the various chromatophores according to the specific environmental stimuli they receive. *Iridophores* are described as reflecting cells that contain guanine, hypoxanthine, adenine, or a combination thereof. They are present in anamniotes and the iris of some birds, but are usually absent from the integument of birds and mammals. *Xanthophores* responsible for the secretion of yellow pigment, and *erythrophores*, containing red pigment, are usually found only in the dermis of elasmobranchs, teleosts, amphibians and reptiles. Vertebrate colour changes result when these different chromatophores interact.

The modern history of pigmentation shows that colouration in vertebrates is the result of interaction not only between pigment-containing cells, but also between these cells and other cells of the skin, such as the cells responsible for forming keratin. This interrelationship can be demonstrated in the human skin. The neural crest cells destined to form melanocytes take up a position at the interface between epidermis and dermis. They extend their branches between the basal cells of the germinativum layer and discharge their pigments into the keratinocytes. Pigmentation of the mammalian epidermis, therefore, is not caused by the melanocyte acting alone, but is the result of an interrelationship between the melanocyte and an associated group of keratinocytes, forming the *epidermal melanin unit.* These units, established during development, are the result of tissue interaction. According to Quevedo (1972) the melanocytes in man exist in a ratio of one to 36 keratinocytes. The maintenance of such a relationship suggests a precise control over mitotic activity.

The melanocytes contain tyrosinase, an enzyme that produces melanin by oxidation of the amino acid, tyrosine. Melanocytes are the only cells capable of synthesizing this material. The pigment is deposited in the cytoplasm within a membrane-bound organelle, the melanosome. Melanin is then transferred to the keratinocytes when these cells phagocytize pieces of the melanocyte dendrites containing the melanosomes. The pigment-containing organelle is transported, and ultimately shed, along with the cornified cells, or catabolized along the way by the keratinocytes.

There is significant variation in the number of melanocytes with regard to sex or race. Skin colour, therefore, appears to depend upon the activity of the pigment cell rather than the number of cells present. This activity, in turn, is genetically determined.

Similar units located in the dermis, called *dermal chromatophore units,* are described for poikilothermic animals. The basic types of chromatophores, previously described for poikilothermic vertebrates, are arranged in such a way in the amphibian that the xanthophores are situated in a top layer, the melanophores form a basal layer, and the iridophores are located in one or more rows in the middle. The dendritic arms of the melanocyte entwine about and between the layers above them. When the frog adapts to a dark background, melanosomes migrate to these dendrites of the melanophores, thus obscuring the other chromatophore types and causing the skin to darken. Migration of the melanosomes out of the dendrites to a more central location exposes the xanthophores and iridophores, causing the animal to lighten.

NERVOUS SYSTEM

In vertebrates there are two integrated systems — the nervous system and the endocrine system. Nervous controls the activities mechanically and endocrine system metabolically. The nervous system is divided into the *central nervous system* (CNS), composed of the *brain* and *spinal cord,* and the *peripheral nervous system,* consisting of sensory and motor nerves. This division is mainly one of convenience, since both act together as one unit. The central nervous system develops from the neural plate of the late gastrula stage. The peripheral nervous system derives partly from the neural plate, partly from neural crest cells, and partly from the thickened ectodermal areas we have identified as placodes.

The basic structural unit of the nervous system is the nerve cells, or *neuron,* consisting of a *cell body and its processes,* the *dendrites* and *axons.* An intricate network of neuron communications carries out co-ordinating functions. Individual nerve cells (neurons) generate messages, which they conduct from one part of the body to another by way of their cell processes. These messages are then chemically transmitted to other nerve cells or to other cells of the body.

The analysis and interpretation of these communications and the attempts at describing them in terms of neural mechanisms mark the frontiers of biological investigation in this decade. Many physiologists believe that the complexities of the nervous system will never be comprehended. They believe that man is at the limits of his understanding and that the inadequacies of his brain will prevent him from completely analyzing and explaining its physiology.

Only fragments of behaviour have been correlated with their neural mechanisms. Perhaps the simplest reflex, present in all vertebrate animals, is most readily understood. A stimulus produces a quick withdrawal movement and involves what is described as a *reflex arc.* This simplest of reflexes, considered to be a monosynaptic (one neuron in contact with one neuron) reflex action, may actually produce many more side effects in the central nervous system than we previously believed. Some of these effects are only beginning to be identified and described.

Although such a simple reflex arc is possible, nervous reactions seldom involve only two neuron chains but are considerably more complex than this. Laterally, billions of nerve cells make contact with one another, forming a maze of interacting fibers. In an attempt to understand basic nerve circuits, the investigator must seek common properties of neural structure and function. To accomplish these ends, the crayfish, the frog, and the cat have been used extensively in the past as investigative animals. However, at the present time, the research carried out on mammals, including man, exceeds that for all other animals.

DEVELOPMENT OF CENTRAL NERVOUS SYSTEM

Development of Neural Tube

By the end of neurulation the neural ectoderm is in the form of a dorsal, hollow neural tube that extends under the surface epithelium of the embryo from its anterior to its posterior end. From the beginning, the neural tube expands in the anterior regions of the body. During gastrulation the notochord assumes a position under the neural plate. As a result of this migration the prechordal plate region lies immediately beneath the presumptive brain areas and the notochordal and somitic mesoderm under the middle and posterior regions. The brain of the organism develops from the expanded anterior region of the neural tube, and the spinal cord develops from the more posterior parts. As a result of further differentiation, induced by the prechordal plate and notochordal and somitic mesoderm, the primary rudiment (the neural tube) gives rise to the central nervous system of the adult.

After the neural tube is established, it undergoes a general organization. At early stages one can distinguish certain concentric layers and longitudinal bands in the walls of both the presumptive brain and neural tube regions. The three concentric layers are (1) an inner layer lining the central canal of the tube, (2) a middle cellular region, the *mantle layer,* and (3) an outer *marginal layer* consisting of fibers from the cells located in the first two layers.

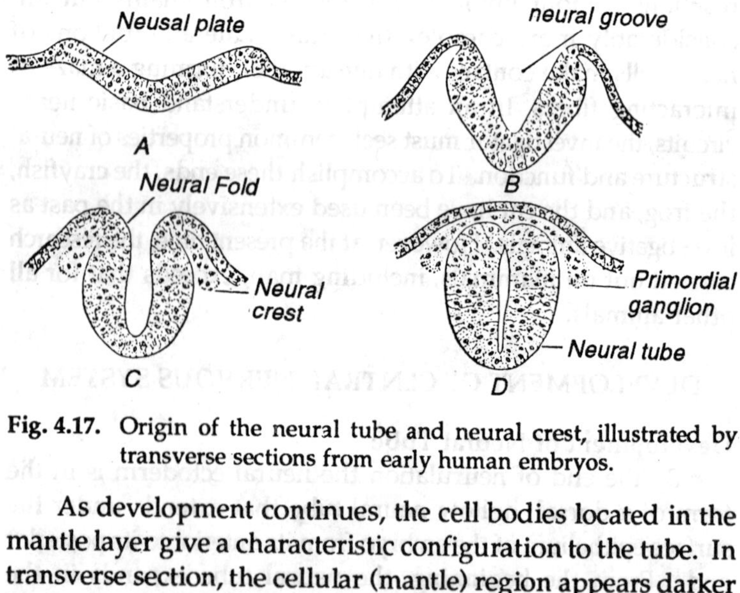

Fig. 4.17. Origin of the neural tube and neural crest, illustrated by transverse sections from early human embryos.

As development continues, the cell bodies located in the mantle layer give a characteristic configuration to the tube. In transverse section, the cellular (mantle) region appears darker because of the closely packed nuclei and is now referred to as the *gray matter* of the tube. The marginal layer lacks cell bodies and contains fibers wrapped with a white, fatlike material, myelin. This layer becomes the *white matter* of the neural tube. The tube grows in diameter by the proliferation of cells from the inner layer and their migration into the mantle layer, and also by the further deposition of myelin around the fibers.

The primitive neural tube can be divided into six longitudinal bands. A relatively thin *roof* (roof plate) and *floor* (floor plate) form the dorsal and ventral areas of the tube. Each lateral wall consists of a dorsal *alar plate* and a more ventral and thicker *basal plate.* Once this basic organization is established, the brain and spinal areas of the cord organize further, both internally and externally, and give rise to the central nervous system unique to the specific organism.

Evidently the neural epithelium at the time of cellular proliferation becomes determined, forming a rather rigid mosaic. Each region of the early neural tube is destined to give rise to a particular functional part of the future cord. For example, removal of the right dorsal half of the neural tube from the chick embryo at 44 hours of incubation results in a normal left side of the spinal cord but the right half is missing.

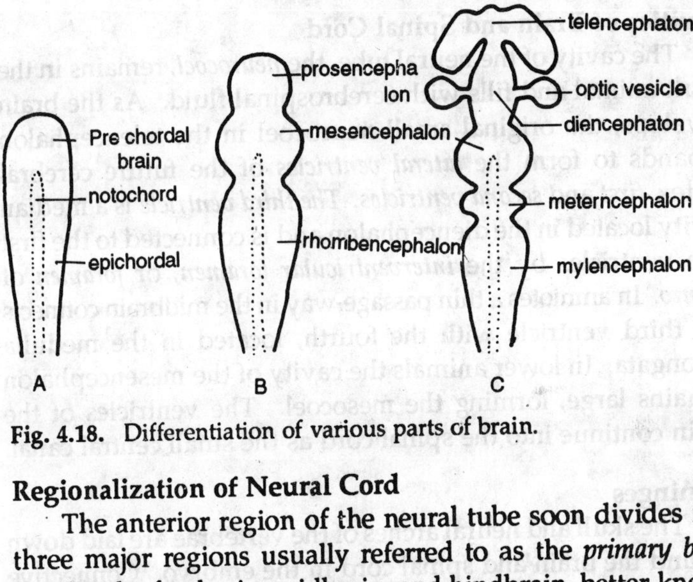

Fig. 4.18. Differentiation of various parts of brain.

Regionalization of Neural Cord

The anterior region of the neural tube soon divides into three major regions usually referred to as the *primary brain vesicles;* the forebrain, midbrain, and hindbrain, better known by their more technical terms as the *prosencephalon, mesencephalon,* and *rhombencephalon,* respectively. These primary vesicles are associated with the three major sense organs. The prosencephalon receives impulses from the olfactory organs, the mesencephalon from the eye, and the rhombencephalon from the ear and lateral-line organs.

The walls of the brain undergo further differentiation and the prosencephalon soon divides into an anterior *telencephalon* from which the olfactory lobe and cerebral cortex form, followed by a *diencephalon,* from which the eyes and thalamus develop. The mesencephalon does not divide further. The rhombencephalon differentiates into the *metencephalon,* which gives rise to the cerebellum and the pons, and the *myelencephalon,* from which the brainstem and upper part of the spinal cord develop. The rest of the tube remains of uniform diameter along its length and is destined to become the spinal cord. As the neural tube develops, it bends and gives rise to the *cephalic and cervical flexures.*

Cavities of Brain and Spinal Cord

The cavity of the neural tube, the *neurocoel,* remains in the adult animal and fills with cerebrospinal fluid. As the brain develops, the original small neurocoel in the telencephalon expands to form the *lateral ventricles* of the future cerebral cortex, *first* and *second ventricles. The third ventricle* is a median cavity located in the diencephalon and is connected to the first two ventricles by the *interventricular foramen,* or *foramen* of *Monro.* In amniotes a thin passage-way in the midbrain connects the third ventricle with the fourth, located in the medulla oblongata. In lower animals the cavity of the mesencephalon remains large, forming the mesocoel. The ventricles of the brain continue into the spinal cord as the small central canal.

Meninges

The skull and neural arches of the vertebrae are laid down around the brain and spinal cord in the embryo. Connective tissue membranes, the *meninges,* wrap intimately around the nervous tissue and offer additional mechanical protection and support for the blood vessels of the central nervous system. The outer membrane is derived from the surrounding mesenchyme; the inner membrane appears to be composed, at least in part, of neural crest cells.

In fish, only one membrane, the *meninx primitiva,* covers the brain and spinal cord. Fine collagen fibers pass through the meninx to the lining of the brain case. Two membranes are present in amphibians, reptiles, and birds: an outer, tough,

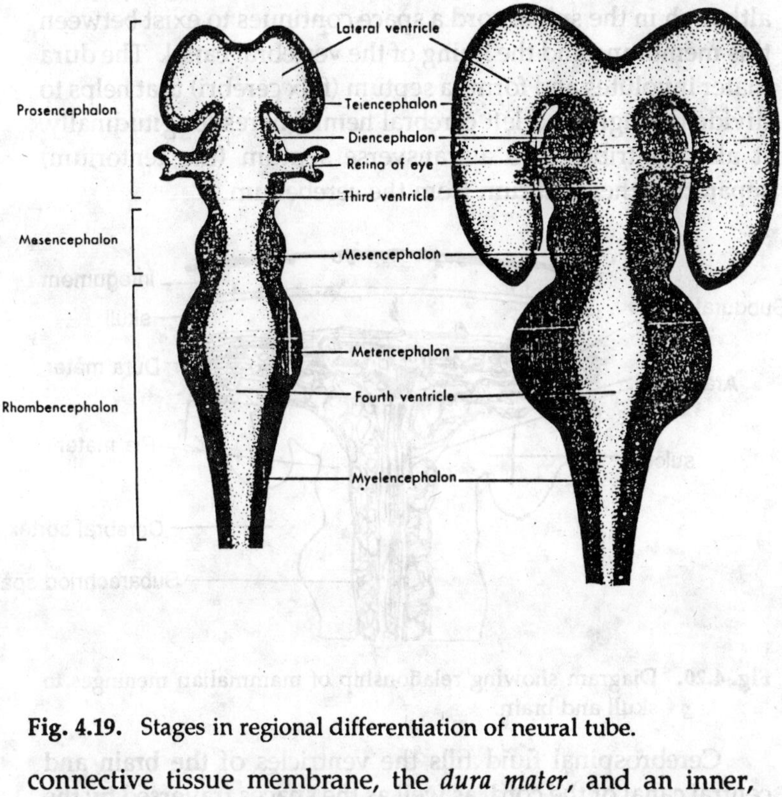

Fig. 4.19. Stages in regional differentiation of neural tube.

connective tissue membrane, the *dura mater*, and an inner, more fragile, vascular, *pia-arachnoid layer*. The *subdural space* separates the two membranes; a second space exists between the dura mater and lining of the brain-case, or neural arch.

In mammals, the pia-arachnoid layer divides, and as a result three membranes are present. There is the outer, tough dura mater, a middle *arachnoid layer*, and an *inner pia mater*. The pia meter is intimately applied to the brain and follows its contours. A space, the *subarachnoid* space, separates the pia mater from the arachnoid and is traversed by delicate fibers that tie the two membranes together. The middle layer, unlike the pia mater, does not dip down into the sulci and fissures of the brain. The outer layer, the dura mater of mammals is closely applied to the lining of the skull, obliterating any space, although in the spinal cord a space continues to exist between this membrane and the lining of the vertebral canal. The dura mater involutes and forms a septum (falx cerebri) that helps to divide the right and left cerebral hemispheres longitudinally. It also contributes to a transverse septum (the tentorium) separating the cerebrum from the cerebellum.

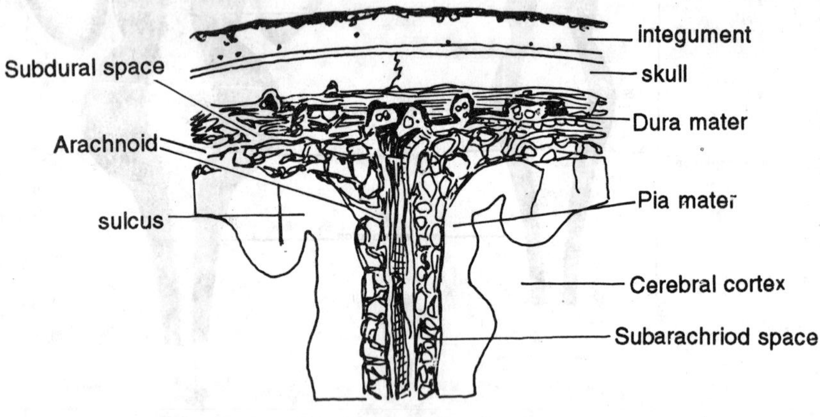

Fig. 4.20. Diagram showing relationship of mammalian meninges to skull and brain.

Cerebrospinal fluid fills the ventricles of the brain and central canal of the cord, as well as the spaces traversed by the connective tissue fibers. The fluid has two functions. It carries the nutrients derived from the blood vessels to the tissue of the central nervous system, and it acts as a cushion to the brain and spinal cord. The *choroid plexuses,* vascular networks located in the roof of the diencephalon and medulla oblongata, secrete the fluid into the third and fourth ventricles. It flows from the ventricles into the subarachnoid space. Here it makes its way into the lymphoid vessels, or arachnoid villi, that connect with certain large venous sinuses under the dura mater. Since secretion of the cerebral fluid occurs continuously, its removal must also go on constantly in order to maintain a stable intracranial pressure.

Basic Embryonic Plan

This basic plan of five cranial divisions plus the spinal cord, along with their cavities and connective tissue membranes, make up the primordium of the central nervous system. It forms early in the development of the organism and is similar in all vertebrate embryos at this early stage. It is far more advanced in the simplest vertebrate, the cyclostome, than in any member of the invertebrate phylum.

Although the nervous system is basically the same in the embryo of all vertebrates and it functions in a similar manner in the adult, the increase in complexity from the lowly cyclostome to the mammal is truly outstanding. The growth of the cerebral cortex brings about an increase in co-ordination and association centers. It is this growth in the cerebral cortex that offers man his great potentialities. According to Romer, "The evolution of the cerebral hemispheres is the most spectacular story in comparative anatomy."

HISTOGENESIS OF NERVOUS TISSUE

Although the presumptive neural tissue at the blastula and gastrula stages is only one cell thick, by the time the neural tube is established its walls consist of a pseudostratified epithelium composed of columnar cells and rounded cells. These cells attach not only to one another, but also at the lumen of the neural tube by special devices known as terminal bars. Watterson (1965) suggests that this attachment at their terminal ends imposes a radial orientation on the neuroblasts and helps to collect similar cells together. All the cells composing the early neural tube are capable of division. They appear similar to one another, but differentiate along two lines. They give rise to the *neuroblasts* that are destined to form the neurons, or else they form the supporting tissue, the *neuroglia*.

Stem-Cells of the Neural Wall

At the beginning of its development, the neural plate is composed of undifferentiated, proliferative epithelium. Its daughter cells enter into two lines of specialization. One path leads toward *nerve cells,* in which irritability and conductivity have become predominant functions; the other course is toward *ependymal* and *neuroglial cells,* which constitute the distinctive supporting tissue of the nervous system.

The embryonic nerve cell is a *neuroblast;* it passes through a bipolar stage, with a process at each end, before reaching a multipolar stage, or immediate precursor of the typical *neuron* of the central nervous system. The *spongioblast* is the forerunner both of the *ependymal cells* and of neuroglial cells known as *astrocytes.* Some spongioblasts are migratory in nature; these differentiate into *oligodendroglia* and into additional astrocytes, as well. It has been claimed that some migratory spongioblasts convert into neuroblasts, but this interpretation is open to doubt.

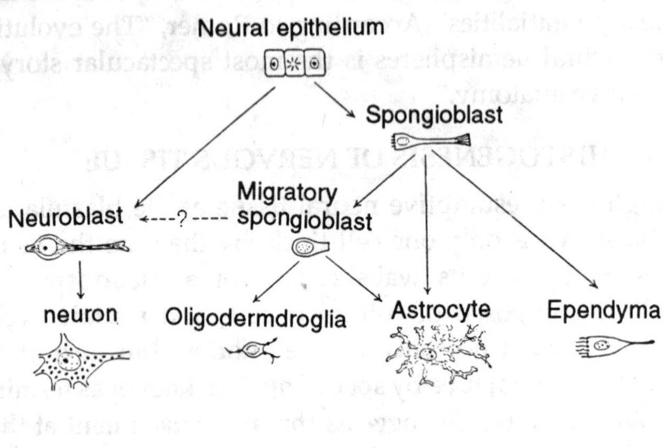

Fig. 4.21. Diagram illustrating the lineage of cells in the neural tube.

Layers of the Neural Wall

The neural plate was originally a single layer of columnar cells, but rapidly becomes many-layered *(C);* in doing this the component cells lose their sharp outlines and seemingly resolve into a compact syncytium *(D)*. Some authorities, however, maintain that the resemblance to a syncytium is an artefact and the constituent cells are always distinct. In the sixth week the neural wall is bounded on the outer and inner surfaces by an *external* and *internal limiting membrane*, respectively, while the cellular elements of the wall are arranged radially *(D)*. At this stage the neural tube is sufficiently organized so that three concentric zones may be distinguished: (1) an inner *ependymal layer*, with its cell bodies abutting on the internal limiting membrane (next the neural canal) and their processes extending peripherally; (2) a middle, nucleated *mantle layer*, derived by proliferation of the innermost cells; and (3) an outer, non-cellular *marginal layer*, into which many nerve processes (nerve fibers) grow.

The ependymal layer, originally the uppermost stratum of the neural-plate stage, not only contains the nucleated bodies of inertly-supporting ependymal cells but also mitotic cells. These are perhaps not special stem-cells, as commonly described; rather, it is now urged that whenever cells of the mantle zone are about to divide they retract and round up, so that they come to lie temporarily in the ependymal zone. The mantle layers makes up the future *gray substance* of the central nervous system; it is predominantly cellular in structure and contains the cell bodies of the neurons and many neuroglial cells. The marginal layer is a 'fibrous' mesh which lacks cells in the early months. It provides a basis into which the processes of nerve cells grow and reach their destinations; thereby neuron is linked with neuron and center with center. It becomes the *white substance* of both the brain and spinal cord. The details of the transformation of neuroblasts into neurons and spongioblasts into ependyma and neuroglia demand further treatment and will occupy the descriptions that follow.

The Differentiation of Neuroblasts

The process by which neural epithelium specializes into neurons is named *neurogenesis*. In accomplishing this end

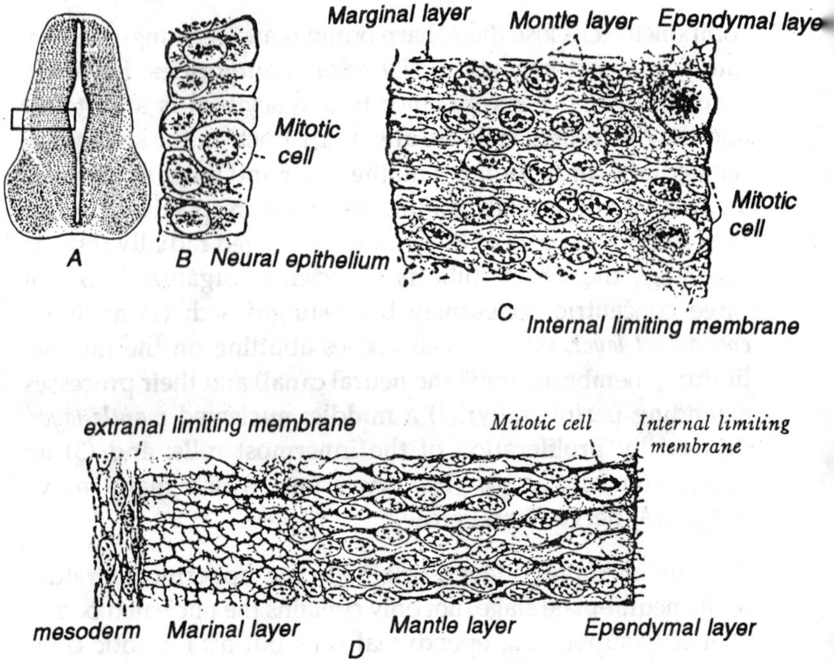

Fig. 4.22. Differentiation of the wall of the neural tube. *A*, Transverse section of a spinal cord; a rectangle explains the relations of the detail-sections, and especially *D*, *B-D*, At stage of the neural plate, 5 mm. and 10 mm.

each neuroblast (and its descendants) sooner or later loses the power of cell division, develops cell processes and converts into a definitive *neuron*. A neuron is a discrete structural and functional unit of nervous tissue; it consists of a nerve cell and all of its processes. The only relation of one neuron to another is that of contact between processes, or between a process and a cell body; this touching constitutes a functional contact known as a *synapse*. Mitosis among neuroblasts ceases during the first year of postnatal life. Thereafter the nervous system matures and enlarges, but the ability to produce new neurons is forever lost. The total number of neurons developed for the use of the nervous system is remarkably constant, and this is true regardless of the size of the individual.

The origin of the nerve fibers as extensions from the neuroblasts is easiest understood in the development of the root fibers of the spinal nerves. The dorsal roots contain sensory, or *afferent fibers* which are processes from ganglion cells located in ganglionic enlargements in these roots. The ventral roots, on the other hand, contain motor, or *efferent fibers* arising from nerve cells located in the neural tube itself. The efferent fibers are the earlier ones to appear.

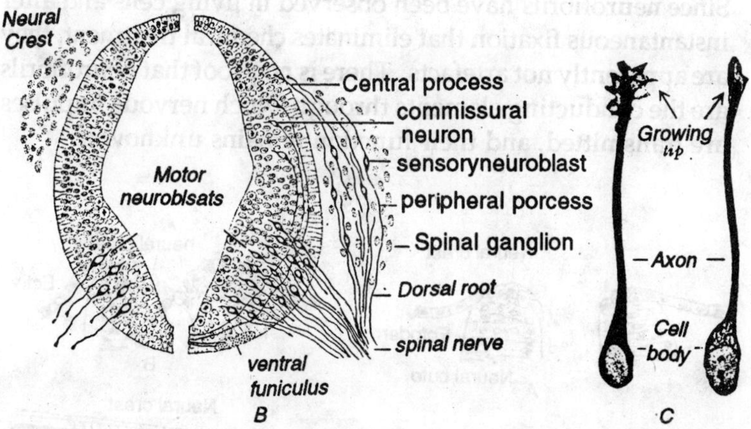

Fig. 4.23. The differentiation and growth of human neuroblasts. *A*, Spinal cord, in transverse hemisection, at 4 mm. (x 225). *B*, Spinal cord, in transverse hemisection, at 5 mm. (X 140). *C*, Two neuroblasts, demonstrating neurofibrils and the enlarged, growing tip (Cajal; X 500).

The Development of Efferent Neurons

Toward the end of the first month neuroblasts separate from the general 'syncytium' in the mantle layer of the ventrolateral wall of the neural tube. The young, bipolar neuroblasts become pear-shaped, and from the slender end of each cell a thin primary process grows out. This extension becomes the chief process, or *axon*, of the nerve cell and is destined to conduct efferent (motor) impulses away from the cell body. Axons penetrate the marginal layer of the spinal cord and, on emerging, straightway combine as a *ventral root* of a spinal nerve. In a similar manner the efferent fibers of the cranial nerves grow out from neuroblasts of the brain wall. The cell bodies of all efferent neurons soon become multipolar by the development of branched secondary processes, the *dendrons* (or *dendrites)*. Within the cytoplasm of even young nerve cells and their secondary processes, fine *neurofibrillae* can readily be demonstrated by selective staining methods. Since neurofibrils have been observed in living cells and after instantaneous fixation that eliminates chemical treatment, they are apparently not artefacts. There is no proof that neurofibrils are the conducting elements through which nervous impulses are transmitted, and their function remains unknown.

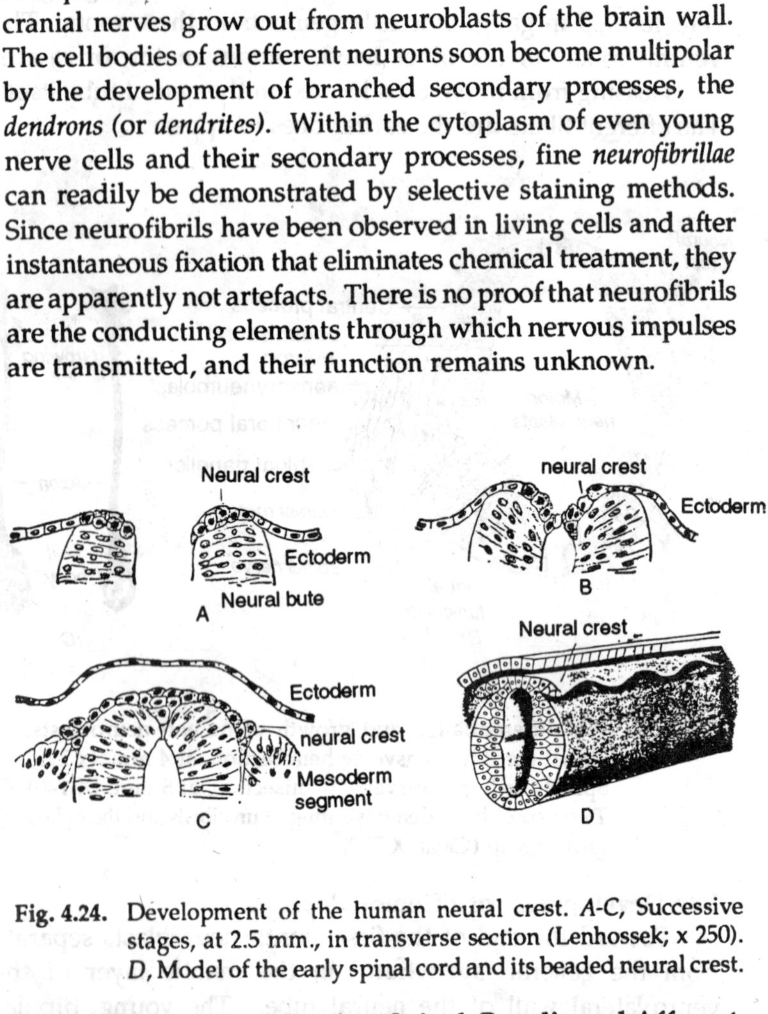

Fig. 4.24. Development of the human neural crest. *A-C*, Successive stages, at 2.5 mm., in transverse section (Lenhossek; x 250). *D*, Model of the early spinal cord and its beaded neural crest.

The Development of Cerebro-Spinal Ganglia and Afferent Neurons

After the formation of the neural plate and groove, a longitudinal ridge of cells appears on each side where the ectoderm and neural plate join. This ridge of ectodermal cells is called the *neural crest*. When the neural folds become a tube and the ectoderm separates from it, the cells of the neural crests overlie the neural tube and extend wedge-like into its seam of closure (B). As development continues the crest-substance separates into right and left linear halves, distinct from the neural tube, which settle to a position between the tube and the myotomes (C). On its arrival in this location the neural crest is a cellular band extending the full length of the spinal cord and far rostrad along the brain wall.

At regular intervals, agreeing with the position of somites, the proliferating cells of the crest give rise to bead-like enlargements, the *spinal ganglia (D)*. The serially repeated ganglia of each side are interconnected for a short time by parts of the originally continuous crest substance, but these bridges soon disappear. In the hind-brain region certain ganglia of the cranial nerves develop also from the crest; they are spaced in relation to the branchial arches.

The cells of the ganglion primordia differentiate into *ganglion cells* and *supporting cells*, groups that are comparable to the neuroblasts and spongioblasts of the neural tube. Each neuroblastic forerunner of a ganglion cell elongates into a spindle-shaped element and, by developing a primary process at each end, transforms into a neuron of the bipolar type. All of the growing processes that are directed toward the neural tube converge into a distinct bundle and thus constitute a *dorsal root*. Each component fiber of the root penetrates the dorsolateral wall of the neural tube, courses in the marginal layer of the spinal cord and, by means of end-twigs, comes in contact with a neuron of the mantle layer. The peripheral processes of the bipolar ganglion cells complete a dorsal spinal root by passing outward and joining the corresponding ventral root; the common bundle, thus formed, constitutes the trunk of a *spinal nerve*. Although bipolar at first, ganglion cells become unipolar in a way not surely understood. Presumably a part of the cell body draws out into a common stem that bears the two processes at its tip. In Figure 4.22 B there can be traced intermediate stages between typical bipolar and unipolar cells.

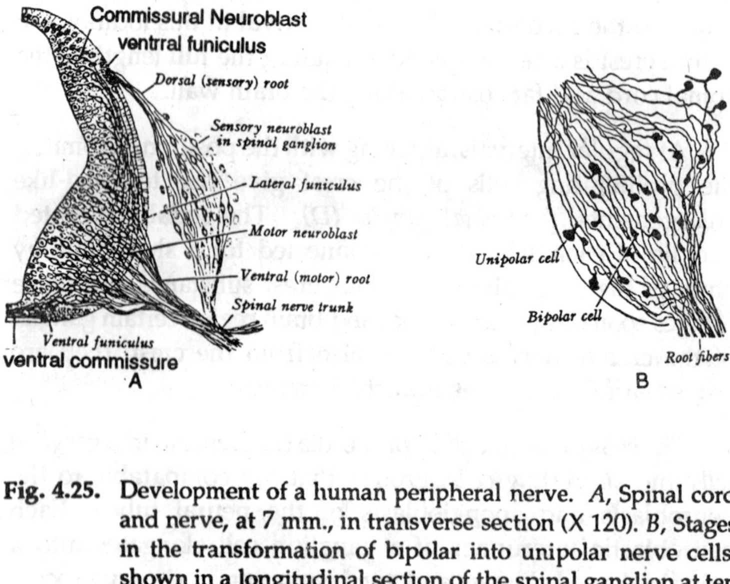

Fig. 4.25. Development of a human peripheral nerve. *A*, Spinal cord and nerve, at 7 mm., in transverse section (X 120). *B*, Stages in the transformation of bipolar into unipolar nerve cells, shown in a longitudinal section of the spinal ganglion at ten weeks.

The Development of Autonomic Ganglia

In addition to forming the cerebro-spinal ganglion cells, certain neuroblasts migrate ventrad and differentiate into cells of the autonomic ganglia. The source of these neuroblasts is disputed, but the evidence seems to favour an origin from the neural crest. The course of differentiation of autonomic ganglion cells differs from that in the cerebro-spinal series inasmuch as the final product is multipolar cells whose chief process is an axon. Functionally, therefore, these ganglion cells are efferent.

Differentiation of the Supporting Elements

Both the primitive neural tube and the early ganglia furnish cells that become non-nervous elements. These are permanent constituents of the brain and spinal cord and of the peripheral nervous system, respectively.

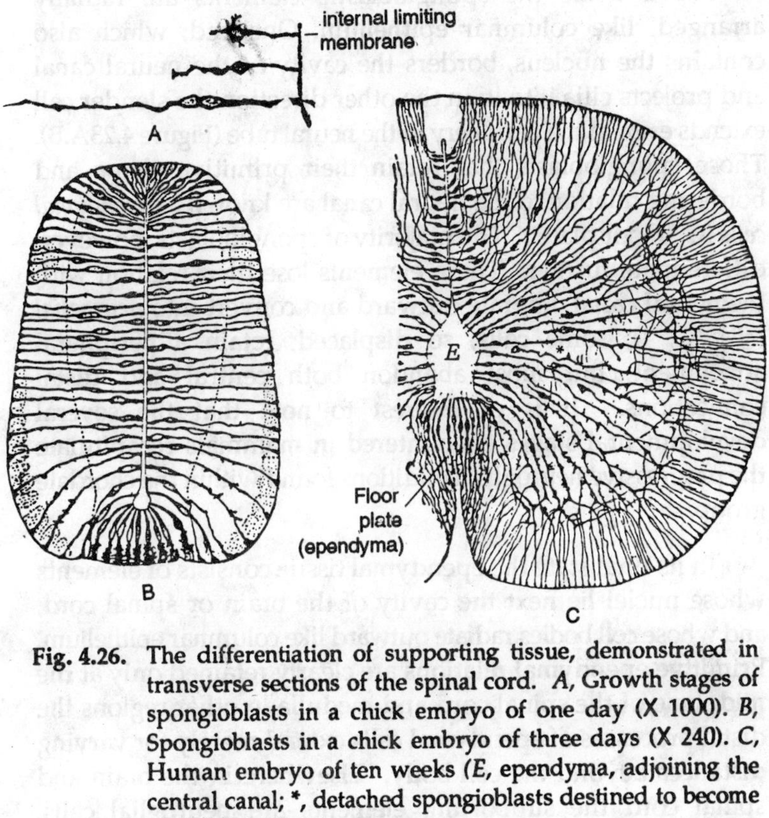

Fig. 4.26. The differentiation of supporting tissue, demonstrated in transverse sections of the spinal cord. *A*, Growth stages of spongioblasts in a chick embryo of one day (X 1000). *B*, Spongioblasts in a chick embryo of three days (X 240). *C*, Human embryo of ten weeks (*E*, ependyma, adjoining the central canal; *, detached spongioblasts destined to become neuroglia). (X 45).

Supporting Elements of the Neural Tube

The brain and spinal cord are given stability by ectodermal, interstitial tissue in the form of ependymal cells, which bound the spinal canal and extend outward toward the periphery, and by neuroglial cells which are more irregularly distributed. A preceding paragraph has described how the spongioblasts originate from the undifferentiated cells of the neural-plate tissue and become more or less altered. The degree and direction of this specialization determine whether they result in ependyma or neuroglia.

For a while the spongioblastic elements are radially arranged, like columnar epithelium. One end, which also contains the nucleus, borders the cavity of the neural canal and projects cilia into it; in the other direction the slender cell extends even to the periphery of the neural tube (Figure 4.23A.B). Those spongioblasts that retain their primitive shape and bordering relation to the neural canal are known as *ependymal cells* (C, 'Floor plate'). The majority of spongioblasts, however, differentiate further. These elements lose their relation with the neural canal, migrate outward and convert into *neuroglial cells* (C, *); some cells, so displaced, retain a peripheral attachment, but most abandon both central and distal connections. It is of interest to note that the several developmental stages encountered in mammals recapitulate the progressive neuroglial conditions found within the chordate group.

In its final state the ependymal tissue consists of elements whose nuclei lie next the cavity of the brain or spinal cord, and whose cell bodies radiate outward like columnar epithelium. Primitive ependymal relations are clearly retained only at the midplane of the spinal cord and medulla in other regions the distal processes of ependymal cells extend merely for varying distances beyond the cell body. Elsewhere in the brain and spinal cord the supporting elements are neuroglial cells, distributed throughout the mantle and marginal layers. They are of two morphological types : (1) *astrocytes,* stellate in shape and with long processes and (2) *oligodendroglia,* with a smaller cell body and fewer, finer processes (C). A third type *microglia* (D), should be mentioned although it would appear that they do not belong developmentally, structurally or functionally with the true neuroglia. In spite of counterclaims, these elements, which are potential ameboid phagocytes, appear late in the neural tube and probably originate from mesenchymal cells. They could, therefore, be appropriately named *mesoglia.*

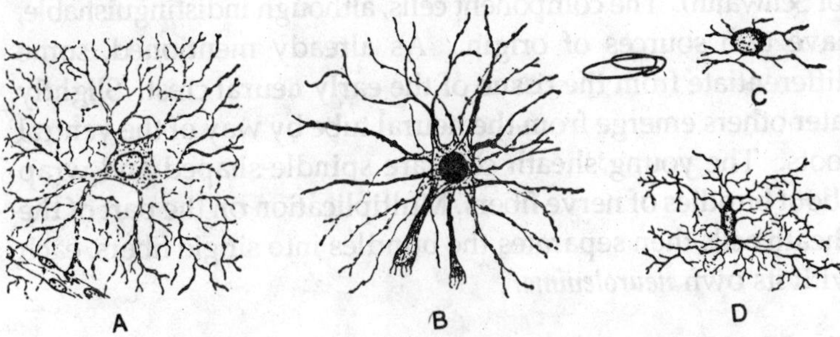

Fig. 4.27. Types of neuroglial cells x 650. *A*, Protoplasmic astrocyte and capillary; *B*, fibrous astrocyte; *C*, oligodendroglial cell; D, microglial cell.

The astrocytes are derived from full-length primitive spongioblasts, from spongioblasts that (because of the thickness of the tube) never connect with the periphery, and from wandering spongioblasts. Astrocytes appear first in the third month. Those occupying the gray substance are named *protoplasmic astrocytes*, another type, *fibrous astrocytes*, develop fibrils within their cytoplasm and are typical of the white substance (B). The oligodendroglia, derived solely from migratory spongioblasts, arise at a later period than astrocytes (C).

Supporting Elements of the Ganglia

The supporting cells of the ganglia at first make up an apparent syncytium, in the meshes of which are found the neuroblasts. The interstitial elements differentiate both into flattened *capsule cells*, which invest the ganglion cells, and into *sheath cells*, which migrate peripherally along the developing nerve fibers and envelop the axons. Still other migratory non-nervous cells, of early neural-crest origin, mingle with the general mesenchyme. Their subsequent history has been explained, at least in part, by properly designed experiments.

The Neurolemmal Sheath

Peripheral nerve fibers are enclosed within a cellular sheath (of Schwann). The component cells, although indistinguishable, have two sources of origin. As already mentioned, some differentiate from the tissue of the early neural crest. Slightly later others emerge from the neural tube by way of the ventral roots. The young sheath cells are spindle-shaped and wrap about bundles of nerve fibers. Multiplication on the part of the sheath cells then separates the bundles into single fibers, each with its own *neurolemma.*

The Myelin Sheath

Between the fourth month of fetal life and the third month following birth a fatty *myelin sheath,* also called a medullary sheath, begins to appear about many nerve fibers. It surrounds the chief process (axon) and, in turn, is enclosed by the neurolemma. The origin of myelin is in doubt. It does not appear until neurolemmal cells are present on a peripheral fiber, and is laid down as units agreement with the extent of those cells. Yet myelin is closely related to the axon and possibly is 'secreted' by it. Probably the origin should be assigned to the co-operative effort of neurolemma and axon. In the central nervous system there are no characteristic neurolemmal sheaths investing the fibers, yet many acquire a myelin sheath. Nevertheless, scattered 'sheath cells' are said not only to be present but also most numerous during the period when myelin is differentiating. Some trace their origin merely to the spongioblastic supporting cells of the neural tube, while others identify them more specifically with oligodendroglia.

Myelination

The *myelinated fibers* (i.e., those with a myelin sheath) have a glistening, white appearance which gives the characteristic colour to the *white substance* of the brain and spinal cord and to most peripheral nerves. Many of the fibers of the central nervous system remain *unmyelinated;* this is true of certain fibers coursing in the white substance, while the portions of all fibers lying within the *gray substance* never acquire myelin. Many fibers in the peripheral nerves are also unmyelinated, yet every peripheral fiber is supplied with a neurolemmal sheath.

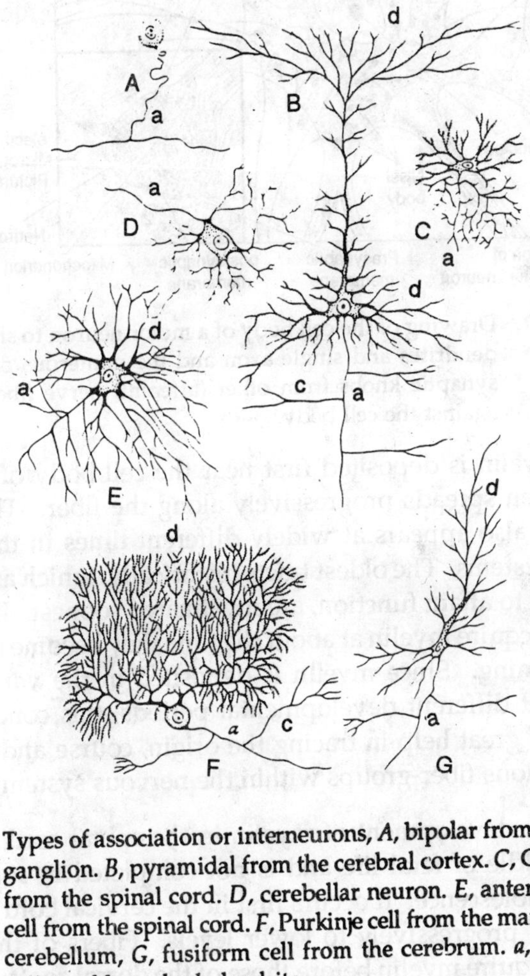

Fig. 4.28. Types of association or interneurons, *A*, bipolar from a spinal ganglion. *B*, pyramidal from the cerebral cortex. *C*, Golgi cell from the spinal cord. *D*, cerebellar neuron. *E*, anterior horn cell from the spinal cord. *F*, Purkinje cell from the mammalian cerebellum, *G*, fusiform cell from the cerebrum. *a*, axon; *c*, collateral branch; *d*, dendrite.

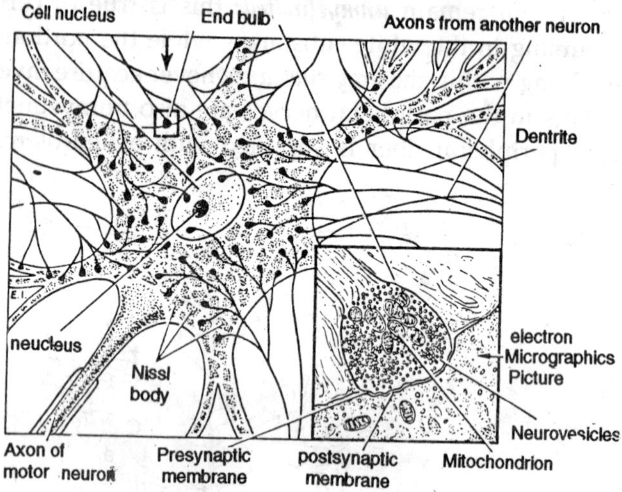

Fig. 4.29. Drawing of the cell body of a motor neuron to show its own dendrites and single axon and the numerous end-bulbs or synaptic knobs from other (afferent) nerve fibers pressing against the cell body.

Myelin is deposited first near the cell body of a neuron and then spreads progressively along the fiber. The myelin sheath also appears at widely different times in the various fiber-systems. The oldest tracts historically, which are also the earliest to attain function, are myelinated soonest. In general, tracts acquire myelin at about the time they become capable of functioning. Since myelin is deposited in the various fiber tracts at different developmental periods, this condition has been of great help in tracing the origin, course and extent of the various fiber-groups within the nervous system.

The development of myelin in the spinal cord begins at the middle of fetal life and is not completed in some fibers until adolescence. It occurs first in the cervical cord and then extends progressively to lower levels. Fibers of the ventral roots acquire myelin before those of the dorsal roots. Tardiest of all are certain descending motor tracts (cortico-spinal; tecto-spinal) which myelinate largely during the first and and second

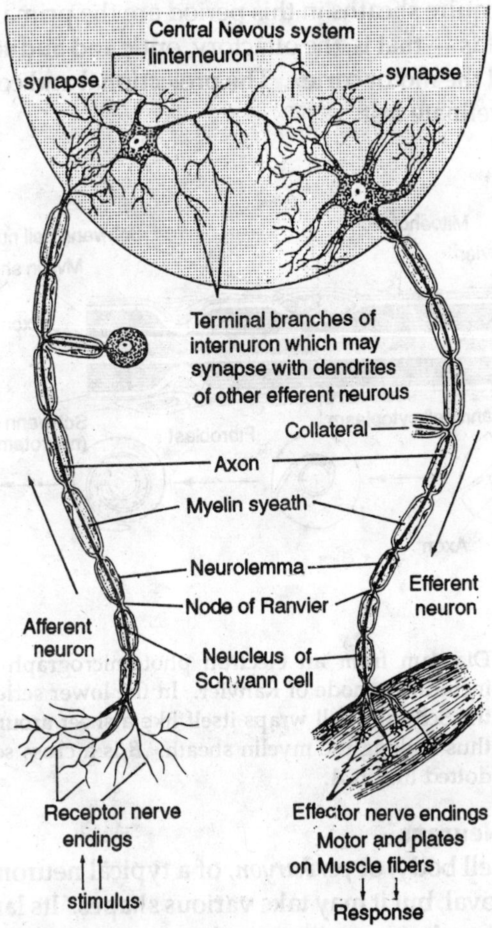

Fig. 4.30. Afferent (sensory) neuron, association or interneuron, and efferent (motor) neuron. So connected they make simplest reflex arc.

postnatal years. The brain begins to myelinate in the sixth fetal month, but progress is slow and only the fibers of the basal ganglia and those that continue the structure of the spinal cord upward possess myelin sheaths at the time of birth. In fact, the brain of a newborn is still largely unmyelinated, so that deposition goes on chiefly from birth through puberty.

First to acquire sheaths in this period are the primary sensory-motor fields — that is, the olfactory, optic and auditory cortical fields and the motor cortex. The projectional and commissural fibers myelinate last.

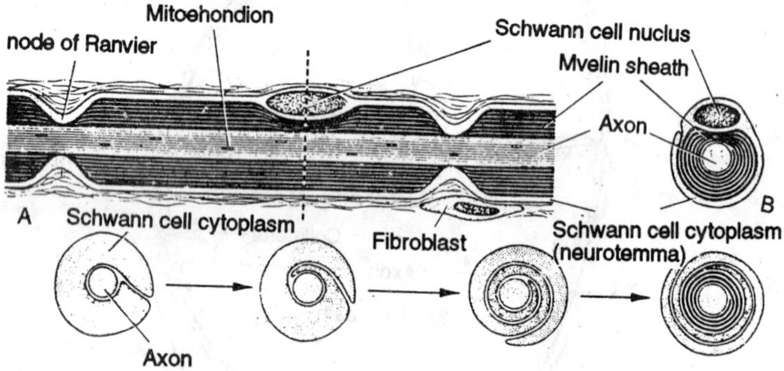

Fig. 4.31. Diagram from an electron photomicrograph of an axon including a node of Ranvier. In the lower series, note how the Schwann cell wraps itself like a sheet around the axon, thus making the myelin sheath. *B* is a cross section at the dotted line in *A*.

Typical Neurons

The cell body, or *perikaryon,* of a typical neuron is usually round or oval, but it may take various shapes. Its large, lightly basophilic nucleus contains a single prominent nucleolus. Basal condensations. *Nissl substance,* are present in the cytoplasm of the cell body and dendrites. The electron microscope reveals these substances to be composed of a dense network of endoplasmic reticulum and RNA granules. As we indicated earlier in the text, RNA is usually associated with protein synthesis and its presence in the nerve cell is believed to be related to the turnover of cytoplasmic material in the axon. The mature neuron is in perpetual growth, since the macromolecules of the neuron undergo continuous degradation and must be replaced. Axoplasm is constantly synthesized in the cell body replenishing the enzyme and structural protein composition of the axon. *Neurofibrils*, cytoplasmic inclusions, are now believed to be artifacts produced in preparing the tissue for histologic sectioning. However, the *neurofilaments* and *microtubules*, seen in the electron micrographs, are considered part of the cytoplasm, although we do not completely understand their function.

Two types of processes extend from the cell body, the *dendritic and axonal* processes. In the past, their identification was based on what was traditionally considered a typical nerve cell, the motor neuron. However, anatomists named the processes one way, neurophysiologists, another way. Today, we realize that no cells in the body differ from one another more than the cells of nervous tissue and that there is no such thing as a typical nerve cell. In an attempt to take into account current functional-anatomical concepts, the conventional terminology was re-examined and a modified nomenclature of vertebrate neuron structure was proposed.

Since the perikaryon (cell body) is considered primarily a growth center, its position on the fiber has little to do with electrochemical functions. Therefore, terminology based on the direction of travel of the nerve impulse, that is, toward or away from the cell body, is meaningless. The new nomenclature refers to a *dendritic zone* of a neuron. In this zone the receptor membrane consists of cytoplasmic branches modified either to receive synaptic endings of other neurons (such as an association neuron), or to respond to stimuli by generating a nerve impulse (as in a sensory neuron). The many dendrites *(dendron, 'tree')* branch in the motor neuron and end in knoblike structures called *gemmules*. On Purkinje cell of the cerebellum, for example, may have as many as 200,000 terminals.

The *axon* is a long extension of cytoplasmic material differentiated to carry nerve impulses away from the dendritic zone. It is synonymous with the general terms "long peripheral fiber" or "axis cylinder." It may be branched and is of uniform diameter. A *neurolemma* (specialized sheath cells) usually wraps around the axon.

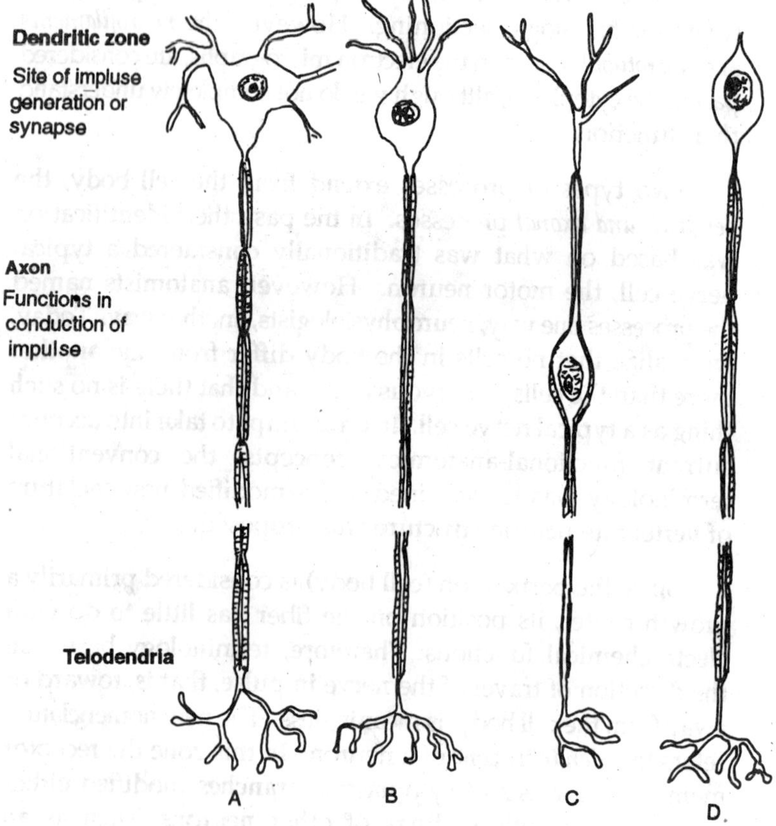

Fig. 4.32. Diagrams of a typical motor neuron, *A*, and three different types of receptor neurons, *B* to *D*.

Axon telodendria are the endings of axon functioning in synaptic transmission and neuro-secretory activity. They may terminate in button-like enlargements, *boutons*. These are in contact with the processes or cell membrane of another neuron, or they terminate at a gland, a special unfolded area of a muscle, or at a blood vessel. According to electron micrographs many small *presynaptic vesicles* are present in the axonal endings or boutons. Anatomists believe that these vesicles are associated with the storage, release, and uptake of transmitter substances and are the only organelles unique to the neuron. The vesicles are morphologically diverse, possibly indicating a variety of chemical transmitters, the most common of which are *acetylcholine* and *norepinephrine*.

Impulses are usually received either by the ends of the dendritic branches or by the membrane of the cell body and transmitted in the direction of the axon. The receptor processes of the sensory nerves are either free or encapsulated. They are either directly in contact with the environment (pain endings) or with another specialized cell that, in turn, is capable of responding to environmental stimuli (rods or cones of the eye, for example). A stimulus generates an impulse in the sensory nerve which carries it to the cord or brain; here the neuron makes contact with association or motor neurons. A nerve fiber itself can conduct an impulse in either direction, but depending on its terminal arborization it can either receive impulses (gemmules) or transmit them (boutons). The synapse, therefore, can conduct the nerve impulse in only one direction.

Except at their point of origin from the cell body and at their terminal fibers, specialized sheath cells, arranged end to end, cover the axons of the peripheral nervous system and most of the axons of the central nervous system. In the peripheral nervous system these cells are derived from the neural crest cells and are known as the *neurolemma*, or *Schwann cells.* Most of the axons also have a sheath of fatty material, *myelin,* inserted between the outer membrane of the Schwann cells and the membrane of the axonal fibers. Those nerve fibers that contain this fatty sheath are referred to as *myelinated fibers;* those fibers that lack the sheath are *non-myelinated*.

The myelinated fibers are usually larger than the non-myelinated and conduct impulses at a faster rate. The post-

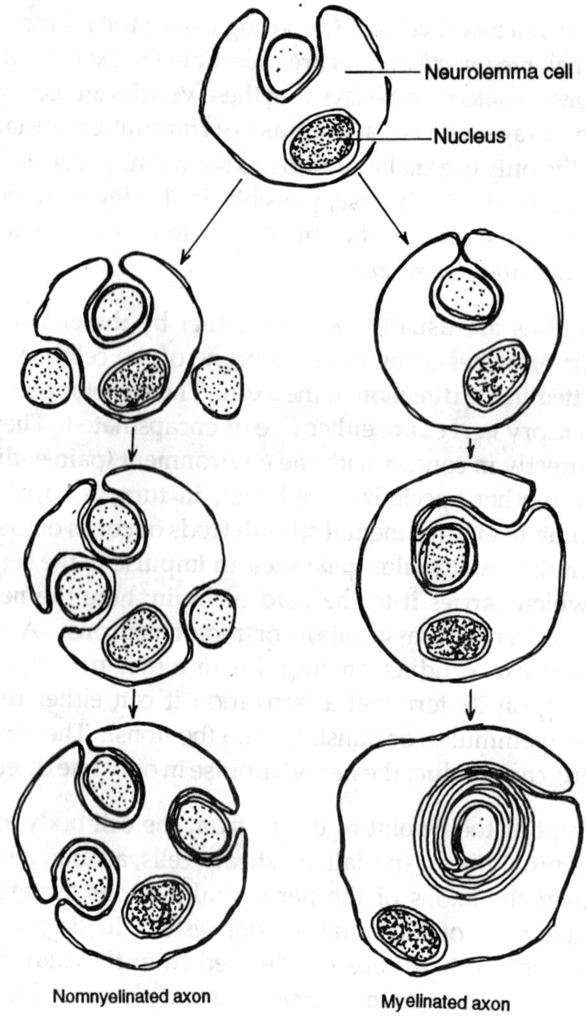

Fig. 4.33. A, Several axons sink into evaginations of the cell membrane (mesaxons) of a neurolemma cell and become enclosed by it. The numerous axons are non-myelinated. B, One axon sinks into the mesaxon. The rotation of the neurolemma cell results in wrapping the axon, "jelly-roll fashion, with double layers of neurolemma cell cytoplasm. The fibers wrapped with such a sheath are said to be myelinated.

ganglionic fibers of the autonomic system, those that pass from the autonomic ganglion to a muscle or gland, are examples of fibers that lack myelin sheaths. The axons of these fibers are usually grouped together with as many as 20 axons in a bundle. Each axon sinks into an invagination of the cell membrane of a neurolemma cell, the so-called *mesaxon A*. As a result, one sheath cell encloses a bundle of axon fibers.

On the other hand, as shown by electron microscopy, a single Schwann cell wraps a myelinated fiber, often as many as 50 times or more, as illustrated. This is sometimes referred to as the "jelly roll" formation of myelin. A cross section of a myelinated nerve fiber reveals a sheath composed of several concentric layers of the Schwann cell's plasma membrane. As the Schwann cell wraps itself around the axonal fiber, the cytoplasm is squeezed out in some manner and the plasma membranes of the Schwann cells are left. These membranes consist of a lipid layer of molecules sandwiched between two protein layers. The protein layers of two adajacent membranes, fuse, freeing the lipid layer. It is this lipid component of the Schwann cell membrane that constitutes the myelin. Myelination begins to develop in the human fetus about the fourth month and may continue until the second or third year after birth.

A constriction, the *node of Ranvier*, marks the point where one Schwann cell ends and another begins. Although the myelin sheath is absent at these nodes, the axonal fiber is not naked at this point. The ends of one Schwann cell overlap another at the nodes and form a protective membrane. Since the cytoplasm is squeezed out from between the layers of the membrane as the Schwann cells rotate around the fiber, most of the cytoplasm accumulates along with its nucleus in the outermost membrane. It is this outer portion of the Schwann cell that is customarily referred to as the neurolemma.

The cells of the central nervous system lack Schwann cells (neurolemma). Instead, one type of glial cell (oligodendroglia cells) provides the myelin around the fibers. The myelinated fibers of the central nervous system, therefore, structurally resemble the myelinated fibers of the peripheral nervous system.

We do not understand the exact function of the myelin sheath, but assume it acts as an insulating material that prevents nerve impulses from jumping from one axon to another. The thickness of the myelin sheath apparently influences the speed of conduction, since the thickest fiber is the fastest conductor. The myelin may also function in a nutritive role for the enclosed axon.

Neurosecretory Cells

Although the neuron functions conventionally as the conductor of nerve impulses, some neurons are modified to secrete. These neurons, referred to as *neurosecretory cells,* are found in the hypothalamus and other parts of the nervous systems of both vertebrates and invertebrates.

Sensory neurons, relaying all possible types of information from other parts of the body, synapse with the dendrites of the neurosecretory cells. Unlike ordinary neurons, the axons of these cells end near a capillary bed. They do not synapse with other neurons nor innervate effector organs. A stimulated neurosecretory cell synthesizes its product and transports it in the form of visible droplets down the axon toward the blood vessel. Here it is either temporarily stored or discharged into the bloodstream, as suggested. The circulatory system carries the product of the neurosecretory cell to some distal site on the body where it produces prolonged effects.

The neurosecretory cells link the nervous system with the endocrine system, the two co-ordinating systems of the body. Transmissions of information to these cells enables them to convert sensory information into a hormonal message that can produce a physiologic response in the organism. Such a circuit is referred to as a *neuroendocrine reflex.*

Synapse

New information resulting from electron microscopic

studies and microelectrode and pharmacologic analysis forced the neuroanatomists to change some of their concepts of the organization and structure of nervous tissue. The nature of the *synapse* morphology is one of these areas. Anatomists no longer talk of *the Synapse* but realize that there is a variety of mechanisms involved in cell-to-cell communication. A synapse is traditionally described as the end-to-end coupling of the axon telodendria of one neuron with the dendritic receptor sites of another neuron. New data reveal, however, that this conventional picture is only one of several relationships that exist between neurons demonstrates some of the diversity of neural linkages that have been described and that do not fall readily into the suggested nomenclature system proposed earlier. The coupling of one cell body with another, dendrite to dendrite, and axon to axon occur frequently. However, cell body to dendrites and cell body to axons have also been described.

Impulses travel down nerve processes until they come to the point where one cell process contacts the cell process or membrane of the cell body or another. This physical junction between the transmitter site of one neuron and the receptor site of another is known as the *synapse*. The two membranes, closely apposed to one another, have only a small but variably sized *synaptic cleft* between them. The impulse passes from one neuron to another by the aid of the transmitted substance believed stored in the vesicles of the many terminal boutons. For example, norepinephrine, one kind of synaptic substance, is associated with a specific type of vesicle with a dense core. Stimulation of nerves known to release norepinephrine causes these vesicles to disappear. The transmitter substance alters the ion permeability of the receptor site by depolarizing its membrane and initiating a new nervous impulse in the second neuron.

Reflex Arc

The quick, automatic withdrawal of a finger when it touches a hot object is the result of a reflex action. The knee jerk is another common illustration. In these examples, a chain of

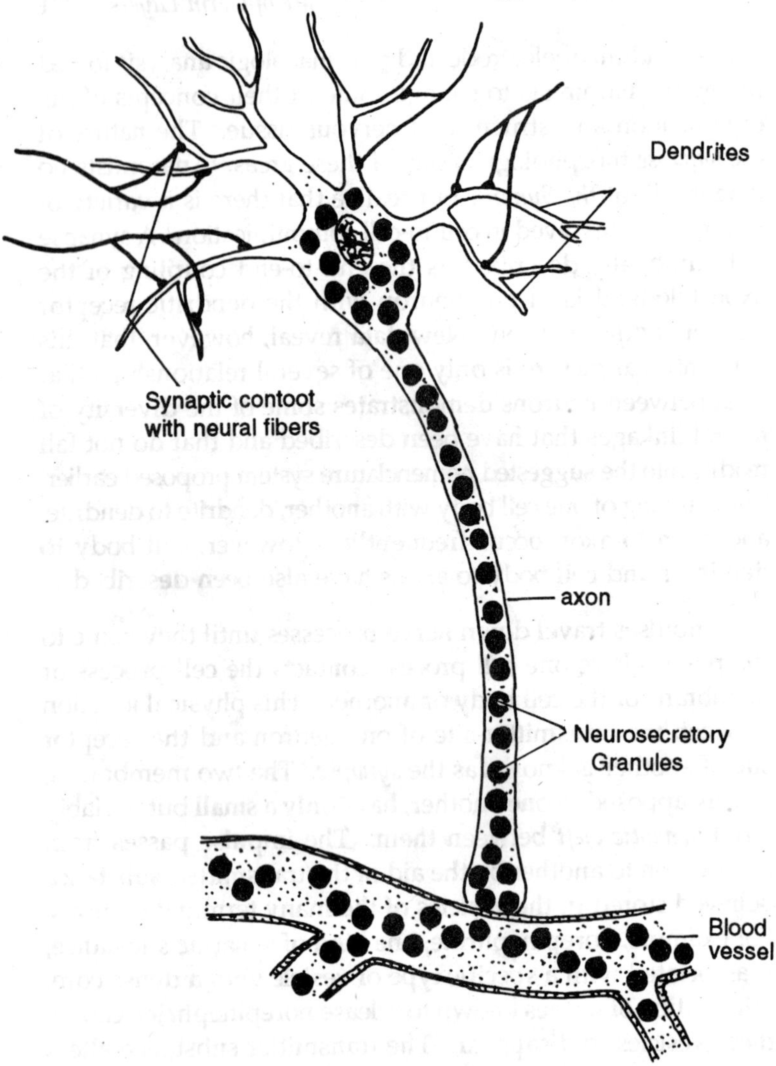

Fig. 4.34. Sensory neural fibers synapse with the dendrites of a neurosecretory cell, stimulating it to synthesize its product. These neurosecretory materials are then transported in the form of visible droplets down the axon toward the blood vessel.

neurons carries the impulse from the point of stimulus, at the finger or patellar ligament to the responding muscle. Such a chain of neurons composes a *reflex arc,* considered the *functional unit* of the nervous system. It first makes its appearance in the human fetus about the fourteenth week of gestation.

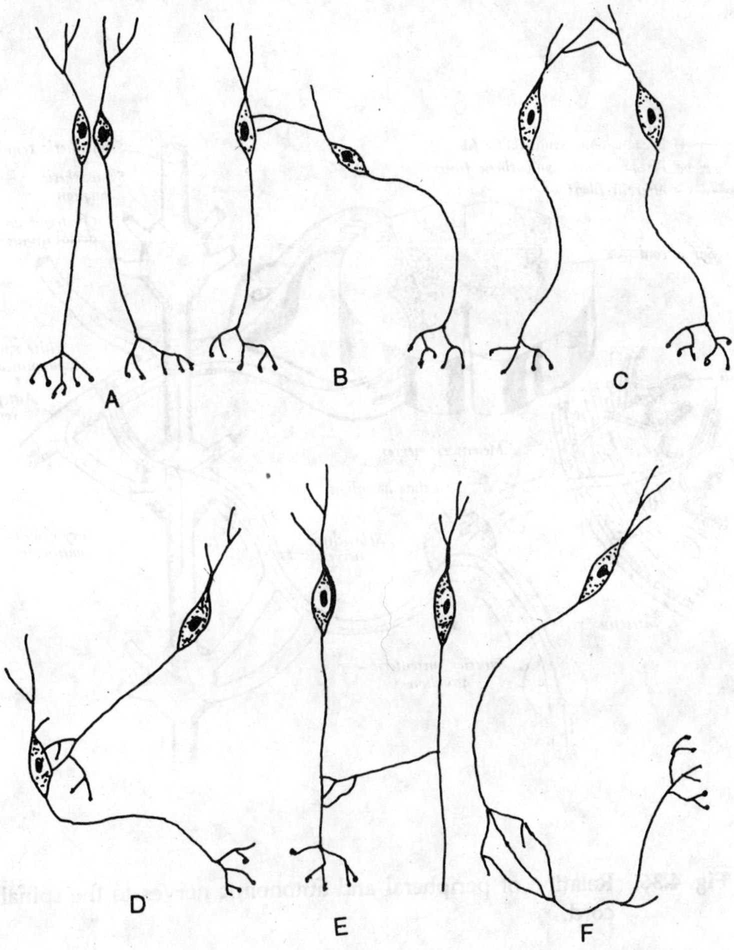

Fig. 4.35. Diagram of various types of synaptic relations between neurons. *A* cell body of one neuron may contact that of a second neuron, *A,* or its dendrites, *B,* or axons, *D.* There can also be dendrite-to-dendrite, *C,* or axon-to-axon, *E,* coupling, *F,* Conventional axon-to-dendrite junction.

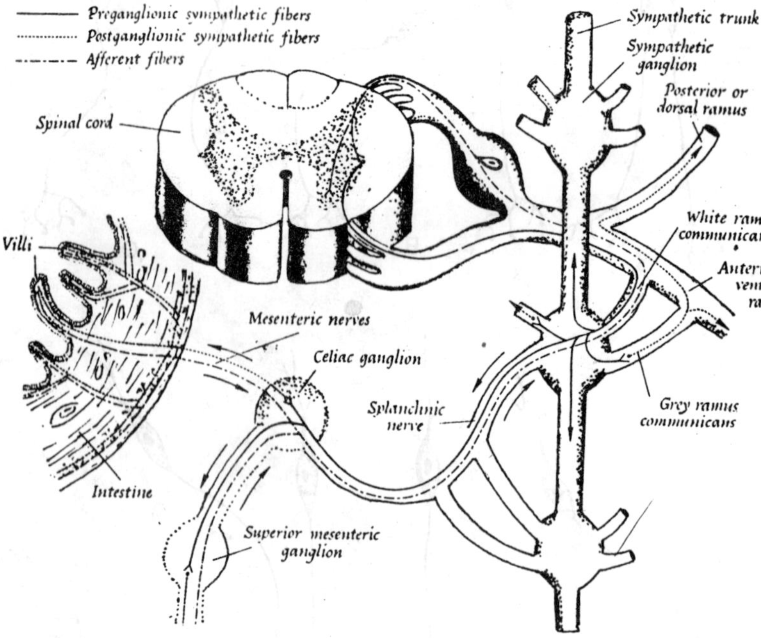

Fig. 4.36. Relation of peripheral and autonomic nerves to the spinal cord.

Although it is possible for the sensory neurons to contact the motor neurons directly, thereby involving a two-neuron chain (monosynaptic reaction), most reflexes are more complicated than this. In general, the simplest reflexes involve at least three neurons. For example, a receptor (dendrite) responds to a heat stimulus and a nerve impulse is generated in the sensory neuron. It passes through the cerebrospinal glands outside the cord and, by way of axonal fibers, enters the spinal cord where it makes contact with many connecting and integrating circuits in the central nervous system. Here analysis and integration of the message occur. The association neurons relay the commands back to the motor neurons. The motor neurons carry the impulse to the effector organ and complete the reflex arc.

In a three-neuron chain reflex, the axonal telodendria of the sensory neurons synapse with several intermediate or association neurons located in the gray matter of the spinal cord. These synaptic connections increase considerably the number of possible pathways. Each association neuron, in turn, receives impulses from several different sensory neurons. The spray of branches at the end of the axonal fiber then allows the association neurons to relay the impulses to several motor neurons. Obviously such interrelationships result in a kind of nerve net.

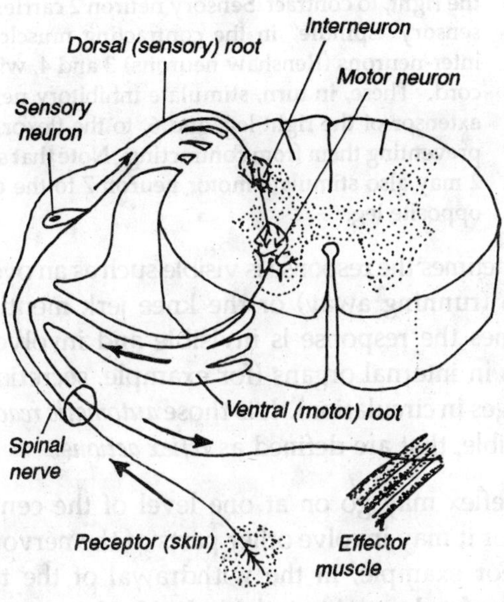

Fig. 4.37. Cross section of spinal cord to show a very simple reflex pathway.

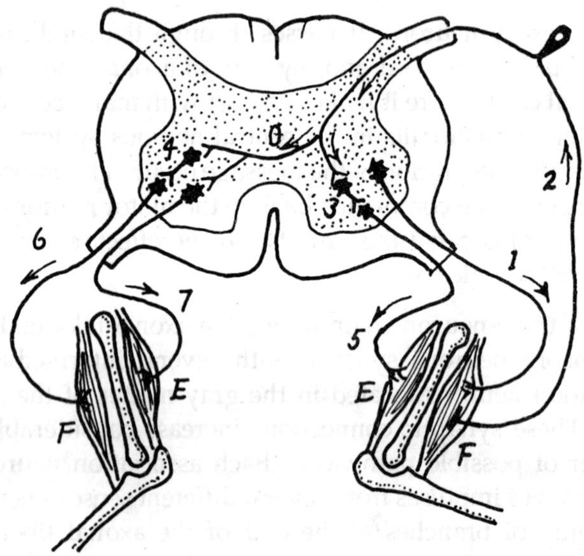

Fig. 4.38. Diagram to show neural pathways in reciprocal inhibition of frog legs. Motor neuron / stimulates F, a flexor muscle on the right, to contract. Sensory neuron 2 carries impulse from sensory "spindle" in the contracting muscle to very short inter-neurons (Renshaw neurons) 3 and 4, within the spinal cord. These, in turn, stimulate inhibitory neurons 5, to the extensor of the right leg, and 6, to the flexor of the left leg, preventing them from contracting. Note that sensory neuron 2 may also stimulate motor neuron 7 to the extensor of the opposite leg.

Sometimes the response is visible such as an overt behaviour reaction (running away) or the knee jerk mentioned above, sometimes the response is invisible and involves regulative changes in internal organs (for example, secretion by glands or changes in circulation). It is those *automatic reactions,* visible or invisible, that are defined as *reflex actions.*

A reflex may go on at one level of the central nervous system or it may involve other parts of the nervous system as well. For example, in the withdrawal of the toe from the stimulus of a sharp object, the individual may move his arms and utter an exclamation of pain as he removes his foot. If both arms are raised, fibers of the intermediate neurons must have crossed to the other side of the cord and relayed the impulse to motor neurons on this side, the opposite side to that in which the stimulus had occurred. Intermediate neurons that cross to the other side of the cord are referred to as *commissural neurons.*

There is evidence that reflex circuitry is laid down early in the embryo and that the nervous system develops orderly pattern of behaviour long before they are actually used. One can demonstrate this by anesthetizing a young embryo at its learning stage. A young amphibian embryo anesthetized before its muscles have functioned remains quiescent during the completion of their development. If the anesthetic is then washed away when the untreated controls demonstrate their ability to swim, the experimental embryos soon recover and join in the vigorous swimming movements. Apparently, the swimming pattern was not learned. Instead, all the neural pathways for reflex swimming movements developed during the anesthetized state. Evidently much of behavioural circuitry is built into the nervous system and is not something imposed upon an established order. The recognition of this fact has caused many researches to turn from a study of psychology to that of neuroembryology in their attempts to understand behavior.

DEVELOPMENT OF PERIPHERAL NERVOUS SYSTEM

The peripheral nervous system consists of paired spinal and cranial nerves that innervate the surface of the body or penetrate to the deeper organs. In the regions of the forelimbs and hindlimbs, several spinal nerves enter the limb buds, establish interconnections, and form a *nerve plexus.* The group of nerves entering the forelimb but make up the *brachial plexus;* those that innervate the hindlimb bud compose the *lumbosacral plexus.*

Formation of Spinal Nerves

The location of the somites on either side of the neural tube in the young neurula determines the position that some of the neural crest cells (future sensory nerve cells) take as they migrate away from the newly closed neural tube. One can demonstrate the importance of the somites by surgically removing them or transplanting additional somites along one side of the body. The opposite unoperated side remains as a control. A corresponding decrease in neural crest cell clusters occurs when somites are removed. Conversely, additional somites induce an increase in numbers of cells clusters when compared with the unoperated side.

There are 31 pairs of spinal nerves in the human embryo, but only 10 pairs in the frog. Each spinal nerve is connected to the spinal cord by way of two roots, *dorsal* and *ventral*, that have different embryonic origins. In amphioxus and some cyclostomes the two roots remain separated, but in all gnathostomes they unite to form a *spinal nerve*. Both sensory and motor elements must be present in the nerves if reflex arcs, the functional units of the nervous system, develop.

The axons of the neuroblasts located in the motor nuclei grow laterally and penetrate the outer wall of the neural tube, the *marginal zone*. Here they emerge as the *ventral*, or *motor*, *root* of a spinal nerve, A. These motor fibers are classified as either *somatic or visceral fibers*. Somatic motor fibers carry impulses to the voluntary muscles of the body, those usually derived from myotomes. Visceral motor fibers, on the other hand, pass deeply into the body of the organism and innervate such internal organs as the heart, stomach and intestine. They maintain those routine functions of the body that are under more or less automatic control. The motor fibers of this visceral system are involuntary in nature and are usually referred to as the *autonomic nervous system*, as compared to the conscious, or willful, nature of the somatic system.

The neuroblasts located in the cerebrospinal ganglia develop into *sensory* neurons, either *somatic* or *visceral,* depending on the organs they innervate. In either case, processes extend from the neuroblast (formerly the neural crest cell) in both directions, out to the receptor and in toward the spinal cord. In the gray matter of the spinal cord they synapse with association neurons. The fibers of the nerve cells collect into bundles outside the spinal cord and give rise to the *dorsal,* or *sensory, root* of the spinal nerve.

The dorsal root unites with the ventral root to form a *spinal nerve,* as can be seen in the section of a 10 mm. pig. A spinal nerve is usually a *mixed nerve;* it can be composed of four different types of nerve processes. It may contain somatic sensory (afferent) and visceral sensory (afferent) fibers, as well as somatic motor (efferent) and visceral motor (efferent) processes.

Growth of Nerves

The factors responsible for directing or guiding certain neurons to specific structures or other neurons have been the subject of much investigation. According to Weiss (1955), there are three phases in the development of nerves; the pioneering phase, the application phase, and the towing phase. In the pioneering phase the cytoplasm protrudes from the side of the nerve as a short thread. The thread lengthens because of the synthesis of protoplasm by the cell body and its flow to the tip of the sprout where ameboid processes develop. During the application phase the free pioneering tip of the fiber attaches itself to a receptor or to an effector cell, located nearby at this early stage. As the embryo grows, the peripheral cell shifts and during the migration "tows" (third phase), the nerve fiber with it. If these peripheral cells migrate outward and become, for example, the big toe of a hindlimb the distance the axonal fiber is towed can be considerable.

We do not understand why the pioneering fiber attaches to a particular cell. The problem of innervation obviously concerns the orientation of the pioneering fibers. Once the application phase is established, additional new fibers from other cells follow the course set by the pioneers, obviously using it as a solid substrate. There is general agreement that nerve fibers follow oriented structural or ultrastructural elements. One hypothesis suggests that the tips of nerves grow into actively growing limb buds by following the stress lines set up as the rapidly dividing mesenchyme cells absorb water. The tendency of nerve fibers to grow along solid objects is referred to as *stereotropism*. The growth of the optic nerve offers a more obvious illustration. The fibers from the inner layer of ganglion cells in the retina converge, pierce the choroid and sclera, and follow the optic stalk as a pathway into the brain. As the fibers grow back into the brain substance, they group together and give rise to the optic nerve.

There is general agreement that growing nerves are subject to contact guidance. Whether additional factors (chemical or electrical) are involved remains a controversial issue. Some investigators explain the migration of nerve fibers toward the source of their stimuli by postulating a difference in electric potential. However, the application of actual electrical currents does not affect either the rate or direction of growth of the fibers. There is little evidence that a fiber can electrically select a particular pathway.

According to Sperry (1965), the modern concept of inherent chemical affinities of the neurons can best explain the selectivity of a growing nerve fiber for a particular pathway. Each of the billions of cells within the nervous system has a specific chemical identification tag dictated by the genetic code. According to this thesis, as the neurons mature and begin to form associations with other neurons, they are attracted to chemically compatible cells or fibers from the millions, or billions, of nerve cells with different chemical identities. As a result numerous neurons having similar chemical identities establish brain pathways. However, chemical affinity most not be the total answer. There are too many different connections that must be made in the total organism for such a simple explanation to apply.

Like most embryonic cells, the developing peripheral nerves appear to be unspecialized as they grow out toward and end organ. There is evidence that a motor nerve makes connection with any muscle cell in its path and becomes specialized after it makes this contact. Motor fibers connect to muscles and glands, and sensory fibers connect to receptors. Evidently these end organs bring about specific chemical changes in the nerves they contact.

Once attained, the specificity of a neuron is irreversible. Sperry (1951) obtained evidence for this specificity when he crossed the cutaneous nerves of the left hind foot to the right hind foot in rats 14 to 26 days old. After the rats recovered from the operation, he found that stimulation of the right foot caused the animal to lick its left foot. No response occurred on the stimulated right side and the proper reactions to the stimuli could not be induced by training. These reactions suggested with the end organs of their receptors and this reaction determined the specificity of these nerves. This information was recorded in the brain.

As development proceeds, the brain irreversibly registers a copy of the non-neural portion of the body. A condensed, point-by-point map of all the body's surfaces, both internal and external organs and muscles, is imprinted on the nervous system. By the time the operation was carried out in the above experiment, the nerves passing to the left side of the animal had contacted the end organs, become specialized as the left sensory or motor leg nerves, and been registered as such in the central nervous system. Therefore, when the left nerve was transplanted to the right side in the postnatal animal and the right side was stimulated, the animal lifted its left foot. This lack of plasticity offers a possible explanation as to why an amputee continues to feel pain in his leg when the stump is stimulated.

Formation of Cranial Nerves

The origin of the cranial nerves is much the same as

spinal nerves, in that motor fibers arise from nuclei located in the gray matter and sensory neurons form from neuroblasts derived either from neural crest cells or ectodermal placodes. There is one difference, however. Although cranial nerves may be mixed nerves, in many cases they are one or the other, either sensory or motor in function. Some biologists believe that cranial nerves at one time contained two roots that did not unite, like the spinal nerves in amphioxus and the cyclostomes. In some cases, the dorsal root became the predominant component and the ventral root degenerated leaving a purely sensory nerve. In other instances, the dorsal root was lost and the ventral root remained as the motor nerve.

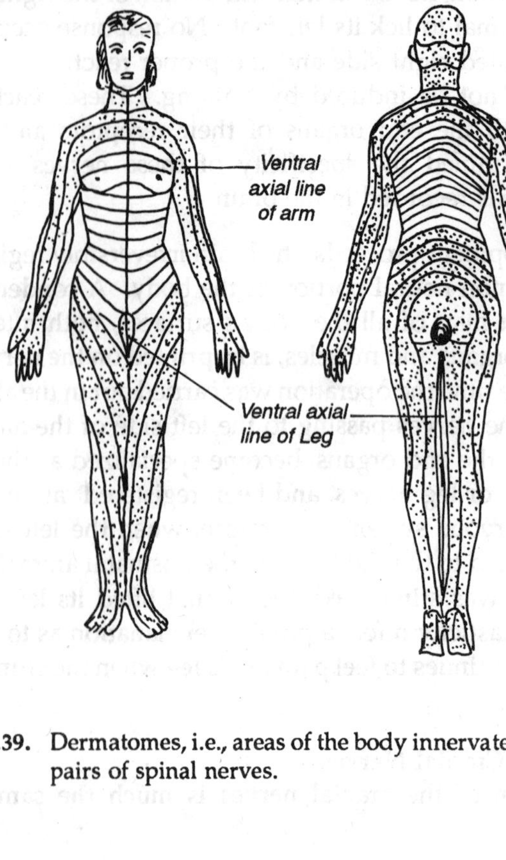

Fig. 4.39. Dermatomes, i.e., areas of the body innervated by successive pairs of spinal nerves.

Segmentation is also lost in the cranial region since some of the cranial nerves represent roots from more than one segment. As an illustration, the hypoglossal nerve (twelfth nerve) forms from the union of ventral roots from several occipital nerves.

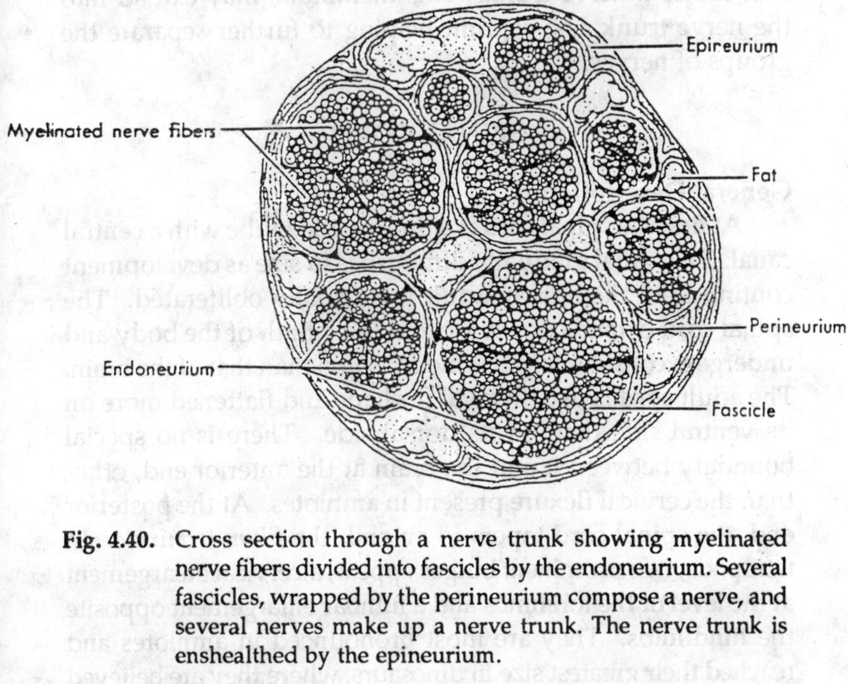

Fig. 4.40. Cross section through a nerve trunk showing myelinated nerve fibers divided into fascicles by the endoneurium. Several fascicles, wrapped by the perineurium compose a nerve, and several nerves make up a nerve trunk. The nerve trunk is enshealthed by the epineurium.

Structure of a Nerve

A spinal nerve consists of axons from the somatic as well as the visceral nervous systems. A cross section of the nerve as seen best reveals its structure. Each nerve axon is surrounded by a sheath of Schwann cells (neurolemma), but may or may not be myelinated. The axons, therefore, are of various sizes ranging from large axons with a thick myelin sheath, fibers of smaller diameter and thinner sheath and thin fibers with no sheath. The large myelinated fibers have a more rapid rate of conduction than do the smaller myelinated fibers. The fibers are arranged in parellel groups and collected into bundles, or *fascicles,* by a connective tissue sheath, the *endoneurium.*

When we refer to a nerve, we are speaking of a group of fascicles bound together by the *perineurium*. This connective tissue sheath not only protects the nerve bundles, but also functions as a diffusion barrier. Several nerves make up a *nerve trunk*. A connective tissue membrane, the *epineurium*, ensheathes a nerve trunk. This membrane may extend into the nerve trunk as a septum helping to further separate the groups of nerves.

SPINAL CORD

General Description

At first the spinal cord is a thick-walled tube with a central canal. This canal gradually diminishes in size as development continues; in the human adult it is usually obliterated. The spinal cord extends down most of the length of the body and undergoes considerably less modification than that of the brain. The adult cord is usually oval in shape and flattened more on its ventral side than on its dorsal side. There is no special boundary between it and the brain at the anterior end, other than the cervical flexure present in amniotes. At the posterior end, the spinal cord tapers to an end, the *filum terminale*. In tetrapods, two prominent bulges appear, a cervical enlargement at the level of the forelimbs and a lumbar enlargement opposite the hindlimbs. They are most pronounced in amniotes and reached their greatest size in dinosaurs, where they are believed to have been much larger than the brain itself. The enlargements are the result of an increase in numbers of cell bodies at the level of the limbs and mark the point of departure from the cord of neurons that pass laterally into these structures.

Early in their development, the neural tube and vertebral column are approximately the same length. However, the vertebral column (cartilage or bone) develops at a faster rate than does the cord. The coccygeal end (posterior) of the cord terminates at the lumbar region in most vertebrates, and the cord, except for the filum terminale, is absent from the posterior regions of the vertebral column. In man the spinal cord terminates at the level of the third lumbar vertebra; in other animals it may be longer but does not usually reach the full length of the vertebral column. As a result of this unequal growth, the spinal nerves in the cervical regions pass laterally into the organs they innervate, but in the posterior part of the cord they are arranged in a more craniocaudad direction.

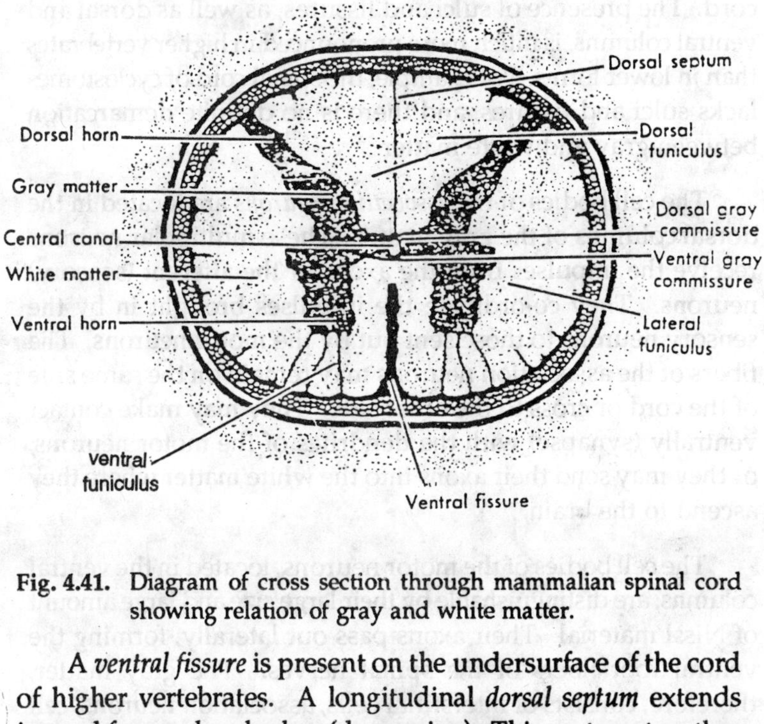

Fig. 4.41. Diagram of cross section through mammalian spinal cord showing relation of gray and white matter.

A *ventral fissure* is present on the undersurface of the cord of higher vertebrates. A longitudinal *dorsal septum* extends inward from a *dorsal sulcus* (depression). This septum, together with the dorsal sulcus and ventral fissure, partially divides the cord into two symmetrical halves as may be Figure 4.37.. The two halves are connected with one another by nerve fibers *(commissures)*.

Gray Matter of Cord

A cross section of the adult cord of amniotes shows that the mantle area of the original neural tube has undergone considerable expansion during development. It forms a "butterfly," or "H-shaped", structure that fills the central part of the section. Anatomists speak of the *dorsal* and *ventral horns* of the gray matter. These are actually cross sections of *columns*, areas of common function, that run up and down the cord. In all vertebrates *dorsal* and *ventral gray commissures* located on either side of the central canal connect the two sides of the cord. The presence of sulci and fissures, as well as dorsal and ventral columns, is much more pronounced in higher vertebrates than in lower forms. For example, the spinal cord of cyclostomes lacks sulci and fissures, and there is no definite demarcation between gray and white matter.

The cell bodies of the *association neurons* are located in the dorsal columns of the gray matter. These multipolar neurons receive the impulses from the axons of the afferent (sensory) neurons. They co-ordinate the impulses brought in by the sensory neurons to those sent out by the motor neurons. The fibers of the association neurons may remain on the same side of the cord or cross to the other side. They may make contact ventrally (synapse) with the dendrites of the motor neurons, or they may send their axons into the white matter where they ascend to the brain.

The cell bodies of the motor neurons, located in the ventral columns, are distinguishable by their large size and large amount of Nissl material. Their axons pass out laterally, forming the ventral root fibers of the spinal nerves. The gray matter, therefore, consists of interstitial cells, association neurons, cell bodies and dendrites of motor neurons, portions of myelinated and non-myelinated fibers, and synapses.

Neural anatomists distinguish functional areas (somatic or visceral) within the dorsal and ventral columns of the gray matter. They identify four such areas on each side of the cord: two areas in the dorsal column where somatic and visceral sensory neurons synapse with association neurons, and two in the ventral area where the cell bodies of the visceral and somatic motor neurons are located. From dorsal to ventral, the areas may be listed sequentially as the somatic afferent and visceral afferent, located in the dorsal column of the gray matter, and visceral efferent and somatic efferent, in the ventral column A. The somatic areas are usually larger than the visceral ones. These general regions persist and can be identified as far forward as the medulla oblongata in higher animals. Anterior to the medulla, this structural plan changes. However, in the amphibian, for example, the entire central nervous system (brain and cord) resembles the cord of higher animals.

White Matter of Cord

Longitudial tracts, referred to as *funiculi* (to differentiate them from the columns of the gray matter) compose the white matter of the cord. Myelinated and some non-myelinated fibers of sensory, motor, and association neurons make up the funiculi. Some transverse fibers are also present running to and from the gray matter. In general, there are no cell bodies and no dendrites in the white matter. The shape of the gray matter divides the surrounding white matter into three general areas on each side of the cord. They are the *dorsal funiculi* located between the dorsal sulcus and dorsal column, the *lateral funiculi* situated between the dorsal and ventral columns, and the *ventral funiculi* between the ventral column and ventral fissure. A white commissure connects the two halves of the cord.

The three funiculi can be further divided into specific *fiber tracts* (fasciculi). A detailed discussion of the fiber tracts of the cord is beyond the scope of this book. One should realize, however, that these tracts connect one part of the cord with another and join the brain with the peripheral system. Their importance increased as the brain evolved and assumed more of the co-ordinating and association functions of the body. The fiber tracts developed to a greater degree in higher animals than in the fish where the trunk is more or less semi-autonomous. They are two-way conduction paths, consisting of ascending axons carrying impulses up the cord and to the brain and descending axons relaying message back to the cord and to the peripheral neurons. Bundles of axons compose the tracts and many tracts make up the funiculi.

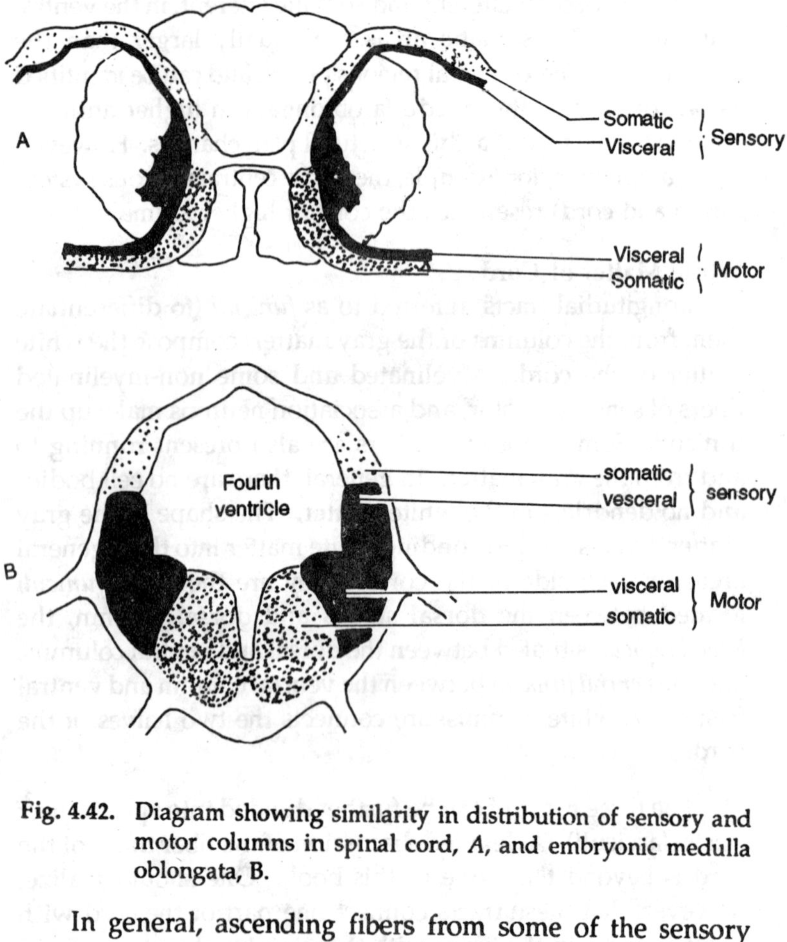

Fig. 4.42. **Diagram showing similarity in distribution of sensory and motor columns in spinal cord, *A*, and embryonic medulla oblongata, B.**

In general, ascending fibers from some of the sensory neurons compose each dorsal funiculus. The fibers enter the cord through the dorsal root and then ascend to a higher level instead of synapsing with an association neuron. Ascending axons from association neurons are also present in this region of the cord. The ventral funiculi carry impulses down the cord and from the brain centers. The cell bodies of these neurons are located in the higher centers in the brain, and the descending fibers synapse either directly with a motor neuron in the ventral column of the cord or with an association neuron, which then contacts the motor neuron. The lateral funiculi consists of both ascending and descending fibers that appear to connect one area of the cord with another.

BRAIN

Myelencephalon

The myelencephalon is the anterior continuation of the spinal cord and is well-developed in all vertebrates. In the adult, it is recognized as the *medulla oblongata* and along with the pons and midbrain compose the *brainstem.* The central canal expands into a triangular structure, the *fourth ventricle,* covered by a thin, highly vascular roof, the *tela choroidea.* Anteriorly, the cerebellum overlaps the medulla. The *posterior choroid plexus* (the mass of capillaries located in the tela choroidea) secretes the cerebrospinal fluid. The white funiculi and gray columns of the cord continue as the walls of the medulla oblongata. The gray matter retains its dorsoventral arrangement of sensory and motor components. Some of the fibers of the medulla cross over to the other side (decussate), thereby integrating the two sides. Consequently, one side of the brain controls the opposite side of the body. Other fiber tracts continue either anteriorly or posteriorly from the medulla. Obviously all tracts between the cord and higher brain areas must pass through the medulla.

Afferent fibers from the posterior regions synapse at two prominent nuclei in the walls of the medulla oblongata. The axons of these cells carry impulses upward to the thalamus and cerebellum. The medulla is the seat of reflex control for such important functions as respiration, heart-beat, and blood pressure. These functions are so necessary for survival that the nuclei of the medulla are referred to as *vital centers.* An injury to this area of the brain often results in death. The reflex centers for certain non-vital functions are also located here, including centers for vomiting, coughing, sneezing, swallowing, hiccuping and salivation. The majority of the cranial nerves are associated with the medulla oblongata. The fifth to twelfth cranial nerves leave or enter the medulla.

Metencephalon

The metencephalon gives rise to the anterior regions of the medulla oblongata, the cerebellum, and the pons. The dorsal part of the metencephalon becomes the elevated and thickened *cerebellum,* which seems to be directly related to the locomotor activity of the organism. The cerebellum does not itself direct activities, but it moderates and co-ordinates those of higher brain centers. It is poorly developed in sedentary forms of vertebrates and most highly developed in birds and mammals. It is a receiving center for sensory and motor information and functions by controlling skeletal muscles. The cerebellum is poorly developed in the cyclostomes, some fishes amphibian and reptiles. In these sluggish and poikilothermic forms, it consists of little more than a shelf of tissue anterior to the fourth ventricle.

The posterior dorsal area of the metencephalon receives information from the vestibular portion of the eighth cranial nerve carrying impulses from the semicircular canals. Also nerves returning from the lateral-line organs give information about vibrations, water currents, and movement of nearby objects. After receiving these sensory data, the cerebellum sorts, correlates, and regulates the body movements. In more active animals (Elasmobranchii, Dipneusti, etc.), irregular-shaped lobes called *auricular lobes,* or *restiform bodies,* project laterally from each side. Their cavities are continuous with the fourth ventricle and they represent the most ancient part of the cerebellum. These lobes are intimately connected with the inner ear and function primarily as organs of equilibrium.

In amniotes, the lateral line components drop out, and as the muscles associated with terrestrial living develop, the proprioceptors increase in importance. The cerebellum of birds is more advanced than in most of the lower animals. It is divided into a middle "worm-shaped" portion, called the vermis (consisting of many horizontal ridges), between two lateral lobes or hemispheres. Its development is associated with the co-ordination of the intricate flight muscles. In the mammal the cerebellum reaches its greatest degree of development. It also receives information about the position of body parts from the organ of equilibrium and on the state of muscle tension from muscle and tendon spindles, types of receptors.

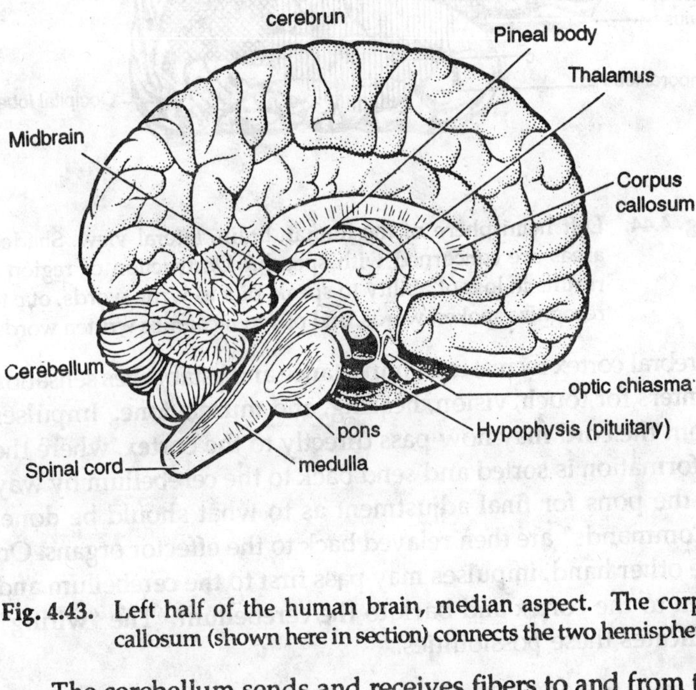

Fig. 4.43. Left half of the human brain, median aspect. The corpus callosum (shown here in section) connects the two hemispheres.

The cerebellum sends and receives fibers to and from the higher brain centers. In the mammal the highly developed cerebral cortex takes over much of the control from the cerebellum. Conspicuous fiber tracts connect the spinal cord and medulla with the cerebellum, and the cerebellum with the cortex. These tracts are located on the ventral part of the myelenecephalon and metencephalon and externally appear as a bulge, the pons (Latin word meaning 'bridge'). The

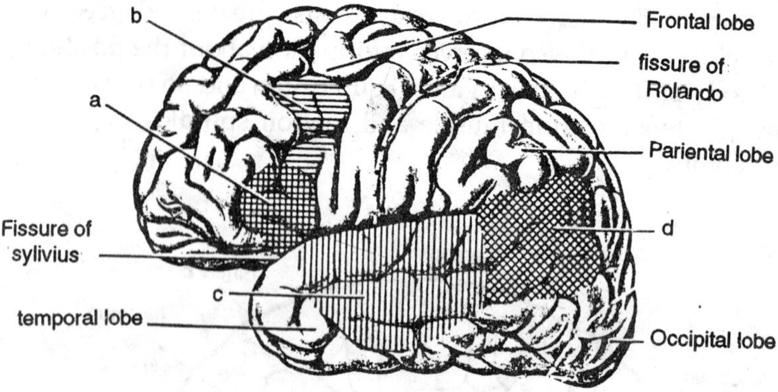

Fig. 4.44. Left hemisphere of the human brain, lateral view. Shaded areas are concerned with speech. Destruction of region *a* results in loss of ability to speak, of *b* to write words, of *c* to recognize spoken words, and of *d* to recognize written words.

cerebral cortex increases in importance and establishes sensation centers for touch, vision, hearing, and muscle tone. Impulses from the cord may now pass directly to the cortex, where the information is sorted and send back to the cerebellum by way of the pons for final adjustment as to what should be done. "Commands" are then relayed back to the effector organs. On the other hand, impulses may pass first to the cerebellum and then to the cortex and back to the cerebellum. The "wiring" indicates these possibilities.

The distribution of gray and white matter in the cerebellum is the opposite of that of the cord. The white matter is concentrated in the center of the cerebellum, and the gray matter, which has increased in amount, forms a convoluted covering over the white matter. The surface of the gray matter has numerous grooves (sulci) and convolutions (gyri). A sagittal section through the cerebellum shows that the white fiber tracts spray out into the gray matter of the cerebellar cortex.

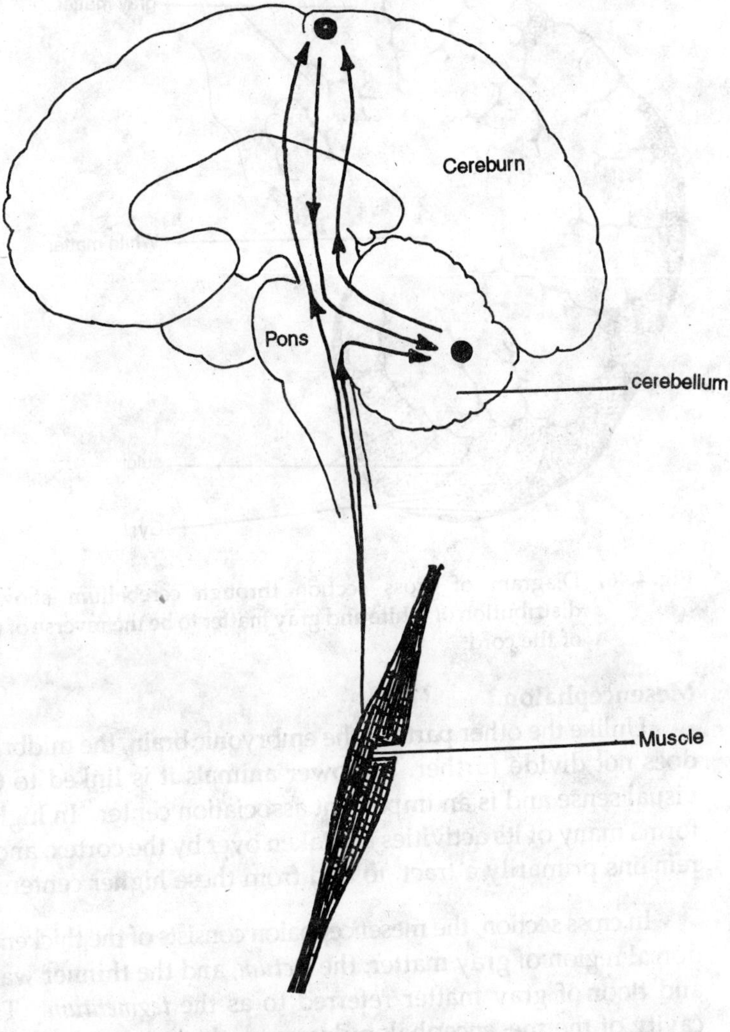

Fig. 4.45. "Wiring diagram" indicating that impulses from spinal cord may pass either-directly to cerebrum or first to cerebellum and then to the higher center. The cerebrum sorts the information it receives and then sends instructions back to the cerebellum as to what should be done.

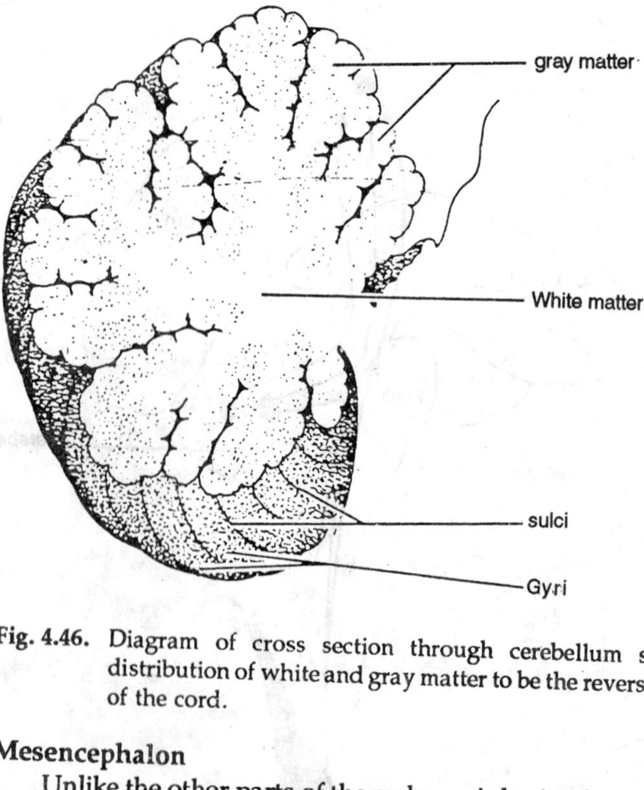

Fig. 4.46. Diagram of cross section through cerebellum showing distribution of white and gray matter to be the reverse of that of the cord.

Mesencephalon

Unlike the other parts of the embryonic brain, the midbrain does not divide further. In lower animals it is linked to the visual sense and is an important association center. In higher forms many of its activities are taken over by the cortex, and it remains primarily a tract to and from these higher centers.

In cross section, the mesencephalon consists of the thickened dorsal region of gray matter, the *tectum*, and the thinner walls and floor of gray matter referred to as the *tegmentum*. The cavity of the mesencephalon decreases during development, forming the narrow *aqueduct*. It communicates posteriorly with the fourth ventricle and anteriorly with the third ventricle. In lower animals the tectum, of roof, an important visual center, appears as a pair of dorsolateral *optic lobes (corpora bigemina*). The fibers of the optic nerve enter the diencephalon and then pass into the roof of the mesencephalon. Fibers from the spinal cord and medulla also terminate in these lobes. The optic lobes are thus important correlating centers for optic and exteroceptive impulses. The tectum is most important in fishes and amphibians, a little less important in reptiles and birds, and least important as a visual center in the mammal. If the cerebral hemispheres are removed from a higher animal, it can no longer function. A frog, on the other hand, after recovering from such an operation, can continue functioning in an apparently normal fashion as long as the brainstem remains intact.

The tectum of mammals is reduced to four prominent swellings known as the *corpora quadrigemina*. The larger anterior pairs are the *superior colliculi* (containing visual receptive centers); the posterior pairs are the *inferior colliculi*) (integrating centers for auditory sense). Although optic and auditory fibers are still sent to the superior and inferior colliculi, the ascending spinal tracts, in general bypass the mesencephalon of mammals and continue anteriorly to the diencephalon, or higher centers. The highly evolved cerebral cortex of mammals takes over the major visual and auditory control, leaving the tectum as essentially a reflex center for light and sound.

Nuclei of the third and fourth cranial nerves are located in the gray matter of the midbrain. Ascending and descending fiber tracts from the rear and from the anterior cerebral cortex form the tegmentum. These are more prominent in mammals than in some of the lower animals, as might be expected, and constitute the major bulk of the mesencephalon. The majority of these fibers originate in the cerebrum and pass by way of the midbrain to the pons, medulla, or spinal cord.

Diencephalon

The diencephalon, or the tween-brain represents an ancient reflex center. It can be divided into a thin roof, the *epithalamus*; the thickened sides, the *thalamus*; and the floor, the *hypothalamus*, each part carrying out *specialized* functions. The third ventricle forms the cavity of the diencephalon. Several dorsal and ventral outgrowths arise from the diencephalon.

Epithalamus

The anterior portion of the roof of the diencephalon resembles the covering of the fourth ventricle in the medulla. It is a thin, non-nervous, vascular membrane that extends into the third ventricle and contains the *anterior choroid plexus.* Along with the posterior choroid plexus, it forms the cerebrospinal fluid, which fills the cavities of the nervous system. In some forms, a fold of the membrane gives rise to the paraphysis, a structure of unknown function.

The posterior portion of the roof of the diencephalon evaginates to form a pair of median stalklike structures, the *anterior parapineal body* and a *posterior epiphysis,* or *pineal structure.* The function of the pineal has long intrigued the biological investigator, but Turner and Bagnara (1971) state that the "time has probably arrived when it is justifiable to consider the pineal as an endocrine gland." There appears to be a decided seasonal interaction between the pineal and the reproductive organs, but other functions are not consistently present among the classes of vertebrates. According to a symposium presented by comparative endocrinologists of the American Society of Zoologists (1970), the pineal may control some basic homoiokinetic mechanism that allows the organism to adjust seasonally to its environment.

Thalamus

The thalamus is an integrating center for fibers that pass to and from the cortex, and it contain several important areas (nuclei). The walls of the diencephalon can be divided into a dorsal and a ventral half. In general, the cells located in the dorsal half are involved with sensory impulses and, next to the cerebrum, reflect the greatest evolutionary change. All sensory pathway, with the exception of the olfactory one, pass to the cerebral cortex of amniotes by way of these tracts. In lower animals the dorsal thalamus is of less importance. The ventral half of the thalamus is concerned with motor impulses.

Hypothalamus

The hypothalamus is of great importance, since it controls involuntary actions relating to sleep, temperature regulation, rate of breathing, genital impulses and metabolism. In the human being it weighs a little more than 7 grams. It is possible to identify four areas.

The *optic chiasma* is located in the anterior parts of the hypothalamus. It is here that the optic nerve fibers cross to opposite sides of the brain before they proceed to terminate in the tectum of the midbrain of the lower animal. In higher animals, they pass to special areas in the gray matter of the cortex. The *tuber cinereum*, believed to be the center of the parasympathetic system, is located behind the optic chiasma. Posterior to this center are the *mammillary bodies*, centers of olfactory function.

The fourth area, the *infundibulum*, grows ventrally between the optic chiasma and mammillary body to meet the ectodermal material from the stomodeal region. Together they give rise to the pituitary gland of the individual. The infundibulum contributes to the neurohypophysis (posterior pituitary and median eminence), and the stomodeal ectoderm contributes to the adenohypophysis, consisting of the anterior and intermediate lobes of the pituitary.

The hypothalamus regulates the function of the anterior pituitary gland, not by way of nerve fibers but by releasing into the portal blood vessels chemicals that stimulate or inhibit its activity. According to present information, the hypothalamus secretes nine regulating or releasing hormones (RH) that affect the pituitary gland. Most of them stimulate the pituitary to both synthesize and release particular hormones. Some, however, inhibit the production of pituitary hormones.

A dual system of hypothalamic control regulates three of the pituitary hormones: the growth hormone, prolactin, and melanocyte-stimulating hormone. One factor stimulates the synthesis and release of the hormone and one inhibits its production. These hormones need a dual regulatory system since their target organs release no negative- feedback products. On the other hand, the thyroid, gonads, and adrenals all produce hormones. The circulating levels of these hormones inhibit the pituitary and hypothalamus secretions by a negative-feedback mechanism. Therefore, only a hypothalamic-releasing factor controls each of these glands.

The hypothalamus also synthesizes oxytocin and vasopressin. The long axonal fibers transport these hormones to the posterior lobe of the pituitary where they are stored until needed.

Telencephalon

Of all the parts of the brain, the telencephalon, or forebrain, evolved the most. Originally a single anterior expansion of the neural tube, it divides into the olfactory lobe and the cerebral hemisphere. In tetrapods, the hemisphere increases in size and separates into two definite lateral swellings, the *cerebral hemispheres*. The olfactory sense is the most important sense influencing behaviour in fish, amphibians, and most reptiles. In primates, on the other hand, the sense of smell is of less importance. The story of the evolution of the telencephalon involves the structural and functional expansion of its cerebral part. This progressive enlargement of the cerebrum in higher groups is diagrammed. The function of the telencephalon shifts from a primarily olfactory one in primitive forms to that of the dominate association and co-ordinating center for all the senses in the mammals. The cerebral hemispheres are responsible for the complex thought, intelligence, and acute sensation associated with man.

The evolutionary history of the telencephalon involves a shift in the location of gray and white matter as compared with that of the spinal cord. The cells of the gray matter move progressively outward as the phylogenetic scale is ascended. The migration of the gray matter outside the white matter occurs slowly and is associated with the growth of a new area, the neopallium. As a result of this reversal of position, the cells of the gray matter occupy a peripheral position in the telencephalon of mammals where there is room for greater surface expansion. The white fiber tracts occupy the central position.

When viewed in cross section, the forebrain of vertebrate embryos consists of two general areas surrounding the *telocoel*, or ventricle; the dorsal roof, or *pallium;* and a thickened floor, the *basal area.* The pallium becomes greatly modified during evolution; it is the primordium of the cerebral cortex. The basal area develops into the *basal nuclei,* essentially equal to the *corpus striatum* of mammals. It forms a part of the brain most concerned with basic instinctive behaviour patterns. Certain cells representing additional sense centers appear in the basal nuclei of the more active fish.

The pallium is thin in the lower vertebrates. Only a small amount of gray matter line the ventricle; the rest of the pallium consists of an outside covering of white fibers. In the primitive vertebrate and in the early embryo of all types, the telencephalon consist of a single cavity. Bilaterally located olfactory bulbs project forward from the forebrain toward the nasal sac. In the cyclostomes the pallium is related entirely to the sense of smell and the basal nuclei is poorly developed.

Even in the cyclostomes, the olfactory bulb of fishes separates from the forebrain and extend toward the olfactory sacs. In the elasmobranchs, for example, a slight longitudinal depression indistinctly divides the single telencephalon into two anteroventral *olfactory lobes* and two posterodorsal *cerebral hemispheres.* The olfactory lobes have only an olfactory function; the hemispheres are probably mostly olfactory, but they may contain a few other centers. Attached to the olfactory lobes are two stalklike anterior *olfactory tracts,* which pass into the rostrum of the fish, terminating in the *olfactory bulbs.* The bulbs are the true anterior ends of the brain. The receptor cells located in the epithelium of the adjacent olfactory sac send their fibers into the olfactory bulb. These fibers form the olfactory, or *first cranial nerve.*

The paired amphibian hemispheres are larger than those of the fish, but it is difficult to differentiate between olfactory lobes and cerebral hemispheres. In these forms the pallium divides into two general regions, a dorsomedial *archipallium* and a lateral *paleopallium.* Although the archipallium continues to be predominantly an olfactory center, the new association center of other sensory impulse first appear here in the amphibians; the paleopallium and the basal nuclei remain olfactory center. The changes that occur in the amphibian correlate with the transition to a terrestrial life. Dependence on the sense of smell begins to diminish, the sensitivity to light, sound, and cutaneous stimulation increases. The shift of additional sense centers from a more posterior area into the cerebral hemispheres represents another trend in the evolution of the mammalian brain. It is also in the amphibians that a few cells from the gray matter begin to move away from the ventricle and into the walls of the pallium. Most of gray matter of amphibians, however, is internal. All of the areas receive fibers from the olfactory center and the thalamus. Fibers that connect the archipallium with the paleopallium and with the basal nuclei also develop, and so these regions are able to correlate olfactory function with a few other simple impulses , possibly gustation. The basal nuclei relays impulses back to the thalamus and to the midbrain. It begins to form an important correlation center that increases in complexity in more advanced groups.

A fourth region, the *neopallium,* differentiates laterally to the archipallium in some reptiles and all mammals. According to Romer, "the evolutionary history of the mammalian brain is essentially a story of neopallial expansion and elaboration."

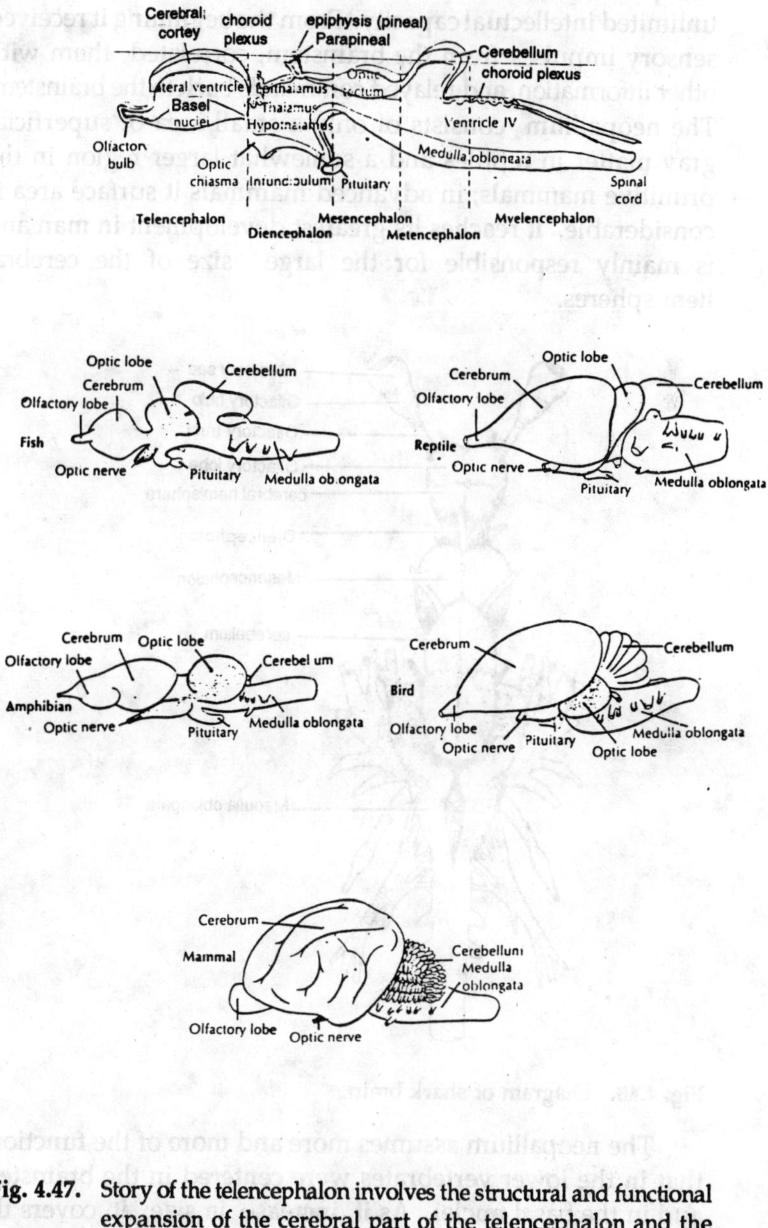

Fig. 4.47. Story of the telencephalon involves the structural and functional expansion of the cerebral part of the telencephalon and the restriction of the olfactory center.

It represents a complex association center that provides unlimited intellectual capacity. From the beginning it received sensory impulses from the brainstem, correlated them with other information, and relayed commands back to the brainstem. The neopallium consists of only a small area of superficial gray matter in reptiles and a somewhat larger region in the primitive mammals; in advanced mammals it surface area is considerable. It reaches its greatest development in man and is mainly responsible for the large size of the cerebral hemispheres.

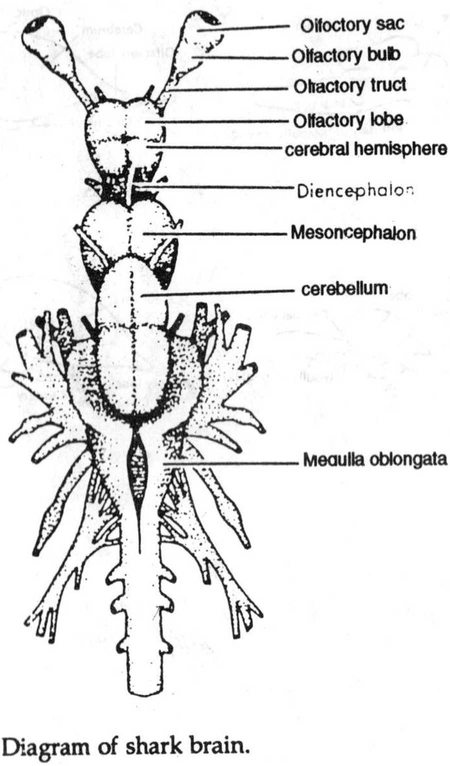

Fig. 4.48. Diagram of shark brain.

The neopallium assumes more and more of the functions that in the lower vertebrates were centered in the brainstem and in the basal nuclei. As it increase in size, it covers the

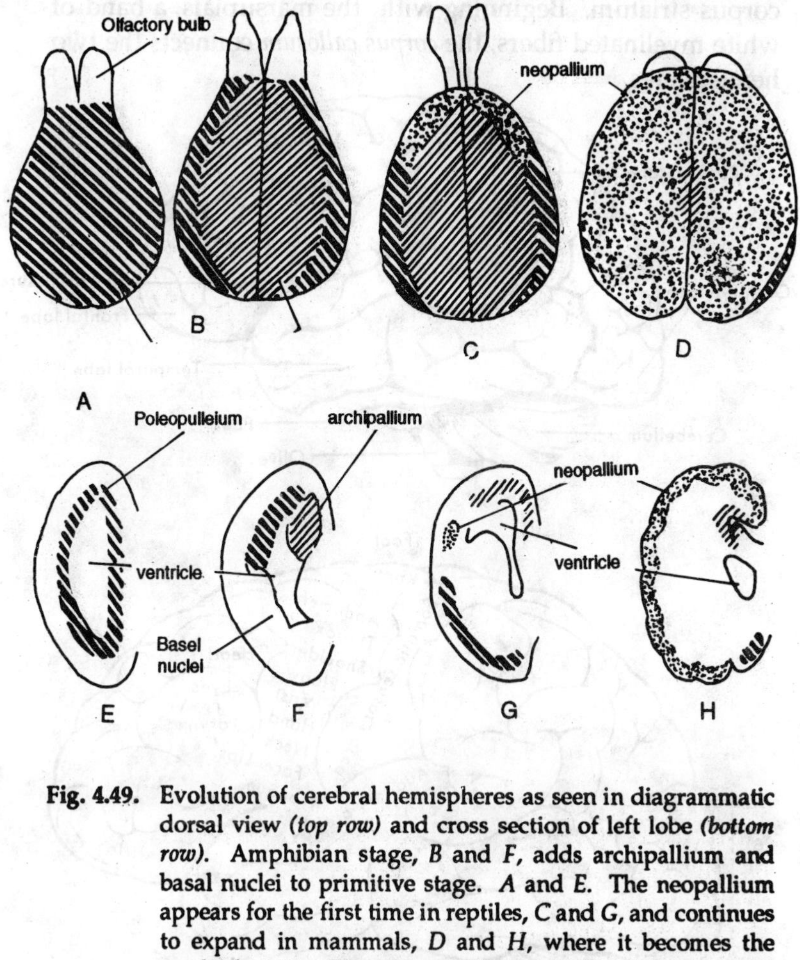

Fig. 4.49. Evolution of cerebral hemispheres as seen in diagrammatic dorsal view *(top row)* and cross section of left lobe *(bottom row)*. Amphibian stage, *B* and *F*, adds archipallium and basal nuclei to primitive stage. *A* and *E*. The neopallium appears for the first time in reptiles, *C* and *G*, and continues to expand in mammals, *D* and *H*, where it becomes the cerebral cortex. The corpus striatum and archipallium are pushed to the interior, and the olfactory lobe diminishes in significance.

other areas, pushing them to the inner part of each hemisphere. In the majority of mammals it overlaps the midbrain and part of the cerebellum. The paleopallium is restricted to a small ventral area, the *piriform lobe*. The well-developed corpus striatum or basal nuclei moves toward the interior of the hemispheres, and the archipallium is forced medially, where it is known as the *hippocampus*. Automatic reactions remain centered in the corpus striatum. Beginning with the marsupials, a band of white myelinated fibers, the *corpus callosum* connects the two hemispheres.

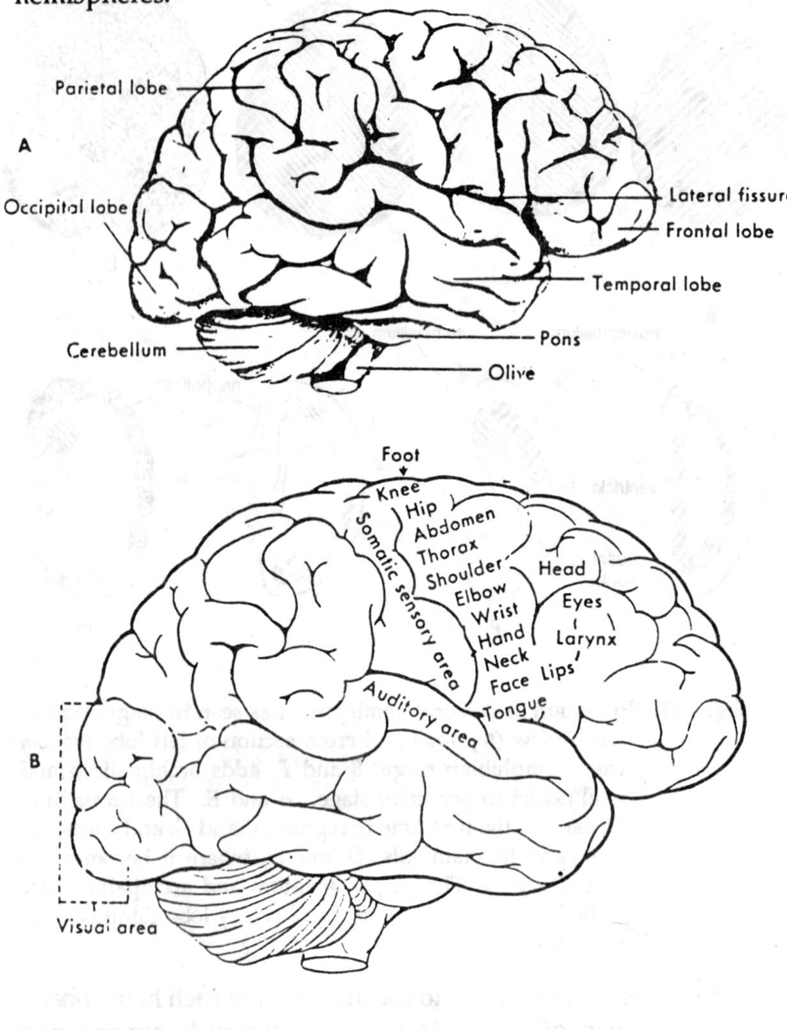

Fig. 4.50. Convolutions of mammalian cerebral cortex divide it into temporal, occipital, frontal, and parietal lobes, *A*. Except for the frontal area, specific functions have been ascribed to specific areas as indicated in *B*.

As expansion continues, the cells of the gray matter composing the neopallium migrate nearer the surface. Here they become concentrated and are known in mammals as the *cerebral cortex.* The walls of the cortex (neopallium) measure in man about 1/2 to 1/12 inch in thickness. As many as six different layers of neurons, with millions of neurons in each layer make up the gray matter of the placental animals cortex. The cells of the various layers of this continues blanket enter in to complex connections with one another by means of their axons and dendrites. It is impossible to comprehend the exact number of correlations that can be made.

The roof of the cerebral hemispheres is smooth in many of the lower vertebrates. In mammals, however, the more rapid increase in number of neurons at the surface of the neopallium as compared to the deeper lying parts, as well as the increase in underlying fiber volume causes the surface to be thrown into convolutions. The folds are referred to as *gyri;* the indentations are *sulci.* The convolutions divide the brain into various general areas. They are identified according to the skull bones surrounding them and are referred to as the temporal, occipital, frontal, and parietal lobes of the cerebral cortex. With the exception of the frontal lobe, specific functions are ascribed to specific areas. For example, the occipital lobe is the center for vision; the rest of the centers are identified.

As a result of the progressive development of the neopallium, the tectum of the mesencephalon loses its co-ordinating function. It now consists primarily of a center that relays information to the new association centers in the cortex. The somatic impulses relayed by the thalamus to the corpus striatum of lower animals are carried through and beyond the corpus striatum by fibers that now terminate in the neopallial surface. The cortex receives all data, integrates and co-ordinates the information, and then relays the commands back to either the cerebellum, the corpus striatum, or the hypothalamus, or by way of the pyramidal tracts to the brainstem and spinal cord.

PERIPHERAL NERVOUS SYSTEM

Autonomic Nervous System

The *autonomic nervous system* is part of the peripheral nervous system and includes both sensory and motor neurons, since it functions reflexively. It is usually further divided into *sympathetic* and *parasympathetic* systems, which have antagonistic functions. In general, the term "autonomic nervous system" refers to the visceral motor fibers that innervate the smooth muscle of the internal organs, that is, those involuntary muscles of the heart, blood vessels, and digestive, urinary, and reproductive organs. The visceral sensory system differs from the somatic sensory system only in that it carries impulses to the cord from the viscera rather than from the body wall. However, the input of any somatic sensory nerve can also produce an automatic motor response. For example, something that an individual sees and finds disturbing can cause his heartbeat rate to increase. The visceral motor system differs distinctly from the somatic motor system; in fact, most authors define the autonomic system as consisting of only these visceral efferent fibers. Sometimes the student has the impression that the autonomic system is an independent system, but it should be considered as a functional division of the whole. Its control is located in the central nervous system and it is intimately bound together with the rest of the nervous system.

Comparison of a Somatic and Autonomic Nervous Systems

Earlier in the chapter we described a typical spinal nerve as being composed of somatic and visceral fibers. Certain functional and structural characteristics help to differentiate the autonomic (visceral) system from the somatic.

When an impulse leaves the somatic motor nuclei in the cord, it passes down an axon to a muscle or gland. Should that muscle be located in the big toe, the axon extends from its cell body in the posterior regions of the cord to the big toe, a distance of perhaps 3 feet. Only *one* neuron is involved between

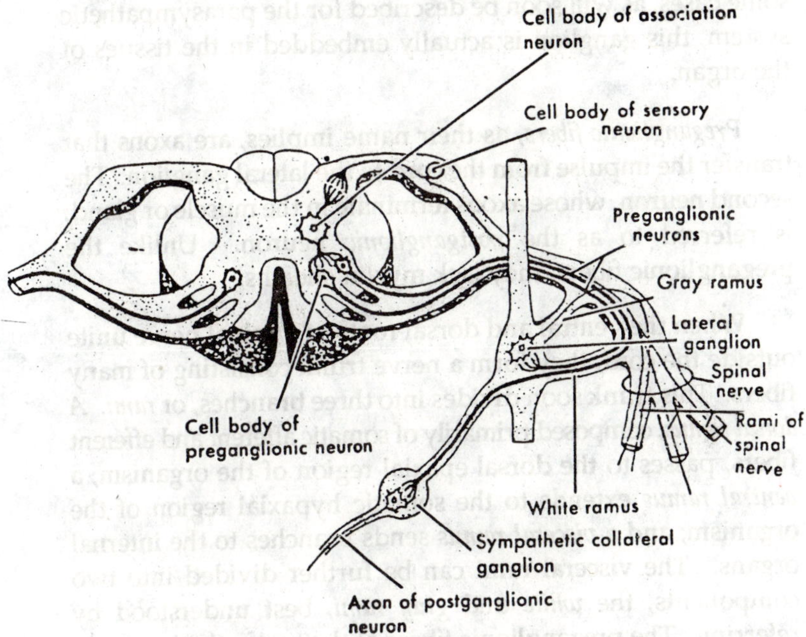

Fig. 4.51. Diagram showing visceral motor pathway from spinal cord to viscera. Two neurons are involved. The cell body of first neuron is located in spinal cord; its axons synapse at lateral ganglion or at sympathetic ganglion. A postganglionic fiber carries impulse to viscera. The myelinated fibers of the preganglionic fiber form the white ramus. The non-myelinated fibers of the postganglionic fiber compose the gray ramus.

the cord and the muscle. In the autonomic system, however, the impulse is conveyed from the cord to the effector organ via two neurons. The cell body of the first neuron is typically located in the visceral motor area of the gray matter, previously described. Its axon may extend to the *lateral ganglia*, lined up in a chain outside the central nervous system on either side of the vertebral column, or it may extend to the *collateral ganglia* located farther away, or even to the organ itself. The ganglia are those described as originating from the neural crest cells. In the ganglion the first neuron may synapse with a second neuron, the axon of which passes to the organ it innervates. In some cases, as will soon be described for the parasympathetic system, this ganglion is actually embedded in the tissues of the organ.

Preganglionic fibers, as their name implies, are axons that transfer the impulse from the cord to the lateral ganglion. The second neuron, whose axons terminate in the muscle or gland, is referred to as the *postganglionic* neuron. Unlike the preganglionic fibers, they lack myelin sheaths.

When the ventral and dorsal roots of a spinal nerve unite outside the cord, they form a nerve trunk consisting of many fibers. This trunk soon divides into three branches, or *rami*. A *dorsal ramus*, composed primarily of somatic afferent and efferent fibers, passes to the dorsal epaxial region of the organism; a *ventral ramus* extends to the somatic hypaxial region of the organism; and a *visceral ramus* sends branches to the internal organs. The visceral rami can be further divided into two components, the *white* and *gray rami*, best understood by referring. The preganglionic fibers of the sympathetic system leave the cord in the ventral root along with somatic motor fibers, but then pass to the lateral ganglia by way of the visceral rami. The myelinated preganglionic fibers make up the white ramus. In the ganglion some of the preganglionic fibers synapse with postganglionic neurons, and these non-medullated (non-myelinated) postganglionic fibers give rise to the gray ramus.

As stated above, all sympathetic fibers do not synapse at the lateral ganglion; some may pass through the ganglia and extend anteriorly or posteriorly in the sympathetic chain before synapsing at levels other than where they originated. A third possibility exists for the preganglionic fiber to pass through the lateral chain ganglia to certain large ganglia located some distance away known as the *collateral ganglia*. The celiac, anterior mesenteric, and posterior mesenteric ganglia compose these collateral ganglia. Regardless of the location of the synapse, any one preganglionic neuron synapses with several postganglionic neurons. The latter, in turn, may then carry the impulse to several different organs. As a result sympathetic responses are widespread and involve many different structures.

Sympathetic and Parasympathetic Divisions

On the basis of function, structure, and location, the autonomic system of higher animals is further divided into two main divisions, the *sympathetic system* and the *parasympathetic system*. Both types of fibers innervate most internal organs. The two systems can be differentiated in several ways.

They differ in their physical relationship to the central nervous system. The preganglionic fibers of the sympathetic system leave the cord in the thoracolumbar regions. For this reason they are sometimes referred to as the *thoracolumbar autonomic system*. However, the preganglionic neurons of the parasympathetic fibers are mostly located in the medulla oblongata. Included are a few fibers to the iris and ciliary body that travel with the third cranial nerve, as well as a part of the facial (seventh), the glossopharyngeal (ninth), and the vagus (tenth) nerves. Also, nerves from the posterior parts of the cord belong to this system and send fibers to the large intestine, bladder, and reproductive tract. As a result, the parasympathetic portion of the autonomic system is referred to as the *craniosacral system*.

The locations of the synapses between preganglionic and postganglionic fibers is a second difference between the sympathetic and para-sympathetic systems. The preganglionic fibers of the parasympathetic system do not synapse with the postganglionic fibers in the lateral or collateral ganglia as described above for the sympathetic system. Instead, they pass directly to the organs they innervate. Here, either in or near the organ itself, synapses occur with the second neuron. Since synapsis occurs some distance from the cord, the axons of the preganglionic neurons of the parasympathetic system are long compared to those of the sympathetic system that synapse just outside the cord. Of course, the postganglionic fibers are short in the parasympathetic system and long in the sympathetic system.

The two systems have opposite effects on the organs that they innervate. On the whole, the sympathetic system alerts the body for activity. It prepares it to meet an environmental challenge by speeding up its heartbeat and respiration rate and stimulating the integumentary glands, hair muscles, and skin capillaries. It depresses those activities not conducive to body activity, such as digestion, excretion, etc. The parasympathetic system, on the other hand, is sometimes referred to as the vegetative system, since it produces a reverse effect. Neither system is entirely stimulative or repressive, but can produce both excitatory or depressant effects depending on the structure under consideration. Under normal circumstances, the organism co-ordinates their effects and maintains a homeostatic state, with either system being able to take over temporarily if the occasion demands it.

The systems produce their effect by liberating certain chemicals, acetylcholine or norepinephrine, at their end organs. These were referred to as a transmitter substances earlier in the chapter and are given the general term of *neurohumors.* The ends of all somatic neurons liberate acetylcholine. It is also released by the preganglionic fibers of both sympathetic and para-sympathetic axons and by the axons of the parasympathetic postganglionic neurons. Such neurons are referred to as *cholinergic nerves.* Epinephrine-like (adrenaline-like) compounds are liberated by the postganglionic endings of the sympathetic system and consequently this system is referred to as *adrenergic.*

CRANIAL NERVES

General Characteristics

The nerves that arise from the ventral surface of the brain and emerge through special openings (foramina) of the skull are referred to as *cranial nerves.* Together with the spinal nerves, they complete the peripheral nervous system. There are 10 pairs present in most anamniotes and 12 pair in reptiles, birds and mammals. The nerves are numbered starting at the anterior end. Roman numerals are often used. The name, number, origin, and distribution of the nerves from fish to man are similar, although, of course, where structures have been added or deleted through evolution, modifications occur. For example, nerves pass to the gills of aquatic animals. When amniotes undergo development, these nerves are either lost or they innervate the structures that evolved in the gill region. Although absent in man, a terminal nerve is present in most vertebrates, anterior to the first cranial nerve. Since it was discovered after the other nerves were numbered, it is usually labeled 0.

Earlier in the chapter, we described four functional components of the spinal nerves. These same general components are present in the cranial nerves. The cell bodies of the cranial motor nerves are located in nuclei in the brainstem. The cell bodies of the sensory neurons collect in the ganglia near the brain. Somatic sensory fibers send twigs to the skin of the head and other somatic structures. Visceral sensory fibers relay information from the membranes of the pharynx, the anterior part of the gut. Somatic motor fibers terminate in structures derived from myotomes, for example, the eye musles, and visceral motor fibers passto the muscles of the gut. However, a special group of nerves, distinguished from the general nerves described above form an additional category of cranial nerves. They are the *special somatic sensory, special visceral sensory,* and *special visceral motor nerves.*

The terms "special" and "special visceral" need further clarifications. The nose, eye, and acoustolateralis organs are defined as special sense organs, distinguished from the simpler receptors. The sensory nerves that pass from these organs to the brain retain this term and are known as the special somatic sensory nerves. A special visceral sensory nerve consists of fibers from the taste organ. The muscles that work the gill bars of fish are a special kind of muscle. They are not derived from myotomes and, although they are formed from lateral plate mesoderm, they differ from the smooth muscle of the gut (of splanchnic mesodermal origin). These gill muscles, referred to as visceral muscles, are striated, and in this respect are similar to skeletal muscles. They are innervated by cranial nerves known as special visceral motor nerves.

Classification of Cranial Nerves

In order to minimize rote memorization of names and functions of the cranial nerves, we advise grouping them in different ways. For example, two of the nerves, cranial nerve 1 (olfactory) and cranial nerve II (optic), are considered atypical nerves, since their embryonic origins do not conform to the usual pattern. These sensory nerves do not possess ganglia. The cell bodies of most sensory nerves are located near the spinal cord or brain, but cranial nerve I does not actual develop from neural tissue. Instead, the olfactory receptor cells, derived from the olfactory placode, send their projections into the olfactory lobe of the brain. These fibers combine to form the olfactory nerve. Cranial nerve II is actually a brain tract, since the retina is an evagination of the diencephalon and is considered part of the brain. The optic nerve forms by a combination of fibers from the ganglionic cells located in the wall of the retina. These cells send their fibers back into other regions of the brain.

It is also helpful to remember that the forebrain arose in relation to the olfactory sense, the midbrain with the visual sense, and the hindbrain with the acousticolateralis sense. Cranial nerves can be grouped therefore. According to the region of the brain with which they are associated. The fibers of the olfactory nerve extend to the telencephalon and the optic fibers extend to the diencephalon and then on to the tectum of the mesencephalon; cranial nerves III and IV arise from the mesencephalon, and nerves V to XII from the medulla oblongata.

The cranial nerves can also be divided into three general groups. Cranial nerves 0 (terminal nerve), 1 (olfactory), II (optic), and VIII (auditory) are all sensory nerves. Motor nerves consist of cranial nerves III (oculomotor), IV (trochlear), VI (abducens), and XII (hypoglossal). The remaining nerves, cranial nerves V (trigeminal), VI (fascial), IX (glossopharyngeal), X (vagus), and XI (spinal accessory), are all mixed nerves.

Specific Cranial Nerves

The following cranial nerves are grouped according to their functional components:

Somatic sensory

0, *Terminal nerve.* This is a small nerve of unknown function that runs parallel to the olfactory nerve but enters the diencephalon. It is apparently sensory in function, since it bears one or more ganglia. It may represent the remnants of an anterior nerve existing in ancestral vertebrates.

Special sensory. These nerves transmit impulses from the special sense organs.

I, Olfactory nerve. This is a sensory nerve that carries impulses from the olfactory sac to the telencephalon. It has no ganglion, but develops from the cells of the epithelial lining of the olfactory sac.

II, Optic nerve. This is a sensory nerve that carries impulses from the eye to the higher regions of the brain. The fibers from each side of the head unite in the floor of the diencephalon where some of them cross to the other side and give rise to the optic chiasma. The fibers then continue in the brain as optic tracts passing to the mesencephalon (in lower animals) or to the cerebral cortex (in higher forms).

VIII. Acoustic nerve. The ganglion giving rise to this nerve divides into two parts in higher animals: the spiral ganglion with fibers passing to the cochlear duct, and the vestibular ganglion that sends fibers to the semi-circular canals and the sacs of the inner ear. The lateral line nerve is closely related to the acoustic nerve. It enters the medulla of fish with nerve VIII, but is usually considered to be more closely combined with nerve VII.

Somatic motor. These nerves carry impulses to the muscles that move the eyes and to the tongue muscles. Apparently they lost their dorsal roots; the remaining ventral roots compose the nerve.

III. Oculomotor nerve. This nerve arises from the mesencephalon and supplies the eye muscles that are derived from the first cranial myotome: the inferior oblique and the superior, inferior, and medial rectus muscles. A few fibers from nerve V usually accompany nerve III as they pass to the smooth muscles of the iris and ciliary body. However, these fibers are usually not considered a part of nerve III.

IV. Trochlear nerve. This nerve arises from neuroblasts in the floor of the mesencephalon just posterior to nerve III; it innervates the eye muscle derived from the second myotome. Its fibers pass upward in the wall of the brain where they then pass dorsally and cross to the opposite side. They emerge dorsally from the brain and pass to the superior oblique muscle of the eye.

VI. Abducens nerve. The abducens nerve originates from the motor nucleus located in the pontine region of the mesencephalon. It innervates the external rectus muscle of the eye, derived from the third myotome.

XII. Hypoglossal nerve. The hypoglossal nerve has several origins. The ventral roots of three or four nerves from the occipital region (back of head and neck) leave the medulla oblongata and merge into one nerve. This nerve innervates the occipital myotomes, which in the fish embryo migrate beneath the gill region to form the hypobranchial muscles. In higher forms, when the gills are lost, this region is remodeled. These muscles are lost in amphibians but contribute to the tongue musculature of amniotes. The hypoglossal nerve therefore innervates the tongue.

Mixed nerves

Mixed nerves all arise from the medulla oblongata and each is related to a gill arch of the fish. As a result, they are sometimes referred to as the *branchial nerves*. When amniotes lose their gills, the nerves innervate the structures derived from the gill area. Each nerve usually includes somatic sensory as well as special visceral sensory fibers and special visceral motor fibers. Cranial nerves V, VII, IX and X are all considered mixed nerves. Originally, each nerve consisted of three main branches, a pharyngeal branch passing inward to the pharynx, a pretrematic branch running in front of the gill slit, and a posttrematic branch running in back of the gill slit.

V. Trigeminal nerve. This is a stout nerve that has a large semilunar (or gasserian) ganglion located outside the hindbrain. It is usually divided into three main branches: the ophthalmic, maxillary, and mandibular nerves. The fibers are primarily sensory from the face, nose, and mouth; they enter the brain at the somatic sensory column of the gray matter of the medulla. Motor fibers pass to the muscles of mastication. Since these muscles are derived from the muscles of the original first branchial arch, the nerves are considered special visceral motor nerves.

VII. Facial nerve. In fish, the facial nerve passes to the second arch and the spiracular gill slit (considered the second gill slit). It contains both somatic and visceral sensory, as well as visceral and special visceral motor fibers. Some of thee fibers become intimately associated with fibers from nerve V and pass along with them to the anterior regions of the head. This nerve is called the facial nerve because in mammals the muscles of the second arch become the muscles of expression and retain the same innervation. There is a sensory ganglion, the geniculate ganglion, housing the cell bodies of the sensory portion of the nerve. The visceral sensory fibers pass to the taste buds and to the lining of the pharynx. The large somatic sensory portion of the nerve passes to a lateral line system.

IX. Glossopharyngeal nerve. In fishes this nerve passes to the first functional gill and therefore contains both special visceral sensory and special visceral motor fibers. Its sensory cell bodies are located in two ganglia, one near the myelencephalon, the superior ganglion, and the other more laterally located, the petrosal ganglion. In fishes it contain both a pretrematic and a posttrematic branch; in mammals this nerve is lost to a great extent. Its visceral motor fibers innervate some of the muscles of the pharyngeal areas and parotid salivary gland; its visceral sensory fibers pass to the taste buds.

X, Vagus, and *XI, spinal accessory nerve.* The vagus is one of the most important nerves. It is a composite nerve formed by the union of several branches that pass to the posterior gill arches, those numbered four to six. When the amniotes lose their gill arches, some of these branchial fibers degenerate, but a few remain and innervate those structures derived from this region, such as the pharyngeal musculature. The main branch of the vagus is visceral and extends backward into the body along the gut where it gives off branches to the stomach, heart, lungs and anterior intestine. It carries the visceral impulses to the medulla oblongata. It also relays visceral motor impulses back to these organs by way of the preganglionic branch of the parasympathetic system. Since the vagus represents a union of several nerves, it has several sensory ganglia as might be expected. A motor portion of the vagus of lower animals divides into a separate nerve. In mammals, it is known as nerve XI, or the spinal accessory nerve.

RECEPTORS

Animals are living in a very fluctuating environment. If an animal is to survive successfully, it must be responsive to the condition of both its internal and external environments and to any changes that occur in them. In the primitive, one-celled animal, this information is easily made known to the active part of the organism, since all cells posses irritability. However, communication between various parts of the multicellular animal becomes more of a problem. As the result of neurulation, the central nervous system (the brain and spinal cord) moves from the surface of the animal to the interior, and the neural crest cells that form the ganglia of the sensory nerves end up in proximity to the brain and the cord. The epidermal ectoderm develops into an integument that covers, insulates, and protects the individual. The central nervous system, therefore, usually cannot be directly stimulated by any change that goes on in or around the organism. A sensory receptor system evolved in the multicellular vertebrate that detects changes in the internal or external environment and transmits the information to the central co-ordinating areas. These areas, in turn, send appropriate instructions for action and organs. If a prompt reaction is necessary to remedy a temporary situation, the message to react is usually sent through the proper nerve to the muscle. On the other hand, if the body needs a slower more generalized adjustment to some situation, the endocrine system functions in producing the response. In general, sense organs are associated with the nervous system.

Sensory receptors are hypersensitive, intermediary structures, located between the environment and central nervous system, that receive and respond to stimuli. They are usually specialized cells that are sensitive to specific chemical or physical stimuli and are associated with nerves that relay the information to the central nervous system. Each receptor responds to one particular kind of stimulus, and depending on the nature of the stimulus receptors are classified as *photoreceptors* (those that respond to visible radiation), *chemoreceptors*, (those sensitive to dissolved chemical materials), *mechanoreceptors* (those stimulated by touch or pressure), *thermoreceptors* (those that react to temperature), *statoreceptors* (those stimulated by changes in body position), and *phonoreceptors* (those stimulated by

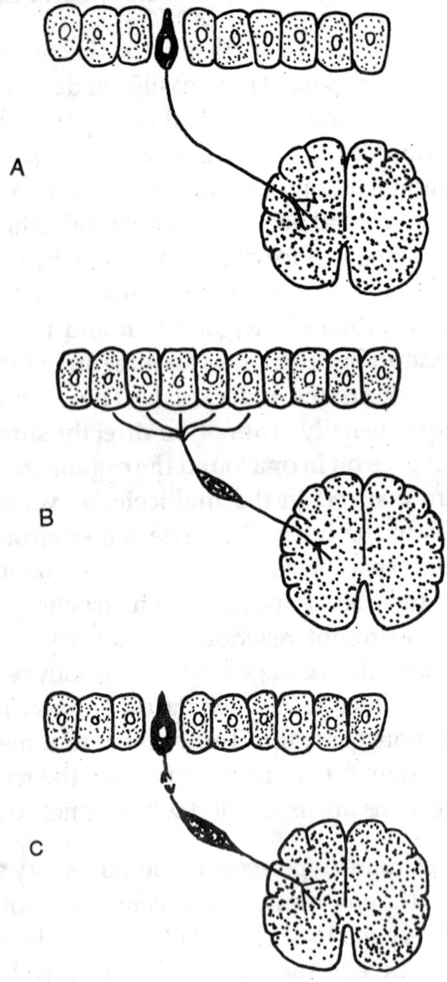

Fig. 4.52. Sense organs can be classified according to their morphological types and levels of organization. At the simplest level, *A*, the epithelial cell serves as both receptor of stimuli and conductor of impulse. At the second level, *B*, the body of the cell migrates inward and cell processes react to the stimuli and transit information. A second cell is inserted at the surface in the third level, *C*, and reacts to stimuli, passing the impulse on to the nerve cell.

vibrations). The sensory receptors develop from ectoderm and their origin may be traced to epidermal ectoderm, neural ectoderm, or neural crest cells. In some instances, a combination of ectodermal types compose the sense organ; often mesoderm, in the form of connective tissue capsules, combines with the ectodermal cell. The receptors are associated with the somatic sensory and visceral sensory nerves. The somatic sensory nerves carry impulses from the cutaneous surface and muscles of the body wall to the central nervous system; visceral sensory nerves relay impulses from the viscera.

The higher centers of the brain interpret the impulses as *sensations* and the individual experiences cold, pain, warmth, sight, sound, etc. The receptor do not perceive anything; they merely refer information to the nervous system where it is sorted and integrated. Obviously most of our insight into sensation is based on man's *reaction to stimuli*, since there is no way of knowing whether the toad or the dog possesses similar sensations. There may be many sensations experienced by lower animals that man is unable to comprehend and, therefore, he has no way of knowing about them. The fact that similar structures (the eye, for example) originate in all vertebrate embryos in a similar manner and that they give rise to organ that are anatomically similar in the adult lead biologists to believe that some sensations, at least, are shared by both lower and higher animals. Those receptor located over the general body are referred to as the *general sensory organs*, although in some cases bare fibers serve as the receptors and the term "organ" is applied from general use. Other receptors concentrate only in special areas of the body and react to very specific stimuli. They are known a the *organs of special senses:* the organs olfaction, sight, hearing and equilibrium.

We classify sense organs according to their morphology and usually recognize at least three levels of organization. These levels also reflect their evolutionary status. At the simplest levels, the epithelial cell, located at the surface of the animal, serves as both receptor of stimuli and conductor of nerve impulses. At the second level, the cell body migrates inward and the modified cell processes react to the stimulus and transmit the information. In the third type of organization, a second cell, inserted in the pathway and located at the surface of the animal, reacts to the stimulus. The generated impulse passes to the nerve cell, which, in turn, transmits it to the brain.

Some organs are classified according to their functions. *Exteroreceptors* are those receptors that receive information about the external environment; *proprioceptors* are those located in the striated muscles and tendons. *Interoceptors* are associated with internal organs.

GENERAL SENSORY ORGANS

The receptors composing the general sense organs are less sensitive to stimuli than those of special senses. They react to touch, pressure, pain, temperature, position, movement, and the viscera. A variety of these stimuli apparently produce the sensation of pain. The most primitive type of sensory cell, found primarily in invertebrates, consists of a modified ectoderm cell located in the superficial layers with long processes extending inward. The vertebrate's olfactory organs (which we shall discuss under the topic of special sense organs) consist of such a simple arrangement. In general, the cell bodies of the general sensory organs do not remain in the superficial ectoderm of the organism but move inward and become associated with the brain and spinal cord. They belong to the second level of organization mentioned above. These cell differentiate from neural crest cells that were brought in at the time the neural tube formed. The cell bodies collect in clusters (ganglia) outside the brain and cord and send processes into the central nervous system as well as to the integument of the organism. These distal processes, modified during evolution, provide a variety of nerve endings usually classified as free (non-encapsulated) or *encapsulated*.

The free type of nerve ending occurs when the peripheral process of the sensory nerve merely ends in a spray of terminal,

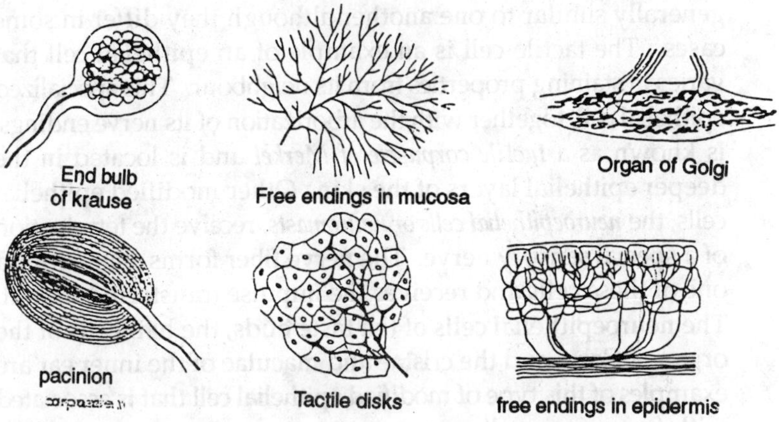

Fig. 4.53 Examples of encapsulated and (non-encapsulated) free nerve endings.

non-myelinated, slender twigs or knoblike swellings. This type of sensory cell is distributed abundantly over the entire body, and its fibrils entwine or embed in epithelial cells, connective tissue, muscle, or serous membranes. The fibril terminate in the epithelium of the cornea, oral cavity, respiratory passages, and kin of all vertebrates. They also appear among the fibers of the connective tissue of the dermis, periostoum, blood vessels, and individual muscle fibers, to name a few example. Properly stained histologic sections of the integument show this type of three ending to penetrate the skin as far as the stratum germinativum layer. Sections through the bronchial epithelium, on the other hand, reveal branches of the nerve wrapped around the cell, with the ends of the fiber actually pushing to the surface. Pain endings are of this type. The sense of touch appears to depend on stimulation of similar endings. For example, nerve endings in hair follicles are important tactile organs. The sensory fibers divide and form a meshlike ending around the hair shaft. Any pressure sufficient to move the hair shaft also stimulates the nerve and adds to the sensitivity of touch.

The epithelial cells associated with the nerve endings are generally similar to one another although they differ in some cases. The tactile cell is an example of an epithelial cell that varies in staining properties from its neighbour. This specialized epithelial cell, together with the arborization of its nerve endings, is known as a *tactile corpuscle of Merkel* and is located in the deeper epithelial layers of the skin. Other modified epithelial cells, the *neuroepithelial cells or neuromasts,* receive the termination of a special sensory nerve. The nerve fiber forms a net around one of these cells and receives the impulse transferred from it. The neuroepithelial cells of the taste buds, the hair cells of the organ of Corti, and the cristae and maculae of the inner ear are examples of this type of modified epithelial cell that is associated with free nerve endings.

When amniotes moved to land, the sensory endings became encapsulated, apparently so that there is a protective adaptation to terrestrial life. The encapsulated endings of the sensory nerve fibers are all similar in that each consists of a terminal nerve ending wrapped with connective tissue. Such structures as the tactile corpuscles of Meisner (touch), end bulbs (temperature), Pacinian corpuscle (tension and pressure), and neuromuscular and muscular-tendon bundles (tension and stretch) are examples of encapsulated nerve endings. They differ from one another in the degree of branching of the nerve, the number of supporting cells present (connective tissue), and the size and shape of the structure. The terminal nerve endings that pass to skeletal muscles and tendons, for example, entwine around a muscle fiber that is encased in a connective tissue capsule. When the muscle fiber stretches, it stimulates the nerve fiber, revealing the position of the body and its parts of the organism. In some instances the connective tissue wrapping around the ends of the nerve is so extensive that the structure can be seen with the haked eye (Pacinian corpuscle).

ORGANS OF SPECIAL SENSES

Organs of Taste

The *taste buds* develop from modified epithelial cells, the neuromasts, mentioned earlier in the chapter. In fish, these cells are distributed over the surface of the body and are innervated by extensions of the cranial nerves. Taste buds are also found buried in the epithelium of the gill rakers and gill arches as well as in the mouth and pharynx of fishes. Materials dissolved in the water stimulate the exposed ends of the cells; taste buds therefore are examples of chemo-receptors that are stimulated by substances at close range rather than at a distance. During their evolution, taste buds lost their generalized distribution over the body of the animal and became localized in special areas. In terrestrial animals they are usually restricted to the tongue, although they may also be present in the palate and pharynx.

The taste buds are usually associated with small elevations of the tongue epithelium, known as *papillae*. Trenchlike depressions separate one papilla from another, and the taste buds are located in the walls of these trenches. The taste buds of man can discriminate between only four taste sensations: acid, bitter, salty and sweet. All other "tastes" are actually olfactory sensations. Fish are believed to discriminate between these same four tastes, but amphibians show no response to substances classified as sweet or bitter. The horny tongues of birds lack taste buds.

Histologically, the epithelial cell differentiates into two types of cells, a *gustatory receptor* cell (neuromast) and a *sustentacular*, or supporting cell. A chemically sensitive hairlike process projects from the apical surface of the elongated neuromast. In mammals, clusters of neuromast cells group together with their supporting cells to form a taste bud. The number of neuromast cells composing a taste bud varies. The supporting cells are interspersed among the neuromast cells and also surround them like staves of a barrel. The hairlike process of each neuromast is in contact with the surface of the epithelium by way of a taste pore.

Nerve fibers that break up into a basketlike arrangement of telodendria (nerve endings) surround the deep ends of the neuromast cells. The cell membrane of the nerve ending is in direct contact with the cell membrane of the neuromast cell. Consequently, when the neuromast cell responds to chemical stimulation, the physiologic changes activate the nerve endings. The nerve then relays the impulse to the central nervous system. Taste buds of all vertebrates are supplied with branches of the eighth, ninth, and tenth cranial nerves. Since these nerves contain visceral sensory fibers, the gustatory organs are considered special visceral receptors. Presumably as a result of the chemical stimulation, the primitive animal moved toward or away from such objects as food, enemies, or a mate.

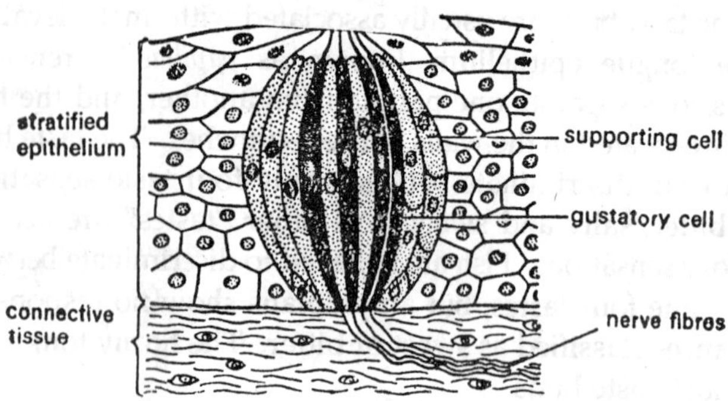

Fig. 4.54. Section of taste bud.

Organs Derived from Placodes

Although most of the epidermal (ectoderm) covering of the embryo develops into the epidermis and epidermal derivatives, a number of other structures are derived, from it. Many of these structures begin their development in the form of *placodes*, plate-shaped thickenings in the epidermal ectoderm. In the early neurula stage the placodes are arranged in an almost continuous horseshoe-shaped swelling around the open cranial neural folds. In most cases the initiation and further development of the placodes depends on induction by the tissues beneath it.

By the time the neural fold closes, the generalized swelling becomes identified as contributing to a specific sense organ. The thickened epidermal ectoderm located anterior to the neural plate (nasal placodes) is stimulated to form olfactory sacs. The optic vesicle contacts the epidermis and induces it to thicken *(lens placode)* and develop into the lens of the eye. An *auditory placode* located laterally to the hind brain gives rise to the ear. *A lateral-line placode* associated with the auditory placode becomes the lateral line sense organs of aquatic animals. Several placodes give rise to the cranial ganglia. These contribute to the fifth and seventh to tenth cranial nerves.

Olfactory Organ

The olfactory or nasal placodes of all vertebrates first appear as oval thickenings on the ventrolateral surfaces of the head of the embryo, anterior to that part of the closed neural tube destined to become the brain. With the exception of the cyclostome, which has a single nasal pouch, the olfactory placodes are bilaterally arranged. They appear in the human embryo as early as the fourth week of gestation. The cells composing the central part of the placode sink inward to form the *olfactory pit;* the walls of the pit soon thicken to become the *olfactory sac.* The opening of the sac to the outside is called the *external naris.*

Some of the cells in the wall of the olfactory sac differentiate into sensory cells. These take the form of rodlike projections ending in sensitive brushes extending to the surface. Processes develop on the proximal end of each cell (end closest to the brain) and grow into that part of the forebrain destined to become the *olfactory bulb.* In the bulb the processes (axons) synapse with the mitral cells, which carry the impulse via the olfactory tracts to the appropriate center in the brain. These processes of the nasal epithelial cells leading from the sac to the telencephalon make up the first cranial nerve, the *olfactory nerve.* Most nerves develop either from the brain itself or from ganglia that are associated with the brain or cord. The olfactory nerve is an exception. It does not develop from nervous tissue at all but from the epithelium of the nasal sacs. It is, therefore, not considered a typical cranial nerve.

In most fishes the olfactory sacs of the embryo develop into similar adult structures. Their primary function is an important one since they act as distance receptors informing the organism of the presence of food some distance away. The olfactory receptors are good examples of extero-receptors. The sacs remain as a pair of pockets in the fish without any communication with the mouth. Each pocket is usually divided into two openings. The water with its dissolved chemical substances flows in one side of the nasal pocket and out the other side. The mucosa of the sac consists of many folds of columnars epithelium thereby increasing its surface area. As the water passes over the sensory end brushes, the dissolved chemicals stimulate the nasal epithelial cells. The proximal processes of the cells conduct the generated impulse to the brain.

The olfactory organ increases in complexity in the Crossopterygii and in the tetrapods. The olfactory sacs in these animals are no longer blind pouches but communicate with the oral cavity by way of an opening, the *internal naris,* or *choana.* When is a cross section of a 10 mm. passing through the head region as from frog it shows the internal naris opening into the mouth cavity. In these animals the nose is also involved in breathing. In the amphibians the nasal sac is a simple elongated structure that opens into the anterior region of the roof of the mouth and functions both on land and in the water. On land it becomes filled with air and special mucous glands keep the epithelial cells moist and capable of functioning. The sensory epithelium of the fish was thrown into folds. In contrast, the olfactory pouch of the amphibian is smooth, and only part of its surface contains the sensory cells.

Figure 4.51. T.S. through eye region of tadpole.

The nasal organs of reptiles and birds are more complex. Although birds retain the reptilian plan of nasal structure, their olfactory organs are of modest size. Birds rely more on their eyes than on their sense of smell to learn about their environment. The olfactory sac divides into two main regions—a small anterior *vestibule* into which the air passes from the external naris, and a larger olfactory chamber opening into the oral cavity by an internal naris more posteriorly placed than in the amphibian. Only the dorsal area of the olfactory chamber contains sensory cells. This sensory epithelium covers the curved, lateral ridges known as the *conchae,* which increase the surface of this area. The walls of most of the chamber, however, function as an air passage and are free of sensory epithelium.

The nasal structures of most mammals increase in size and may occupy as much as half of the skull space (dog). As in the reptile, the olfactory area divides into a dorsal sensory area and a non-sensory air passage. The development of a secondary roof to the oral area, the secondary and soft palates, moves the openings of the internal nares more posteriorly in the mouth. The chonchae, formed from coiled scrolls of the ethmoid, maxilla, and nasal bones, are highly developed and covered with sensory epithelium. The incoming air passes through the scrolls and is warmed and filtered in the process. The nasal cavities of eutherian mammals also communicate with cavities (sinuses) of adjacent bones, lightening the skull.

Except for some mammals, turtles, and crocodiles, a second sensory area appears in tetrapods, the *vomeronasal organ,* or *organ of Jacobson.* It forms from another part of the olfactory sac that retains its sensory function. It detects olfactory sensations from the food in the mouth. This "organ" is either

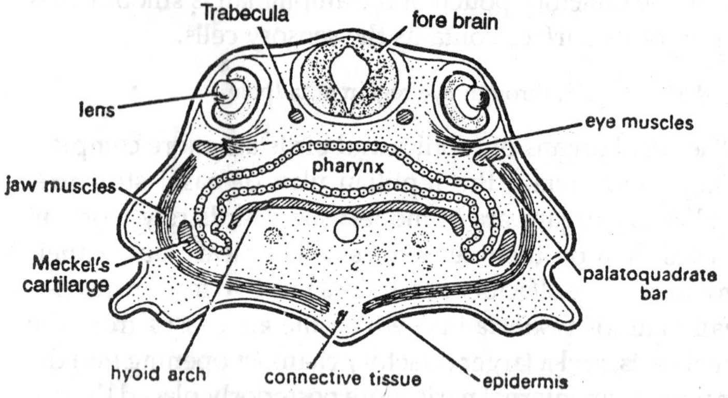

Fig. 4.55. Diagram of structure of neuromast

a groove covered with olfactory epithelium, or else it exists as a separate pocket. In either case, it usually communicates with the oral cavity by way of the internal naris. Jacobson's organ is particularly well-developed in reptiles. It may consist of a medioventral, club-shaped pouch whose opening into the oral cavity is associated with the internal naris, or it may exist as a completely separated pouch with its own opening. In lizards and snakes, the cleft, flicking tongue also serves as an olfactory organ. When drawn into the mouth, the tips of the tongue are inserted in to the vormeronasal pocket. Chemical particles adhering to their surface pass into solution and stimulate the sensory epithelium of the pocket. The pouch is absent in primates but is present in other mammal a functional structure.

Acousticolateralis Placodes

A lateral series of placodes, located on either side of the neural tissues of the head, develop into the lateral-line system in fishes and aquatic stages of amphibians. The cells composing this system in modern vertebrates are sensitive to mechanical stimulation. Apparently they evolved from the neuroepithelial cell (neuromast) considered common in ancestral types. A group of stereocilia and a long motile cilium extend from the

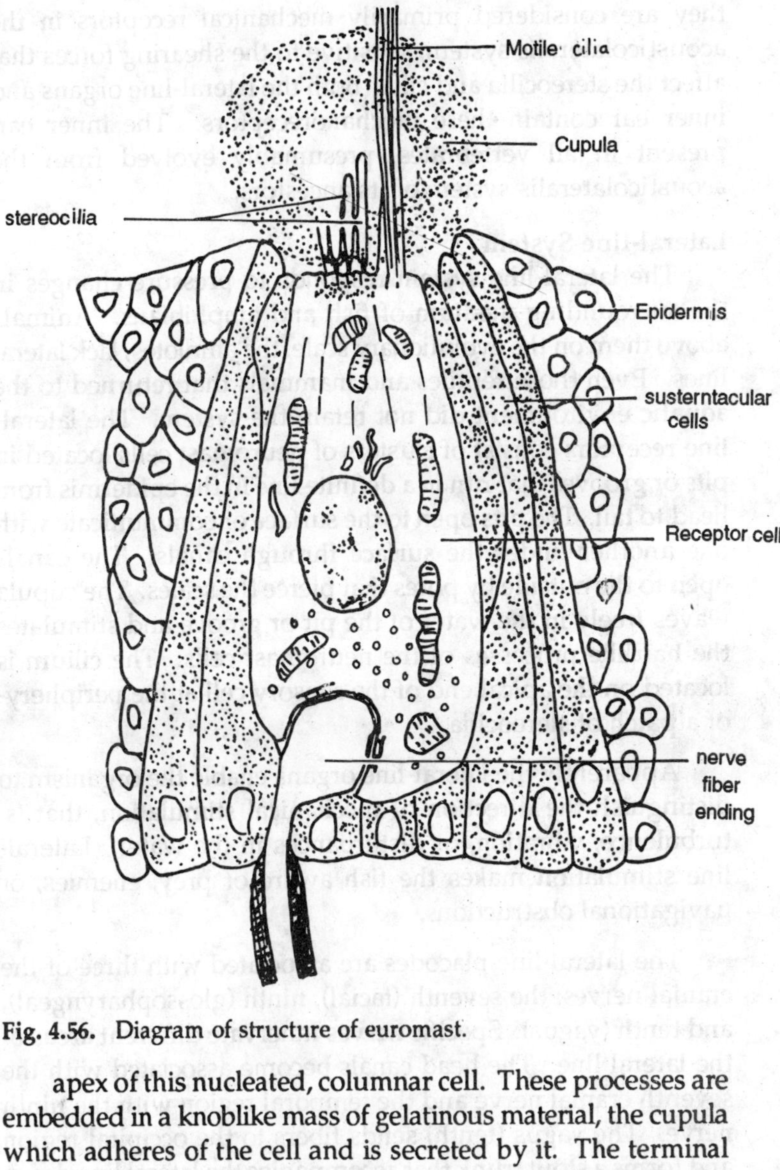

Fig. 4.56. Diagram of structure of euromast.

apex of this nucleated, columnar cell. These processes are embedded in a knoblike mass of gelatinous material, the cupula which adheres of the cell and is secreted by it. The terminal endings of the associated neurons wrap around the neuromast and receive the impulse of the stimulated receptor cell. Although some neuromast cells respond to chemical and thermal stimuli, they are considered primarily mechanical receptors in the acousticolateralis system, sensitive to the shearing forces that affect the stereocilia and cilia. Both the lateral-line organs and inner ear contain these mechanoreceptors. The inner ear, present in all vertebrates, presumably evolved from the acousticolateralis system of its ancestors.

Lateral-line System

The lateral-line system responds to pressure changes in the surrounding medium of fish and amphibians. Animals above them on the evolutionary scale, the amniotes, lack lateral lines. Even those reptiles and mammals that returned to the aquatic environment did not retain the system. The lateral-line receptors consist of clusters of neuromast cells located in pits or grooves that run in a definite line in the epidermis from head to tail. The pits open to the surface or communicate with one another under the surface through canals. The canals open to the outside by pores that pierce the scales. The cupula waves freely in the water of the pit or groove and stimulates the hairlike processes of the neuromast cells. The cilium is located on the apical end of the sensory cell at the periphery of a patch of stereocilia.

Apparently the lateral-line organs enable the organism to distinguish the direction of mechanical stimulation, that is, turbulence, vibration, or disturbances in the water. Lateral-line stimulation makes the fish aware of prey, enemies, or navigational obstructions.

The lateral-line placodes are associated with three of the cranial nerves: the seventh (facial), ninth (glossopharyngeal), and tenth (vagus). Specific nerves innervate different areas of the lateral line. The head canals become associated with the seventh cranial nerve and the temporal region with the ninth nerve. The vagus (tenth) sends fibers to the occipital region and forms a stout trunk that accompanies the lateral line down the length of the animal's body.

Ear

The mammalian ear is composed of three regions: *external ear, middle ear,* and *inner ear.* All vertebrates have inner ears, but the middle and external ears are relatively recent additions. The middle ear first appears in the amphibians and is an adaptation to aerial living. As the animal changed from an aquatic to a terrestrial existence, one of the gill pouches became the middle ear chamber and one of the bones of the gill apparatus became adapted to transmit vibrations from the outside to the receptors of the inner ear. The external ear merely collects and directs sounds to the middle ear where they are magnified. Some reptiles have an external ear, but it is most developed in the mammals.

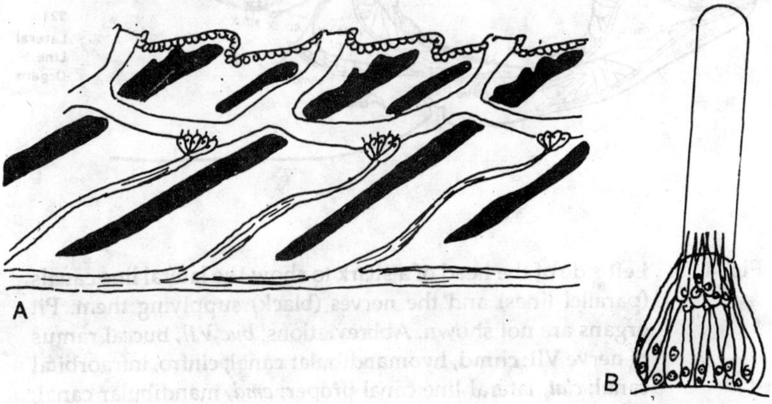

Fig. 4.57. Lateral lines of fish. *A*, Part of the main lateral line canal of the perch *(Perca)* showing three pores to the surface and three neuromast organs. *B*, Diagram of a single neuromast organ showing the sensory processes and the gelatinous cupula.

Embryologic development of the inner ear suggests that it evolved from that part of the lateral line innervated by the eighth cranial nerve. The otic placodes located in the epidermis lateral to the hindbrain (rhombencephalon), on each side of the head, arise in much the same manner as other components of the lateral-line organs. The epidermis thickens to form a plate; it sinks more deeply into the loose mesenchyme cells of the head region than do the lateral-line placodes and becomes the otic sac. It may or may not retain its connection with the surface. The development of the inner ear of the bony fish is an exception. In these animals a solid plate of cells on the inner surface of the epidermis sinks downward and later becomes hollowed out.

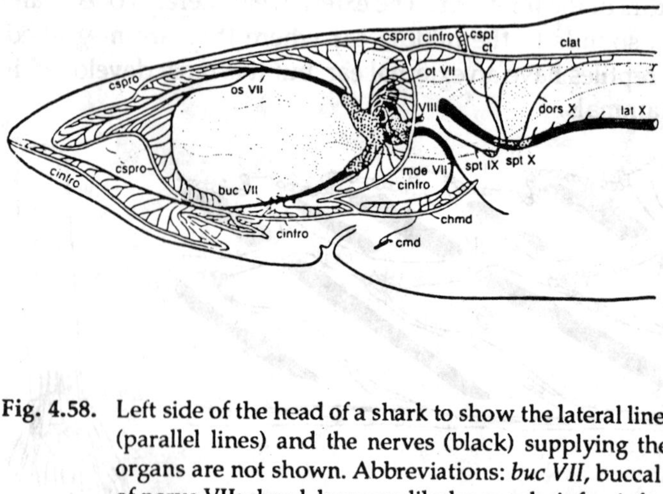

521 Lateral Line Organs

Fig. 4.58. Left side of the head of a shark to show the lateral line canals (parallel lines) and the nerves (black) supplying them. Pit organs are not shown. Abbreviations: *buc VII*, buccal ramus of nerve VII; chmd, hyomandibular canal; cinfro, infraorbital canal; *clat*, lateral line canal proper; *cmd*, mandibular canal; *cspro*, supraorbital canal; *cspt*, supratemporal (or occipital) canals; *ct*, temporal canal; *dors* X, dorsal ramus of nerve X; *tat* X, lateral line ramus of nerve X; mde *VII*, external mandibular ramus of nerve VII; os *VII*, superficial ophthalmic ramus of nerve VII; ot *VII*, otic ramus of nerve VII; *spt* IX, supratemporal ramus of nerve IX; *sp* X, supratemporal ramus of nerve X. The brain is stippled in and the ganglia of the cranial nerves shown by a pattern of small circles.

Although the exact mechanism varies with the species, the otic sac is converted into the *otic vesicle*, or *otocyst*. This completely closed structure differentiates into the inner ear, composed of the organ of equilibrium and the organ of hearing.

In the fish the organ of hearing is under-developed and the ear functions primarily in equilibrium. Since olfactory function was of primary importance in the primitive animal, auditory function in the ancestral vertebrates was perhaps of little value and may have been completely absent. However, the ability of the ear to function as an organ of equilibrium is more fundamental and remains relatively unchanged from fish to man. Hearing becomes an important part of the sensory system at the tetrapod level and increases in complexity as the evolutionary scale is ascended. The saccule of the fish, for example, is only a small structure, whereas in man the lagena, or organ of hearing develops from its posterior end.

Neuroblast cells differentiate from the otic vesicle and form a club-shaped mass on the medial wall. These cells are destined to become the acoustic, or auditory, ganglion. Processes develop on the cells of the ganglion and penetrate the rhombencephalon, giving rise to the root of the eighth nerve. In adult animals this nerve divides into two main branches, the *vestibular nerve,* which passes to the anterior regions of the organ of equilibrium, and the *cochlear nerve,* which services the posterior part of the organ of equilibrium and the cochlea. The fact that neuromast cells provide the sensory receptors for both the lateral-line organs and inner ear offers additional evidence of their close phylogenetic relationship.

Ear as organ of equilibrium. The otic vesicle develops into a pear-shaped structure with the pointed end oriented toward the dorsal surface of the head. The pointed area becomes the *endolymphatic duct;* the expanded portion of the otic vesicle gives rise to the *membranous labyrinth.* The labyrinth undergoes irregular growth and soon divides into two chambers, a dorsal *utricule* connected to a more ventral *saccule* by way of a narrow *sacculoutricular duct.* The endolymphatic duct is attached either to this duct or to the saccule. Further expansion and constriction of the utricule give rise to the rudiments of the three *semicircular canals,* first present in the human embryo as early as the sixth week. In the adult of all vertebrates, these rudiments become three distinct canals, each of which is oriented in a different plane (two vertical and one horizontal) and is open at both ends into the utricule. They are referred to as the *anterior, posterior,* and *horizontal semicircular canals.* In lower vertebrates, the saccule forms a pocketlike structure, the lagena, at its posterior end. This elongates in reptiles and birds and expands in mammals into the *cochlear* duct. The utricule, saccule, semicircular canals, and lagena compose the inner ear and should be differentiated from the auxiliary middle- and outer-ear structures, which develop from different primordia.

The thin walls of the membranous labyrinth consist of flat epithelial cells. The medioventral area, however, thickens, and the cells develop into patches of supporting cells and sensory cells known as the *maculae.* In the adult organ they are distributed in a horizontal plane on the floor of the utricule and in a vertical plane on the inner wall of the saccule. These spots of sensory epithelium, known as the *utricular macula* and *saccular macula,* respectively, are associated with the branches of the auditory nerve. The fibers of this nerve grow between the epithelial cells early in development and arborize about the bases of the sensory cells. Stimulation of these sensory cells offers information about the tilt of the head, and the individual is made aware of static position. An expansion at the end of each semicircular canal, referred to as the ampulla, also contains sensory areas, termed *cristae.*

The hollow, irregularly shaped membranous labyrinth, consisting of the utricule, saccule, and semicircular canals, fills with fluid. The *endolymph,* which is more viscous than water, resembles the interstitial fluid. Supporting cells and hair cells with long processes compose the maculae and cristae (the actual receptors for the sense of equilibrium). The hair cells are histologically similar to neuromast cells and respond to pressure or movements of the liquid in which they reside. The supporting cells are believed to secrete a jellylike substance, the *cupula,* in which the hairlike projections of the sensory cells of the cristae rest. Various concretions (otoliths) or grains of sand are associated with the gelatinous covering of the sensory cells of the macula. As the animal moves about, the weight of the otoliths, which rest on the sensory epithelium, stimulate different hair cells. This stimulation in turn causes an increase in the rate of impulses fired to the central nervous system by way of the vestibular branch of the eighth cranial nerve. The stimulation of these slightly elevated patches of cells furnishes the animal information about its turning movements. The endolymph also surges about in the tiny semicircular canals as the animal changes positions, and the movement of the fluid stimulates the sensory hairs of the cristae, providing additional information about the activity of the animal.

As mentioned above, the original placode (induced to form by the rhombencephalon) sinks into the mesenchyme of the head area and gives rise to the otic vesicles. This vesicle develops into the labyrinth. Depending on the species, the mesenchyme (derived from sclerotomes) is in turn induced by the otic vesicle to differentiate into skeletal tissue. It is of harder texture immediately surrounding the membranous labyrinth than elsewhere and forms a capsule about the ear known as the *cartilaginous* or *osseous labyrinth*. A space exists between the membranous labyrinth (derived from ectoderm) and the osseous labyrinth (composed of mesoderm). It fills with *perilymph*, a fluid very similar to the endolymph of the membranous labyrinth.

Ear as Organ of Hearing

Middle and outer ear. The organ of equilibrium remains relatively unchanged in tetrapods, but hearing becomes an important additional sense. How well fishes hear remains a controversial topic, but, in general, investigators believe that the sensory maculae act as possible phonoreceptors. Skeletal structures of the head transmit sound waves from the water to the inner ear of fishes. The air bladder also may function as a transmitter of sound vibrations. However, since air waves are relatively faint, there is a need for some kind of device to amplify the vibrations in land animals. The development of the *middle ear* from the anterior gill pouch (the spiracular gill cleft) occurs in terrestrial animals. This first pharyngeal pouch does not break through to the outside to form a typical gill cleft; instead, it remains as an expanded lateral cavity, the primordium of the *middle ear*, or *tympanic cavity*. The addition of mesoderm to the thin membrane left between the pouch and the anterior ectoderm converts it to the *tympanic membrane*, or eardrum. Since the gill pouches are outpocketings of the pharynx, the internal lining of the tympanic membrane is endoderm. Its outer surface, however, is derived from superficial epidermal ectoderm. The resulting tympanic membrane thins and responds sensitively to air waves. The remnant of the attachment of the gill pouch to the pharynx becomes narrowed and remains as the *eustachian tube*.

Mesenchyme condensations give rise to the primordium of the transmitting apparatus of the middle ear. In amphibians and reptiles a piston-like bone, the columella (which evolved from the hyomandibular bone of fish to be discussed later), crosses this pouch from the eardrum to a membrane-covered opening in the ear capsule, the *oval window* (fenestra ovalis). One end of the columella (stapes) rests against this membrane. It picks up vibrations received by the ear-drum and transfers them internally to the fenestra ovalis. As a result, the liquid of the internal ear is set in motion on the other side of the membrane, stimulating the various sensory spots. A second membrane-covered opening, the *round window* (fenestra rotunda), is located at the other end of the perilymphatic duct. It functions as a release valve. In mammals pressure waves, introduced to the perilymph by the action of the stapes, travel up the scala vestibuli and then across to the scala tympani (dorsal and ventral ducts of the original perilymphatic space) and back toward the middle ear. The membranous fenestra rotunda bulges into the middle ear cavity dissipating the pressure waves.

In mammals, three *auditory ossicles* (malleus, incus and stapes) replace the columella and lead from the tympanic membrane to the fenestra ovalis. The mammal also adds a cartilaginous *external ear (pinna)* to help collect and concentrate air waves. In many animals, muscles attach to the pinna and move the ear about and orient it in the direction of sound. A short canal, the *external auditory meatus*, leads from the outside to the tympanic membrane. Wax glands are usually present in the walls of this canal.

Development of cochlear duct. The lagena, present in all vertebrates beginning with the fish, develops in higher animals into the basic structure of hearing, the *cochlear duct* with its *organ of Corti.* The evolution of the cochlea involves the *lagena*, the *perilymphatic duct*, and the *organ of Corti.* The lagena first appears as on endolymphatic papilla in the amphibian saccule. It expands into a long fingerlike extension in the crocodiles and birds, but in mammals it is so long that it must coil in order to fit within the ear capsule. In man, the cochlea makes 2 $^{3}/_{4}$ turns. Since the lagena is an extension of the saccule, endolymph also fills its cavity. This duct, formed by the extended lagena, is the *cochlear duct*, or *scala media*.

The *organ of Corti*, named after the Italian anatomist who first described it, is the true receptor organ for the sense of hearing. It extends down the length of the cochlear duct and its cells discriminate the sounds of different pitch. In the human embryo it develops as early as the third month as a local thickening of the floor of the cochlear duct. Between the third and fifth months the epithelial cells differentiate into the hair cells of the organ of Corti and establish the adult structure. The mature structure basically consists of a series of sensory hair cells (neuromast cells) intermingled with supporting cells. A gelatinous membrane, the *tectorial membrane*, extends over them and is in contact with the stereocilia of the hair cells. This membrane may be compared to the cupula of the neuromast cells previously described as composing the lateral-line sense organs. A basilar membrane, capable of vibrating, forms the floor of the organ. The roof of the cochlear duct (scala media) is known as the *vestibular*, or *Reissner's membrane.*

The *perilymphatic duct* is the name given to the space between the cartilaginous capsule and lagena portions of the membranous labyrinth. It is present in amphibians, reptiles, birds, and mammals. It consists of a well-defined channel that follows one side of the cochlear duct from its origin at the membranous labyrinth to the end of the coil and then turns and follows the other side back. A portion of the perilymphatic duct, therefore, lies above and below the cochlear duct. This relationship of the cochlear duct to the perilymphatic canals can best be seen in a cross section of the duct.

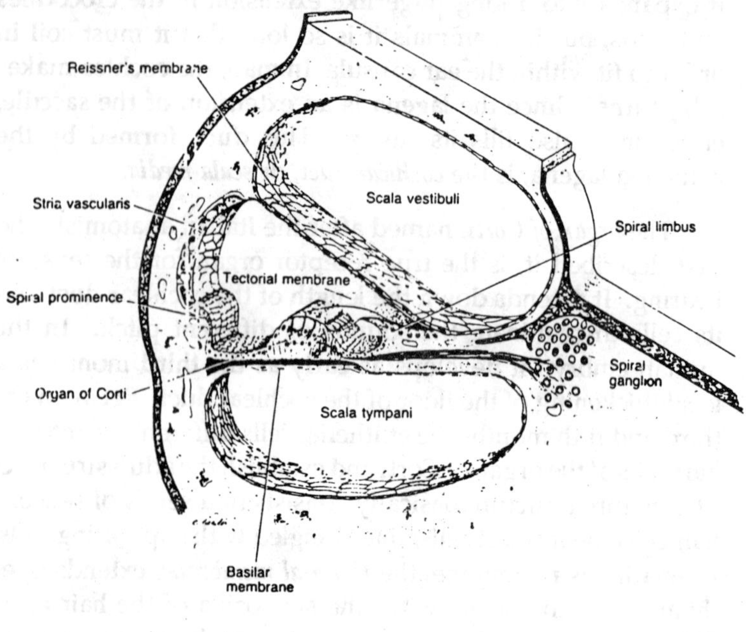

Fig. 4.59. A much enlarged and schematic section through one turn of a mammalian cochlea to show the organ of Corti.

As a result of its relationship with the perilymphatic ducts, three main longitudinal chambers referred to as *scalae* compose the cochlea proper. The upper chamber is the *scala vestibuli* and the lower one is called *scala tympani;* both form from the perilymphatic ducts and fill with perilymph. In between the two chambers is the *scala media,* or *cochlear duct,* filled with endolymph and the only part derived from the membranous labyrinth. The scala media ends blindly at the apex of the coil, but the scala vestibuli and scala *tympani* communicate with one another at the apex, the *helicotrema,* through a small junction. They are, therefore, continuous with one another. The cochlear duct is attached to the sides of the bony labyrinth but is surrounded above and below by the perilymphatic ducts. Delicate strands of tissue cross these spaces and help anchor the fragile organ of Corti in place.

In the mammal the external ear directs the sound toward the tympanic membrane. Sound vibrations that impinge on this membrane cause it to vibrate. These vibrations are magnified and transmitted inward by the three auditory ossicles. The stapes vibrates against the fenestra ovalis at the same frequency as the original sound. The ascending scala vestibuli is on the other side of the fenestra ovalis and a pressure-change is produced in the perilymph when the stapes moves the window in and out. Since the perilymph is enclosed by a rigid bony casing and cannot be compressed, as stated above, the membranous fenestra rotunda bungles outward into the middle ear cavity dissipating the pressure.

Pressure changes initiated in the scala vestibuli (when the stapes vibrates against the fenestra ovalis) are believed to travel up the scala vestibuli to its junction with the scala tympani at the helicotrema and then move down this canal. However, many vibrations in the perilymph are transmitted across the membrane separating the scala vestibuli from the cochlear duct. This causes in the cochlear duct a disturbance of the endolymph, which displaces the basilar membrane toward the scala tympani. This displacement (caused by waves in the perilymph) alters the contact of the hair cells with the tectorial membrane. This change in relationship acts as a stimulus for the hair cells.

Terminal branches of the auditory nerve contact the hair cells. Presumably the stimulation of the hair cells is translated into a nerve impulse that passes to the medulla oblongata by way of the cochlear branch of the auditory nerve. Clinical and experimental evidence shows that the regionalization of response by the organ of Corti enables the individual to discern pitch. That part of the organ of Corti closest to the membranous labyrinth discriminates high tones; the receptors for low tones are located near the apex of the duct.

Eye

The eye is an exceedingly complex structure well-developed in even the lowest vertebrate group. There are few clues as to how it evolved. It is a sensitive detector of light waves and differs from the other sense organs in that it develops from the wall of the brain. Its basic structure is the same from cyclostomes to mammals, although adaptive modifications allow it to function better in one ecologic niche than in another. It basically consists of a spherical *eyeball,* located in an *orbit,* protected by cartilage or bone, and connected to the brain by an *optic nerve.* The wall of the adult eyeball consists of three concentric coats: an outer *sclera,* a middle *uvea,* and an inner *sensory retina.* The eye also contains a crystalline lens and a transparent cornea (anterior portion of the sclera), which focus the light rays on the retina. Three cavities make up most of the eyeball: the *anterior* and *posterior chambers* and the *vitreal cavity,* which contain transparent viscous substances that allow the light rays to pass through to the retina. The eye results from a series of inductions. It is derived from three main groups of tissues in the embryo. The retina forms from neural ectoderm, the lens develops from epidermal ectoderm, and the tunic of the eye develop from the surrounding mesenchyme.

Early Development

Early in development, when the brain is still in the process of differentiation, lateral evaginations from the wall of the forebrain push toward the skin ectoderm. In the human embryo, the evaginations appear as early as the middle of the third week. These protrusions of the brain are the beginning of the *optic vesicles.* They become bulblike at their distal ends, where they are in contact with the surface of the head: each contains a cavity, the *opticoel,* continuous with the cavity of the brain (*prosocoel*). A constriction soon separates the proximal end of the vesicle from the brain; the resulting connection between the two structures becomes the *optic stalk.*

The underlying chorda-mesodermal roof induces the optic vesicles to form. As a result of this influence, the eye develops in a step-by-step manner, with specific tissue interactions necessary before the next stage of the development process can proceed. For example, mesenchyme cells must surround the evaginations of the brain for normal development to occur. Also, as the lateral evaginations from the diencephalon expand outward, they must contact the epidermal ectoderm for the retina to differentiate. If no contact is made, the presumptive neural retina becomes pigmented rather than sensory.

When the optic vesicle contacts the epidermal ectoderm, the vesicle flattens and the distal wall invaginates, with obliteration of the opticoel and formation of a double-layered *optic cup.* At first the wall of the optic vesicle is only one cell thick. However, the cells of the indented wall of the optic cup undergo mitosis and become the sensory *retina,* the photosensitive lining of the eye. Pigment granules invade the thin outer layer and form the *pigmented layer* of the retina. It absorbs any light that passes through the retina layer, thereby preventing scattering of the light and blurring of the image. The resulting structure is a double-layered cup attached to the diencephalon of the brain by a narrow, hollow, optic stalk. By the time the eye completes its differentiation, the retina and the pigmented layer fuse form one layer.

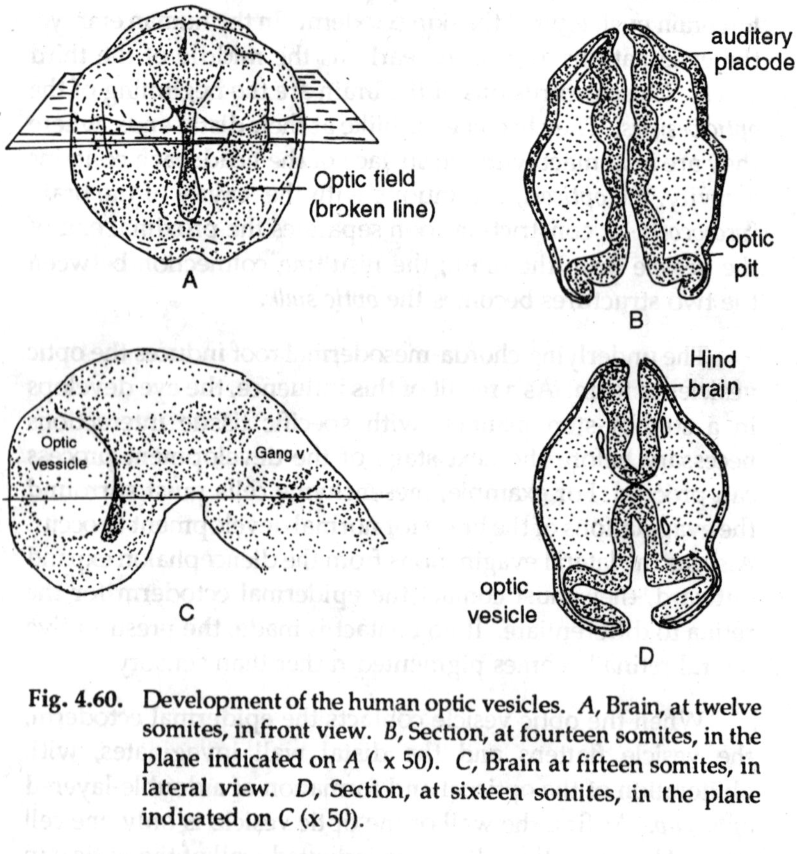

Fig. 4.60. Development of the human optic vesicles. *A*, Brain, at twelve somites, in front view. *B*, Section, at fourteen somites, in the plane indicated on *A* (x 50). *C*, Brain at fifteen somites, in lateral view. *D*, Section, at sixteen somites, in the plane indicated on C (x 50).

The optic vesicle changes into the optic cup by an invagination that begins at the lower edge of the optic vesicle near the optic stalk and proceeds upward. As a result of the differential invagination, the ventral rim is not as well-developed as the rest of the cup and a groove remains, the *choroid fissure*. The indentation that gave rise to the cup continues along the ventral surface of the optic stalk. As a result, during early development of the eye a troughlike structure extends from the ventral rim of the cup toward the brain. The central artery follows the trough of the choroid fissure and enters the optic

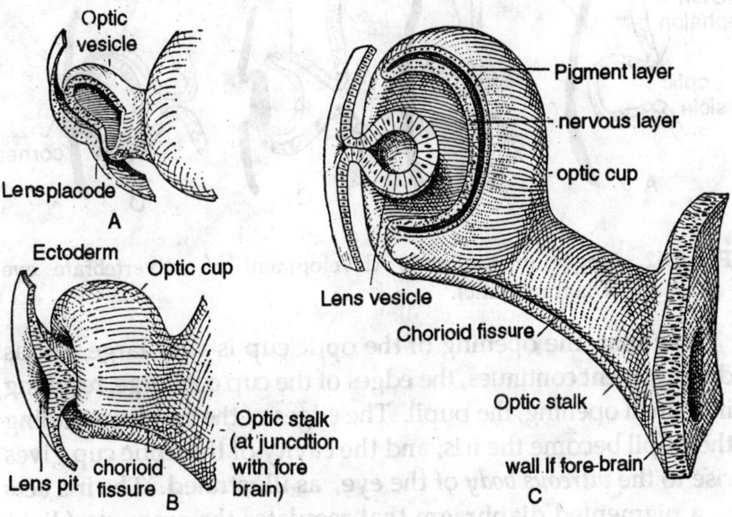

Fig. 4.61. Human optic primordia, shown as models in side view X 100. The lens is sectioned and the optic cup in *A* and C has been partly cut away. *A,* At 4.5 mm,; B, at 5.5 mm,; C, at 7. 5 mm.

cup where it is known as the *hyaloid artery*. A network of capillaries forms from the artery around the developing lens. This step appears necessary in the early stages for differentiation of the lens. By birth the hyaloid artery regresses, leaving the *hyaloid canal* in its place in the vitreous body. The edges of the choroid fissure eventually fuse and a small tube forms within the already tubular optic stalk. The central artery and vein remain in this canal and become the main blood vessels of the retina. When the optic nerve forms, the axons of the ganglionic cells converge to the walls of the optic stalk and follow its lumen into the brain. The walls of the optic stalk become the sheath of the optic nerve, enclosing the axons and the central artery and vein a well.

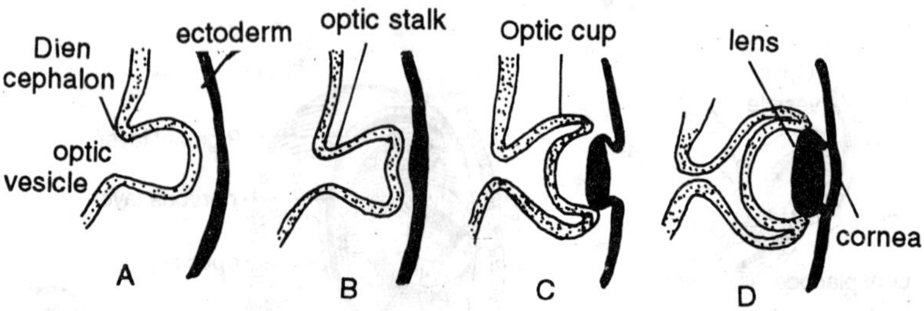

Fig. 4.62. Stages in the early development of a vertebrate eye (Diagrammatic).

At first the opening of the optic cup is very large, but as development continues, the edges of the cup converge, resulting in a small opening, the pupil. The edges of the cup surrounding the pupil become the iris, and the cavity of the optic cup gives rise to the *vitreous body* of the eye, as illustrated. The iris acts as a pigmented diaphragm that regulates the amount of light falling on the retina.

Formation of Lens

The general pattern of lens induction is believed to be the same for all vertebrates. First endoderm and then mesoderm induces the epidermal ectoderm to form a lens placode. When the lateral protrusions of the brain approach the surface of the embryo during the formation of the optic cup, the retina becomes the third inductor. The underlying nerve ectoderm induces the epidermal ectoderm to thicken into the lens placode and to form a lens. The inductive influence occurs even through porous sheets of material experimentally placed between the two layers of cells. The results of RNA-staining suggest that the optic vesicle changes chemically during the induction process. For example, its cytoplasm becomes less basophilic. At the same time, the presumptive lens ectoderm shows a gain in basophilia, as if "something" were transferred to it.

The influence of the optic vesicle on the epidermis can be further demonstrated by surgical removal of the lens placode. The surrounding epidermal cells move into the empty space

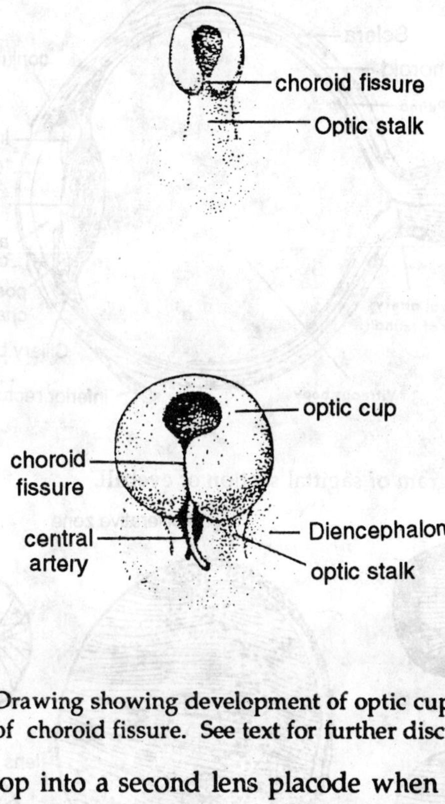

Fig. 4.63. Drawing showing development of optic cup and formation of choroid fissure. See text for further discussion.

and develop into a second lens placode when they come in contact with the optic vesicle. The optic vesicle, therefore, induces other areas of the epidermis to form a lens, and determines the exact position of the lens in relation to the eye cup. If the optic cup is removed, either no lens forms at all or it is defective. This ability of the epidermal ectoderm to respond to the optic cup is a clear example of an induction process and has been extensively studied.

The induced lens placode, under the continued influence of the retina, either bends inward or separates itself as a mass and reassembles to form a lens vesicle. In either case, the lens

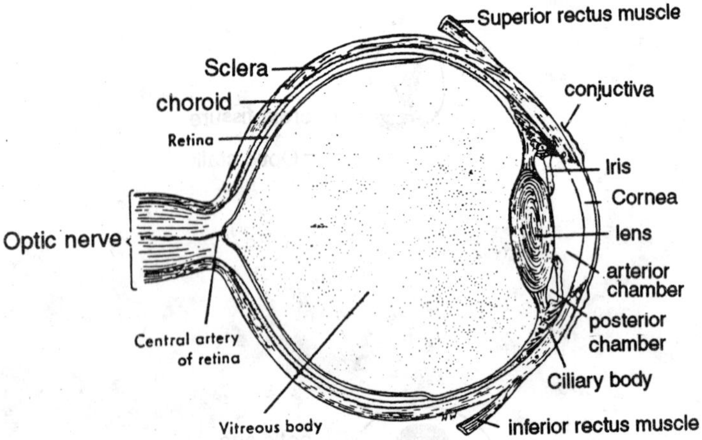

Fig. 4.64. Diagram of sagittal section of eyeball.

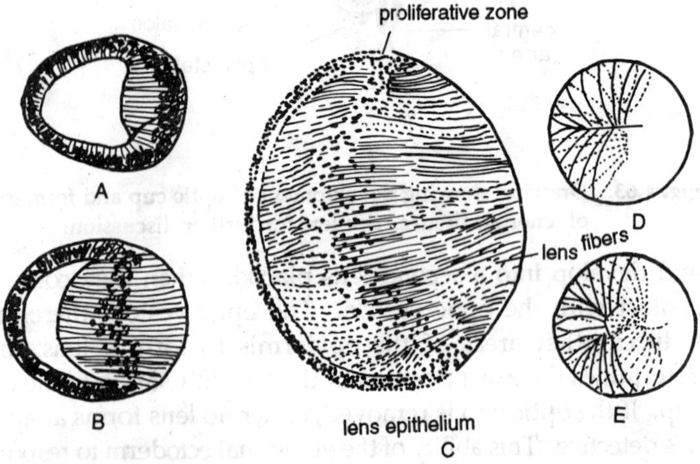

Fig. 4.65. Differentiation of the human lens. *A-C*, Sections through the lens of embryos at 10 mm., 13 mm. and 20 mm. *D, E*, Diagrams of the formation of sutures; *D* represents a stage when sutures are linear, like the adult dogfish lens (front suture, horizontal; back suture, vertical); E shows a four-pointed star of the early adult.

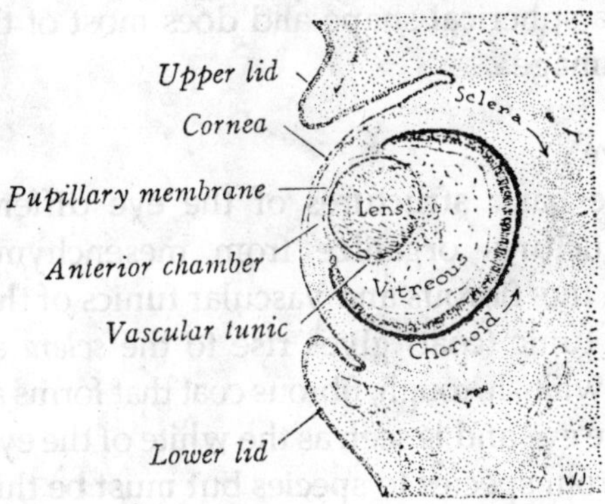

Fig. 4.66. Human eyeball and eyelids, at two months, in longitudinal section.

eventually occupies a position in the pupil of the iris almost filling the optic cup during early development. Once the presumptive lens separates from the overlying ectoderm, it differentiates further. The cells of the inner portion of the lens elongate, lose their nuclei, and transform into longer fibers, arranged in an orderly fashion. The outer layer of the lens becomes its epithelial capsule.

The lens takes its place at the mouth of the cup as a small crystalline structure, capable of either moving forward and backward (in anamniotes) or changing shape (amniotes). This difference in the mechanism of lenticular adjustments is correlated with the kind of adaptations the animal must make to its environment. For example, the ability to change the shape of the lens gives the animal a greater range of focus necessary for aerial vision. In cyclostomes the lens is not attached but the pressure of the vitreous humor and the cornea keep it in its proper position. In other forms, a ring of fibers hold the periphery of the lens in place. The ability of the eyes to focus light on the retina is known as *accommodation*. It is accomplished in general, in land animals by the curved surface of the cornea focusing the image on the retina and the lens bringing it into sharp focus. This is not the case in fish. The fish lens has a spherical shape and does most of the work of focusing the image.

Tunics of Eye

While the basic structures of the eye differentiate, its accessory structures organize from mesenchyme. These structures are the fibrous and vascular tunics of the eye. The outer fibrous tunic (coat) gives rise to the *sclera* and *cornea.* The sclera is a whitish, tough, fibrous coat that forms a protective covering to the eye and is seen as the white of the eye. It varies in thickness from species to species but must be thick enough to withstand the intraoptic pressure. The adult sclera of some vertebrates is entirely cartilaginous. In several species of fish, birds, and reptiles, sclerotic bones, or ossicles, are embedded in the sclera near the cornea and give additional strength to the tunic. The tendons of the six optic muscles that move the eye attach to the surface of the sclera and the fibers of the optic nerve perforate its posterior surface at the *lamina cribrosa.*

In the anterior portion of the eye the sclera combines with the epidermal ectoderm to form the transparent, circular cornea. The epidermis overlying the area of the lens and optic cup is induced to form the anterior epithelium of the cornea. Mesenchyme cells condense around the brain, optic stalk, and optic cup. These cells develop into the dura mater of the brain and optic stalk and the sclera of the eyeball. Those mesoderm cells that lie under the epidermal ectoderm (anterior epithelium of the future cornea) become transparent and contribute to the cornea. The cornea, therefore, is a composite structure, derived from both ectoderm and mesoderm.

It is at the corneal surface that much of the refraction of light occurs in terrestrial forms as pointed out above. Irregularities in its curvature bring about an impairment of vision in the human eye known as *astigmatism.* The cornea is an avascular structure, nourished by the secretions of the eye and able to repair itself rapidly if injured. It must be kept moist at all times. The location of many pain receptors in the cornea makes it an extremely sensitive structure.

The *uvea,* the inner tunic of the eye (the middle layer between retina and sclera), is derived from mesoderm and includes the *choroid coat,* the *ciliary body,* and the *iris.* The choroid coat is pigmented and extremely vascular, providing the nourishment for the eye. It contains arteries and veins of various sizes as well as capillaries. It also reflects light back through the retina in several species, a phenomenon that makes their eyes shine in the dark. The area of the choroid coat responsible for such eyeshine is known as *tapetum lucidum.*

The uvea differentiates into the connective tissue of the ciliary body, the fibers of the ciliary muscle, and the connective tissue of the iris. The iris (as well as the ciliary body) has a dual composition. This structure present in all vertebrates, receives contributions from both the choroid coat and the non-nervous part of the retina. The presence of pigment in this layer is responsible for the colour of the eyes. If no pigment is present at all, the blood vessels of the iris give the eye a pink colour, characteristic of albino eyes. Blue eyes occur when the pigment is present only in the inner portion of the retinal layer. The colourless tissue covering the iris interferes with light rays in such a manner that the iris appears blue. The presence of dark pigment in both the choroid and retinal layers produces brown or black eyes. However, the eyes of newborn children, even those of the black races, do not reflect their true future colour and are blue. The addition of pigment later to the layers of the iris gives rise to its true adult colour.

Muscles are present in the iris, arranged in circular and radial patterns. Contraction of these muscles results in decreased pupil size. The iris, therefore, regulates the amount of light that passes to the lens and on to the retina. In dim light, the pupil opens wide, but in bright light it narrows. These muscles of the iris are derived from its retinal portion and are one of the very few examples of a muscle originating from neural ectoderm rather than the usual embryonic source of muscle, the mesoderm.

The ciliary body also receives contributions from the optic cup. It is a ring-shaped structure that encircles the equatorial zone of the lens and is located near the juncture of sclera and cornea. It functions in accommodation and may or may not actually contact the lens. When the muscles of the ciliary body contract in reptiles and birds, processes on the ciliary body push against the lens and cause it to thicken and shorten its focal length. The lens is a very elastic structure in mammals. It is anchored to the ciliary region by certain zonary fibers that hold the lens under tension and, therefore, flat. Contraction of the zonary muscles brings the ciliary body closer to the lens, the fibers relax, and the elastic lens becomes more rounded. This allows the mammal to accommodate for closer objects.

The eyeball consists of a series of cavities filled with transparent liquids of various consistencies. The *anterior chamber* of the eye is the space between the cornea and the iris; the *posterior chamber* is the small area between the iris and the lens. These two cavities are continuous with one another through the pupil. They are filled with a thin watery liquid known as the *aqueous humor* secreted by epithelial components of the ciliary body. This liquid is similar to the blood serum and nourishes the lens, iris, and cornea and produces the turgor of the eyeball. Drainage of the fluid into a series of sinuses that make up the canal of Schlemm, located in the angle of the iris, regulates the intraocular pressure. A thick gelatinous substance known as the *vitreous humor* fills the principal cavity of the eyeball, the space between the lens and the retina. The photosensitive cells of the retina secrete the material during the development of the eye.

Histogenesis of Retina

As demonstrated by the manner in which it is formed, the retina is an extension of the brain. The origin of the rods and cones, the two types of cells sensitive to light, can be traced to the ependymal cells of the neural tube and to the presumptive neural ectoderm on the surface of the blastula. A knowledge of their development explains why light must pass through several layers of the adult retina before reaching the neurosensory photoreceptors. These receptors are located in the outermost part of the retina, that layer closest to the choroid coat and farthest from the pupil. In the early embryo the ependymal layer lines the neural cavity. As the optic cup forms, this layer becomes adjacent to the pigmented layer of the cup. As a result, the rods and cones are located away from the surface exposed to the vitreous body and away from the source of light.

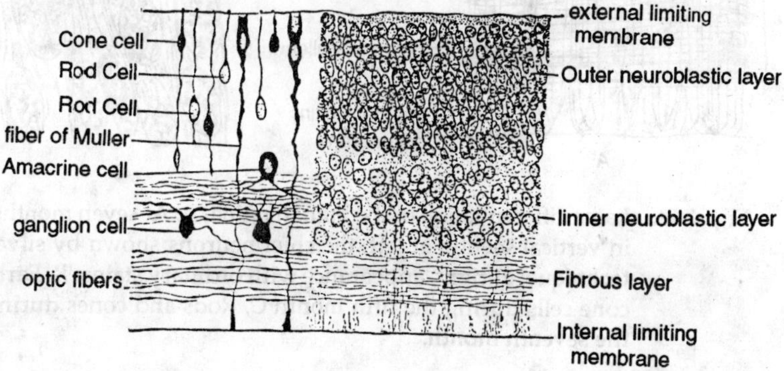

Fig. 4.67. Early differentiation of the nervous layer of the human retina, shown in a vertical section at three months. At left, Cajal's analysis of the component element after silver impregnation; at right, the appearance with ordinary stains.

Three layers of cells make up the retina of vertebrates. Listed in order, the processes of the rod and cone cells are located next to the pigmented cells; the cell bodies of these photoreceptor cells make up the outer nuclear layer. This is followed by the *bipolar neurons,* with the layer of *ganglion cells* situated nearest the vitreous humor. The plexiform layers separate one cell layer from another and consist of cell processes

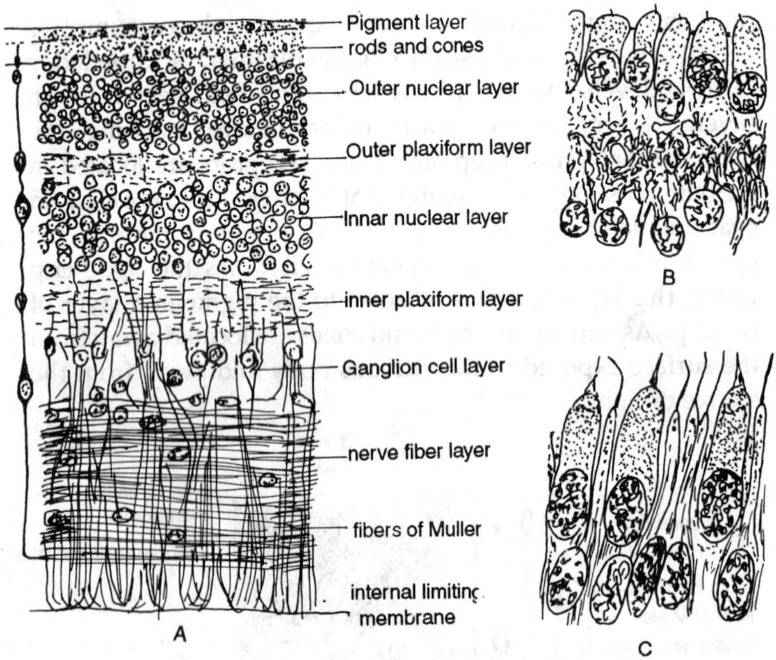

Fig. 4.68. Later differentiation of the human retina. *A*, At seven months, in vertical section at left, the chief neurons shown by silver technique; at right, appearance with ordinary stains. *B*, Early cone cells during the fifth month C, Rods and cones during the seventh month.

synapsing with one another. The *outer plexiform layer* is the area of synapse between the rods and cones and the bipolar cells; the *inner plexiform layer* consists of the synapses between the bipolar cells and the ganglion cells. The distribution of rods and cones in the vertebrate retina differs among the various species. Nocturnal vertebrates, for example, usually have more rods than cones, and the retina of diurnal vertebrates contains more cones. A depression, known as the *fovea*, often marks the site of sharpest vision in certain teleosts, birds, and primates.

The rods and cones differ from one another in shape, number, sensitivity to light, and specific visual pigment

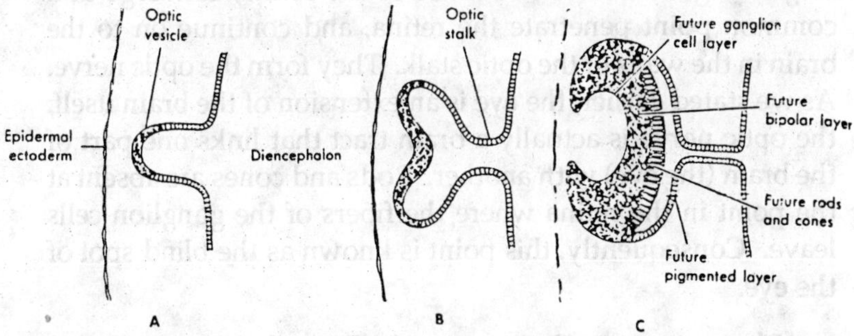

Fig. 4.69. ***A*, Lateral extension of the brain, the optic vesicles, grow outward until they contact the epidermal ectoderm. *B*, Optic vesicles flatten and invaginate, forming the double-walled cuplike structure, the optic cup. *C*, Invaginated layer of optic cup, the future retinal layer, thickens and invaginates until it is adjacent to future pigmented layer. Light must pass through the depth of the retina before it reaches the receptor cells, the rods and cones.**

contained in the receptor. Vertebrate rods are slender, filamentous cells, sensitive to degrees of darkness, are more numerous than cones (at least in the human eye), and contain in their outer segments a specific reddish purple pigment called *rhodopsin.* The cones on the other hand, are shorter and thicker, are sensitive to colour, and contain the visual pigment called *iodopsin.* The visual pigment consists of the protein opsin and a particular configuration of an aldehyde of vitamin A. When stimulated by light, the retinaldehyde is converted to an unstable form.

The impulse generated as a result of this chemical transformation passes on to the processes of the bipolar cell that connect with the rods and cones on one hand and with the associated processes of the ganglion on the other. It appears that several rods can synapse with one bipolar cell. The long processes of the ganglion cell pass to the brain and relay the impulse generated by the light. These long processes (axons), originating on the inner surface of the retina, converge at a common point penetrate the retina, and continue on to the brain in the walls of the optic stalk. They form the optic nerve. As we stated earlier, the eye is an extension of the brain itself; the optic nerve is actually a brain tract that links one part of the brain (the eye) with another. Rods and cones are absent at the point in the retina where the fibers of the ganglion cells leave. Consequently, this point is known as the blind spot of the eye.

The optic nerve fibers pass to the floor of the diencephalon and cross to the opposite side of the brain at the X-shaped optic *chiasma*. There is complete *decussation* (crossing of fibers) in all classes except mammals. In these animals, decussation is incomplete and only the fibers from the medial half of the retina cross to the other side. This leaves about one third to one half of the fibers uncrossed in man. This incomplete decussation is associated with the development of *stereoscopic vision*. Physiologists believe that two duplicate mental pictures form in most animals. In mammals however, the development of stereoscopic vision enables each brain hemisphere to build up a half-picture of the total image. Interconnections between the two cerebral hemispheres allows the individual to "see" the total picture.

In all non-mammalian vertebrates the optic fibers continue in the walls of the thalamus to the roof of the midbrain (mesencephalon). Here they terminate in the optic lobes of the mesencephalon where the primary visual center is located. When the cerebral hemispheres assume their prominence as a co-ordinating and association center in mammals, the mesencephalon loses its importance and no longer acts as a visual center. As a result, most of the optic fibers in mammals are no longer sent to the mesencephalon. Instead, the optic impulses in man are shunted from the thalamus to the new visual centers of the cerebral cortex. It is in the cerebrum that the sensation of sight occurs.

Accessory Structures of Eye

Certain accessory structures protect the eye or aid it in its function. Folds of developing skin, consisting of ectoderm and mesoderm, arise above and below the cornea and give rise to the eyelids of tetrapods. Most fish lack a movable eyelid. The blinking of the eyelids of tetrapods keeps the cornea moist and clean and prevents it from becoming dry and opaque. A transparent third eyelid, the *nictitating membrane,* is present in birds and some reptiles and mammals. It passes from the inner angle of the eye to the outer surface. Rows of hairs, the eyelashes and eyebrows, furnish additional protection to the eye.

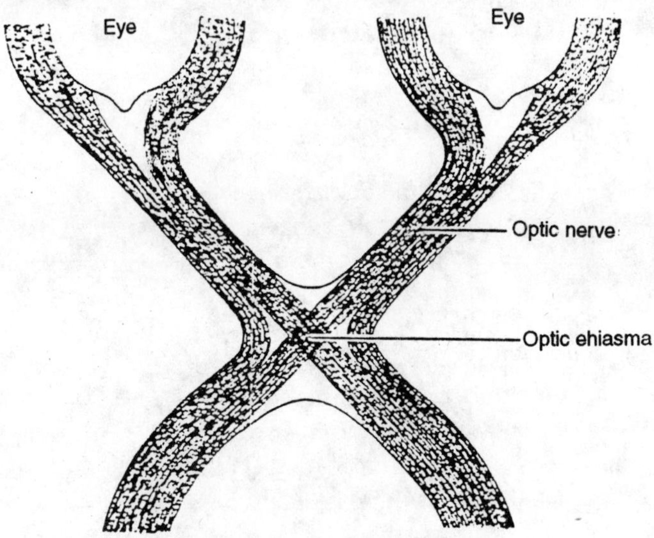

Fig. 4.70. Nerve fibers of optic nerve cross to the other side of the brain at the x-shaped optic chiasma. In mammals there is incomplete decussation and about one third to one half of the fibers remain uncrossed. Duplicate images are created on each side of the brain and stereoscopic vision is especially well-developed.

In land animals, lacrimal glands (tear glands) secrete a watery fluid that keeps the cornea moist and clean and provide nourishment to this avascular structure. Epithelial buds form on the undersurface of the eyelid and during development converge as a group to one area of the eyeball. In most animals, they become branched and form an acinous gland that secretes constantly. A lacrimal duct drains the secretions into the nasal passageway.

5

Mesodermal Derivatives in Vertebrates

In vertebrates the mesoderm is differentiated either from embryonic ectoderm or from endoderm. It appears in the form of mesodermal somites. The somites, when first formed, are masses of mesodermal cells with a small cavity in the middle. The cells are arranged radially around the central cavity. With further development, the shape of the somites changes; they become extended in the dorsoventral direction and flattened mediolaterally. In the vertebrates with discoidal cleavage, this leads to the elevation of the dorsal parts of the embryo above the general level of the blastodisc.

The flattening of the somite is accompanied by a change in the shape of its central cavity (the *myocoel*); instead of being spherical, the cavity becomes a narrow vertical slit. An inner wall and an outer wall become clearly distinguishable, corresponding to the parietal and visceral layers of the lateral plates. The inner wall of the somites becomes very much thicker than the outer wall.

The fate of the inner and the outer wall of the somites is completely different. The outer wall contributes to the formation of the connective tissue layer of the skin, and is therefore called the dermatome. The inner wall produces skeletogenous tissue and the voluntary striated muscles of the body. The skeletogenous tissue develops from the lower edge of the inner wall of the somite, and this part of the somite is therefore called the sclerotome. The sclerotome breaks up into a mass of mesenchyme cells. The cells migrate into the spaces surrounding the notochord and the spinal cord, envelop these organs and later differentiate into cartilage, thus forming the bodies and the neural arches of the vertebrae. The hemal arches in the tail region and the ribs are of the same origin.

The dorsal part of the inner wall of the somite is the source of somatic muscles, in the vertebrate's body. It is therefore called the myotome. The cells of the myotome rearrange themselves so that they become elongated in a longitudinal direction. These longitudinally elongated cells differentiate subsequently into the striated muscle fibers. Originally each myotome becomes a muscle segment, separated from the one anterior and the one posterior to it by a connective tissue layer, the vertical myocomma. At one time in its development the whole somatic musculature of the vertebrate consists of such segments arranged in linear order.

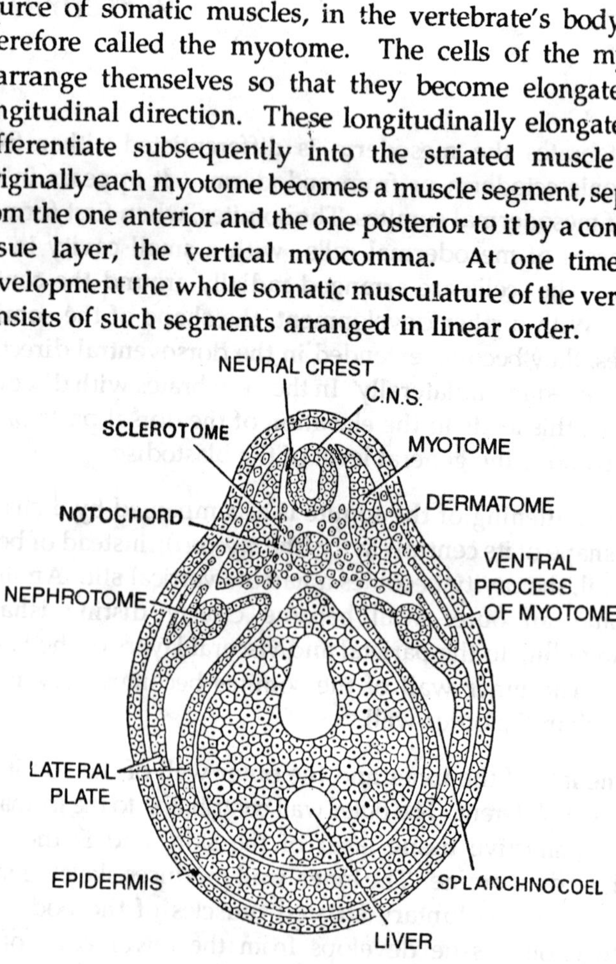

Fig. 5.1. Frog. T.S. of young tadpole through trunk region.

In the lower vertebrates, the myotome is the largest part of the somite, the sclerotomes being small and rather inconspicuous. In the Amniota, however, the sclerotomes are much larger. Only the upper edge of the inner wall of the somite adjoining the dermatome becomes the myotome. The size of this part increases rapidly; the myotomes grow downward to assume the same position, lateral to the neural tube and the notochord, that they occupy in the amphibians right from the start.

Developing as they do from the somites, the myotomes are originally dorsal in position. In the course of the subsequent development, the muscle segments spread downward in the space between the skin on the outside and the somatic layer of the lateral plate on the inside, until the muscle segments on the right and the left side meet ventrally. It is in this stage, with minor alterations, that the somatic muscles persist in the fishes and in the aquatic larvae of amphibians. The segmentation of the lateral and ventral muscles, as well as of the dorsal muscles, is directly derived from the segmentation of the mesodermal mantle in somites.

In terrestrial vertebrates, the primitive segmentation of the somatic muscles is more or less obliterated in connection with a change of locomotion. The segmented lateral bands of muscle are adapted to locomotion by lateral inflections of the body and the tail. The locomotion by means of two pairs of legs requires a completely different organization of the muscle system; as a consequence, only traces of the original muscle segmentation can be discovered in the terrestrial vertebrates.

In *Amphioxus,* muscle segments *continue* anteriorly almost to the tip of the snout. With the development of the brain and the skull, in vertebrates, the head region is exempted from the process of propelling the body by lateral inflections. The somatic muscles become superfluous in the head region. Nevertheless, mesodermal somites are formed in the head region of the embryo. The muscle segments derived from the somites lying posterior to the ear (the postotic somites) may persist and be linked up with the muscle segments of the neck region. The somites lying anterior to the ear vesicle are always very transitory structures: a part of the cells of these somites, however, differentiates as somatic muscle. The muscles derived from this source do not serve for locomotion but are the six pairs of oculomotor muscles: the four rectus muscles and the two oblique muscles on each side.

THE AXIAL SKELETON: VERTEBRAL COLUMN AND SKULL

In most vertebrates, the axial skeleton passes through three phases in its development. In the first phase, the supporting system of the body is represented by the notochord. In the second phase, cartilage develops partly in direct connection with the notochord, partly independently of the latter. This condition is preserved in the adult state of contemporary cyclostomes and elasmobranch fishes (we are not concerned with the question of whether this condition is primitive or secondary). In the remainder of the fishes and in tetrapods, the cartilaginous skeleton is later replaced or supplemented by the bony skeleton. The bony skeleton and its relationship to the cartilaginous skeleton are amply dealt with in courses on comparative anatomy of vertebrates and may, therefore, be entirely left out of consideration. We will consider here only the following aspects in the development of the skeleton;

1. The origin of cells giving rise to the cartilaginous skeleton.
2. The arrangement of the parts of the early cartilaginous skeleton in relation to the other organ rudiments of the embryo.
3. The dependence of the formation of the skeletal parts on the adjoining structures.

The material for the cartilaginous axial skeleton in the body and tail of vertebrates is derived from the sclerotomes.

The sclerotomes, being parts of the somites, are segmental in origin, but once they become transformed into mesenchyme the segmental arrangement is largely lost, and the mesenchyme spreads out as a continuous sheath along the notochord, enveloping the spinal cord above and the caudal artery and vein below the notochord in the caudal region. In this continuous mesenchymal sheath, nodules of cartilage, known as the arcualia, appear later in close apposition to the external surface of the notochord. Typically the arcualia appear in double pairs; one pair is formed dorsolaterally to the notochord and another pair ventrolaterally. The dorsolateral arcualia grow out dorsally alongside the spinal cord and unite above the spinal cord to form the neural arch. The ventrolateral cartilages in the caudal region grow downward and unite underneath the caudal vein to form the hemal arch. In the cervial and thoracic region they give rise to lateral outgrowths—the rudiments of the ribs. At the same time, the proximal parts of all four cartilages spread out around the surface of the notochord, contributing in varying degrees to the formation of the body of the vertebra.

The cartilaginous neural and hemal arches do not bear a simple relationship to the somites (or sclerotomes); in many groups of vertebrates there are two series of cartilages (two pairs of dorsal elements and two pairs of ventral elements) developed in each mesodermal segment. In other cases, although there is only one set of cartilages formed, these are situated intersegmentally at the junction of two myotomes (the myotomes retain the original segmentation of the somites). The way in which the original cartilage rudiments co-operate in the formation of the definitive vertebrae will not concern us here.

The fact that the cartilaginous (and later bony) vertebral column replaces the notochord functionally suggests that the location and the arrangement of the vertebral cartilages should be dependent on the notochord. In the absence of the notochord, the cartilaginous axial skeleton is very irregular. Cartilage, however, is not completely absent in those sections of the body which do not have the notochord. The notochord is thus not indispensable for the formation of axial cartilages. The irregularity of the latter could be a secondary effect of the extirpation of the notochord, seeing that other systems, such as the spinal cord and segmented muscles, are greatly distorted as a result of the failure of the embryo to stretch normally.

In some experiments in which the spinal cord instead of the notochord was removed from salamander embryos, it was found that the axial cartilages were either completely absent or reduced to insignificant vestiges. This could not be the result of damage to the sclerotomes, since in other experiments very extensive defects of the somites, including the sclerotomal region, were completely restored at the expense of the remaining fragments of the somites. The results of extirpation experiments were corroborated by the transplantation of pieces of spinal cord into an incision on the lateral surface of the somites of another embryo. It was found that complete and well-developed neural arches were formed in association with the grafted spinal cord.

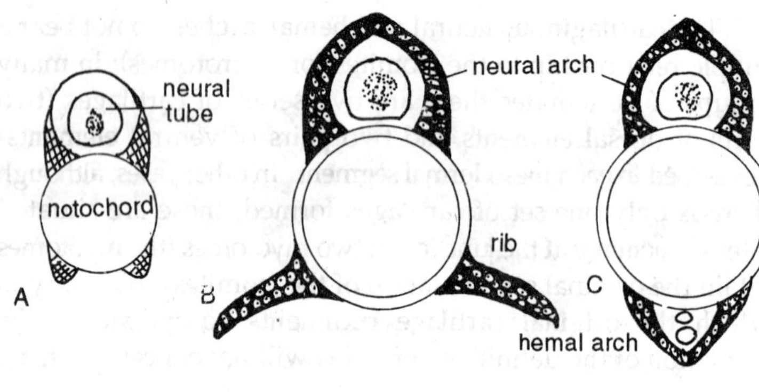

Fig. 5.2. Diagram of the dorsal and ventral pairs of arcualia *(a)* giving rise to the neural arch and the ribs in the trunk *(b)* and to the neural arch and the hemal arch in the tail *(c)*.

The segmentation of the neural arches has not so far been considered. In normal development there is a definite relationship between the segmentation of the longitudinal dorsolateral muscles, the segmentation of the spinal nerves and ganglia, and the segmentation of the vertebral skeletal elements, in particular that of the neural arches. There is a pair of spinal nerves with spinal ganglia corresponding to each muscle segment, the ganglia being situated opposite the median surface of the muscle segment. The neural arches alternate with the spinal ganglia and are thus situated intersegmentally with respect to the myotomes, so that each neural arch (and subsequently each vertebra) is connected to two consecutive muscle segments. The blood vessels also bear a relationship to this common pattern: an intersegmental artery arising from the dorsal aorta along each myocomma, close to the vertebral column.

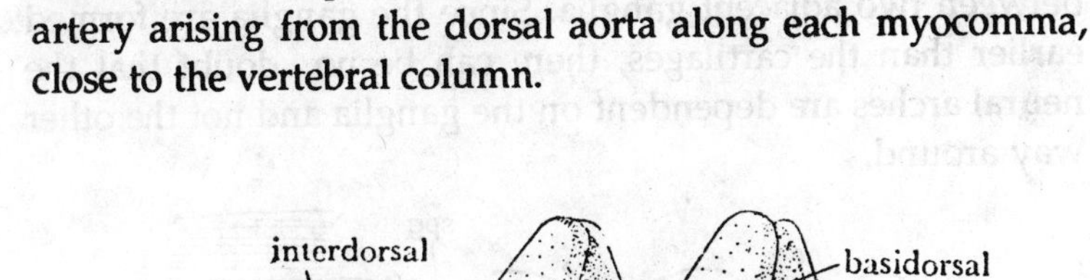

Fig. 5.3.

Experimental evidence is available to show that this whole system of metamerically arranged parts is originally dependent on the segmentation of the muscle rudiments. In a salamander embryo at the tailbud stage, the anterior somites are somewhat broader than the posterior ones, so that if a block of several somites in the branchial region is removed and replaced by a block of somite from the posterior trunk region, more somites may be fitted into the wound than had been cut out. Thus somites 7 to 12 could be substituted for somites 3 to 5. As a result, the number of muscle segments on the operated side was increased in comparison with the normal. It was found that the number of spinal nerves and spinal ganglia was increased also, but not necessarily in strict correspondence to the number of muscle segments, one nerve sometimes supplying more than one muscle segment. On the other hand, the number of neural arches was found to be strictly in accord with the number of spinal ganglia, a bar of cartilage always appearing between two adjacent ganglia. Since the ganglia are formed earlier than the cartilages, there can be no doubt that the neural arches are dependent on the ganglia and not the other way around.

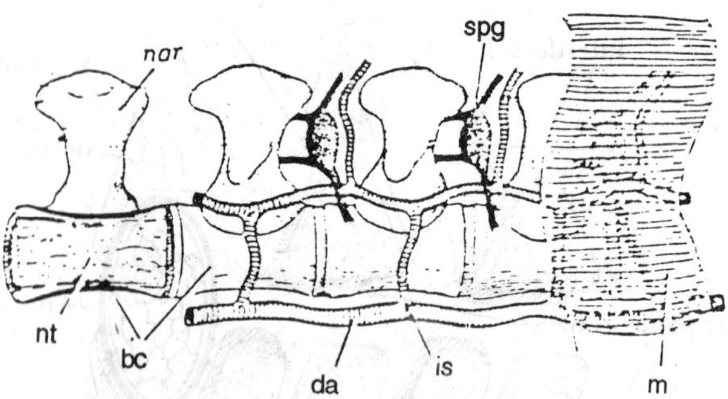

Fig. 5.4. The relation between the muscle segments *(m)*, the spinal ganglia *(spg)*, the intersegmental arteries *(is)* and the vertebrae in a larva of *Ambystoma*, *bc*, Bony cylinder of centrum of vertebra; *da*, dorsal aorta; *nar*, neural arch; *nt*, notochord.

The whole chain of reactions would then appear to be as follows. The neural crest cells which are produced along the whole length of the neural tube become aggregated opposite the median surfaces of the somites, and these aggregations become the rudiments of the spinal ganglia. Next, the cells of the skeletogenic mesenchyme, produced by the sclerotomes, spread out over the notochord and neural tube, but they are apparently repulsed by the spinal ganglia and thus, instead of forming a continuous sheet of cartilage enclosing the spinal cord, they give rise to a series of disconnected element (the dorsal arcualia) alternating in position with the spinal ganglia. This is obviously not the complete picture, as it does not account for the cases in which two pairs of cartilages appear between each consecutive pair of spinal ganglia. It is thus likely that some structures other than the spinal ganglia, the neural tube and the notochord take part in determining the position of early cartilaginous rudiments.

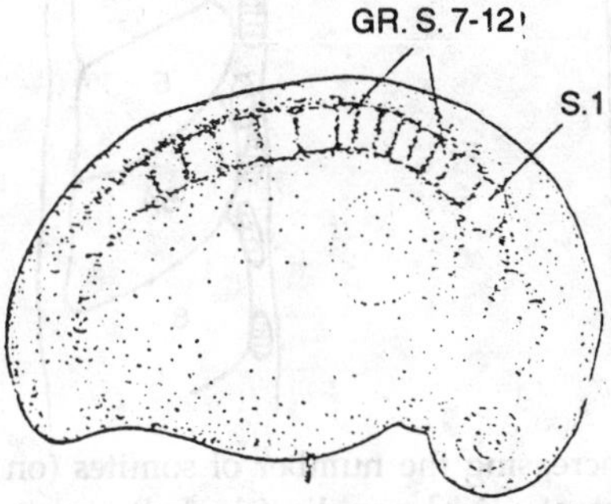

Fig. 5.5. *Ambystoma punctatum embryo* with somites 7-12 transplanted in place of somites 3-5.

The neural portion of the cranium in vertebrates in part bears the same relationship to the notochord and to the neural tube as does the axial skeleton in the posterior parts of the body. But there is little, if any, trace of a segmentation in the development of the cranium, and important parts of the cranium are quite peculiar in this respect.

In the lower vertebrates at the earliest stages of development, the cartilaginous cranium consists of several independent rudiments. These are: (1) The trabeculae; (2) the parachordals; (3) the capsules of the sense organs—the nose, the eye and the ear.

The trabeculae (or trabeculae cranii) are a pair of elongated cartilages which appear in the most anterior part of the head, in front of the hypophysis. The trabeculae lie ventrally and

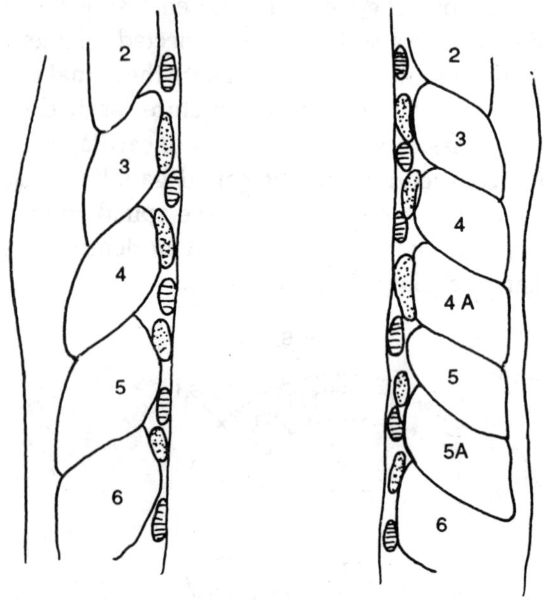

Fig. 5.6. Result of increasing the number of somites (on right side). The number of spinal ganglia (shaded) and neural arches (stippled) increased, as compared with control side (left).

ventrolaterally to the diencephalon and telencephalon, and their upper edges are wedged in between the brain on the inside and the rudiments of the nose and the eye on the outside. It has been established both by observation and experiment that the mesenchyme from which the trabeculae are developed comes from two different sources. The anterior part of the trabeculae is formed of neural crest cells migrating forward and downward anterior to the eye cup. The posterior part of the trabecula is mesodermal origin and is derived from the prechordal plate mesenchyme.

The parachordals, or parachordal cartilages, are derived from the mesenchyme produced by the sclerotomes of the somites in the head region. This mesenchyme spreads out on both sides of the notochord and eventually chondrifies in the form of two longitudinal rods, situated alongside the notochord and ending anteriorly at the same level as the notochord, that is, just posteriorly to the infundibulum and the rudiment of the hypophysis. The parachordals are very similar in origin and position to the rudiments of the cartilaginous vertebral column, but they lack the segmentation of the latter (possibly because the cranial ganglia, owing to the greater breadth of the neural tube in the head region, lie much farther laterally and away from the region in which development of the parachordals is taking place).

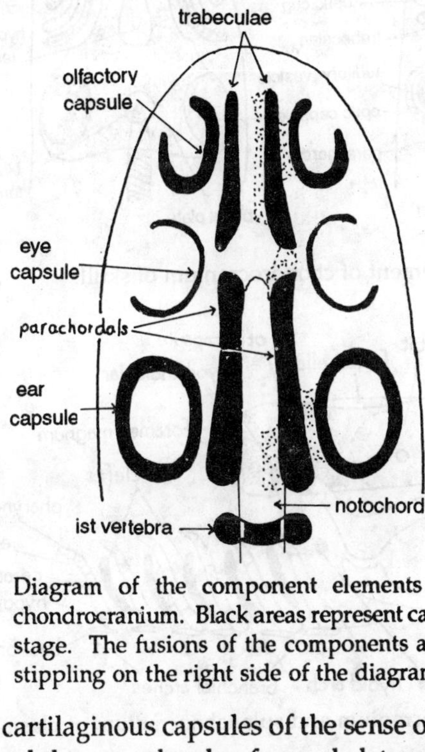

Fig. 5.7. Diagram of the component elements of the vertebrate chondrocranium. Black areas represent cartilages in the initial stage. The fusions of the components are indicated by the stippling on the right side of the diagram.

The cartilaginous capsules of the sense organs—the nose, the eye and the ear—develop from skeletogenic mesenchyme accumulating around the surface of the epithelial parts of these organs. The source of the mesenchyme may not be the same in all three cases. It is fairly certain that the ear capsule is formed by mesenchyme derived from the same sclerotomes as give rise to the parachordals. In fact, the ear capsules are formed in close proximity and even in continuity with the parachordals, and strands of mesenchyme have been observed leading from the rudiments of the parachordals to the site in which the cartilaginous ear capsules start to develop.

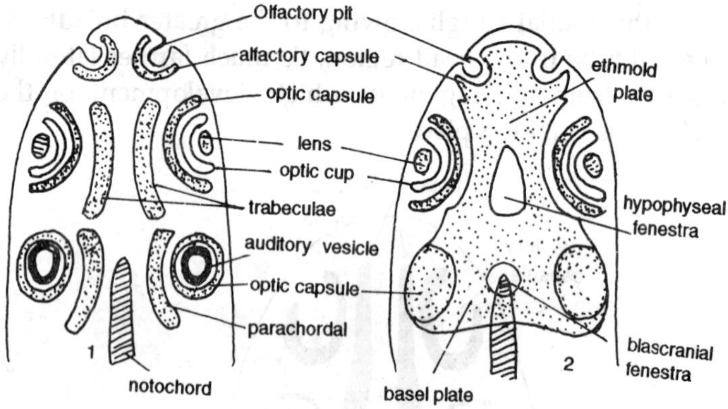

Fig. 5.8. Development of chondrocranium of skull.

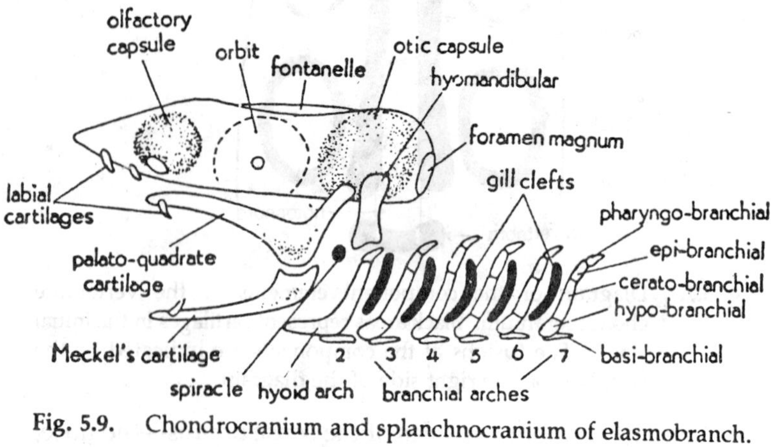

Fig. 5.9. Chondrocranium and splanchnocranium of elasmobranch.

It has been claimed that cartilage of the nose capsules is derived from the epithelial nose rudiment itself. It would seem that this statement needs further corroboration. On the other hand, we have seen that the nasal placodes develop in very close proximity to the anterior transverse neural fold. The neural fold being the source of neural crest cells, it would not be very astonishing if the neural folds could produce the nasal cartilages as well as the epithelial parts of the olfactory organ.

In most vertebrates, the eyebrow is surrounded first by a connective tissue capsule, the sclera, which becomes cartilaginous is later stages. The source of the cells forming the sclera has not yet been established.

The capsules of all three sense organs are dependent in their development on the epithelial parts of the organs. This has been clearly shown in the case of the ear capsule and is very probably true in the case of the olfactory capsule and the sclera of the eye, as these capsules fail to be formed if the epithelial parts of these organs are removed.

The parachordals, through spatially intimately associated with the notochord, show a high degree of independence from the latter organ. In experiments in which the notochord rudiment was removed, the development of the base of the skull does not seem to be affected to any extent. In particular, there is no foreshortening of the posterior part of the head, in sharp contrast to the shortening and stunting of the trunk and tail region in the absence of the notochord.

The further development of the cartilaginous skull (chondrocranium) is characterized by the enlargement and fusion of the initially formed cartilages. The right and left trabeculae fuse across the midline underneath the forebrain, and their posterior ends fuse with the tips of the parachordals. The parachordals envelop the notochord, particularly its dorsal side, and so give rise to the basal plate of the skull. At the point where the infundibulum and the hypophysis are situated, the cartilages leave a ventral opening (the hypophyseal fenestra) which persists for a long time and is closed only much later by cartilage or bone on the ventral side. The posterior ends of the parachordals grow upward and eventually fuse above the medulla, thus enclosing the foramen magnum of the skull. The nose capsule and the otic (auditory) capsule become firmly

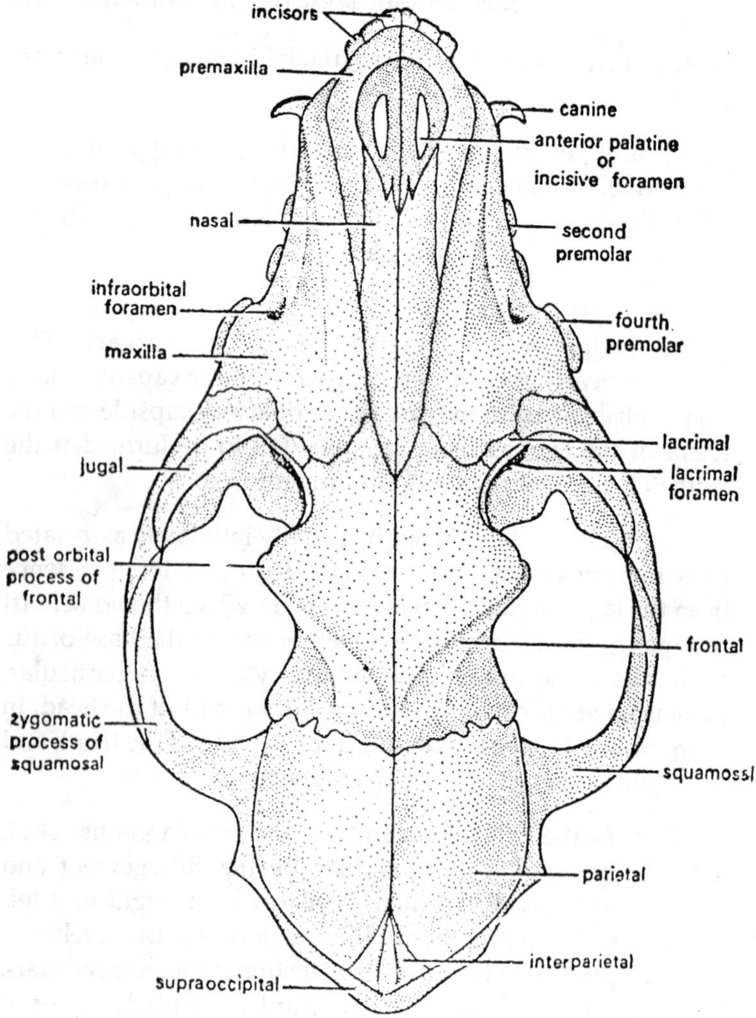

Fig. 5.10. Skull (Dorsal view).

joined to the trabeculae and the parachordals respectively, thus contributing to the formation of the lateral walls of the chrondrocranium. Gradually the lateral edges of the cartilaginous skull grow upward, and in the more primitive vertebrates (cyclostomes, many fishes and amphibians) form a roof over the dorsal surface of the brain. In teleost fishes and

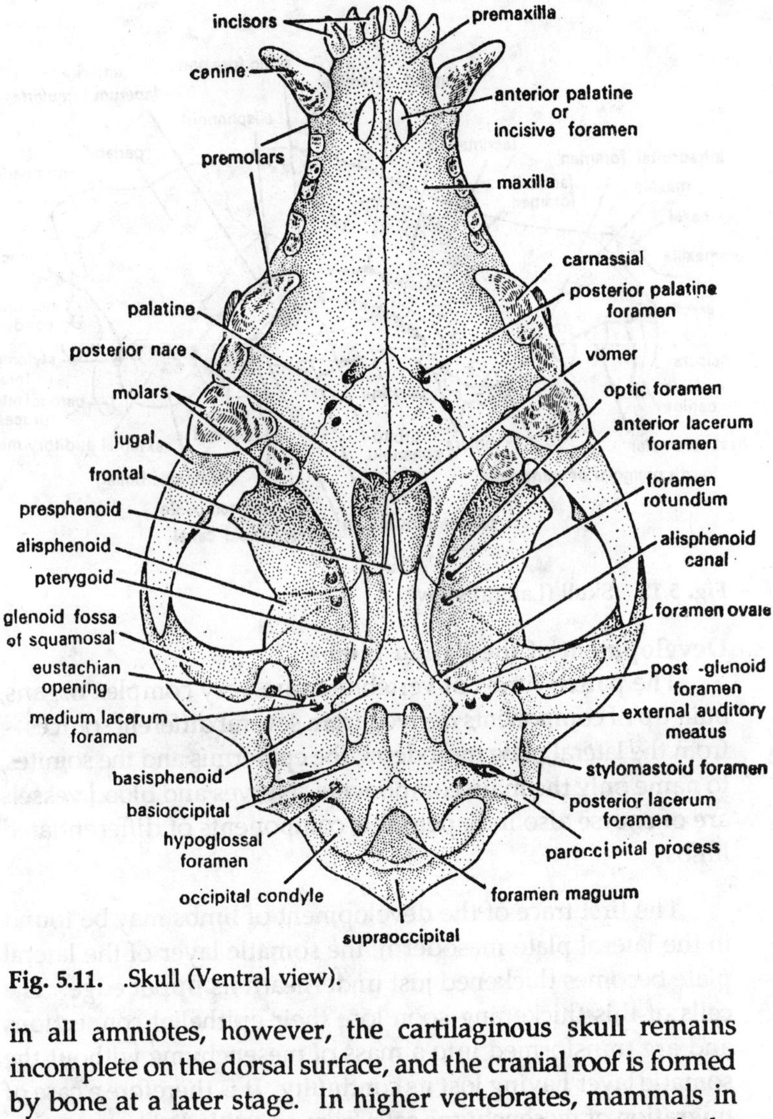

Fig. 5.11. .Skull (Ventral view).

in all amniotes, however, the cartilaginous skull remains incomplete on the dorsal surface, and the cranial roof is formed by bone at a later stage. In higher vertebrates, mammals in particular, the initial stages in the development of the cartilaginous skull may be speeded up in such a way that the trabeculae and parachordals are fused right from the start— a condition which is achieved in lower vertebrates secondarily.

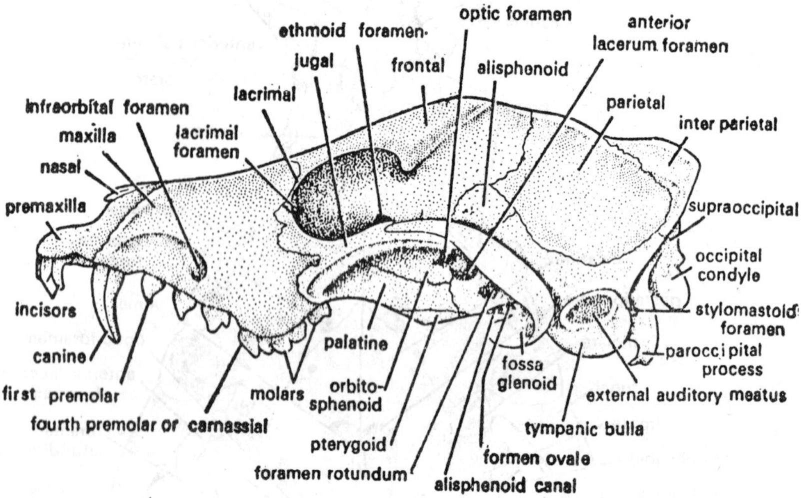

Fig. 5.12. Skull (Lateral view).

Development of the Paired Limbs

The paired limbs of vertebrates are very complex organs, built up of components derived from several different sources—from the lateral plate mesoderm, the epidermis and the somites, to name only the main components. Nerves and blood vessels are of course also indispensable components of differentiated limbs.

The first trace of the development of limbs may be found in the lateral plate mesoderm; the somatic layer of the lateral plate becomes thickened just underneath its upper edge. The cells of this thickening soon lose their epithelial connections and are transformed into a mass of mesenchyme without the somatic layer having lost its continuity. It is therefore a case of migration of mesenchyme cells from an epithelial layer, rather than that of the breaking up of epithelium into mesenchyme. The mesenchyme accumulates between the remaining lateral plate epithelium and the epidermis and soon becomes firmly attached to the inner surface of the epithelium. The thickening of the lateral plate mesoderm and the subsequent formation of a mass of mesenchyme under the epithelium may coincide rather closely with the position of the two pairs of limbs, that is, they may appear in two disconnected regions—just behind the branchial region and just in front of the anus. This is the case in the amphibians. In other vertebrates, however, the thickening and the mesenchyme accumulation may spread far beyond the actual region of limb development. In fishes the early limbs rudiments are more elongated anteroposteriorly in the earlier stages than in the later stages of development. In the amniotes the thickenings and the mesenchyme gatherings are continuous throughout the whole length of the body in the form of horizontal ridges—the Wolffian ridges. However, the most anterior and the most posterior parts of the ridge are thicker than the intermediate part, and it is only these anterior and posterior parts that develop progressively, giving rise to the forelimbs and hindlimbs. The intermediate part of the Wolffian ridge later disappears.

The epidermis over the mesenchyme mass becomes slightly thickened and bulges outward. This happens over the Wolffian ridges as well, but in the intermediate parts of the ridge the epithelial thickening disappears together with the gathering of mesenchyme. In the regions where the fore-and hindlimbs are to develop, the protrusion, consisting of a thickened epithelial covering and of an internal mass of densely packed mesenchyme, increases and becomes the limb-bud.

Of the two components contributing to the formation of the limb-bud, the mesoderm is determined as such at an early stage shortly after the closure of the neural tube. Pieces of lateral plate may be cut out in this stage and transplanted under the epidermis on the flank or on the head. The local epidermis will then become the epithelial component of the limb-but, and a limb will develop heterotopically. The presumptive epidermis of the limb in the same stages, that is, before a limb-bud has been formed, does not possess any special properties and, if transplanted alone, will not give rise to a new limb. Epidermis from any part of the body is able to co-operate with the presumptive limb mesoderm in forming a limb-bud. This can be shown by removing the epidermis in the limb region and then covering the wound with a flap of epidermis taken from any part of the body.

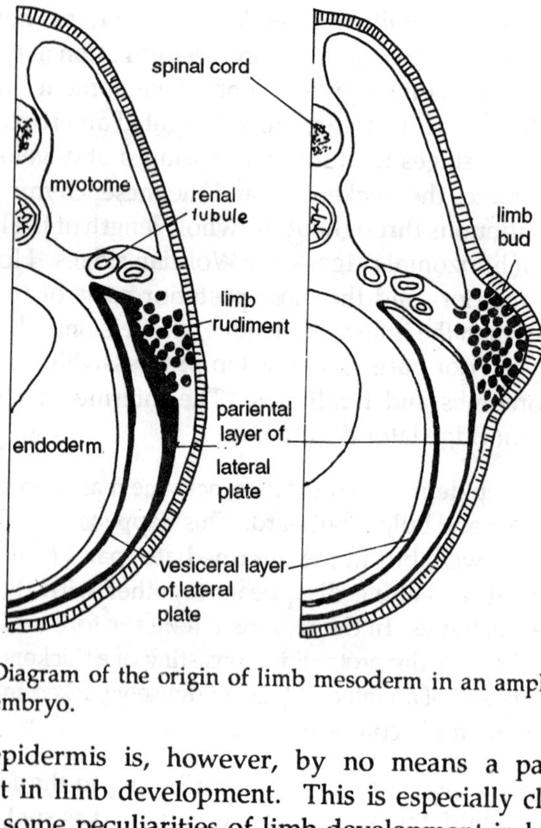

Fig. 5.13. Diagram of the origin of limb mesoderm in an amphibian embryo.

The epidermis is, however, by no means a passive component in limb development. This is especially clearly shown by some peculiarities of limb development in higher vertebrates. In the amniotes, the limb-bud becomes slightly flattened at an early stage, and an epidermal thickening develops along the edge of the flattened bud. The thickening is in the form of a very sharply defined ridge and sometimes (in reptiles) even takes the form of a solid fold of the epidermis. In cross section the ridge looks like a nipple. It is referred to as the ectodermal apical ridge.

The cells of the ridge differ from the ordinary epidermal cells not only in their arrangement but also in their physiological properties. It was found that they contain more ribonucleic acid and more glycogen, and they differ very conspicuously from surrounding epidermal cells in their high content of the enzyme alkaline phosphate. All these biochemical properties may be taken as indications of active metabolism.

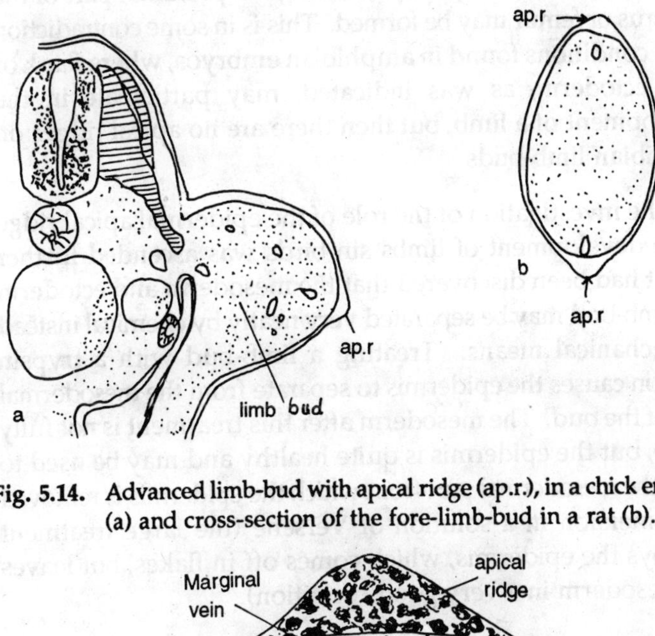

Fig. 5.14. Advanced limb-bud with apical ridge (ap.r.), in a chick embryo (a) and cross-section of the fore-limb-bud in a rat (b).

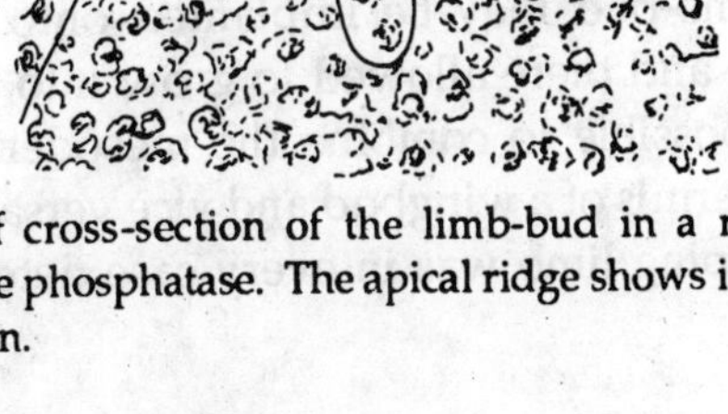

Fig. 5.15. Part of cross-section of the limb-bud in a rat stained for alkaline phosphatase. The apical ridge shows intense positive reaction.

The apical ectodermal ridge is indispensable for the normal outgrowth of the limb rudiment. If the ectoderm of the apical ridge of a wingbud of a three-day chick embryo is removed without causing damage to the underlying mesoderm, the distal parts of the wing fail to be formed, although the proximal parts develop quite normally. If the ectoderm covering the limb-bud in a chick embryo is removed and replaced by epidermis from another part of the body, the latter is found to be incapable of developing an apical ridge. The result is that the development of the whole of the distal part of the limb is suppressed. The girdle may, however, develop normally, and a short piece of cartilage representing the proximal part of the humerus or femur may be formed. This is in some contradiction to the conditions found in amphibian embryos, where flank or head ectoderm, as was indicated, may participate in the development of a limb, but then there are no apical ridges on amphibian limb-buds.

The investigation of the role of the epidermal apical ridge in the development of limbs sin birds was ascended further after it had been discovered that the mesoderm and ectoderm of a limb-bud may be separated very neatly by chemical instead of mechanical means. Treating a limb-bud with a trypsin solution causes the epidermis to separate from the mesodermal core of the bud. The mesoderm after this treatment is not fully viable, but the epidermis is quite healthy and may be used to cover the mesodermal part from which the epidermis is removed by immersion in a solution of Versene (the large treatment destroys the epidermis, which comes off in flakes, but leaves the mesoderm in a very good condition).

After the mesoderm of a limb-bud is covered by epidermis of a different origin, the two stick firmly together, and such a composite limb-bud may be implanted onto the flank of a third embryo and there allowed to grow into a limb. In this way it was possible to combine the mesoderm of a legbud with the epidermis of a wingbud and vice versa. The structure of the developing limb was in every case determined by the origin of the mesodermal component; it was wing if the mesoderm was taken from a wingbud, and a leg if the mesoderm was that of a posterior limb rudiment. The origin of the ectoderm did not affect the nature of the developing limb. This experiment stresses the leading part played by the mesoderm in limb development.

The next experiment emphasizes the importance of the epidermis. In a "wingless" mutation in fowls, the forelimb-bud appears approximately at the same time as the normal limb-bud but fail to grow and produce a limb, though parts of the limb girdle may be present. It was noticed that the epidermis covering the wingbud in the affected embryos does not have an apical ridge. The experiment was therefore undertaken of combining the mesoderm of a normal wingbud with the epidermis of an embryo of the wingless strain. The result was as had been expected; in the absence of an apical ridge the bud did not grow and no distal parts of the wing were formed.

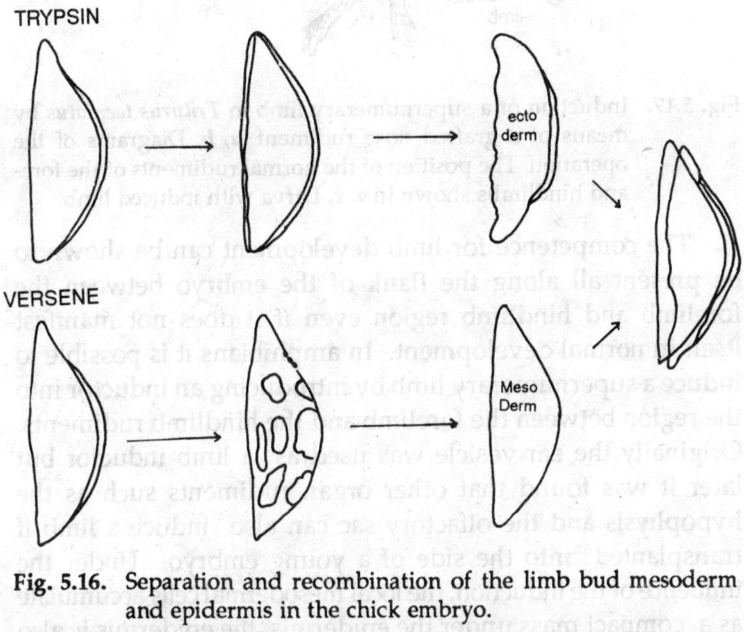

Fig. 5.16. Separation and recombination of the limb bud mesoderm and epidermis in the chick embryo.

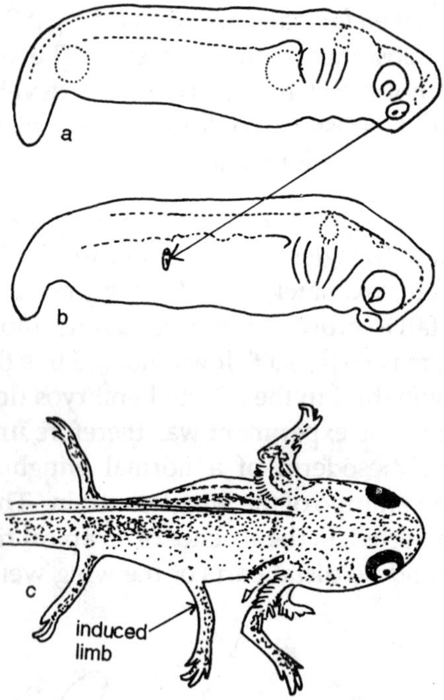

Fig. 5.17. Induction of a supernumerary limb in *Triturus taeniatus* by means of a grafted nose rudiment. *a, b,* Diagrams of the operation. The position of the normal rudiments of the fore- and hindlimbs shown in *a. c,* Larva with induced limb.

The competence for limb development can be shown to be present all along the flank of the embryo between the forelimb and hindlimb region even if it does not manifest itself in normal development. In amphibians it is possible to induce a supernumerary limb by introducing an inductor into the region between the forelimb and the hindlimb rudiments. Originally the ear vesicle was used as a limb inductor but later it was found that other organ rudiments such as the hypophysis and the olfactory sac can also induce a limb if transplanted into the side of a young embryo. Under the influence of the induction, the local mesodermal cells accumulate as a compact mass under the epidermis, the epidermis is also made to react, and an additional limb develops. Depending on whether the induced limb lies nearer to the normal forelimb or hindlimb region, it may resemble a forelimb or a hindlimb in its structure.

Experiments on limb induction suggest that in normal development, as well, there must be some factor determining which part of the mesoderm competent for limb development actually produces a limb rudiment. Limb-buds are often induced together with other structures when "spino-caudal" inductors are introduced into embryos in the gastrula stage. By transplanting the presumptive mesoderm together with the notochord into the lateral plate region of an embryo in the neurula stage, it is chord into the lateral plate region of an embryo in the neurula stage, it is possible to induce the lateral plate mesoderm to develop into kidney (normally produced by the stalks of the somite. In these experiments it has often been observed that additional limb-buds developed together with the kidney tubules. It follows that the determination of the limb mesoderm occurs in conjunction with the determination of other parts of the mesodermal mantle. The kidney in itself cannot, however, induce a limb.

Differentiation in the Limbs

After the limb-bud has grown so far that its length exceeds its breadth, the differentiation of the subordinate parts of the limb gets in. We have already noted the slight flattening of the limb-bud. Now the distal portion of the bud becomes flattened even more, and at the same time it becomes distinctly broader than the proximal part of the limb rudiment. The flattened and broadened distal part is the hand (or foot) plate. The edge of the plate is initially circular, but soon it becomes pentagonal, the projecting points indicating the rudiments of the digits. While the tips of the digit rudiments keep growing out farther, extensive necrosis occurs in the intervening sections. Mesodermal and ectodermal cells (including those of the apical ridge) in these sections die and are consumed by macrophages. As a result the digits become separated by distinct incisions.

The five digits appear simultaneously in all amniotes with pentadactyl limbs, but in amphibians, especially in urodeles, the first two digits appear earlier and digits 3, 4 and 5 are formed one after another on the posterior edge of the limb. Where less than five digits are present in the adult limb, or more than five in cases of hyperdactyly, this condition is reflected in the structure of the hand (foot) plate. In the early limb rudiments, the future flexor surface is ventral and the future extensor surface is dorsal, but as the limb elongates a rotation takes places so that the flexor surface is turned posteriorly, and eventually it may even face in a posterodorsal direction. The preaxial edge of the limb, which is originally anterior, is then turned downward. With the elongation of the limb, it becomes bent at the elbow joint (or knee joint). A less pronounced flexion develops at the base of the carpus (or tarsus). The three main sections of the limbs thus become recognizable externally.

Concurrently with changes in the external appearance of the limbs, differentiation occurs in the interior of the limb rudiments. The mesenchyme cells which are closely packed in a young limb-bud become segregated into areas in which the mesenchyme is lying more loosely and into other areas in which the mesenchyme cells are crowded. The latter are the rudiments of the skeletal parts of the limb. The concentrated masses of mesenchyme in due course become converted into procartilage, and then, by further deposition of intercellular matrix into cartilage. Whereas in the initial stage of mesenchyme concentration large sections of the limb skeleton are represented by a common mass of mesenchyme, in the procartilage stage individual elements of the skeleton are laid down as separate units; these may fuse together later. The differentiation of the limb skeleton generally proceeds in a proximodistal direction, though some deviations from this order are of fairly general occurrence. In amphibians, the first skeletal part to become recognizable is the stylopodium (humerus or femur). Parts of the zeugopodium (radius and ulna in the forelimb, tibia and fibula in the hindlimb) are laid down next, and the autopodium differentiates considerably later. The girdle rudiments appear after the stylopodium, but earlier than the autopodium. In higher vertebrates the girdle tends to be developed simultaneously with the proximal elements of the limb. In the autopodium, the proximodistal sequence is upset by the larger skeletal elements, the metacarpals and metatarsals differentiating more rapidly than the smaller elements, namely, the carpals and the tarsals. In the digits, however, the proximal phalanges are laid down earlier than the distal ones.

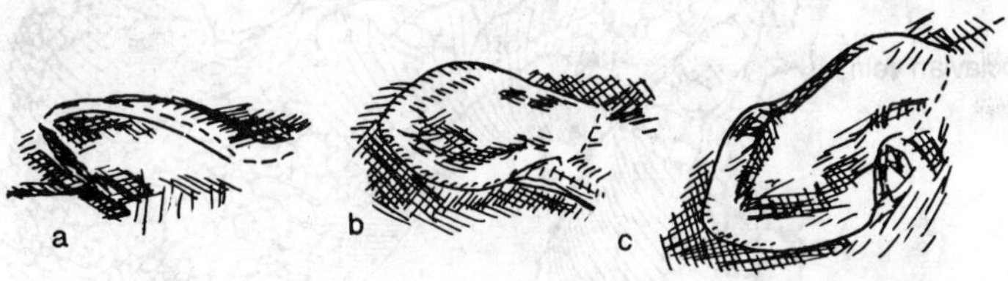

Fig. 5.18. Rotation of the developing forelimb in the lizard.

The blood vessels appear in the limb-bud at an early stage, and although the pattern of the arteries, veins and capillaries is too variable to deserve much attention here, one blood vessel may be mentioned. It is situated along the edge of the hand (or foot) plate, just underneath the ectodermal apical ridge, and possibly it is responsible for supplying nutriment to this rapidly growing area of the limb.

The pattern according to which the various constituent parts (bones, muscles, blood vessels, nerves) are arranged in each limb is asymmetrical — proximal and distal ends, dorsal and ventral surfaces, anterior and posterior sides of the limb being different from one another. The first indication of this asymmetrical pattern can be noted when the tip of the limb-bud grows out in a slanting posterior direction instead of growing straight out away from the side of the body.

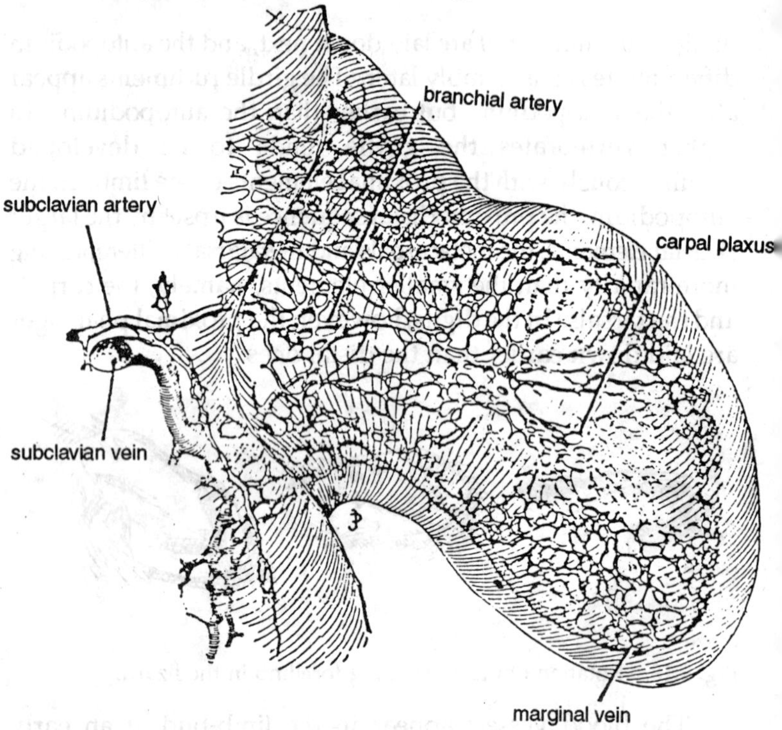

Fig. 5.19. Reconstruction of the blood vessels in a limb rudiment of a pig embryo.

Special experiments have been carried out to find out how the asymmetry of the limb, and thus the basic pattern of its differentiation, is determined. Experiments were originally carried out on the embryos of the salamander *Ambystoma punctatum*. Forelimb rudiments were transplanted at different stages after the end of neurulation in such a way that either one limb rudiment axis or two or all three were inverted (disharmonious) with respect to the host's body. The inversion of the proximodistal axis could only be carried out with the mesodermal part of the limb rudiment, but this is not of importance, as it is the mesoderm that is the carrier of the limb determination. The transplantations were done either *orthotopically*, that is, in place of a normal limb rudiment, or *heterotopically*, on the flank. The experiments showed that the three axes of the limb rudiment were not determined simultaneously. In the earliest rudiments immediately after neurulation, the anteroposterior axis is already fixed; limb rudiments, transplanted disharmoniously, with this axis inverted, had the tip of the limb-bud growing forward instead of backward, and the limbs later had their postaxial (ulnar) side placed anteriorly. At the same time, the inversion of the dorsoventral axis of the transplanted limb rudiment could still right itself; that is, the original upper part of the rudiment developed into the ventral (palmar) surface of the limb and the original lower part of the rudiment developed into the dorsal surface of the limb. This shows that the pattern of the limb differentiation was not yet fixed in respect to its dorsoventral axis and that this pattern could be imposed upon the limb rudiment by the host, that is, by the parts which were in connection with the limb rudiment in its new position. In a later stage of development when the tail rudiment begins to elongate, the dorsoventral axis of the limb was also found to be determined; if the limbs were transplanted in an inverted position, with the dorsoventral axis of the limb rudiment the inverse of the dorsoventral axis of the host, the limb developed with an abnormal orientation, its plantar surface facing upward. The surroundings could no longer change the pattern of limb differentiation in respect to the dorsoventral axis, as well as in respect to the anteroposterior axis.

At the same time that the dorsoventral axis of the limb rudiment is determined, the proximodistal axis may still be inverted without impairing the normal development of the limb. Only in a later stage, when the limb-bud begins to be visible from the outside, does the proximodistal axis show signs of being determined, and if the limb mesoderm is transplanted with the axis inverted the limb shows abnormalities in its development. (A limb cannot actually grow inward into the body because it would then be no longer in contact with the epidermis, and this contact is indispensable for the differentiation of parts of the limb). In cases when the transplanted limb rudiment grows out in a disharmonious orientation to the host, it is often observed that a kind of regulation occurs by means of the formation of a second limb-bud, whose orientation and differentiation is harmonious with the host's body. The appearance of the second limb-bud is the result of a sort of splitting of the original rudiment and is due to an influence of the host on the graft. This influence is not strong enough to invert the axial structure of the transplanted rudiment as a whole, but is sufficiently strong to divert a part of the cells of the transplanted rudiment and to cause them to take on the axial structure of the host. If the original disharmonious limb were now to degenerate (as sometimes happens), the host would be in possession of a normal set of limbs.

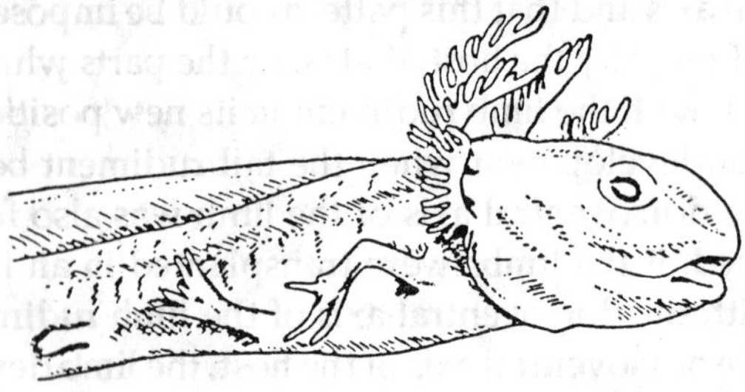

Fig. 5.20. Heterotopic transplantation of the limb rudiment in the late tailbud stage, with inverted dorsoventral axis. The palmar surface of the transplanted limb (partially reduplicated) facing upward.

The splitting of a limb rudiment (or any other organ rudiment) to produce two similar rudiments is known as reduplication. That reduplication is possible shows that the cells of the rudiment are not determined each to fulfill a definite part in the developing organ, even though the axial pattern of the rudiment as a whole is determined. From this it is further inferred that the determination of the axial pattern is not a matter of determining what each cell or group of cells has to do in the process of development of the organ, but it is a matter of polarity, of a heteropolar structure of the rudiment as a whole. This immediately recalls the polarity of the egg in the early stages of development, and some similarity between the rudiment of an organ and the early egg as a whole is indeed shown. Both possess the ability to develop a number of subordinate parts, each within its own scope of action (the whole animal in the case of the egg, one organ in the case of an organ rudiment), and without these subordinate parts being represented by discrete particles in the initial system. Just as the egg can be split mechanically into parts and each part will develop into a miniature whole, so an organ rudiment, in a suitable stage of development, may be split mechanically into halves, and each half will develop into a complete organ. We have already mentioned similar results in the case of the eye rudiment. The same applies to the limb rudiment. The limb rudiment up to the stage of the limb-bud formation may be cut in two and each half transplanted separately, or the two halves left in place and kept apart by inserting a piece of extraneous tissue between them. Each will develop into a whole limb. The two limbs resulting from such a splitting may later grow to the normal size. Splitting of limbs may sometimes occur accidentally in young amphibian larvae developing in nature, or it may possible be caused by the pressure of folds of the amnion in higher vertebrates, and abnormalities will be the result: limbs that are completely or partially reduplicated. If the splitting of the rudiment is due to a mechanical cause rather than being spontaneous, the two halves retain the same axial structure (polarity), thus being replicas of each other.

The limb girdles normally develop in intimate connection with the limbs themselves. The mesodermal material for the girdle is derived from the peripheral parts of the mesenchyme mass, the central part of which becomes the limb-bud. However, in their determination the girdles are partially independent of the limbs. The limb girdles may develop independently of the limbs, if the development of the limb itself is suppressed in some way or other, as by the removal of the limb-bud. In limb induction experiments, limb girdles sometimes are found where the limb itself fails to develop. The development of the girdle is in no way dependent on the existence of an interaction with the epidermis of the skin, as is that of the limb itself.

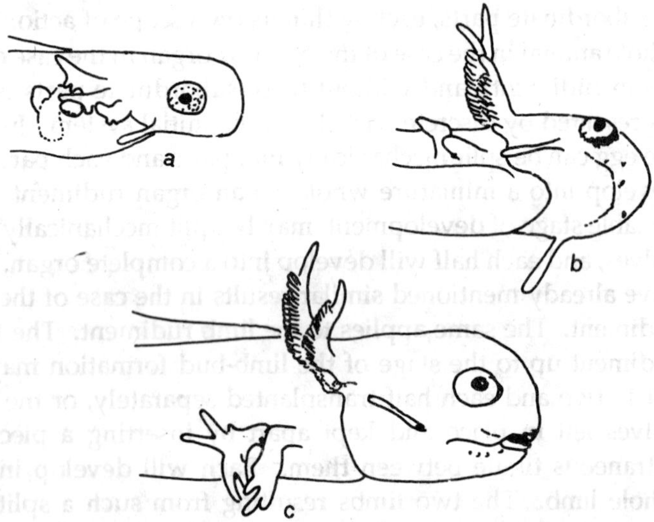

Fig. 5.21. .Splitting of the forelimb rudiment, with a strip of extraneous tissue (T) inserted across it (a) to produce two limbs (b, c).

In birds, the limb girdle may also develop fairly well even if the distal part of the limb is absent as a result of a failing co-operation between mesoderm and ectoderm, as in cases where the epidermis over the limb-bud lacks an apical ridge.

If the girdle (shoulder girdle or pelvic girdle) develops in the absence of the limb, it may be normal in its peripheral parts, but the fossa for the articulation with the humerus or femur does not develop in the absence of at least the proximal part of the stylopodium (humerus or femur). The thickening of the girdle in the region of the articulation is also lacking in the absence of a limb. In so far as these parts are concerned, the amphibian girdles are dependent on the limbs. In bony fishes, in which the cartilaginous part of the shoulder girdle is reduced to a small skeletal element in the region of articulation, the bony cleithrum can develop independently of the pectoral fin (Balinsky, unpublished).

The mesenchyme of the limb-bud does not contain all the mesodermal cells used for the development of a limb; after the limb-bud is formed it receives an additional supply of mesodermal cells from the lower edges of the myotomes. These cells produce the muscles of the limb or at least a part of these. In fishes it is possible to observe muscle buds being formed at the lower edges of the myotomes. These muscle buds are protrusions of the cell masses of the myotomes. The muscle buds push downward and outward until they enter the limb-buds. In the limb-bud itself, the muscle buds derived from different myotomes fuse into a common mass of cells (myoblasts) from which the muscles of the limb subsequently develop. In the higher vertebrates, beginning with amphibians, muscle buds are not found, although the migration of individual cells from the myotomes into the limb rudiments may take place.

In the fishes, if the lateral plate mesoderm of the limb region is transplanted to near the midventral line on the abdomen, it gives rise to fins which are devoid of muscles, thus proving that the lateral plate mesoderm is not capable of producing the limb muscles. Muscles are developed, however, if a part of the somite is included in the graft. In amphibians, on the other hand, when the rudiment of the limb, consisting of lateral plate mesoderm with or without epidermis, is transplanted heterotopically, it will develop into a complete limb with muscles. This, however, does not definitely exclude the participation of the myotome material, since under experimental conditions some regulation could have taken place, just as half the rudiment may produce a whole limb in cases or reduplication.

The nerve supply to the developing limbs has been dealt with in the section on the differentiation of the nervous system.

DEVELOPMENT OF THE URINARY SYSTEM

The excretory organs in vertebrates are essentially aggregates of uriniferous tubules, connected originally at their proximal ends with the coelomic cavity by ciliated funnels (the nephrostomes) and communicating to the exterior by a system of ducts which, in the lower vertebrates, open into the cloaca. Both the tubules and the ducts are of mesodermal origin, and they develop from the stalks of the somites (the nephrotomes). In the most primitive vertebrates, such as the cyclostomes, and also in the Gymnophiona one uriniferous tubule develops from the nephrotome in each mesodermal segment.

The nephrotome, prior to the development of a uriniferous tubule, is a strand of cells connecting the somites to the lateral plate mesoderm. The cells become separated into the parietal and visceral layer, and the cavity, which we may call the *nephrocoele,* is for a while continuous with both the myocoele and the definitive coelom between the two sheets of the lateral plate mesoderm. The connection of the nephrocoele with the cavity of the somite is soon obliterated, but the connection with the cavity of the lateral plate persists in the more primitive type of vertebrate excretory organs and becomes the nephrosome. The dorsolateral wall of the nephrotome becomes drawn out into a hollow tube, the cavity of which is an extension of the nephrocoele. The tube, which is in open connection with the coelomic cavity, becomes the uriniferous or renal tubule. The distal (outward) ends of the most anterior tubules (that is, the tubules formed in the anterior part of the trunk) soon turn backward and then fuse with each other, thus giving rise to the common excretory duct, known as the *pronephric duct*.

It is an essential feature in vertebrates that the renal tubules are associated with bunches of fine blood vessels (the *glomeruli),* through the endothelial walls of which the blood plasma containing excretory products is filtered into the uriniferous tubules or into the coelom in the immediate vicinity of the nephrostomes, so that the nitrogenous waste products may be carried through the tubes and the excretory ducts and removed from the body. Either the mass of blood vessels is invaginated into the wall of the renal tubule, which enlarges to contain the

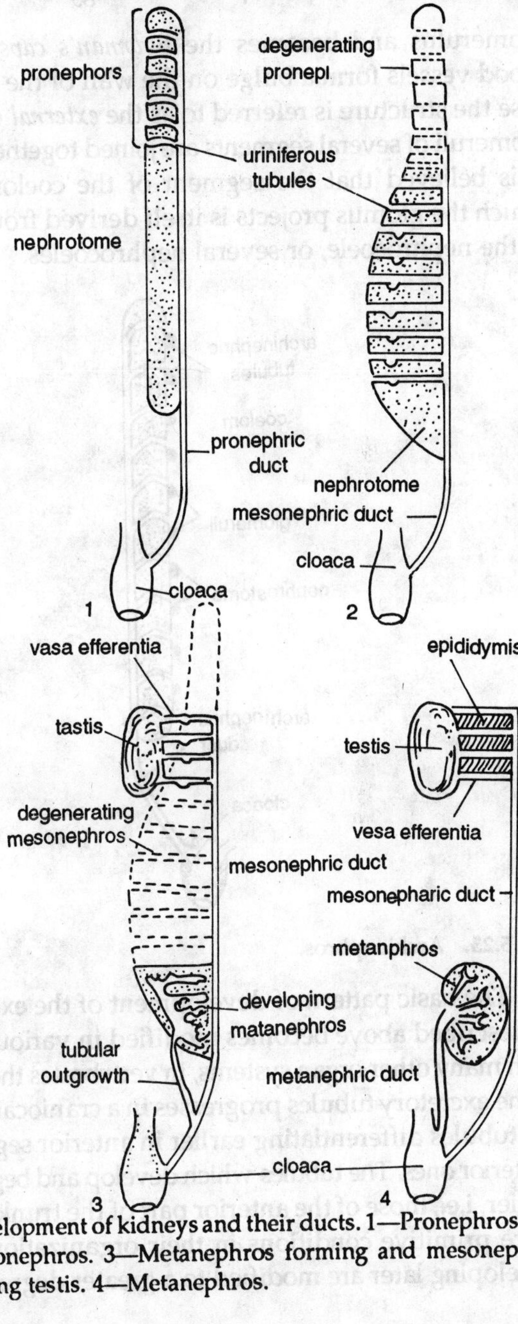

Fig. 5.22. Development of kidneys and their ducts. 1—Pronephros. 2—Mesonephros. 3—Metanephros forming and mesonephros joining testis. 4—Metanephros.

glomerulus and becomes the *Bowman's capsule,* or else the blood vessels form a bulge on the wall of the coelom. In this case the structure is referred to as the *external glomerulus,* or, if glomeruli of several segments are joined together, as the *glomus.* It is believed that the segment of the coelomic cavity into which the glomus projects is itself derived from an expansion of the nephrocoele, or several nephrocoeles.

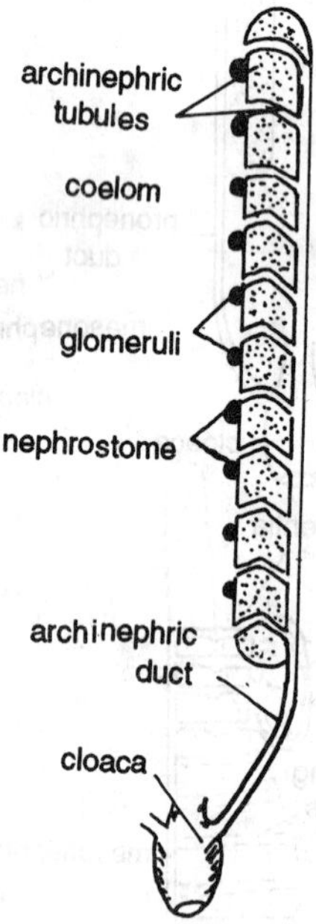

Fig. 5.23. Archinephros.

The basic pattern of development of the excretory organs as described above becomes modified in various degrees. As with many other organ systems, in vertebrates the development of the excretory tubules progresses in a craniocaudal direction, the tubules differentiating earlier in anterior segments than in posterior ones. The tubules which develop and begin functioning earlier, i.e., those of the anterior part of the trunk, tend to show more primitive conditions in their organization, while those developing later are modified to a greater degree. In the most primitive vertebrates, the cyclostomes and some fishes, the differences between the anterior and posterior parts of the excretory system are only gradual, but in the higher vertebrates we may distinguish more or less clearly three sections of this system : the *pronephros*, the *mesonephros* and the *metanephros*.

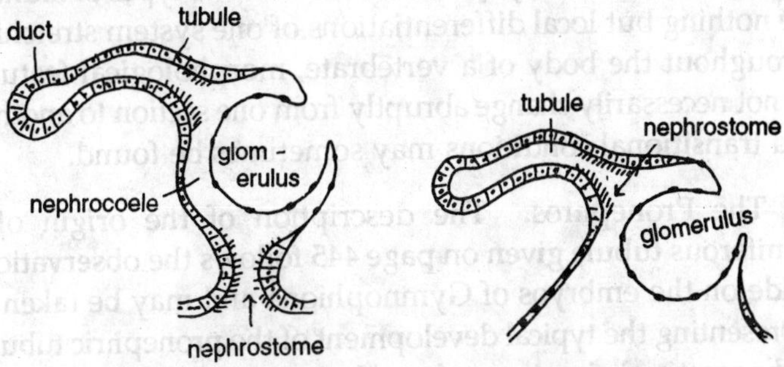

Fig. 5.24. Diagrams of two main types of an excretory unit, with internal glomerulus (left) and external glomerulus (right).

The most anterior nephrotomes give rise to the pronephros, the mesonephros develops from those of the midtrunk region, and the metanephros is derived from the nephrotomes of the posterior part of the trunk. The resulting three types of kidneys can thus be distinguished as anterior, middle and posterior with respect to their place of origin. The pronephros is the first to develop, the mesonephros second, and the metanephros, if present, appears latest of all.

The three types of kidneys differ also in the way in which they are produced by the stalks of the somites, in the presence or absence of the nephrostomes, in the position of the glomerulus and in the origin and connections of the excretory ducts. In the pronephros, the nephrostomes are well-developed and the glomeruli tend to be replaced by a common glomus (see previous description), though in some primitive forms, such as the Gymnophiona, each tubule of the pronephros may be supplied with a separate glomerulus. In the mesonephros, a separate glomerulus is, as a rule, intercalated in the course of each uriniferous tubule. Nephrostomes are present at first but may disappear later. In the metanephros, the nephrostomes are not formed at all, and the uriniferous tubules begin with the Bowman's capsules.

Further differences in the development of the tubules and ducts can best be described separately in relation to each section of the excretory system. Since all three types of kidneys are nothing but local differentiations of one system stretching throughout the body of a vertebrate, morphological features do not necessarily change abruptly from one section to another, and transitional conditions may sometimes be found.

The Pronephros. The description of the origin of a uriniferous tubule given on page 445 follows the observations made on the embryos of Gymnophiona and may be taken as representing the typical development of the pronephric tubules in the most archaic groups of vertebrates. The strictly segmental origin of the tubules is an essential feature of this development. In amphibians, the pronephric tubules are formed from a common mesodermal thickening appearing beneath the third and fourth somites (in salamanders) or second, third and fourth somites (in frogs). Nevertheless, the number of the pronephric tubules corresponds to the number of segments participating in the development of the pronephros.

The rudiment of the pronephros can be traced back in amphibians to the neurula stage. By means of vital staining, the presumptive material of the pronephros was found to lie in the mesodermal mantle just outside of the edge of the neural plate, posterior to the middle of the embryo. In the neurula stage, mesoderm of this region is capable of self-differentiation when transplanted heterotopically. At the same time, however, other parts of the mesodermal mantle also possess the ability to develop into renal tubules, as for instance when an inductor (the notochord) is transplanted into the lateral plate region of the embryo. The presumptive somites, when isolated from the notochord and cultivated *in vitro* (surrounded by a coat of skin epidermis for protection), may develop into renal tubules even though their normal destiny would be a different one (the development of muscle). After the end of neurulation (closure of the neural tube), the competence for the development of the pronephric tubules is restricted to the presumptive material of the pronephros. From this stage onward the pronephric rudiment cannot be replaced, and the removal of the rudiment leads to the absence of the pronephros.

The sequence of events during the transformation of the nephrotomes into the pronephric tubules, in frogs, salamanders and higher tetrapods, is not so clear as in the lower vertebrates. It is carried out by means of a rearrangement of cells, that is, by morphogenetic movements. The end result is, however, the same; several pairs of pronephric tubules are formed; they open by means of the nephrostomes into the coelomic cavity and fuse distally to form the pronephric duct.

The formation of the pronephric duct, by fusion of the distal ends of the pronephric tubules, is a very important phase in the development of the excretory system, as this duct not only serves the pronephros but is also instrumental in providing pathways for the outflow of urine from the mesonephros and metanephros and, in the males, for the passage of spermatozoa.

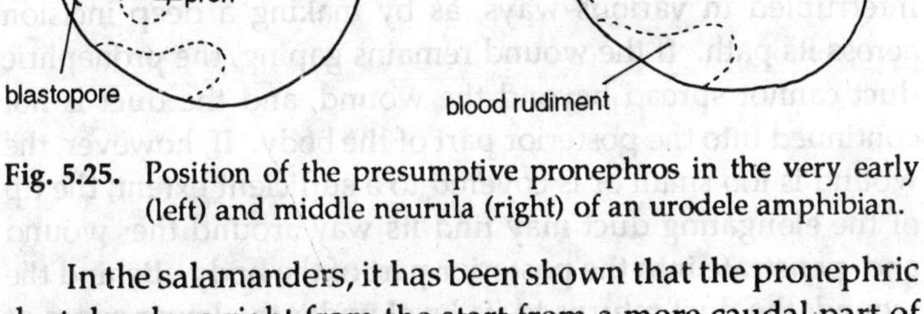

Fig. 5.25. Position of the presumptive pronephros in the very early (left) and middle neurula (right) of an urodele amphibian.

In the salamanders, it has been shown that the pronephric duct develops right from the start from a more caudal part of the mesoderm than the pronephric tubules; while the two

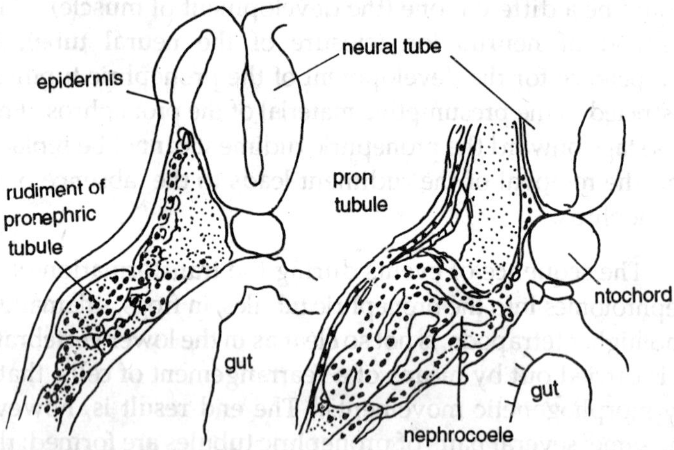

Fig. 5.26. Early development of a pronephric tubule in the frog.

Once they have been formed, both the pronephric tubules and the pronephric duct elongate very considerably. The pronephric tubules, as a result of their elongation, are thrown into numerous loops and eventually form a more or less spherical body, consisting of tangled and interwined tubules. The pronephric duct, on the other hand, remains straight, and while it elongates, its posterior free end pushes itself backward along the lower ends of the somites. This backward movement of the tip of the pronephric duct ends when the duct reaches the cloaca and fuses with its wall, while the lumen of the duct opens into the cavity of the cloaca.

The backward elongation of the pronephric duct may be interrupted in various ways, as by making a deep incision across its path. If the wound remains gaping, the pronephric duct cannot spread beyond the wound, and the duct is not continued into the posterior part of the body. If, however, the wound is too small or is covered to a sufficient extent, the tip of the elongating duct may find its way around the wound and penetrate into the posterior part of the body. Behind the wound, the duct returns to its level under the lower edges of the somites and eventually reaches the cloaca. This experiment shows that the elongation of the pronephric duct is largely independent of the surrounding parts, although the latter do seem to direct the duct by furnishing a suitable path for its movement. The same is borne out with the utmost clarity by some further experiments in which embryos in the closed neural tube stage bisected transversely, one half reversed, and the two parts caused to heal together with opposite dorsoventral orientation. When the pronephric ducts reached the level of the operation in their backward elongation, the tip of the duct changed its direction and struck a new path across the lateral body wall, until it reached the somites of the inverted posterior part of the embryo. It then elongated along the edge of the so mites and reached the cloaca as usual.

The development of the pronephric duct shows a great similarity to the development of the lateral line. In both cases the migration of the posterior end of the organ rudiment along the length of the body is due to intrinsic tendencies of the cells of the rudiment, but the organs and tissues with which the rudiments come in contact influence the path taken by the migrating cells.

The kidneys of the most primitive living vertebrates, the cyclostomes and elasmobranch fishes, in their development and organization in the adult, show features characteristic of the pronephros, although they may perhaps be more correctly described as representing a stage in which the differentiation into pronephros and metanephros has not yet taken place. The typical pronephros is a functional kidney of the larval stages of bony fishes and amphibians. In frog tadpoles, the pronephric canals form a bulky convoluted mass on both sides in the anterior part of the trunk at the level of the forelimbs, on the median aspect of the pectoral girdle. In frog tadpoles approaching metamorphosis, the pronephros gradually degenerates; the glomus shrinks, the neprhic tubules are resorbed and the anterior portions of the pronephric ducts are also resorbed. The function of excretion is taken over by the mesonephros.

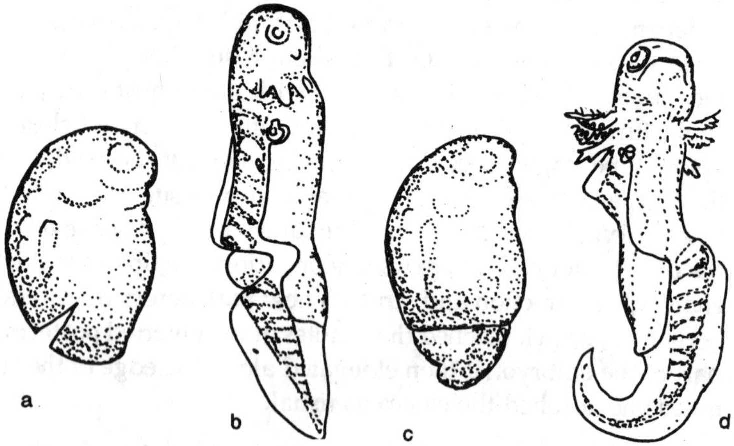

Fig. 5.27. Pronephric duct reaching the cloaca by an abnormal path after transection of the embryo (a, b) and after rotation of the posterior part of the embryo through 180 degree (c, d).

In embryos of amniotes, the pronephros develops in the anterior trunk region, but it is not functional at any stage. In the human embryo, about seven pairs of rudimentary pronephric tubules are formed. These soon degenerate, but not before they give rise to the pronephric ducts, which remain after the pronephric tubules have disappeared.

The Mesonephros. The mesonephric tubules are derived from the nephrotomes, as in the case of the pronephros, but their interrelation is by no means so simple. Only in the more archaic groups of vertebrates (selachians, Gymnophiona, ganoids) are the renal tubules formed directly from the nephrotomes. In most amphibians and in all higher vertebrates, the mesodermal cells of the nephrotomes dissolve into a mass of mesenchyme stretching on each side of the body along the dorsal edge of the lateral plates from the level where the pronephros ends to the pelvic region. This elongated mass of mesenchyme is known as the nephrogenic cord or nephrogenic tissue. The mesonephric tubules are developed from the nephrogenic tissue by a secondary aggregation of the mesodermal cells.

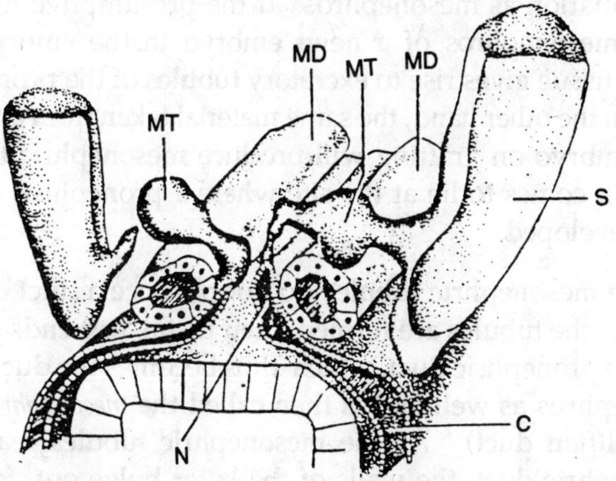

Fig. 5.28. Relation of the mesonephric tubules to other mesodermal parts in the embryo of a sturgeon. The rudiment of the mesonephric tubule (MT) is shown as still connected to the somite (S) on one side, and just separated on the other. C, coelom; N, nephrostome; MD, mesonephric duct.

The aggregating cells began by forming epithelial vesicles which stretch and elongate to become tubes. One end of such a tube becomes connected to the pronephric duct, while the other end becomes invaginated to form the Bowman's capsule; or, in rarer cases, it opens into the coelomic cavity by a nephrostome while a Bowman's capsule is formed higher along the course of the tubule. The Bowman's capsules become supplied by small branches from the dorsal aorta. The number of the mesonephric tubules does not correspond to the number of segments, several tubules being developed in the region of each segment. Even if at first only one tube is formed in each segment, it soon gives rise by budding to secondary and tertiary tubes, and so on. If present at first, the nephrostomes may disappear later. In the frogs, the nephrostomes lose their connection with the rest of the nephric tubule but open secondarily into the veins (connecting the veins to the coelom).

The nephrogenic tissue is not at once determined for

differentiation as mesonephros. If the presumptive material of the mesonephros of a newt embryo in the embryo, the grafted tissue gives rise to excretory tubules of the pronephric type. On the other hand, the same material taken from a tailbud stage embryo on grafting will produce mesonephric tubules even if it comes to lie at the site where a pronephros should have developed.

The mesonephric tubules do not produce a duct of their own. As the tubules are formed, their distal, free ends join up with the pronephric duct, which thus becomes the duct of the mesonephros as well and is then called the *mesonephric* duct (or Wolffian duct). As the mesonephric tubules reach the mesonephric duct, the walls of the latter bulge out, forming collecting ducts. The mesonephric tubules, especially the ones formed later, open into these collecting ducts rather than into the mesonephric duct itself.

The development of the mesonephros has also been found to be dependent on the pronephric duct in another way. The nephrogenous tissue develops into the mesonephric tubules only if stimulated to this by the pronephric duct. In the preceding section, experiments have been mentioned in which the penetration of the pronephric duct into the posterior half of the body was prevented by placing an obstacle (in the form of a gaping wound) in its path. When the operation is successful—that is, if the duct does not reach the region where the mesonephros normally develops—the nephrogenous tissue fails to form the renal tubules or forms only poorly developed tubules. Apparently some stimulus (induction) from the pronephric duct is necessary for the normal development of the mesonephric tubules from the nephrogenous tissue.

The most anterior mesonephric tubules degenerate later in many animals, so that the kidney becomes shortened at its anterior end and somewhat more compact. In many fishes and in amphibians, the posterior part of the mesonephros shows some traits found in the metanephros: complete absence of nephrostomes, joining of the nephric tubules to large collecting ducts—out growths of the mesonephric duct. For this reason, some authors prefer to regard the kidney in these animals as a joint mesonephros and metanephros, giving if the special name of *opistonephros.* In fishes and amphibians, the mesonephric (or opistonephric) kidney is the excretory organ of the adult animal.

In reptiles and birds, on the other hand, the mesonephros functions only during the embryonic period of development and loses its excretory function at the time of hatching. In the mammals, the placenta takes on the task of removing excretory products from the blood of the embryo and passing them into the blood of the mother, from where they are removed by the maternal kidneys. In mammals such as the pig, in which the connection between the fetal and maternal tissues is not very close, the mesonephros is still active as an excretory organ, and a certain quantity of urine eventually reaches the cavity of the allantois and the amniotic cavity. In mammals having a very close connection between the fetal and maternal tissues (as in rodents with a hemochorial placenta), there is no sign of excretory activity in the mesonephros, which is thus a rudimentary organ like the pronephros.

The Metanephros

The metanephros develops from the posterior part of the nephrogenic cord, the part adjacent to the cloaca. There is no trace of a relation of the tubules to the nephrotomes, although the nephrogenic tissue is primarily derived from the latter. As in the case of the mesonephros, the metanephros does not develop a duct of its own but uses the mesonephric duct as a means of removing the urine that it excretes. However, the connection between metanephros and *mesonephric* duct is established not directly but by means of a special outgrowth or branch of the duct. Before the metanephros starts differentiating, a bud is formed on the mesonephric duct a small distance in front of the point where the duct joins the cloaca. The bud elongates in the direction of the posterior part of the nephrogenic tissue. The duct formed in this way becomes the *ureter* and the bud from which it develops is the *ureteric bud.* Having reached the nephrogenic tissue, the end of the ureter begins to branch, the branches later becoming the collecting tubules of the kidney. The point at which the branching begins expands to form the renal pelvis. The nephrogenic mesenchyme accumulates around the tips of the collecting tubules and in due course differentiates into the renal tubules with their glomeruli. No trace of nephrostomes can be discovered in the metanephros at any stage.

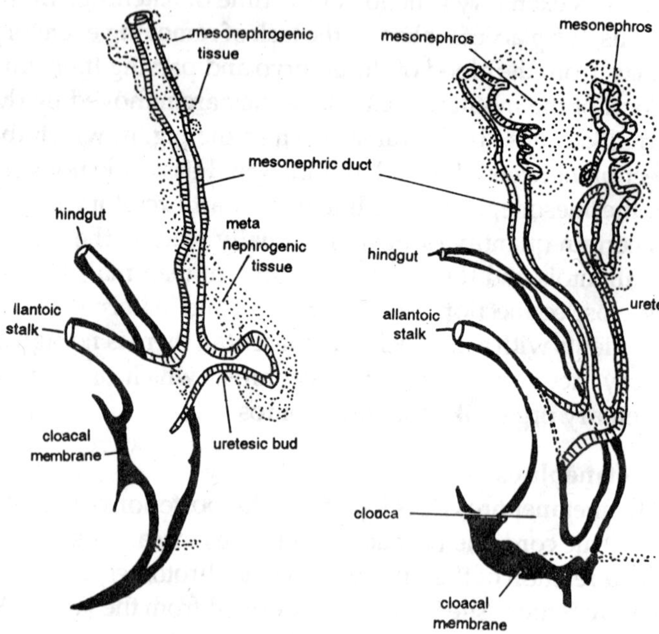

Fig. 5.29. Development of the ureter and the metanephros in a rabbit embryo.

As in the mesonephros, the conversion of the metanephrogenic tissue into a system of renal tubules is dependent on a stimulus from the excretory duct—in this case from the ureter and its branches, which develop from the ureteric bud. In experiments on chick embryos, in which the backward growth of the pronephric duct is prevented by destroying its growing tip, no mesonephric duct develops in the posterior part of the body and naturally no ureteric bud is formed. The result is that no metanephric kidney is formed on the operated side of the embryo.

A confirmation of results gained by operation may be found in certain defects caused by mutations.

The metanephros is the functional kidney in the postembryonic life of reptiles, birds and mammals. In lizards, the fully differentiated metanephros largely retains the position of the metanephrogenic part of the nephrogenic cord: the functional kidney lies at the level of the hindlimb and even projects into the base of the tail. In birds and mammals, however, at an early stage the metanephric kidneys become displaced in an anterior direction. The metanephros shifts past the posterior end of the mesonephros and eventually comes to lie at about the same level as the latter: in the lumbar region, in mammals, and filling the space underneath the synsacrum and the pelvic girdle in birds. Concurrently with the displacement of the kidneys, the ureters elongate to a very great extent as compared with their length at the time when the ureteric bud first contacts the metanephrogenic tissue. The excretory ducts of the kidney in most vertebrates acquire close relations with the reproductive organs. These relations and the modifications of the excretory ducts in this connection will be dealt with together with the development of sexual organs.

DEVELOPMENT OF THE HEART

The heart in vertebrates develops from the mesoderm forming the ventral edges of the lateral plates in the pharyngeal region of the body. It will be most convenient to describe first the development of the heart in animals such as the amphibians and to consider later the development of the heart in higher vertebrates.

The Heart in Lower Vertebrates

During gastrulation in the amphibians, the sheet of mesoderm advances forward from the blastopore, penetrating between the ectoderm and the endoderm. The rate of movement of the mesoderm is greatest in the dorsal region of the embryo, intermediate laterally and least ventrally. In the neurula stage, the dorsal and dorsolateral parts of the mesodermal mantle have reached the head region of the embryo; ventrally there remains an approximately triangular area in which there is no mesoderm intervening between ectoderm and endoderm. This triangle, roughly corresponding to the oral and pharyngeal region of the embryo, has its apex posteriorly and the broad base anteriorly. The posterior part of the area is the site of development of the heart, and it is later filled in by the mesoderm participating in the formation of the heart. The most anterior part of the area remains free of mesoderm, and here the mouth breaks through after a fusion between the endoderm and the ectoderm has taken place.

The presumptive material of the heart is found in the edges of the mesodermal mantle bordering the mesoderm free area on the right and on the left. By vital staining in the neurula stage, it has been shown that the presumptive material for the heart is located in the edge of the mesodermal mantle, rather high on the flank on each side, where it adjoins the parts of the neural folds which become the hindbrain portion of the neural plate.

If the mesoderm of this area is stained with nile blue sulphate or neutral red (by lifting a flap of ectoderm and mesoderm and applying a piece of cellophane soaked in stain to the *inner* surface of the mesodermal layer), it can be observed that the mesoderm shifts in later stages in a ventral direction and eventually comes to lie midventrally in the area where the heart can be seen to differentiate. The particles of vital stain remain in the cells long enough for it to be ascertained that the stained material on each side of the embryo gives rise to exactly one half of the heart rudiment, the other half being derived from a similar area of mesoderm on the other side of the embryo.

Parts of the free edge of the mesodermal mantle may be excised in the neurula stage and cultivated in a saline solution. To prevent the mesoderm from disintegrating, it is sometimes isolated together with a flap of ectoderm, the ectoderm then closes into a vesicle with the mesoderm inside. The mesodermal cells under these conditions differentiate into muscle tissue, and this begins to pulsate rhythmically—an unequivocal indication that the developed muscle tissue is cardiac muscle, as only this type of muscle is capable of autonomous rhythmical contraction. It has been claimed that in some experiments the heart rudiments explanted in neurula and even late gastrula stages can even produce pulsating tubes having an obvious resemblance to normal hearts.

It is rather remarkable that there is a definite discrepancy between experiments using the vital staining technique and explantation experiments. While vital staining indicates that the material going into the formation of the heart lies in the upper part of the mesodermal mantle, heart differentiation has been observed in explanted pieces taken from different parts of the free edge of the mesodermal mantle: from the upper part from the middle part and from the ventral part. It would seem that mesoderm in a wide region of the early embryo has a potentiality for heart development. From this wider area the potentiality for heart development is later restricted to the actual material participating in the formation of the heart in normal development.

In these early stages, the development of the heart is in some way dependent on the endoderm. By cutting through the ectoderm and the mesodermal mantle and removing the whole of the endoderm, it is possible to produce newt embryos which consist of ectoderm and mesoderm only. The embryos survive the operation quite satisfactorily, but their term of life is limited because they do not possess the food supply normally

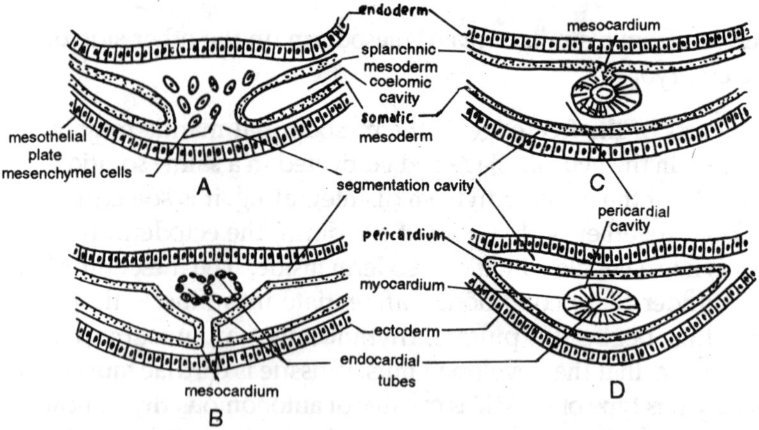

Fig. 5.30. Development of Heart.

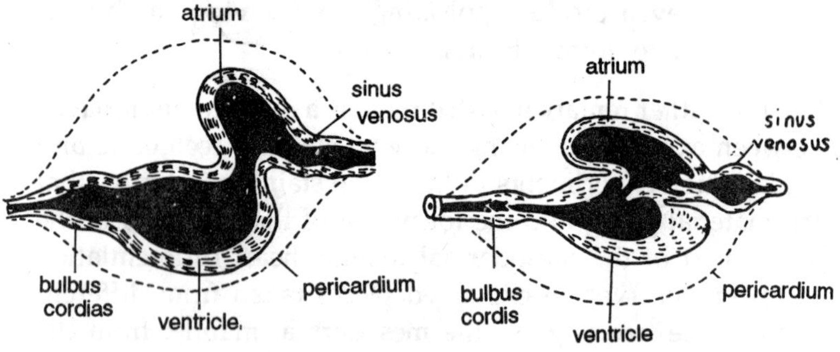

Fig. 5.31. Differentiation of various regions of heart.

contained in the yolky cells of the endoderm. The endodermless embryos show various defects in the ectodermal and mesodermal organs owing to the lack of inductive influences emanating from the endoderm. One of such defects is the complete absence of the heart, although the mesodermal layer, from which the heart rudiment is derived, may remain intact.

The dependence of heart development on the endoderm and also on other adjacent tissues has been tested in hanging drop cultures, the techniques used being similar to those used in studying the induction of nervous tissues. Pieces of mesoderm

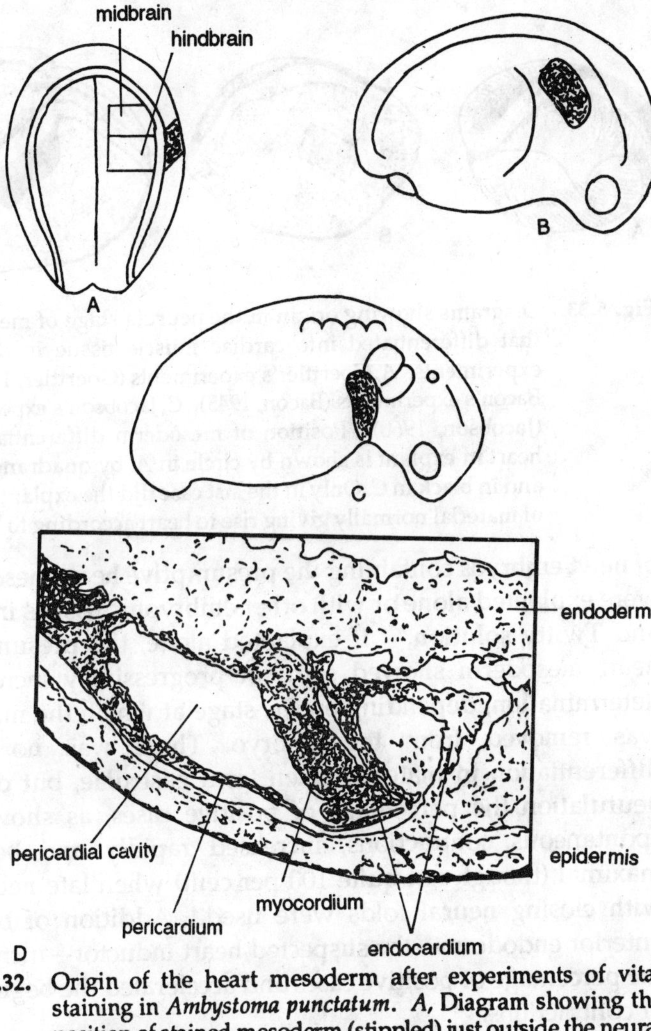

Fig. 5.32. Origin of the heart mesoderm after experiments of vital staining in *Ambystoma punctatum*. *A*, Diagram showing the position of stained mesoderm (stippled) just outside the neural fold, in the hindbrain region of a neurula. *B*, Embryo in early tail bud stage with stained mesoderm in the ear region. *C*, Embryo in late tail bud stage with the heart mesoderm shifted to the ventral position, where the heart rudiment is actually formed. D, Cross-section through the heart region of an embryo in which the heart mesoderm was stained on the right side in the neurula stage (as in *A*). The stained material (stippled) on the right half of the heart includes all layers: endocardium, myocardium and pericardium.

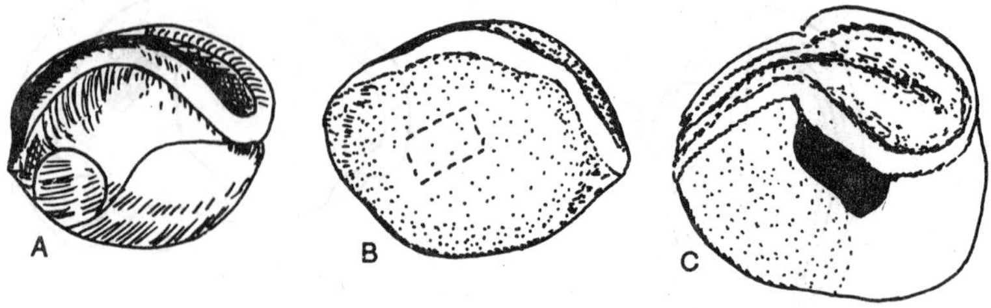

Fig. 5.33. Diagrams showing origin in the neurula stage of mesoderm that differentiated into cardiac muscle tissue in different experiments. *A*, Goerttler's experiments (Goerttler, 1928). *B*, Bacon's experiments (Bacon, 1945). *C*, Jacobson's experiments (Jacobson, 1960). Position of mesoderm differentiating as heart in explant is shown by circle in *A*. by quadrangle in B and in black in C. Only in the last case did the explant consist of material normally giving rise to heart according to Wilens.

of newt embryos containing the presumptive heart mesoderm were explanted alone or with other embryonic tissues in "Niu and Twitty solution." If explanted alone, the presumptive heart mesoderm showed signs of progressively increasing determination, depending on the stage at which the material was removed from the embryo. There was no heart differentiation in material taken from gastrulae, but during neurulation the percentage of positive cases, as shown by spontaneous contractions, increased rapidly and became maximal (though not quite 100 per cent) when late neurulae with closing neural folds were used. Addition of bits of anterior endoderm—the suspected heart inductor—increased the percentage of positive cases and accelerated the beginning of contractions.

Instead of bits of intact endodermal tissue, an extract could be added to the hanging drop, the extract being prepared by homogenizing embryonic tissues and passing them through a cephadex column with large particle size, so that small bits of cytoplasmic organoids (up to the size of pigment granules) could get through. The extract was found to be able to promote heart differentiation, thus suggesting that in this case, as in others, the induction is mediated by soluble substances, although the action of fragments of cytoplasmic membranes in this process has not been ruled out. Anterior epidermis was also able to stimulate heart development but its effect can be considered as only a supporting factor, since hearts fail to develop in endoderm free embryos when the epidermis is present. Posterior endoderm on the other hand had no effect, and the transverse neural fold seems to exercise even a depressing effect on heart development.

After the end of the neurulation, the free edges of the mesodermal mantle gradually converge toward the middle of the mesoderm free area and become thickened in the heart region, foreshadowing the formation of the heart rudiment. At this stage it can be noticed that between the free edges of the mesodermal mantle covering from the right and the left there lie a number of loose cells, similar to mesenchyme cells in their structure. These cells are derived from the ventral edge of the mesodermal mantle. They are the rudiment of the endocardium, that is, the endothelial lining of the cavity of the heart. The endocardial cells soon accumulate in the midline as a longitudinal strand and eventually become arranged in the form of a thin-walled tube. The lumen of the tube is the cavity of the heart. The endocardial tube bifurcates at both end; at the anterior end its two prolongations are the ventral aortae, and at the posterior end it receives the two vitelline veins—the first venous blood vessels to reach the heart. All these vessels are at first similar to the endocardial tube in that they consist only of a thin layer of endothelium, produced by the mesenchyme cells joined together.

While the endocardial tube is being formed, the edges of the mesodermal mantle, which at this stage may also be called the edges of the lateral plate mesoderm, close in along the midline and fuse with each other. The fusion first occurs under the endocardial tube, between it and the ectoderm. Very soon, however, the visceral layer of mesoderm envelops the endocardial tube on the dorsal side as well. By fusion of

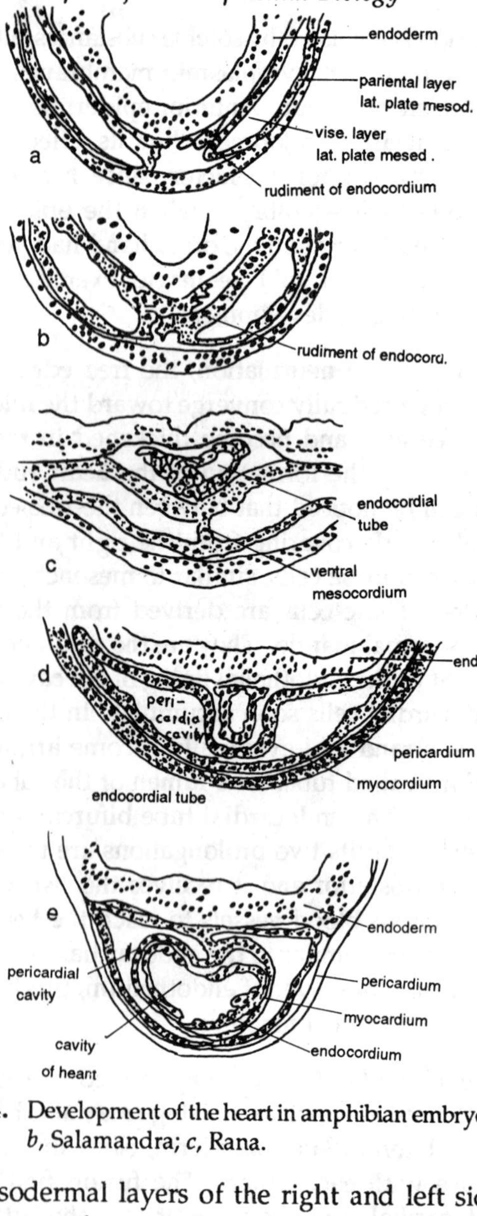

Fig. 5.34. Development of the heart in amphibian embryos. *a,d,e,* Triturus: *b,* Salamandra; *c,* Rana.

the mesodermal layers of the right and left side, epithelial partitions are formed above and below the endocardial tube. In analogy to the dorsal and ventral mesentery, these are called the *dorsal* and *ventral mesocardium.* The ventral mesocardium is of a very ephemeral existence; it is very soon perforated, and the coelomic cavities of the right and left side become continuous underneath the endocardial tube. The dorsal mesocardium persists longer but is also dissolved at a later stage. The coelomic cavities expand in the heart region to form the pericardial cavity. The pericardial cavity is initially only a part of the general body cavity, the coelom. It becomes completely separated from the remainder of the coelom, largely owing to the failure of the coelom to develop in the region of the branchial pouches, which lie immediately dorsal to the heart rudiment. Posteriorly the connection of the pericardial cavity with the rest of the coelom becomes occluded by the developing liver, to which the posterior end of the endocardial tube becomes closely connected. A connective tissue wall developing in this position (at the anterior boundary of the liver) is the *septum transversum.*

From the above description, it will be clear that the pericardial cavity is lined by lateral plate mesoderm. The parietal layer of this mesoderm persists as the epithelial wall of the pericardial cavity, or the pericardium proper. The visceral layer adheres to the endocardial tube. This layer differentiates as muscle tissue and thus gives rise to the myocardium of the heart.

The heart is at first an almost straight tube and does not show a sub-division into its various chambers. Later the tube becomes inflected in a very characteristic way. Starting from behind, the tube first runs forward, then bends downward and to the right and eventually again to the left, upward and forward. The heart thus becomes coiled in the shape of an S. The degree of the twisting in higher vertebrates is greater than in lower ones, so that in the former the tip of the second inflection comes to lie well posterior to the tip of the first inflection. The tubular heart rudiment becomes constricted in some places and dilated at others and is thus subdivided into its four main parts. The sinus venous lie posteriorly; the atrium develops at the tip of the first inflection of the heart; the descending part, from the first to the second inflection, becomes the ventricle: and the part going forward from the second inflection becomes the course arteriosus.

Before this subdivision is performed, however, the functioning of the heart starts; it begins to pulsate at a regular rhythm. The pulsations of the heart start very early in the development of the embryo, even before the peripheral blood vessels are ready to receive the blood stream.

As the presumptive heart rudiment is being formed, its capacity for performing the further stages of development independently of the norms surroundings increases perceptibly. If the heart rudiment is excised after the end of neurulation and transplanted into an abnormal position or allowed to develop in vitro, enclosed in a vesicle of skin, the differentiation goes much farther than in the experiments referred to above. Not only pulsating muscle tissue develops but a cardiac tube is formed, and the tube becomes inflected just as does the heart in normal development. The rudiment of the heart in this stage, however, is by no means strictly determined in all its parts. The left or right half of the heart rudiment may be excised, and the remaining half then develops into a complete whole. What is more, half of the heart rudiment may be explanted or transplanted, and it still develops into a complete heart. The ability of a half of the heart rudiment to form a whole heart may be used to produce two hearts in the same embryo. For this an incision should be made lengthwise before the two halves of the heart rudiment unite in the middle. Inserting a piece of extraneous tissue, a somite for instance, or leaving the wound to gape may prevent the halves of the heart rudiment from coming together. Two complete pulsating hearts develop under these conditions. The reverse can also be done: a complete heart rudiment may be superimposed on the intact presumptive heart mesoderm of a host embryo. The two rudiments then fuse into one whole and produce a normal heart.

The main points emerging from the foregoing will now be summarized:

1. The heart develops in a very anterior position, in the pharyngeal region. The position of the heart in the thorax (as in an adult tetrapod) is thus secondary and is due to a displacement of the organ in later development.
2. The heart develops from a paired rudiment, uniting in the midtive secondarily.
3. Each half of the heart rudiment is able to differentiate even before the two halves fuse.
4. Two heart rudiments may fuse into one organ of normal structure.
5. The functioning of the heart begins at an early stage of development.

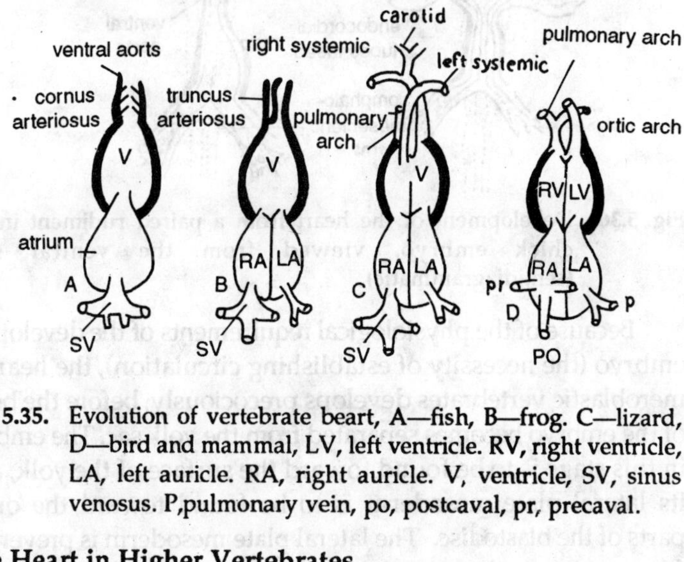

Fig. 5.35. Evolution of vertebrate beart, A—fish, B—frog. C—lizard, D—bird and mammal, LV, left ventricle. RV, right ventricle, LA, left auricle. RA, right auricle. V, ventricle, SV, sinus venosus. P,pulmonary vein, po, postcaval, pr, precaval.

The Heart in Higher Vertebrates

With all this in view, it is easy to understand the development of the heart in the vertebrates having yolky eggs and partial cleavage.

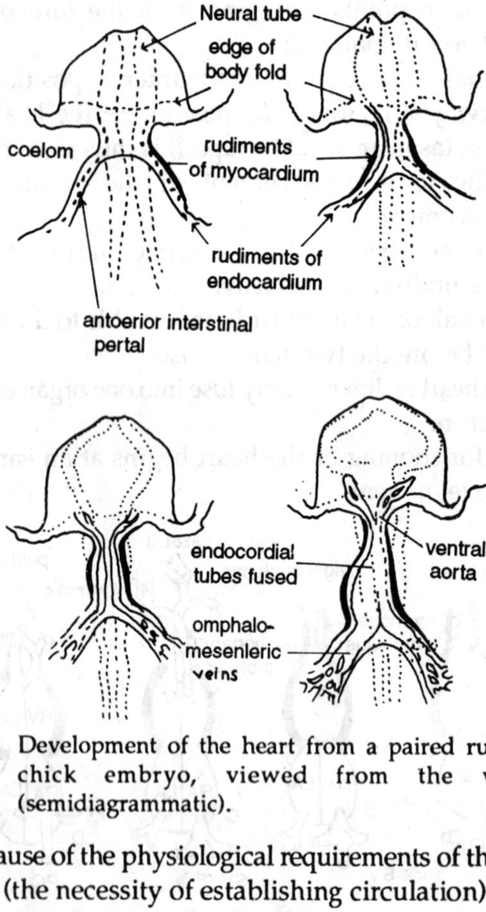

Fig. 5.36. Development of the heart from a paired rudiment in the chick embryo, viewed from the ventral side (semidiagrammatic).

Because of the physiological requirements of the developing embryo (the necessity of establishing circulation), the heart in meroblastic vertebrates develops precociously, before the body of the embryo becomes separated from the yolk sac. The embryo in this stage is to be found toward the surface of the yolk, and its lateral plate mesoderm is to be found toward the outer parts of the blastodisc. The lateral plate mesoderm is prevented from uniting on the ventral side of the embryo by the intervening yolk. As a result, the two halves of the heart rudiment begin differentiating independently of one another. Two endocardial tubes are actually formed; each becomes invested by the myocardium and surrounded laterally by the pericardial cavity. When the body folds undercut the anterior end of the embryo and the foregut becomes separated anteriorly from the yolk sac, the right and left heart rudiments become able to meet in the midline under the pharynx. The two endo-cardial tubes come to lie alongside each other and soon fuse into one tube in the cardiac region, while in front and behind the heart the endothelial tubes remain separate, thus leading to a state described earlier for the lower vertebrates. The visceral walls of the right and left pericardial cavities also meet and fuse above and below the endocardial tubes, forming a single pericardial cavity. The single endocardial tube is now completely surrounded by the myocardium. In view of the experimental results mentioned above, the fusion of two heart rudiments to form one single organ is not at all surprising.

Of all the organs of a vertebrate, the heart is the one which starts its definitive function earliest. It is also greatly dependent in its development on function. It is essential for the development of the heart that a blood stream should actually flow through it. It is true that a heart of a typical shape with recognizable parts will develop even in complete isolation, as in the explantation experiments, but such an isolated heart soon stops developing further. Also, a heart in its normal position is arrested in its development if in some way or other (interruption of the afferent blood vessels) it is deprived of circulation. The degree of the heart's development and growth appears to be dependent on the volume of blood passing through it or on the size of the animal which the heart supplies with blood. Heart rudiments have been transplanted reciprocally between large and small species of salamanders and it was found that the hearts of small species grew beyond their usual size in large hosts, and the hearts of large species were undersized in small hosts. In every combination the transplanted heart grew to approximately the same size as the host heart should have grown.

In the vertebrates developing a pulmonary circulation, the heart becomes separated to a greater or lesser degree into a right half, carrying blood to the lungs, and a left half, receiving the blood from the lungs by way of pulmonary veins and sending it to the rest of the body. The first indication of this separation is found in the lungfishes (Dipnoi). In the amphibians, the atrium becomes subdivided by a partition wall arising between the points of entry into the atrium of the sinus venosus and pulmonary vein. After the separation is completed, the right atrium receives blood from the sinus venosus and the left atrium from the lungs by way of the pulmonary vein. Both atria, however, pour out their blood into the ventricle, which remains undivided.

The Mammalian Heart

The most elaborate transformations are found to occur in the development of the heart in mammals. The task of directing the blood into its various channels is complicated not only by the presence in the adult state of a double circulation the systemic and the pulmonary circulation) but also by the necessity of providing for a rapid change from fetal to postnatal circulation. The subdivision of the heart into right and left halves is dependent on the co-operation of several structures and carried out in stages.

The atrium is separated into right and left halves first by a primary partition wall (septum primum), which is similar to that found in amphibians. As soon as this partition wall subdivides the cavity of the atrium, however, it is perforated by an opening allowing for intercommunication between the right and left atrial cavities. Consequently, the left atrium is not deprived of a blood flow (on which, as we have seen, the development of the heart is dependent in a high degree) during the intra-uterine life, when the amount of blood passing through the lungs is very small.

At a later stage, a second partitions wall (septum secundum) develops in the atrium just to the right of the first, but this remains incomplete, a large foramen ovale remaining open in its lower part. The foramen ovale does not quite coincide in position with the opening in the primary partition wall;

nevertheless, blood may pass through both openings so long as the blood pressure in the right atrium is greater than that in the left atrium, as it actually is during fetal life.

After birth, however, when the pulmonary circulation is fully established, the pressure in the left atrium becomes higher than that in the right atrium. The primary partition wall becomes pressed against the secondary wall, and as a result the previously existing passage becomes closed. Thus the atria become finally separated. Eventually the primary and secondary partition walls of the atrium fuse together.

The atrium originally communicates with the ventricle by way of the atrio-ventricular opening, which is a narrowed part of the heart tube. The separation of this single opening into two is performed by means of connective tissue outgrowths developing on the dorsal and ventral edges of the opening. The outgrowths, known as the *endocardial cushions,* fuse in the middle and are also joined by the lower edge of the primary interatrial partition wall; as a result two canals are left, one leading from the right atrium into the right ventricle and the other leading from the left atrium into the left ventricle. The endocardial cushions also participate in the formation of the atrio-ventricular valves.

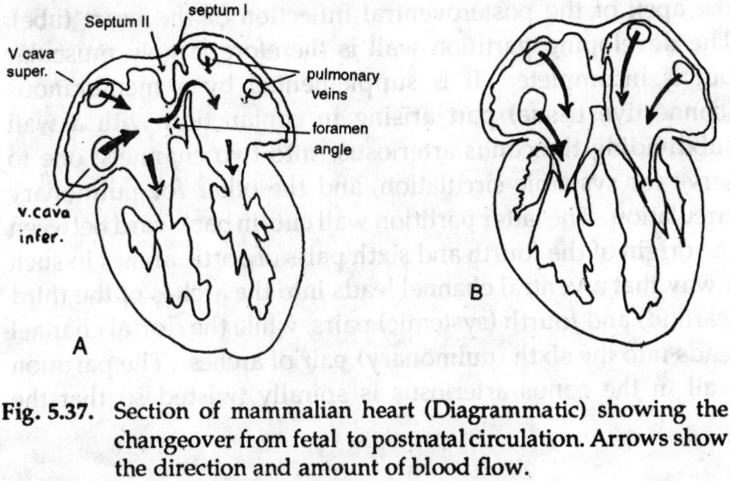

Fig. 5.37. Section of mammalian heart (Diagrammatic) showing the changeover from fetal to postnatal circulation. Arrows show the direction and amount of blood flow.

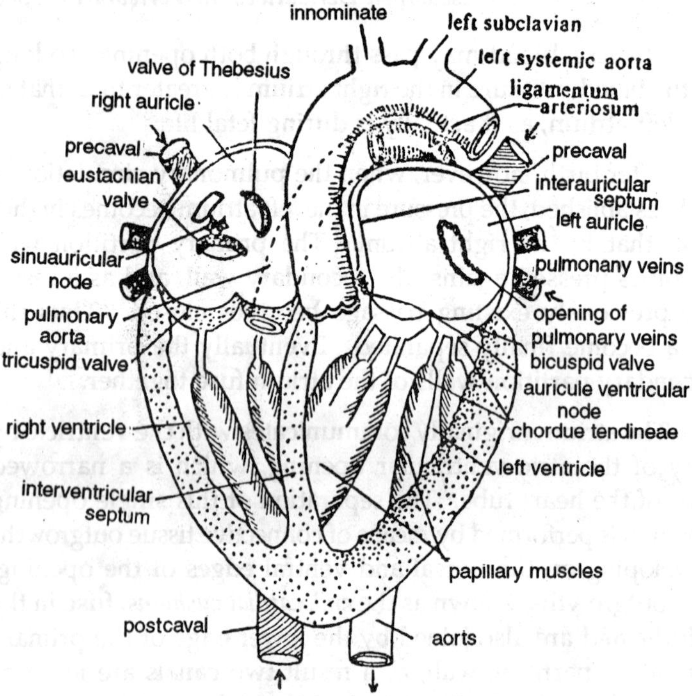

Fig. 5.38. Heart of mammal (ventral dissection).

The ventricle becomes partially separated into left and right halves by an indentation of the myocardium at its tip (at the apex of the posteroventral inflection of the heart tube). The developing partition wall is therefore largely muscular but is incomplete. It is supplemented by a membranous (connective tissue) part arising in conjunction with a wall subdividing the conus arteriosus into two channels: one to serve for systemic circulation, and the other for pulmonary circulation. The latter partition wall cuts in backward between the origin of the fourth and sixth pairs of aortic arches in such a way that a ventral channel leads into the arches of the third (carotid) and fourth (systemic) pairs, while the dorsal channel leads into the sixth (pulmonary) pair of arches. The partition wall in the conus arteriosus is spirally twisted so that the ventral channel becomes connected to the left ventricle and the dorsal channel to the right ventricle.

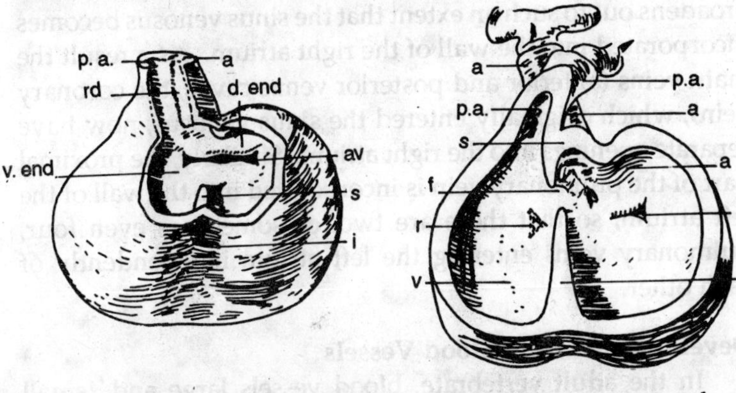

Fig. 5.39. Subdivision of the atrio-ventricular canal in mammals, a, Aorta; *a.v.*, atrio-ventricular foramen: *d.end*, dorsal endocardial cushion: *f*, interventricular foramen; *i*, indentation of ventricular myocardium; *l*, left atrio-ventricular foramen; *p.a.*, pulmonary artery; *r*, right atrio-ventricular foramen; *rd*, ridge separating the conus arteriosus in two channels; *s*, parts of interventricular septum; *v. end*, ventral endocardial cushion; *v*, cavity of right ventricle.

It will be seen that the blood entering the right atrium from the sinus venosus, including the blood from the placental circulation is directed into the pulmonary arch (except for the volume of blood passing from the right into the left atrium via the foramen ovale). The pulmonary blood vessels, during the fetal life, cannot take all this blood, neither is there any necessity to have the blood pass through the lungs, as they are as yet not serving as respiratory organs. The greater part of the blood passing through the right half of the heart is therefore returned to the systemic circulation by way of the arterial duct (ducts arteriosus), which connects the pulmonary arch to the dorsal aorta. After parturition, the arterial duct becomes closed by a violent contraction of its muscular wall, and all the blood from the right part of the heart is forced into the blood vessels of the lungs. The double circulation thus becomes finally established.

It is a further peculiarity of the mammalian heart that the opening between the sinus venosus and the right atrium broadens out to such an extent that the sinus venosus becomes incorporated into the wall of the right atrium. As a result the main veins (anterior and posterior venae cavae, the coronary vein), which originally entered the sinus venosus, now have separate openings into the right atrium. Similarly, the proximal part of the pulmonary vein is incorporated into the wall of the left atrium, so that there are two, or sometimes even four, pulmonary veins entering the left atrium independently of each other.

Development of the Blood Vessels

In the adult vertebrate, blood vessels large and small permeate almost all parts of the body. Supernumerary parts of the body, whether they are grafts or results of induction, become vascularized, and the blood vessels, which in this case are additional to the ones produced in normal development, link up with host arteries and veins and are included in the host's blood circulation. It would appear that blood vessels are attracted to penetrate any parts which are in need of a blood vessel supply. Furthermore, it appears that a suitable situation for blood vessels is duly reserved in the pattern of organization of any organ. It must be noted, however, that in vertebrates the blood vessels are invariably situated in spaces occupied by connective tissue or its derivatives. Where capillaries seemingly penetrate into epithelium, this is actually tantamount to the invasion of channels or other spaces between epithelial cells by connective tissue from which the endothelial cells of the blood vessels are derived.

The rudiments of the blood vessels are laid down as aggregations of mesenchyme cells. The cells participating in the formation of blood vessels are called angioblasts. They are probably always of mesodermal origin. We have seen that in the vascular area of amniotes the walls of the blood vessels are developed from the blood islands in conjunction with the development of the first blood cells. Similar, though more concentrated, blood islands occur in vertebrates with holoblastic cleavage (in particular in amphibians) on the ventral side of the abdomen. In the amniote body proper and in all areas except for the ventral body wall in the anamniotes, the blood vessels develop independently of blood corpuscles. The aggregations of angioblasts become arranged in the form of a flat epithelium surrounding a cavity. The epithelium is the endothelium of the blood vessel; the outer layers of the blood vessel walls are differentiated much later. A student of the anatomy of the adult vertebrates is accustomed to find the blood vessels in the form of tubes of a constant diameter over relatively long stretches. The first blood vessels laid down in the embryo are only rarely in the form of straight tubes. Over large areas the blood vessels are initially laid down in the form of a network. The further development of individual canals in the network depends on the amount and the direction of blood flow. Those channels which happen to come in the line of the greatest blood flow become increased in diameter, develop the connective tissue and the muscular layers and become arteries or veins. The channels that receive less blood remain in the form of capillaries or degenerate and disappear completely. Another form in which early blood vessels appear is as sinuses — extensive irregular spaces surrounded by endothelium. Large sections of some of the larger veins appear in this form. Some of these sinuses later acquire a more regular tubular form; others remain as such in the adult animal— for instance, the postcardinal "vein" becomes a tubular blood vessel in urodeles but remains essentially in the form of a sinus in the dogfish.

Once the network of endothelial tubes — the rudiments of the blood vessels — has been established, new blood vessels continue to be formed by sprouting and outgrowth of those already present. Lateral branches may form on already existing capillaries, or a capillary may become interrupted and each free end grow out in a new direction. Two outgrowth from different blood vessels may contact each other and fuse, thus establishing a new channel for circulation.

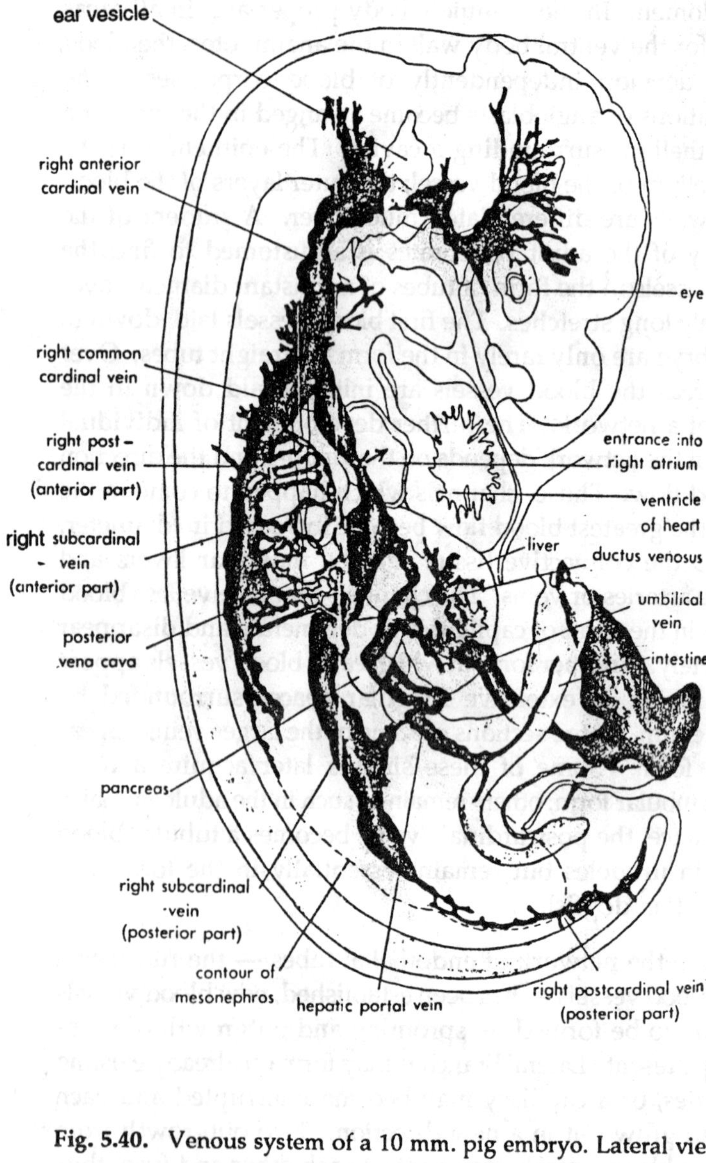

Fig. 5.40. Venous system of a 10 mm. pig embryo. Lateral view.

As a result of the extreme plasticity of the blood vessel system, the eventual arrangement of arteries, veins and capillaries in any part of the body is largely dependent on the amount and direction of blood flow in the part of question. In their first formation the embryonic blood vessels also appear in conjunction with other rudiments, suggesting a dependent mode of differentiation, though the dependence is not of a functional nature since the blood vessels develop prior to the establishment of circulation. If the heart rudiment in an amphibian embryo is removed before circulation starts, the main blood vessels continue to develop for sometime, until the embryo dies. The exact position of the blood vessel rudiments is probably determined by a process similar to an induction, though this has not been shown conclusively.

Certain locations in the embryo offer preferential conditions for the development of blood vessels. These locations are : (1) between the visceral mesoderm and the endoderm (2) around the kidneys, especially the pronephros and the mesonephros. Close networks of capillaries are always found in these two locations, and they are the site of development of some of the major blood vessels.

The heart obviously belongs to the first group. We have seen that the determination of the heart occurs at an early stage of development, and that as a consequence it shows a high degree of autonomy in its further differentiation. It is noteworthy, nevertheless, that the heart does not develop in embryos from which the endoderm has been removed. The blood vessels which form the (paired) continuation of the heart tube anteriorly and posteriorly also develop in conjunction with the endodermal parts. The two posterior vessels are the vitelline veins, which collect the blood from the network on the surface of the gut and, in amniotes, from the network on the yolk sac. The anterior prolongations of the heart tube are the ventral aortae, which become connected along the partition walls between the endodermal pharyngeal pouches with the dorsal aortae, a pair of blood vessels developing on the dorsal side of the endodermal gut. The vessels conveying blood from the ventral aortae to the dorsal aortae are the aortic arches.

Fig. 5.41. Outgrowth of new capillaries, establishing new channels of circulation.

The aortic arches develop in a cranio-caudal sequence, and the first to appear is the mandibular arch which ascends between the edge of the mouth and the first (spiracular) pharyngeal pouch. The second aortic arch develops between the first spiracular and the second (first true branchial) pouch, and subsequent aortic arches develop between the more posterior branchial pouches. Six aortic arches are laid down in the embryos of all vertebrates except for some very archaic forms (cyclostomes and some selachians), whose number of gill clefts is larger than six. The dorsal aortae become extended forward into the head and backward throughout the length of the trunk and to the base of the tail. In the trunk region the two dorsal aortae later fuse into one unpaired dorsal aorta, while in the branchial and head region the two aortae remain separate. In the trunk, the dorsal aorta gives off at the level of each myocomma an intersegmental artery. Larger arteries convey blood from the dorsal aorta to the viscera, to the limbs and, in embryonic stages, to the yolk sac (vitelline arteries) and to the allantois (umbilical arteries).

The blood vessels developing in conjunction with the excretory organs are the *cardinal veins*. The first rudiments of these appear, in Amphibia, in the form of a venous sinus around the pronephros. Very soon prolongations of this sinus

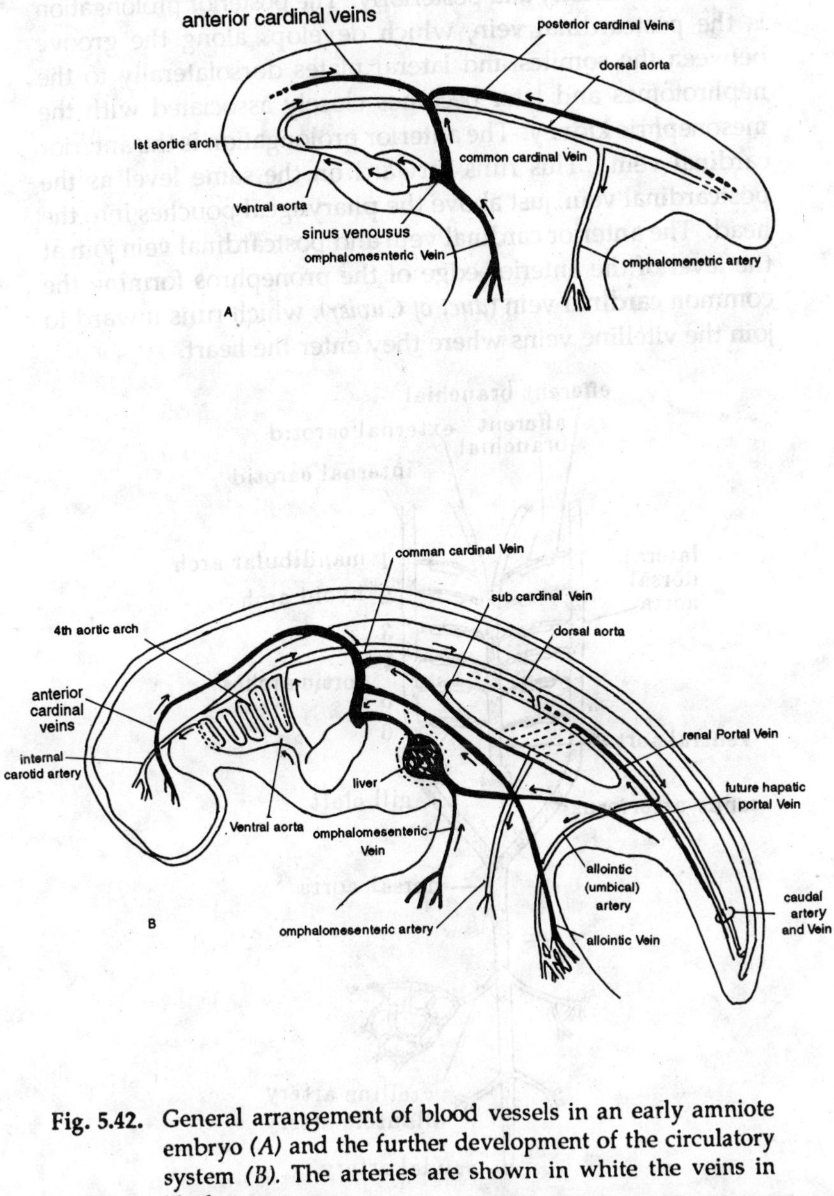

Fig. 5.42. General arrangement of blood vessels in an early amniote embryo *(A)* and the further development of the circulatory system *(B)*. The arteries are shown in white the veins in black.

are found anteriorly and posteriorly. The posterior prolongation is the postcardinal vein, which develops along the groove between the somites and lateral plates dorsolaterally to the nephrotomes and later becomes closely associated with the mesonephric kidney. The anterior prolongation is the anterior cardinal vein. This runs forward on the same level as the postcardinal vein, just above the pharyngeal pouches into the head. The anterior cardinal vein and postcardinal vein join at the level of the anterior edge of the pronephros forming the common cardinal vein *(duct of Cuvier)*, which runs inward to join the vitelline veins where they enter the heart.

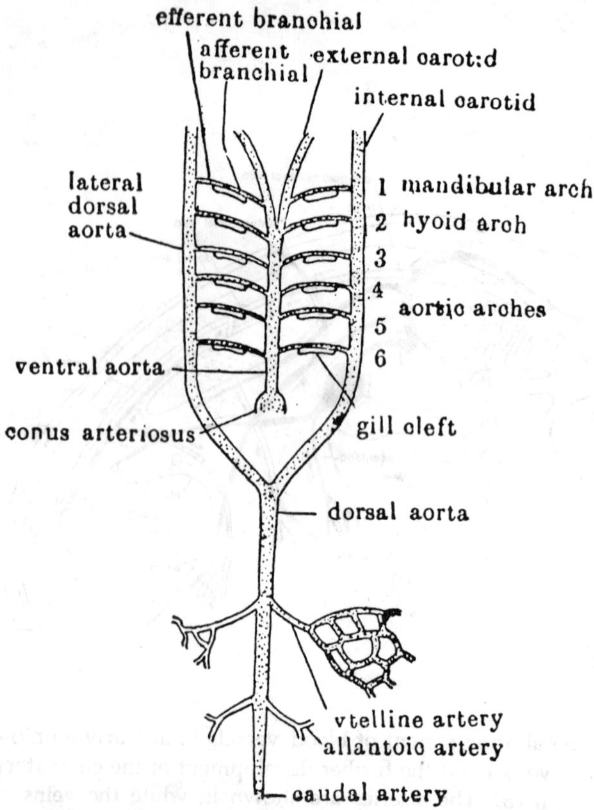

Fig. 5.43. Embryonic aortic arches and arteries.

At a later stage, a second pair of veins develops in tetrapods on the median side of the mesonephric kidney — the *subcardinal veins.*

This basic pattern of the main blood vessels is laid down with great regularity in embryos of all classes of vertebrates, but it becomes greatly modified as development goes on, especially in the higher vertebrates.

In the fishes and in the amphibian larvae, a number of aortic arches from the third onward supply blood to the gills. The blood flow from the ventral portion of each aortic arch involved is directed into a network of capillaries which develop in the gill lamellae attached to the gill septum or in the filaments of the external gills. From these, the blood then returns by way of collecting vessels into the upper part of the arch and through it to the dorsal aorta. The middle part of the aortic arch becomes greatly reduced in diameter. It becomes lost in the capillary network in the branchial septum but may not be completely interrupted, though the amount of blood which passes through it is small compared with the volume of blood going through the gills.

In amphibians which lose their gills on metamorphosing into the adult stage, the connection between the ventral and dorsal portions of the aortic arches is reinstated by the enlargement of the narrow blood vessels connecting them during the larval stages. The capillary system in the gills is reduced and disappears, and the aortic arches again assume the shape of simple tubes leading from the ventral aorta to the dorsal aorta.

Of the six pairs of aortic arches, the anterior two pairs becomes degenerate to a greater or lesser extent, but even in the higher vertebrates remnants of the ventral portions of these arches take part in the formation of the branches of the external carotid in the ventral region of the head. The anterior portions of the ventral aortae, which gave rise to the first and second aortic arch, become the external carotids.

In the terrestrial vertebrates (including amphibians after metamorphosis), the third pair of aortic arches become the carotid arches. The anterior parts of the dorsal aortae now serve to forward the blood from the carotid arch to the dorsal part of the head and to the brain. This blood vessel becomes the *internal carotid artery.* The sections of the ventral aortae between the points of origin of the fourth and the third (carotid) aortic arch carry blood both to the carotid arches (now the internal carotids) and to the external carotids. They thus become the common carotids.

The fourth pair of aortic arches becomes the main channel for the blood flow from the heart to the dorsal aorta and hence to the body. These are the systemic pair of arches (in amphibians and reptiles).

The fifth pair of aortic arches degenerates in all tetrapods except in urodeles, where they carry part of the blood to the dorsal aorta. The general direction of blood flow in the dorsal aorta is from the anterior to the posterior end. This direction of blood flow is patent from the level at which the fourth (systemic) arches join the dorsal aortae. However, the anterior prolongations of the paired dorsal aortae serve for supplying blood to the head. This blood enters the dorsal aortae mainly through the carotid arches; the anteriorly directed flow of blood starts therefore from the point of entry of these arches. Between the level where the carotid arches join the dorsal aorta to the level where it is joined by the systemic arches, there is an area of blood stagnation. As in other cases, where a blood vessel is devoid of blood flow, the sections of the dorsal aortae between the carotid and the systemic arches tend to become atrophied; the vessels become attenuated or even completely interrupted. In some urodeles and reptiles this section of the dorsal aorta, known as the *carotid duct,* persists as a narrow vessel in the adult stage, but in most terrestrial vertebrates the duct, though existing in the embryo becomes completely reduced and the anterior sections of the dorsal aortae, now becoming a part of the internal carotid artery, are completely separated from the remainder of the dorsal aorta.

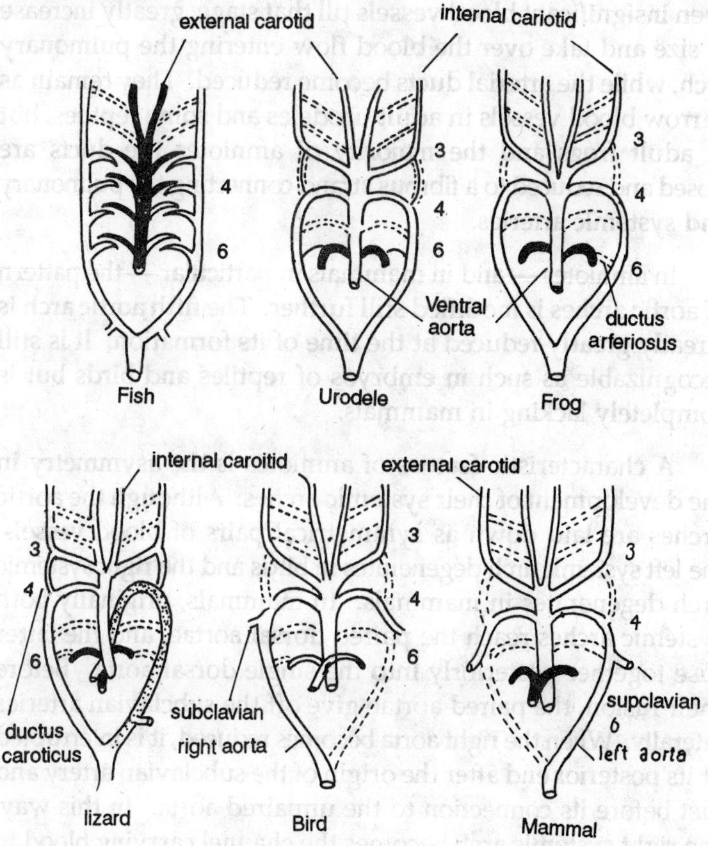

Fig. 5.44. Modifications of aortic arches in vertebrates. 3—carotid arch. 4—systemic arch. 5—abortive. 6—pulmonary arch.

The sixth pair of aortic arches gives off blood vessels to the lungs (and to the skin in amphibian). It is thus the *pulmonary arch*. The upper part of the pulmonary arch, connecting the base of the pulmonary artery to the dorsal aorta, is known as the *arterial duct* (ductus arteriosus or ductus Botalli). In the larvae of amphibians and in the embryonic stages of amniotes, the whole of the pulmonary arch is a fairly large vessel carrying blood from the ventral aorta to the dorsal aorta. This condition changes drastically, however, when the lungs start functioning as respiratory organs. The pulmonary arteries, which had been insignificant blood vessels till that stage, greatly increase in size and take over the blood flow entering the pulmonary arch, while the arterial ducts become reduced. They remain as narrow blood vessels in adult urodeles and some reptiles, but in adult frogs and the majority of amniotes the ducts are closed and reduced to a fibrous strand connecting the pulmonary and systemic arteries.

In amniotes — and in mammals in particular — the pattern of aortic arches is modified still further. The fifth aortic arch is already greatly reduced at the time of its formation. It is still recognizable as such in embryos of reptiles and birds but is completely lacking in mammals.

A characteristic feature of amniotes is the asymmetry in the development of their systemic arches. Although the aortic arches are laid down as symmetrical pairs of blood vessels, the left systemic arch degenerates in birds and the right systemic arch degenerates in mammals. In mammals, originally both systemic arches reach the paired dorsal aortae, and the latter fuse together posteriorly into the single dorsal aorta. Before their fusion, the paired aortae give off the subclavian arteries laterally. When the right aorta becomes reduced, it is interrupted at its posterior end after the origin of the subclavian artery and just before its connection to the unpaired aorta. In this way, the right systemic arch becomes the channel carrying blood to the right subclavian artery. In later stages, the distinction between the aortic arch and the right subclavian artery is lost, and the whole vessel is then known as the right subclavian artery. It is, therefore, not entirely homologous to the left subclavian artery, which starts, as in lower vertebrates, from the left paired dorsal aorta.

The section of the paired ventral aorta between the branching of the common aorta and the origin of the right common carotid artery becomes what is known as the *innominate artery*.

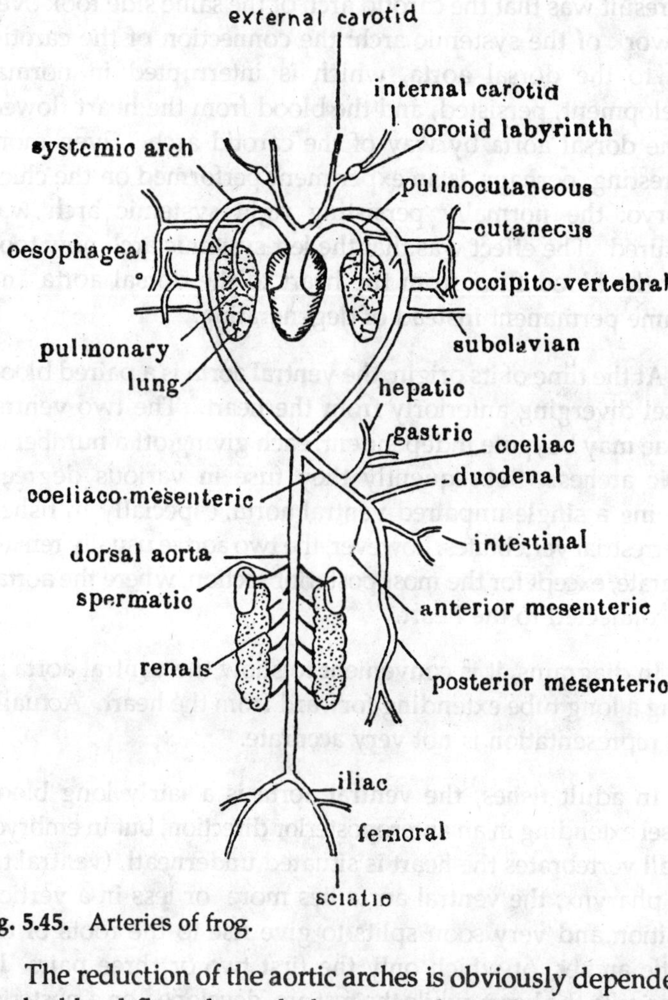

Fig. 5.45. Arteries of frog.

The reduction of the aortic arches is obviously dependent on the blood flow through these vessels, in the same way as the blood flow has been found to mold the development of vessels from the original capillary networks. The following experiments prove this. In frog tadpoles, all four (third to sixth) aortic arches persist up to metamorphosis, and only then the transformations occur which lead to the adult condition. In the tadpole of the clawed toad, *Xenopus laevis*, the systemic arch on one side was destroyed shortly before metamorphosis. The result was that the carotid arch of the same side took over the work of the systemic arch: the connection of the carotid arch to the dorsal aorta, which is interrupted in normal development, persisted, and the blood from the heart flowed to the dorsal aorta by way of the carotid arch. Even more interesting, perhaps, is an experiment performed on the chick embryo: the normally persisting right systemic arch was ligatured. The effect was that the left systemic arch now took over the blood flow from the heart to the dorsal aorta and became permanent instead of degenerating.

At the time of its origin, the ventral aorta is a paired blood vessel diverging anteriorly from the heart. The two ventral aortae may be quite independent, each giving off a number of aortic arches. Subsequently they fuse in various degrees, forming a single unpaired ventral aorta, especially in fishes. In terrestrial vertebrates, however, the two aortae usually remain separate, except for the most posterior section, where the aortae are connected to the heart.

In diagrams, it is convenient to show the ventral aorta as being a long tube extending forward from the heart. Actually this representation is not very accurate.

In adult fishes, the ventral aorta is a fairly long blood vessel extending in an anteroposterior direction, but in embryos of all vertebrates the heart is situated underneatl. (ventral to) the pharynx; the ventral aorta lies more or less in a vertical position and very soon splits to give rise to the roots of the aortic arches, of which only the first two or three pairs lie anterior to the heart, while the rest are developed on a posterior extension of the aorta. The pulmonary arteries, when these develop, may be described as being on the extreme backward prolongation of the ventral aortae; when the last aortic arches area formed, the pulmonary arteries shift upward and can then be seen as originating from these last (sixth) aortic arches.

A further important change in the arterial system in higher vertebrates accompanies the separation of the heart into right and left halves. The pulmonary arteries acquire a separate connection to the right ventricle, while the systemic and carotid arteries are linked up to the left ventricle. The separation occurs from the branching points of these arteries backward toward the heart. The ventral aorta becomes split lengthwise by a partition cutting in at the point where the pulmonary arches branch off from the systemic arches, the latter carrying the carotid arteries with them.

In fishes, the heart remains in the pharyngeal region, but in terrestrial vertebrates the heart is shifted into a more posterior position in the thoracic cavity of the animal. The shift is, in part, owing to the emancipation of the heart from its connection with the gills and, in part, is a result of the development of the neck as a distinct region of the body. The posterior position of the heart causes a considerable elongation of some of the blood vessels conveying blood to the head. The carotid arteries are the ones affected most. There is, however, some variety in the way that the elongation is achieved. In the amphibians, mammals, birds and most reptiles, the carotid arch is carried backward together with the heart (and the other aortic arches). It is then the external and internal carotids which become elongated. In snakes and some lizards, on the other hand, the carotid arches remain in the original position, just posterior to the head, and it is the common carotids which elongate. The systemic and pulmonary arches in all terrestrial vertebrates move into the thoracic cavity together with the heart.

The disappearance of the branchial respiratory apparatus, and the development of the neck, separate parts of a previously closely knit system and bring them in widely different positions in the body. The ganglia of the cranial nerves originally connected to the branchial apparatus and developing in part from epibranchial placodes in the head, inside the brain case. The remnants of the visceral skeleton (hyoid and branchial arches) are to be found in the neck as the hyoid bone and the cartilages of the larynx and trachea. The aortic arches lie in the thoracic cavity, and close to them also lies the thymus, another derivative of the branchial pouches.

The vitelline veins on their way to the heart pass through the region in which the liver is developed at a later stage. When the rudiment of the liver is formed it envelops the vitelline veins. These break up into a system of hepatic sinusoids permeating the liver lobules and thus give rise to the hepatic portal system. The veins in front of the liver and leading to the heart become hepatic vein, and the parts of the veins caudal to the liver give rise to the *hepatic portal vein.*

In amniotes, the development of the vitelline veins is asymmetric, the right vein tending to degenerate posteriorly and the left vein disappearing, in its most anterior part, in front of the liver. Through a system of anastomoses, however, the continuity of the blood flow is preserved, and a single blood vessel may eventually be built from parts of the two vitelline veins. The parts of the vitelline veins collecting blood from the yolk sac of the embryo disappear at the time of birth, but a branch from the vitelline veins—the mesenteric vein—takes over the drainage of blood from the intestinal tract and becomes the main component of the hepatic portal vein of the adult animal.

In embryos of reptiles and birds, the vitelline veins are not completely broken up into hepatic sinusoids as they pass the liver but preserve a large unpaired channel, the *ductus venosus,* by means of which blood passes through the liver and reaches the sinus venosus. After hatching, however, the ducts venosus is occluded, and all the blood of the portal vein is forced into the system of hepatic sinusoids.

The blood vessels of the cardinal system undergo extremely complicated transformation in the course of ontogenetic development and of evolution in vertebrates. The anterior cardinal veins are least changed; they become the internal jugular vein, and the common cardinal veins become the anterior venae cavae. In some mammal, including man, a transverse anastomosis is formed between the anterior cardinal veins of the right and left side, whereupon the connection of the left anterior cardinal to the heart disappears, and all the blood from the head, neck and forelimbs is directed into the right common cardinal vein which becomes the single anterior vena cava.

The transformations of the veins in the posterior part of the body are very much more drastic: the postcardinal veins, which are the main veins of the posterior trunk region in the fishes; and in the early embryos of all vertebrates, are gradually reduced in the embryos of tetrapod.

The first stage in this reduction, as it occurs during the embryonic development in amniotes, is the formation of a venous network on the ventromesial side of the mesonephric kidneys. This plexus gives rise to veins lying in the same position—the *subcardinal veins.* The subcardinal veins become linked to one another by anastomoses in the midline.

Initially, the subcardinal veins join the postcardinal veins anteriorly and use the latter as a channel for blood outflow. At a later stage, however, a connection become established between the network of the subcardinal veins, with their associated plexuses, and the system of veins in the liver. This opens a more direct route to the heart for the blood collected from the kidneys into the subcardinal vein.

At first, the blood from the subcardinal veins passes through the hepatic sinusoids in the tissue of the liver, but as more and more blood goes through, the sinusoids enlarge and develop into one straight channel, passing through the liver directly to the sinus venosus. This channel is the anterior portion of the posterior vena cava. (As result of the reduction of the part of the hepatic tissue, the posterior vena cava later

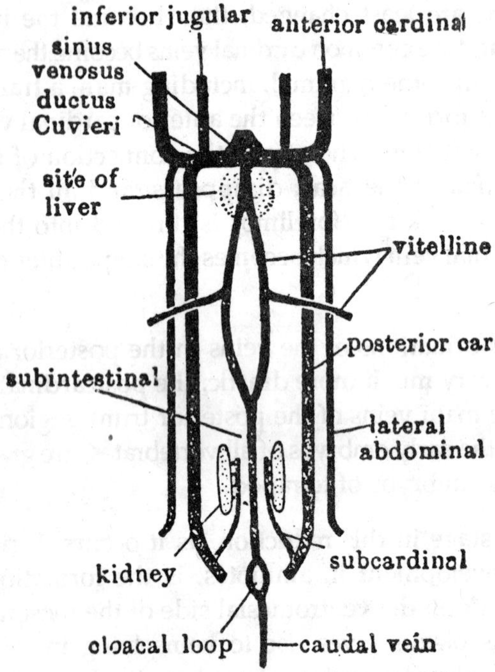

Fig. 5.46. Primitive embryonic veins.

comes to lie on the edge of the liver.) The middle portion of the posterior vena cava is formed by the joining together of smaller vessels belonging to the subcardinal vein system. Eventually, the posterior vena cava, through a system of intermediate smaller vessels, establishes a connection with the iliac veins which collect the blood from the hindlimbs. The posterior vena cava is then established in its final form as it is found in adult amniotes.

The postcardinal veins, in the meantime, lose their significance as channels conveying blood from the posterior part of the body, and their anterior halves leading to the ducts of Cuvier degenerate, leaving behind only insignificant remnants. The posterior parts of the postcardinal veins remain intact a renal portal vein, in amphibian throughout life, and in

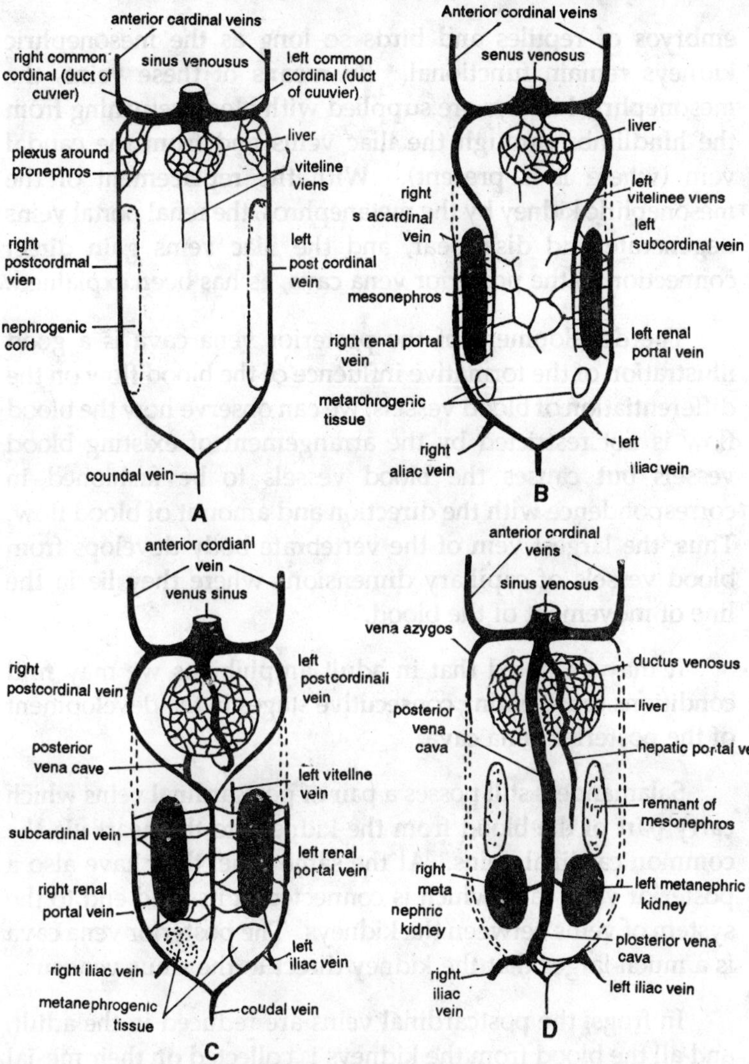

Fig. 5.47. Stages in the transformations of the postcardinal veins in embryos of terrestrial vertebrates (diagrammatic). *A*, Early stage, slightly later than, *A*, the vitelline veins have already given rise to the hepatic sinusoids. *B*, Stage corresponding to that in allantoic fig. 5.42, B[1] veins not shown. C, First appearance of posterior vena cava formed from a connection between the right subcardinal vein and the venous system of the liver. D, Posterior vena cava fully developed.

embryos of reptiles and birds so long as the mesonephric kidneys remain functional. By means of these veins, the mesonephric kidneys are supplied with blood returning from the hindlimbs through the iliac veins and from the caudal vein (where it is present). With the replacement of the mesonephric kidney by the metanephros, the renal portal veins degenerate and disappear, and the iliac veins gain direct connection to the posterior vena cava, as has been explained.

The development of the posterior vena cava is a good illustration of the formative influence of the blood flow on the differentiation of blood vessels: we can observe how the blood flow is not restricted by the arrangement of existing blood vessels but causes the blood vessels to be fashioned in correspondence with the direction and amount of blood flow. Thus, the largest vein of the vertebrate body develops from blood vessels of capillary dimensions where they lie in the line of movement of the blood.

It may be noted that in adult amphibians we may find conditions representing consecutive stages in the development of the posterior vena cava.

Salamanders still posses a pair of postcardinal veins which carry part of the blood from the kidneys to the heart via the common cardinal veins. At the same time, they have also a posterior vena cava which is connected at its hind end to the system of veins between the kidneys. The posterior vena cava is a much larger that the kidney direct to the sinus venosus.

In frogs, the postcardinal veins are reduced in the adult, and all the blood from the kidneys is collected on their mesial aspect into the posterior vena cava. The latter, however, is not yet linked up with the hind legs. This last stage in the development of the posterior vena cava is completed only in the amniotes.

In addition to the hepatic portal vein and the inferior vena cava, a third system of veins becomes associated with the liver in amniotes. The allantoic (or umbilical) veins go toward the heart in the body wall and enter the common cardinal veins. Before joining the latter, the allantoic veins pass very close to the liver. As the liver grows, it makes contact with the allantoic veins, whereupon a connection is established between these and the system of hepatic sinusoids in the liver. More and more blood from the allantoic veins enters the liver, until the main channel into the common cardinal veins becomes reduced and obliterated, so that all the blood from the allantois (from the placenta in mammals) goes to the heart by way of the liver. The right allantoic vein later degenerates in all amniotes, the left remaining as the only functional one.

The large amount of blood entering the liver from the left allantoic vein cannot be accommodated in the hepatic sinusoids and eventually makes for itself a broad direct channel which joins the posterior vena cava near the point where it leaves the liver at its anterior end. This direct channel is referred to by students of mammalian (especially human) development as the ductus venosus, although it is not homologous to the ductus venosus of reptiles and birds, where it is a section on the pathway of the vitelline veins. After birth and the cessation of placental circulation, the ductus venosus degenerates and becomes transformed into a strand of connective tissue (the *ligamentum venosum),* as often happens with large blood vessels when they cease to be functional.

DEVELOPMENT OF THE REPRODUCTIVE ORGANS

The gonads in vertebrates develop from the upper edge of the visceral layer of lateral plate mesoderm in the posterior half of the body. The first rudiment of the gonad appears as a thickened longitudinal strip of mesodermal epithelium lining the body cavity immediately lateral to the dorsal mesentery. This thickening is called the germinal ridge. Initially the ridge is produced by the mesodermal cells, assuming a high columnar shape, but soon the cells become arranged into a compact mass several cells thick, and the ridge protrudes into the coelomic cavity. Still later, the connection between this cell mass and the peritoneal wall becomes constricted laterally, and the gonad remains suspended from the peritoneal wall only by a double layer of peritoneum—the *mesorchium* or *mesovarium*.

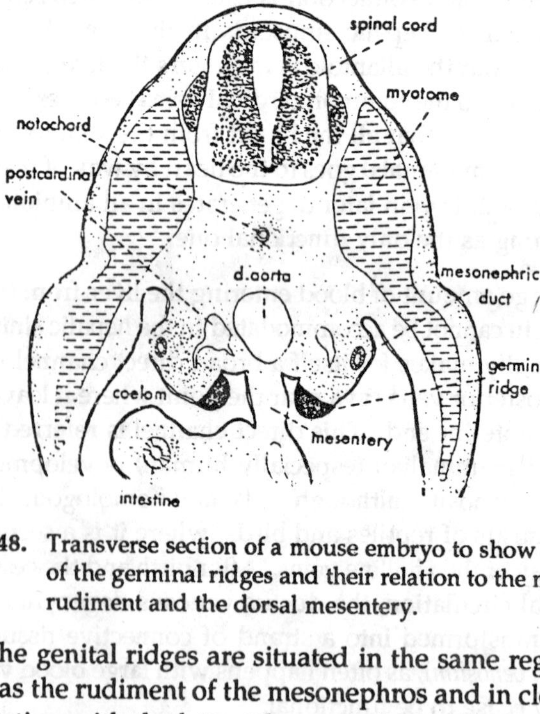

Fig. 5.48. Transverse section of a mouse embryo to show the position of the germinal ridges and their relation to the mesonephric rudiment and the dorsal mesentery.

The genital ridges are situated in the same region of the body as the rudiment of the mesonephros and in close spatial association with the latter, the genital ridge being medial to the mesonephros. Although in the later development of vertebrates the genital apparatus becomes associated with the excretory system, the initial stages of the two organ systems are completely independent of each other.

What has been said so far about the formation of the gonad rudiment refers only to its gross morphology. A more detailed microscopic examination of the germinal ridge epithelium discloses that it consists of two types of cells. The majority of cells are very similar to the other cells of the peritoneal epithelium, even though they may become columnar

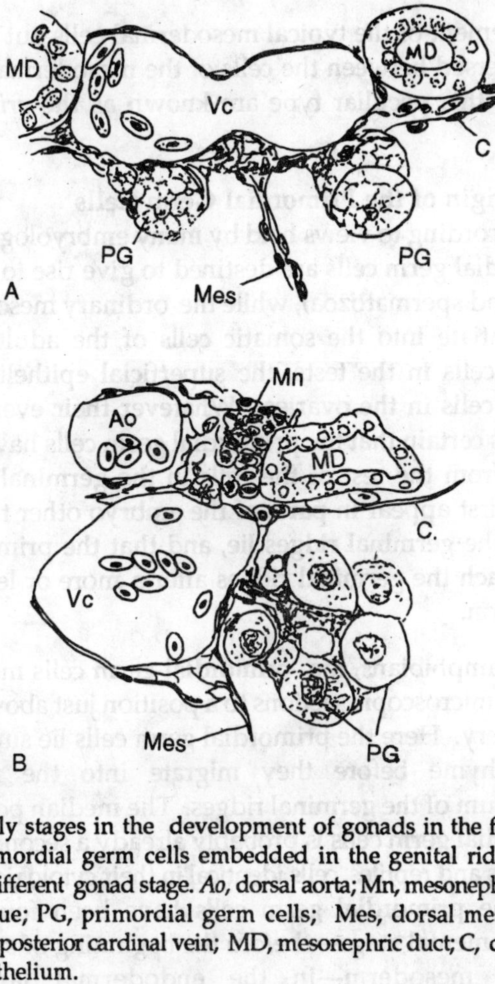

Fig. 5.49. Early stages in the development of gonads in the frog. *A*, Primordial germ cells embedded in the genital ridges. *B*, Indifferent gonad stage. *Ao*, dorsal aorta; Mn, mesonephrogenic tissue; PG, primordial germ cells; Mes, dorsal mesentery; Vc, posterior cardinal vein; MD, mesonephric duct; C, coelomic epithelium.

in the initial stages of the development of the ridge. The cells of the second type have a very different appearance. They are much larger than the ordinary mesodermal cells, they contain large, more or less vesicular nuclei and their cytoplasm differs from that of the surrounding cells in its staining properities amphophil instead of acidophil in birds) and, in amphibians, in its higher yolk content. The shape of the cells is nearly spherical, and they do not participate in the epithelial arrangement of the typical mesodermal cells but appear to be interspersed between the cells of the mesodermal epithelium cells of this peculiar type are known as the *primordial germ cells.*

The Origin of the Primordial Germ Cells

According to views held by many embryologists, only the primordial germ cells are destined to give rise to the gametes (eggs and spermatozoa), while the ordinary mesodermal cells differentiate into the somatic cells of the adult gonad: the Sertoli cells in the tests, the superficial epithelium and the follicle cells in the ovaries. Whatever their eventual fate, it appears certain that the primordial germ cells have a different origin from the rest of the cells in the germinal ridges, that they first appear in parts of the embryo other than those in which the germinal ridges lie, and that the primordial germ cells reach the germinal ridges after a more or less extensive migration.

In amphibians, the primordial germ cells may be traced back in microscopic sections to a position just above the dorsal mesentery. Here the primordial germ cells lie surrounded by mesenchyme before they migrate into the mesodermal epithelium of the germinal ridges. The median position of the primordial germ cells is probably already a secondary one. In the birds and reptiles, cells identical in their cytological properties with the primordial germ cells have been found in early developmental stages—before the beginning of segmentation of the mesoderm—in the endodermal layer of the extraembryonic part of the blastoderm. Similarity of appearance cannot, however, be considered as proof that the cell found in the extra-embryonic endoderm are the same cells a appear later in the germinal ridges.

To test whether the primordial germ cells are derived from the cells in the extraembryonic endoderm, the latter may be destroyed prior to their migration. In the chick, the cells in question occupy a crescentic area in front of the head end of the embryo. This area can be destroyed in various ways, as for instance by cauterizing it with a hot needle or irradiating it with radium emanation. As the crescentic area lies well beyond the embryo proper, the development of the latter is unimpaired. At the time when the germinal ridges are formed, no primordial germ cells are to be found in and around the germinal ridges, although the area of the germinal ridges had not been tampered with.

The experiment shows that the peculiar cells of the extraembryonic endoderm are actually the primordial germ cells, which are later incorporated into the gonad.

This conclusion is further supported by the following experiment: if the primordial germ cells are exposed to very weak irradiation with x-rays, while still in the extraembryonic endoderm, they are damaged to a greater or lesser extent but not killed at once. some of the damaged cells succeed in reaching the site of gonad formation and are to be found there for a short time. Then all the primordial germ cell perish, and there remains a gonad consisting of mesodermal cells only. The gonad may grow and develop further without showing any signs of producing generative cells. Such a gonad, devoid of generative elements, may be called a sterile gonad.

As to the means by which the primordial germ cell travel from the extraembryonic endoderm to their later position in the germinal ridges, it is supposed that in bird they do so largely through the blood vessel system. The cells have been observed to leave the endodermal epithelium, to move slowly into the space between endoderm and mesoderm, and then to penetrate into the blood vessels of the area vasculosa. The primordial germ cells may actually be seen in the blood stream. Although most of them eventually reach the germinal ridges, some go astray and are found in various parts of the body, where they eventually degenerate.

Further experimental work on the origin of primordial germ cells has been carried out in amphibians. According to

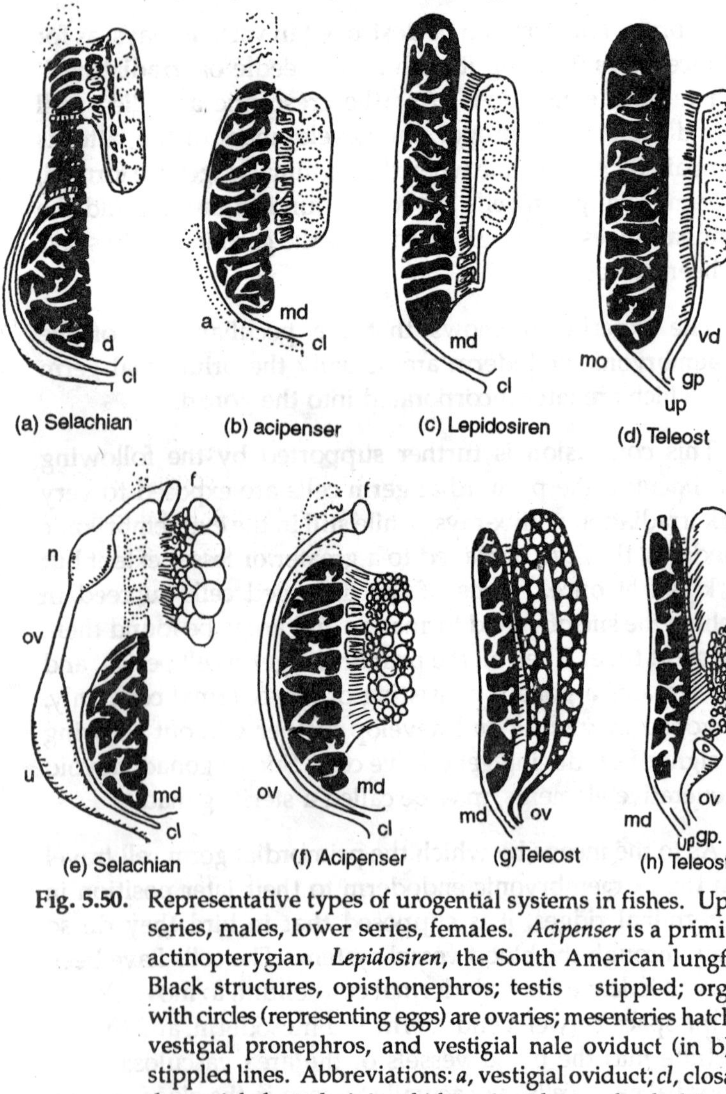

Fig. 5.50. Representative types of urogential systems in fishes. Upper series, males, lower series, females. *Acipenser* is a primitive actinopterygian, *Lepidosiren*, the South American lungfish. Black structures, opisthonephros; testis stippled; organs with circles (representing eggs) are ovaries; mesenteries hatched; vestigial pronephros, and vestigial nale oviduct (in b) in stippled lines. Abbreviations: *a*, vestigial oviduct; *cl*, closaca; *d*, special duct draining kidney in advanced selachians; *f*, open funnel of oviduct; *gp*, genital papilla and pore of telecots; *l*, gland in anterior region of selachian kidney analogous to epididymis; *md*, archinephric duct as kidney duct; *n*, shell gland; *ov*, oviduct; *u*, selachian ovisac, or "uterus" *up*, urinary pore of teleosts; *vd*, sperm duct of teleosts, analogous to ductus deferens of amniotes. In male selachians, as in amniotes,

the testis tends to take over the original duct for sperm transport, but in other fishes, this duct continues to drain the kidneys, and in teleosts a separate sperm duct *(vd)* develops. Specialized methods of egg transport are present in teleosts.

the view held by the majority of embryologists, the cytoplasmic material included in the primordial germ cells in amphibians can be traced back as far as the uncleaved egg. Before cleavage begins, cytoplasmic inclusions of a particular kind can be found in the subcortical layer at the vegetal pole of the egg. The inclusions are very rich in RNA. During cleavage, these inclusions move upward along the cleavage furrows and surround the nuclei of a group of endodermal cells lying right in the middle of the mass of yolky endoderm. The cells receiving this particular cytoplasmic substance are the primordial germ cells, and the substance of the cytoplasmic inclusions has been called the germinal plasm.

By irradiating the vegetal pole of newly fertilized frog's eggs with ultraviolet light, it is possible to destroy the "germinal plasm" and in this way to greatly reduce the number of primordial germ cells formed in later stages. By removing the endoderm in the neurula stage, the embryos are completely sterilized . After the completion of neurulation, the primordial germ cells migrate first to the dorsal region of endoderm, to the dorsal crest, the part of the endodermal wall lying above the cavity of the gut. From here the primordial germ cells move into the mesenchyme of the dorsal nesentery and eventually into the germinal ridges.

The area containing primordial germ cell can be transplanted in the late neurula stage from one embryo to another and if the tissues of the donor embryo can be distinguished from the tissue of the hot it can be shown that the eggs and spermatozoa are actually derived from the grafted material. In one experiment, the mensoplantation was made between two subspecies of the clawed that *Xenopus laevis* differing, among other things, in the properties (size, colour) of their eggs. Operated female embryos were reared to the adult stage and laid eggs which in colour and size clearly showed the properties of the donor species. Thus the primordial germ cells transplanted with a section of endoderm at the end of neurulation not only found their way into the gonads but also preserved intact their genetic characters, in spite of the fact that during their development and maturation in the gonads they were surrounded by somatic cells of a different subspecies.

Surprisingly, it appears that the primordial germ cells in urodele amphibians have an entirely different origin from those in frogs and also from those in amniotes. They develop not from the endodermal cells but from cells originally located in the posteroventral parts of the lateral plate mesoderm.

It has been stated earlier that in salamander embryos it is possible to remove the whole of the endoderm in the neurula state. The endoderm from another embryo may then be inserted into the empty ecto-mesodermal shell. If the operation is carried out heteroplastically, and if the two species are chosen in such a way as to allow the cells of each to be distinguished by some peculiarity, then the derivation of the primordial germ cells, or any other cells, from the endoderm or the mesoderm respectively can be determined with certainty. The experiments were carried out with the purpose of investigating the origin of the primordial germ cells using *Triturus cristatus* as one of the components and *Triturus alpestris* or *Ambystoma mexicanum* as the other. The eggs of the latter two species contain in their cytoplasm considerable amounts of pigment granules, while in the eggs of *Triturus cristatus* pigment granules are either completely absent or very few in number. The operation is a rather difficult one, but a sufficient number of operated embryos survived. The result was a very clear-cut one: all the primordial germ cells found in the operated embryos had the specific properties (pigmentation) of the species to which the ectoderm and mesoderm belonged and not of the species to which the endoderm belonged. The primordial germ cells were, in this experiment, embedded in germinal ridges consisting of cells of the same species.

In another series of experiments, the lateral plate mesoderm was transplanted in the same heteroplastic combination and it was found that after such operations numbers of primordial germ cells develop from the graft. The primordial germ cells, although mesodermal, do not arise from cells in the germ ridges, but, as in the case of endodermal primordial germ cells, they migrate from their site of origin (ventrolaterally in the posterior part of the body) to the area above the dorsal mesentery and then into the germ ridges.

The conclusion seems inevitable that the primordial germ cells are derived from different germinal layers in the urodele amphibians and in the frogs. In many invertebrates, however (*Ascaris, Sagitta,* some insects), the primordial germ cells are known to be distinguishable in very early cleavage stages even before the germinal layers become segregated. The primordial germ cells can thus be considered as not belonging to any of the germ layers. This is in contrast to the somatic parts of the gonads and the auxiliary parts of the sexual organs (ducts, etc.) which are mesodermal.

As to the mechanism by which the primordial germ cells in the amphibians reach their destination, it is ameboid movement of the cells themselves. There is no reason to suppose that blood vessels play any part in their transport.

No experimental investigations have been carried out on the origin of the primordial germ cells in human embryos, but careful observations show that these are first found in the endodermal epithelium of the yolk sac in the vicinity of the allantoic stalk, and that from there the germ cells migrate into the adjoining mesenchyme and eventually take up their position in the germinal ridges.

In mice, on the other hand, investigations comparable to those done on birds and amphibians have been carried out. As in man, the primary germ cells first occur in the endoderm and later migrate into the germ ridges. These cells contain large quantities of alkaline phosphatase in early embryonic stages when this enzyme is absent from other cells of the embryo. By using a special histochemical stain for alkaline phosphatase on sections, therefore, it is possible to distinguish the primary germ cells and to follow their migration from the gut into the dorsal mesentery and then into the germinal ridges. Originally there are less than a hundred primary germ cells in a mouse embryo, but during their migration the cells increase in number by repeated mitosis, so that eventually 5,000 or more cells are present. The primary germ cell are extremely sensitive to x-rays, and if the female mice are irradiated between the eighth and twelfth day of gestation with a dose of 400r, all or almost all the germ cells are killed off before they reach the germinal ridges. The germinal ridge develop nevertheless and differentiate into gonadal tissue which is, however, completely sterile—devoid of sex cells or their predecessors.

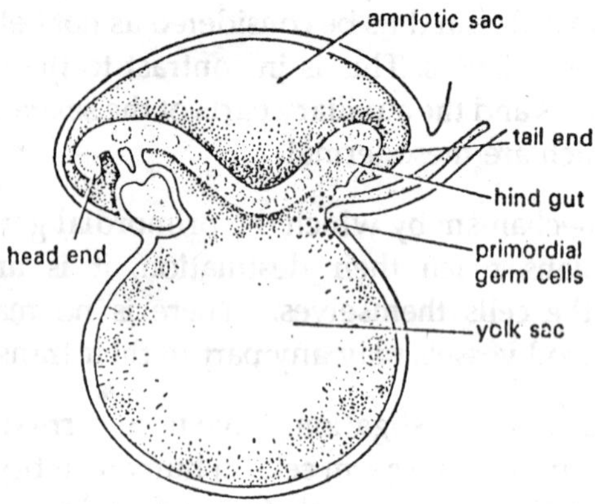

Fig. 5.51. A 4-week old human embryo showing the original position of the primordial germ cells.

Differentiation of the Gonads

After arriving in the germinal ridges, the primordial germ cells become embedded in the epithelium of the germinal ridge. At the same time, the germinal ridge epithelium becomes convert toward the coelomic cavity and, on it dorsal side, forms a hollow which is filled initially with loose undif-

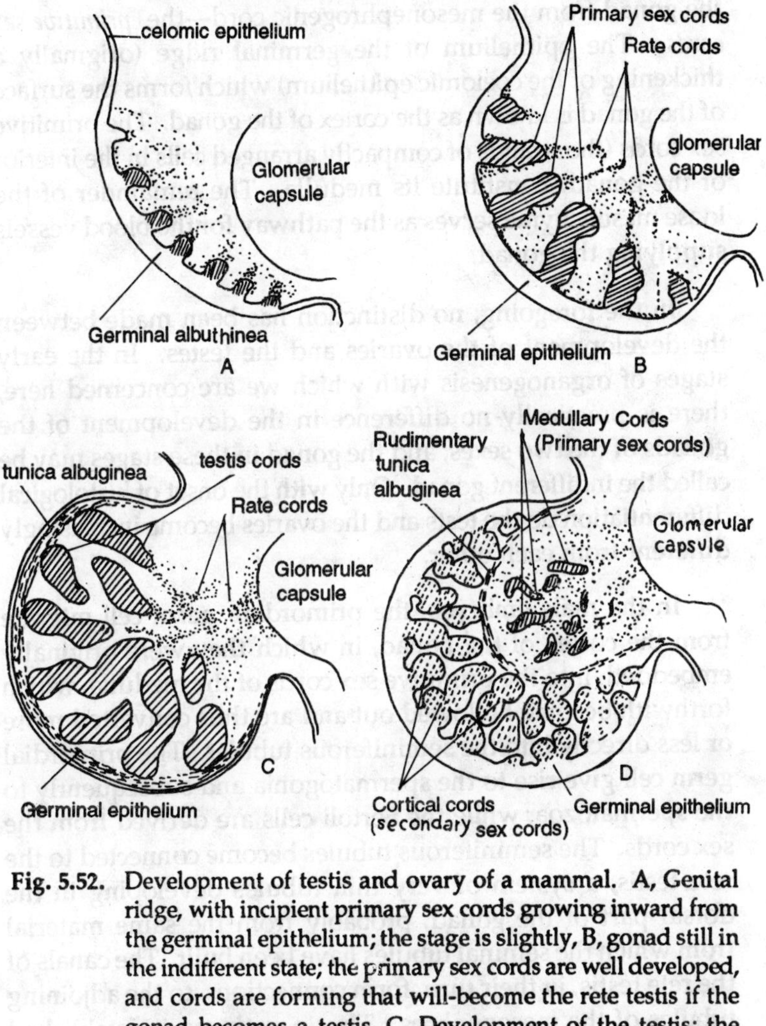

Fig. 5.52. Development of testis and ovary of a mammal. A, Genital ridge, with incipient primary sex cords growing inward from the germinal epithelium; the stage is slightly, B, gonad still in the indifferent state; the primary sex cords are well developed, and cords are forming that will-become the rete testis if the gonad becomes a testis. C, Development of the testis; the germinal epithelium degenerates and is replaced by a sheath around the testis; testis tubules are rete continue development. D, Early development of an ovary, with reduction of primary sex cords and rete rudiments, but, on the other hand, great development externally of secondary cords in which eggs develop.

ferentiated mesenchyme. Subsequently, this loose mesenchyme is partly replaced by compact strands of cells migrating into the gonad from the mesonephrogenic cord—the *primitive sex cords.* The epithelium of the germinal ridge (originally a thickening of the coelomic epithelium) which forms the surface of the gonad is known as the cortex of the gonad. The primitive sex cords (the strands of compactly arranged cells in the interior of the gonad) constitute its medulla. The remainder of the loose mesenchyme serves as the pathway for the blood vessels supplying the gonad.

In the foregoing, no distinction has been made between the development of the ovaries and the testes. In the early stages of organogenesis with which we are concerned here, there is practically no difference in the development of the gonads of the two sexes, and the gonad in these stages may be called the indifferent gonad. Only with the onset of histological differentiation do the tests and the ovaries become increasingly different from each other.

In the male embryos, the primordial germ cell migrate from the cortex of the gonad, in which they were originally embedded, into the primitive sex cords of the medulla, which forthwith become hollowed out and are thus converted more or less directly into the seminiferous tubules. The primordial germ cell give rise to the spermatogonia and subsequently to the spermatozoa, while the Sertoli cells are derived from the sex cords. The seminiferous tubules become connected to the rete testis, a system of very thin tubules developing in the dorsal part of the gonad, probably from the same material from which the seminal tubules have been built. The canals of the rete testis, in their turn, form connections to the adjoining tubules of the mesonephros. Thus a pathway is established from the male gonad, via the mesonephros, to the mesonephric duct, which in vertebrates serves as the outlet for the sperm. While the medulla of the testis becomes its functional part, the cortex undergoes reduction and is converted into a thin epithelial layer covering the coelomic surface of the testis.

In female embryos, the medulla of the gonad becomes

reduced, the primary sex cords are resorbed, and the interior of the mature gonad is filled with loose mesenchyme permeated by blood vessels. The primordial germ cells remain embedded in the cortex, which greatly increases in thickness. During the growth of the gonad, the masses of cortical cells become split up on the inner surface of the cortex into groups and strands of cells surrounding one or several primordial germ cells. These then become the primary oocytes which enter in the initial stages of oogenesis. The primordial germ cells lying nearer the surface of the cortex retain their undifferentiated state longer and serve for the production of eggs at a later stage in the female's life.

Special features in the development of gonads in some groups of animals will now be moted.

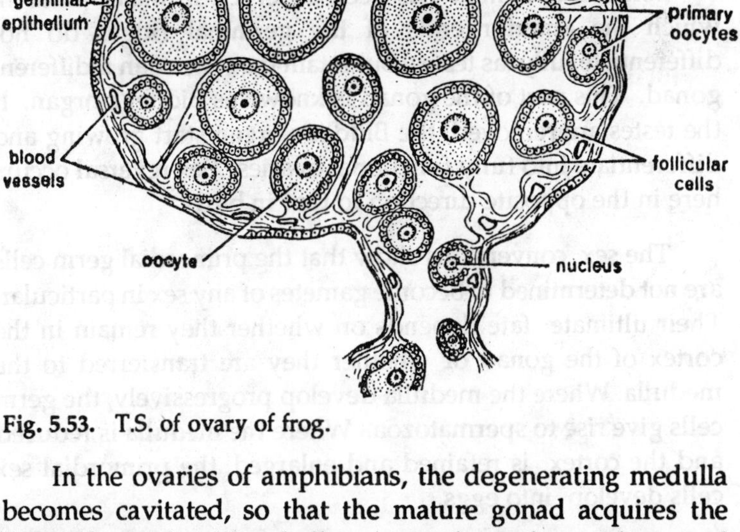

Fig. 5.53. T.S. of ovary of frog.

In the ovaries of amphibians, the degenerating medulla becomes cavitated, so that the mature gonad acquires the form of a sack with a hollow space (or rather with several hollow spaces) in the interior. The mature eggs bulge into these internal cavities. When ovulation occurs, however, the external surface of the gonad becomes ruptured, and the eggs are shed not into the internal cavity but into the coelom.

In birds, with some rare exceptions, only the left ovary becomes fully developed. From the start, more primordial germ cells enter the left germinal ridge, and only here the ovarian cortex acquires a typical structure. The gonad on the right side remains in the state of an indifferent gonad, with the primary sex cords not reduced. The primordial germ cells in the right gonad eventually degenerate. However, if the left gonad is removed by operation, the right gonad enters into a stage of compensatory growth and differentiates progressively. The right gonad then acquires the general structure not of an ovary but of a testis and even starts to secrete male hormones. Under the influence of the male hormones, the bird shows some external features of the male sex (especially in colour and shape of the feathers) even though it was and remains a genetic female.

Something similar has been observed in male toads in which the anterior ends of the germinal ridges do not differentiate fully (as testes) but retain the state of an indifferent gonad. This part of the gonad is known as Bidder's organ. If the testes are removed, the Bidder's organ start growing and differentiate into fully functional ovaries. Sex reversal occurs here in the opposite direction to that in birds.

The sex conversions show that the primordial germ cells are not determined to become gametes of any sex in particular. Their ultimate fate depends on whether they remain in the cortex of the gonad or whether they are transferred to the medulla. Where the medulla develop progressively, the germ cells give rise to spermatozoa. Where the medulla is reduced and the cortex is retained and enlarged, the primordial sex cells develop into eggs.

Development of the Genital Duct

An essential part of the genital organs are the genital ducts by means of which the sex cells (ova and spermatozoa) are passed to the exterior or to an other location where fertilization can take place.

The channels by which the spermatozoa are conveyed from the testes are the mesonephric ducts. The seminiferous tubules are connected to the mesonephric ducts through the rete testis and a number of mesonephric tubules which become the epididymis. In fishes (except for the bony fishes, whose gonaducts are not comparable with those of other vertebrates) and in amphibians, the mesonephric ducts serve simultaneously for the passage of urine from the mesonephros and of the spermatozoa from the testes. There may be, however, a certain degree of segregation between the genital and the excretory functions: the anterior part of the kidney and the corresponding part of the mesonephric duct serving mainly for reproductive functions, while the posterior part of the mesonephros carries out the main part of excretion (in selachians and in urodeles). Nevertheless, both the spermatozoa and the urine pass eventually into the cloaca through the same duct.

In amniotes, the mesonephros ceases to be an excretory organ and the only part remaining functional is that which in the males develops as epididymis. The mesonephric duct becomes the vas deferens, serving only for reproduction.

In embryos of the female sex, the development of the mesonephric duct starts in the same way as in male embryos. In anamniotes, if functions throughout life as the pathway for urine. In amniotes, the mesonephric duct is laid down in connection with the development of the pronephros and mesonephros. When the mesonephros ceases to function as an excretory organ, the mesonephric duct degenerates.

In addition to the mesonephric ducts, a second pair of genital ducts, the *paramesonephric* (or Mullerian) *ducts* are formed in all vertebrates (again with the exception of bony fishes). At

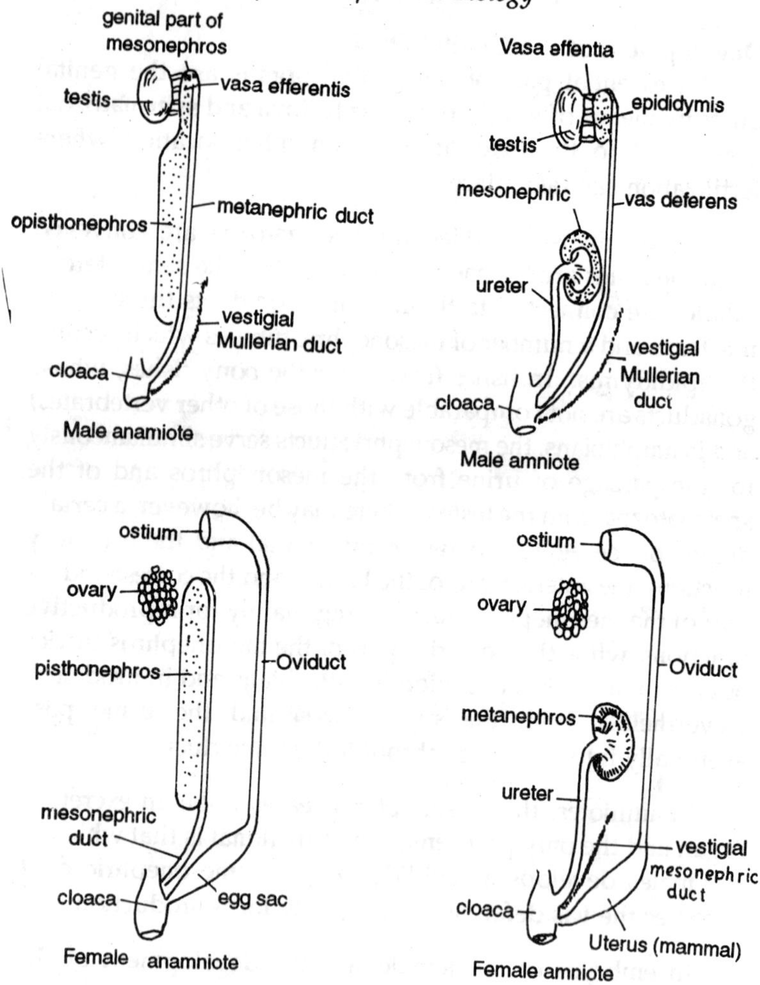

Fig. 5.54. Urogential organs and their ducts.

their anterior ends, the paramesonephric ducts open by a funnel-shaped *ostium tubae* into the coelomic cavity, and in adult females the paramesonephric ducts serve as channels for the eggs, that is, as oviducts. Although functional only in females, the paramesonephric ducts develop initially in embryos of both sexes. The development starts at the anterior end with the ostium tubae. In lower vertebrates (selachians, amphibians) the ostium tubae is the surviving funnel (nephrostome) of one of the pronephric tubules, which remain intact after the rest of the pronephros degenerates and is resorbed (there are indications that two or more pronephric nephrostomes may fuse to form the ostium). The deep end of the ostium starts growing in a backward direction retroperitoneally and reaches the mesonephric duct. It then continues its growth alongside the mesonephric duct until it reaches the cloaca and fuses with the cloacal wall.

In higher vertebrates, no nephrostomes remain after the degeneration of the pronephros, but the coelomic epithelium becomes thickened in the area where the pronephros was connected to the coelom. This thickening, which is elongated in a craniocaudal direction, sink in, forming a longitudinal groove, and later it rolls into a tube. The tube remains open and in communication with the coelom anteriorly, but is completely closed posteriorly, where its free end is imbedded in the retroperitoneal connective tissue. At this stage, the rudiment of the paramesonephric duct become essentially similar to the initial stage of the duct in anamniotes, where it starts growing from a remnant of the pronephric nephrostome. The opening of the tube anteriorly, of course, becomes the newly formed tubae. There follows a backward growth of the paramesonephric duct alongside the mesonephric duct and its connection to the cloaca.

The details of the growth of the paramesonephric ducts are not very well-known, and it appears that they vary to some extent in different animals. It is questionable to what extent the mesonephric duct contributes to the development of the paramesonephric duct, if at all. It has been found, however, that if the mesonephric duct is incomplete because its outgrowth was interrupted in early stages, the paramesonephric duct, developing at a very much later stage, grows backward only as far as the posterior end of the mesonephric duct. It appears that the mesonephric duct in some way provides a path for the subsequent growth of the paramesonephric duct, even if it does not itself contribute material for the elongation of the latter.

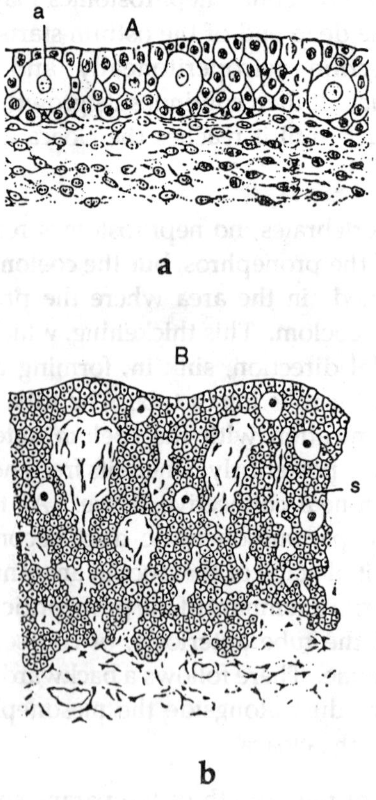

Fig. 5.55. Early stage in mammalian gonad development. *A*, A section of the germinal epithelium of the gonad at an early in different stage; primordial germ cells are present *(a)*, but sex cords are not developed. *B*, A somewhat later stage, with primary sex cords *(s)* growing inward into the epithelial layer.

The paramesonephric duct initially develops in both male and female embryos, although it become functional in the latter sex only. In this way a most peculiar situation arises, namely, that embryos of both sexes possess a double set of genital ducts: the mesonephric duct, which in the male sex function as the vasa deferentia, and the paramesonephric ducts, which become the oviducts in the female sex. This condition of the genital ducts can be called the indifferent stage in respect to sex. In chick embryos this condition prevails up to the ninth day of incubation, and in a human embryo to the age of about two and one-half months. After this age, the ducts of the opposite sex regress. The paramesonephric ducts in male embryos of amniotes disappear practically completely. The mesonephric ducts in female embryos are reduced to a lesser degree: in birds, the ducts persist as rudimentary structures even in the adult state, and in mammals and man small remnants of mesonephric ducts may also remain. In birds, the right paramesonephric duct become reduced in correspondence with the reduction of the right gonad, while the left duct becomes the functional oviduct.

The progressive development of the genital ducts, corresponding to the sex of the embryo, and the reduction of the genital ducts of the opposite sex are the result of the action of sex hormones which start being secreted by the gonads when these emerge from the indifferent state and differentiate into either testes or ovaries. The hormones are secreted not by the germ cells (future eggs and spermatozoa) but by the somatic cells of the gonads. The female hormones are believed to be secreted by cells of the cortex; the male hormones are produced by some of the cells of the medulla (the future interstitial cells).

A proof that the differentiation of the genital ducts is under hormonal control has been provided by experiments in which sex hormones were injected into embryos of birds or mammals, or sex hormones were added to water in which the larvae of amphibians were being reared. The result of one such experiment performed with chick embryos. The injection of a female hormone into a male embryo caused the retention and hypertrophy of the paramesonephric ducts (compare with normal male of same age; Injection of a male hormone into a female embryo caused the paramesonephric ducts to disappear almost completely and the mesonephric ducts to increase greatly in size. In these experiments the gonads were not affected, so that the original sex of the treated embryo could be easily ascertained. Essentially similar results were produced in numerous experiments on other animals.

The results of sex hormone injections may be checked by experiments of castration of early embryos. In birds, castrated embryos, female as well as male, retain both paramesonephric and mesonephric ducts and remain thus in an indifferent condition as regards the genital duct system. Castrated mammalian embryos of both sexes tend to develop according to the female sex pattern: their paramesonephric ducts develop progressively, but the mesonephric ducts degenerate. This is interpreted as showing that in mammals it is the male hormone which mainly causes differentiation of the secondary sex characters, while the female characters do not need special stimulation for their development.

The dependence of the development of sexual characters on hormone has been the subject of very numerous investigations, which could not be reviewed here. For further information on this subject the student is referred to Burns, 1961.

Differentiation of the Cloacal Region and the External Genitalia in Mammals

In amphibians, reptiles and birds, the ducts carrying excretory products (the mesonephric ducts and ureters) as well as the genital ducts carrying eggs and sperm (the oviducts and vasa deferentia) open into the posterior section of the endodermal gut-the cleva. The cloacal opening thus serves, in these animals, for the exit of feces, urine and the sex cells. In mammals, the cloaca becomes subdivided so that the channel for feces is separated from the pathway serving for the conveyance of urine and sex products.

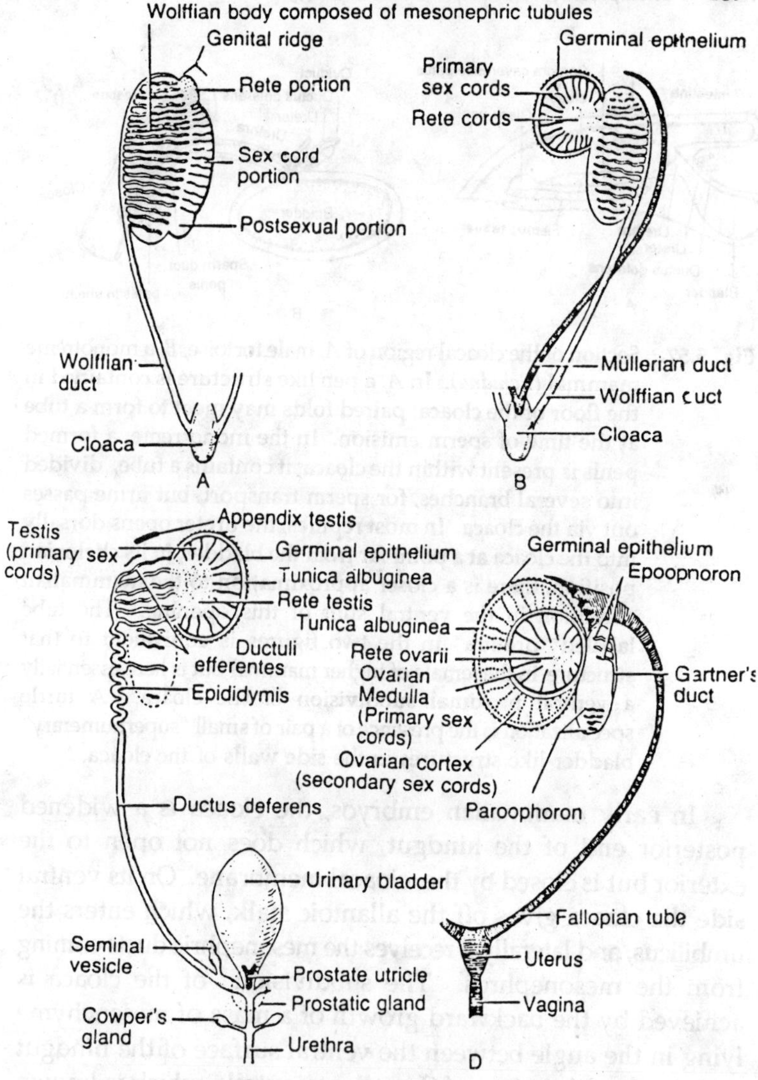

Fig. 5.56. Embryonic development of the genital system of amniotes. *A*, Early indifferent stage, showing sex cords, developing rete testis, mesonephric kidney, and archnephric duct (wolffian body, wolffian duct). *B*, Somewhat later stage, in which the embryonic oviduct (mullerian duct) has appeared. *C*, The adult male. *D*, Adult female.

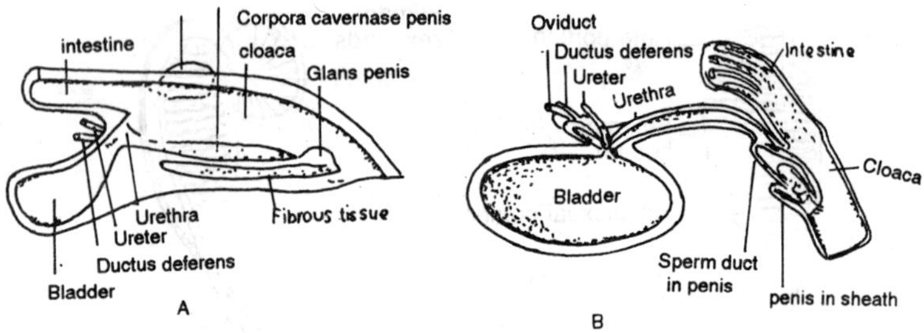

Fig. 5.57. Section of the cloacal region of *A*, male tortoise, *B*, a monotreme mammal *(Echidna)*. In *A*, a pen like structure is contained in the floor of the cloaca; paired folds may meet to form a tube at the time of sperm emision. In the monotreme, a formed penis is present within the cloaca; it contains a tube, divided into several branches, for sperm transport, but urine passes out via the cloaca. In most reptiles, the ureter opens dorsally into the cloaca at a point far from the bladder; in its chelonian position, there is a closer approximation to the mammalian condition in the ventral shift of this opening. The tube labeled "urethra" in the two figures is equivalent to that structure in the female of higher mammal but is here essentially a ventral proximal subdivision of the cloaca. A turtle specialization is the presence of a pair of small "supernumerary" bladder-like structures in the side walls of the cloaca.

In early mammalian embryos, the cloaca is a widened posterior end of the hindgut, which does not open to the exterior but is closed by the cloacal membrane. On its ventral side the cloaca gives off the allantoic stalk, which enters the umbilicus, and laterally it receives the mesonephric ducts coming from the mesonephros. The subdivision of the cloaca is achieved by the backward growth of a mass of mesenchyme lying in the angle between the ventral surface of the hindgut and the dorsal surface of the allantoic stalk, which is known as the *urorectal* or *cloacal septum*. As the urorectal septum extends backward the cloaca is split into a dorsal and a ventral part. In the process of this subdivision, the openings of the mesonephric ducts become assigned to the ventral part, which becomes the urogenital sinus, while the dorsal part becomes the rectum. The urorectal septum eventually reaches the cloacal

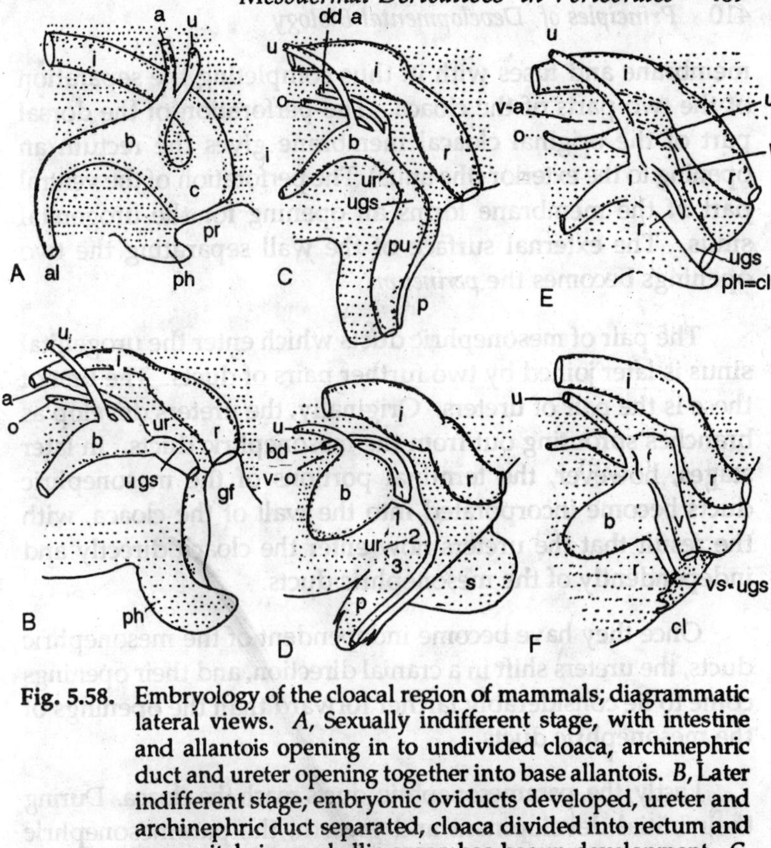

Fig. 5.58. Embryology of the cloacal region of mammals; diagrammatic lateral views. *A,* Sexually indifferent stage, with intestine and allantois opening in to undivided cloaca, archinephric duct and ureter opening together into base allantois. *B,* Later indifferent stage; embryonic oviducts developed, ureter and archinephric duct separated, cloaca divided into rectum and urogenital sinus; phallic organ has begun development. *C,* Early stage of development of male, and *D,* adult male structure. In contrast to female, the oviduct disappears (broken line in *D),* the archinephric duct becomes the ducts deferens, and an extensive urethra includes, in addition, (1) the base of the allantois, (2) the urogenital sinus, and (3) a duct traversing the penis. *E,* Early stage of development of female, and F, adult female structure; disappearance of archinephric duct (broken line in *E),* development of bladder from base of allantois with short urethra distal to it, differentiation of uterus and vagina from embryonic oviduct, development of phallic organ as clitoris. Abbreviations; *a,* archinephric duct; *al,* staslk of allantois; *b,* bladder; *c,* cloaca; *cl* , clitoris; *dd,* ductus deferns; *gf,* genetial fold; *i,* intestine; *o,* oviduct; *p,* penis; pb,phallic organ; *pr,* proctodeum; *pu,* penile urethra; *r,* rectum; *s,* scrotum; *u,* ureter; *ugs,* urogenital sinus; *ur,,* urethra; *ut,* uterus; *v,* vagina; *v-a,* vestige of archinephricduct; *v-o,* vestige of oviduct; *vs,* vestibulum.

membrane and fuses with it, thus completing the separation of the two parts of the cloaca. The perforation of the dorsal part of the original cloacal membrane gives the rectum an opening to the exterior (the anus). The perforation of the ventral part of the membrane forms an opening for the urogenital sinus. The external surface of the wall separating the two openings becomes the *perineum.*

The pair of mesonephric ducts which enter the urogenital sinus is later joined by two further pairs of ducts. The first of these is the pair of ureters. Originally, the ureters develop as branches sprouting out from the mesonephric ducts. In later stages, however, the terminal portions of the mesonephric ducts become incorporated into the wall of the cloaca, with the result that the ureters now enter the cloaca directly and independently of the mesonephric ducts.

Once they have become independent of the mesonephric ducts, the ureters shift in a cranial direction, and their openings come to lie considerably farther forward than the openings of the mesonephric ducts.

Lastly, the paramesonephric ducts reach the cloaca. During their initial development and growth, the paramesonephric ducts are situated laterally to the mesonephric ducts. The posterior ends of paramesonephric ducts, however, pass over the mesonephric ducts and contact the urogenital sinus mid-dorsally, between the mesonephric ducts. The terminal sections of the paramesonephric ducts closely approach one another and fuse partially.

The urogenital sinus is continuous anteriorly with the stalk of the allantois. After the ureters have become separated from the mesonephric duct and have moved to a more anterior position, the part of the urogenital sinus with the openings of the ureters, together with the adjacent part of the allantoic stalk, expand to form the *urinary bladder.* The canal leading from the bladder to the umbilicus later closes and degenerates.

The portion of the urogenital sinus immediately following the bladder becomes narrow and develops as the *urethra*.

The external opening of the urogenital sinus flanked on two sides by elongated thickenings — the genital folds. Anteroventral to the opening of the urogenital sinus, the genital folds join and form a median outgrowth, the *genital tubercle,* which is the rudiment of the *phallus.*

Just as did the gonads in the earliest stages and the genital ducts somewhat later, the cloacal region with its associated structures, as described so far, represents an indifferent condition which does not allow for the distinction of the two sexes. When the secretion of sex hormones from the differentiating gonads begins, the urogenital sinus and the rudiments of the external genitalia undergo specific changes together with the changes occurring in the genital ducts.

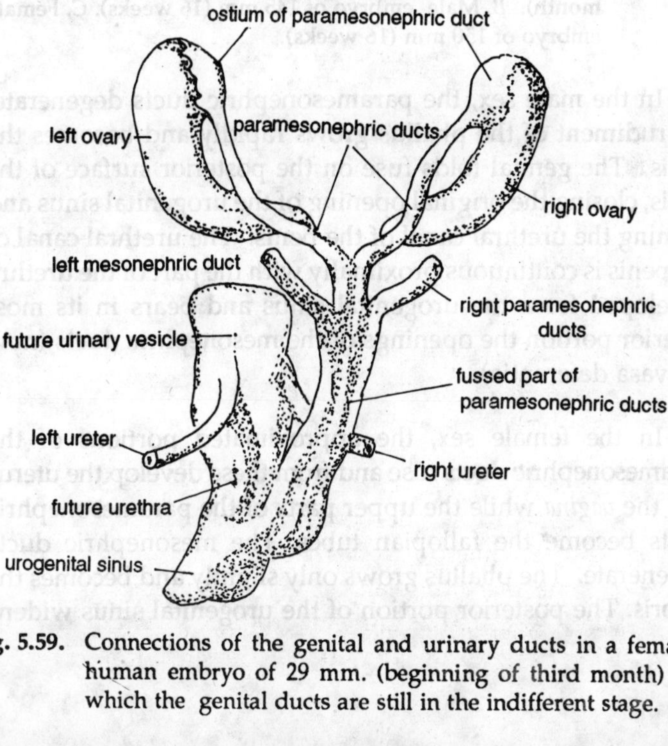

Fig. 5.59. Connections of the genital and urinary ducts in a female human embryo of 29 mm. (beginning of third month) in which the genital ducts are still in the indifferent stage.

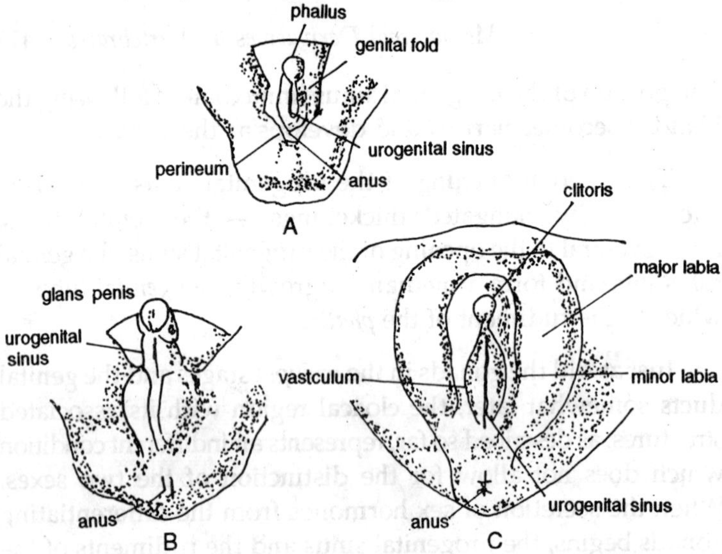

Fig. 5.60. Development of external genitalia in human embryos. *A,* Indifferent stage in an embryo of 29 mm. (beginning third month). *B,* Male embryo of 145 mm (16 weeks). *C,* Female embryo of 150 mm (16 weeks).

In the male sex, the paramesonephric ducts degenerate, the rudiment of the phallus grows rapidly and becomes the penis. The genital folds fuse on the posterior surface of the penis, closing the original opening of the urogenital sinus and forming the urethral canal of the penis. The urethral canal of the penis is continuous proximally with the part of the urethra developed from the urogenital sinus and bears in its most anterior portion the openings of the mesonephric ducts (now the vasa deferentia).

In the female sex, the approximated portions of the paramesonephric ducts fuse and from these develop the uterus and the *vagina* while the upper parts of the paramesonephric ducts become the fallopian tubes. The mesonephric ducts degenerate. The phallus grows only slightly and becomes the clitoris. The posterior portion of the urogenital sinus widens out to the exterior, so that the short urethra and the vagina open independently into the vestibulum, which is the widened posterior part of the urogenital sinus. The genital folds do not fuse and remain as the minor labia.

6

Endodermal Derivatives in Vertebrates

The relation between the archenteron and the definitive alimentary canal is very different in vertebrates with complete cleavage from that in those with incomplete (meroblastic) cleavage. The archenteron in *Amphioxus* is originally lined with presumptive endoderm, presumptive mesoderm and presumptive notochord cells. When the notochord and the mesoderm are segregated from the endoderm, the latter closes the gap on its dorsal side, and the resulting cavity becomes the cavity of the alimentary canal. In holoblastic vertebrates— cyclostomes, ganoid fishes, lungfishes, and amphibians — the presumptive notochord and mesoderm also participate in the lining of the archenteron, forming its roof. This roof is segregated from the endoderm and the endodermal cavity which is closed by the fusion of the free edges of the endodermal layer.

The resulting endodermal cavity consists of three unequal portions. The anterior portion is dilated and lined by a relatively thin endodermal epithelium. This part is usually referred to as the *foregut*. The following portion is called the *midgut*. The cavity of the midgut is narrower. Dorsally it is lined by a rather thin epithelium, but the ventral wall consists of a mass of large cells containing abundant yolk; the wall here is therefore very thick. The most posterior part of the cavity which adjoins the blastopore may be distinguished as the *hindgut*. The fate of the various parts of the endodermal lining of the fore-and midgut in amphibians has been elucidated by the method of

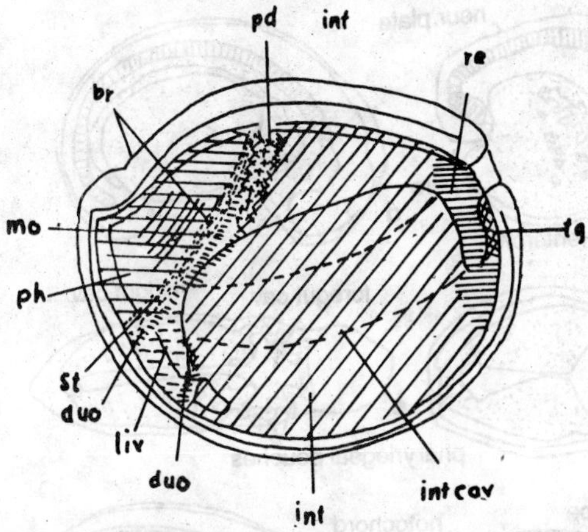

Fig. 6.1. Fate map of the endodermal organs in the neurula of a frog. The embryo is represented as being cut in half in the median plane and viewed from side of the cut surface, br, branchial pouches, duo, duodenum, int, small intenstine, int cav, region in which the cavity of the intetine will later develop, liv, liver; mo, mouth, pd, dorsal pancreas; ph, pharynx; re, rectum, st, stomach; tg, postanal gut.

local vital staining. To stain the inner surface of the gut, cuts were made through the body wall (in some experiments through the neural plate) and pieces of agar soaked in vital stain were applied to the endoderm from the inside. After the stain was taken up by the endodermal cells, the agar was removed; the wound healed easily, and the embryos continue to develop apparently quite normally.

In the neurula stage the ventral wall of the foregut becomes flattened out and later even folded slightly upward. The cavity is thus subdivided into two pocketlike recesses. The larger anterior one, lying immediately underneath the brain, gives rise to the cavities of the mouth and of the branchial region. The posterior pocket, bordering on the mass of the yolk-laden cells of the midgut, is known as the *liver diverticulum* (or hepatic diverticulum), although in addition to giving rise

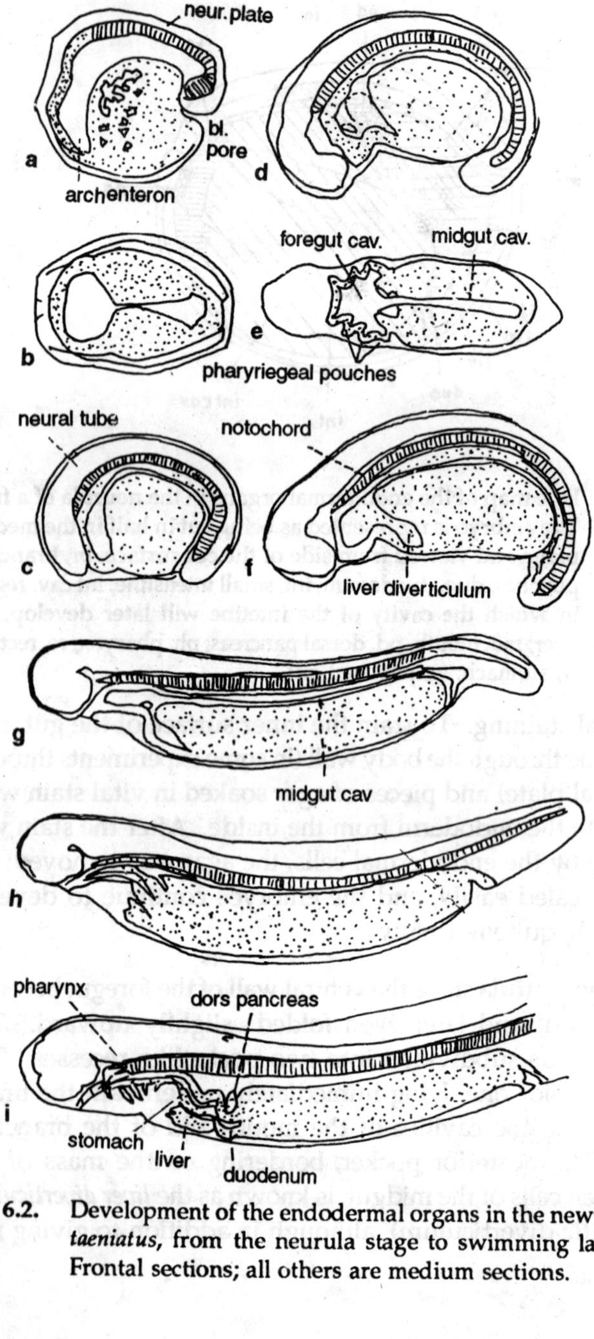

Fig. 6.2. Development of the endodermal organs in the newt *Triturus taeniatus*, from the neurula stage to swimming larva. *b, e,* Frontal sections; all others are medium sections.

to the rudiments of the liver and the pancreas it also participates in the formation of the stomach and the duodenum. It is rather remarkable that in frogs the liver diverticulum is extended downward and backward until the endoderm is perforated and the cavity of the gut opens into the space between the endoderm and mesoderm on the ventral side of the embryo. The significance of this perforation is not known.

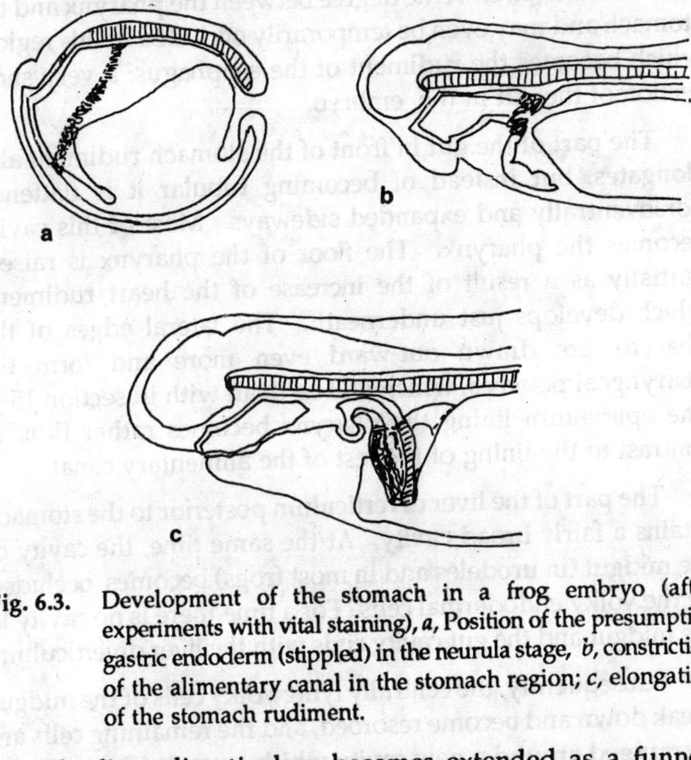

Fig. 6.3. Development of the stomach in a frog embryo (after experiments with vital staining), *a*, Position of the presumptive gastric endoderm (stippled) in the neurula stage, *b*, constriction of the alimentary canal in the stomach region; *c*, elongation of the stomach rudiment.

The liver diverticulum becomes extended as a funnel-shaped invagination farther and farther in a posteroventral direction, drawing into it the rudiments which were originally located in the wall of the broadened cavity of the foregut (such as the rudiment of the anterior part of the duodenum, stomach and esophagus). These rudiments become re-shaped in the form of a narrow tube. This can best be shown by considering the development of the stomach in the frog embryo.

In the neurula stage, the presumptive material of the stomach is arranged in the form of a narrowring, the ventral part of which lies just in front of the live diverticulum, the dorsal part occupies the roof of the gut just at the mouth of the midgut, and the right and left parts obliquely cross the lateral walls of the foregut. This ring contracts and at the same time broadens until eventually the rudiment of the stomach becomes more or less barrel-shaped, with a fairly narrow cavity. The tube is constricted to an extreme degree between the pharynx and the stomach and may even be temporarily occluded in this region, which becomes the rudiment of the esophagus; a very short section of the gut in the embryo.

The part of the gut in front of the stomach rudiment also elongates, but instead of becoming tubular it is flattened dorsoventrally and expanded sideways. Most of this cavity becomes the pharynx. The floor of the pharynx is raised, partially as a result of the increase of the heart rudiment, which develops just underneath. The lateral edges of the pharynx are drawn out-ward even more and form the pharyngeal pouches, which will be dealt with in section 15-3. The epithelium lining the pharynx becomes rather thin, in contrast to the lining of the rest of the alimentary canal.

The part of the liver diverticulum posterior to the stomach retains a fairly broad cavity. At the same time, the cavity of the midgut (in urodeles and in most frogs) becomes occluded by the yolky endodermal cells. For a time there is no cavity in the midgut and the gut cavity ends with the liver diverticulum.

Subsequently, the centrally lying yolky cells of the midgut break down and become resorbed, and the remaining cells are rearranged around a new cavity which extends in a posterior direction from the liver diverticulum. The live diverticulum thus becomes incorporated into the main channel of the alimentary canal.

This mode of development of the alimentary canal of the amphibians was first described for urodeles, and the findings were confirmed in all essential points by work done on frogs.

In some frogs, however, as in the clawed toad *Xenopus laevis*, the midgut cavity is not completely closed at any stage and later expands to form the intestinal cavity.

At the time when the midgut becomes occluded, the hindgut does not lose its cavity and the latter persist as the cavity of the cloaca. The dorsal wall of the hindgut becomes extended into the tail rudiment as a postanal gut. The postanal gut has only an ephemeral existence and is soon broken up and disappears. The hindgut also gives rise in amphibians to the urinary bladder, which develops as a ventral evagination of the gut in late stages of larval life.

As the main portions of the alimentary canal begin to take shape, the canal as a whole becomes twisted in a characteristic way. At an early stage, the stomach has already assumed a slanting position. Subsequently, the posterior end of the stomach is shifted to the left, while the adjoining part of the duodenum comes to lie transversely, going from left to right. The distal part of the duodenum is then bent in such a way that it leads to the anterior end of the intestine, which is more or less in a dorsal position. The alimentary canal thus performs a complete spiral revolution, which may be referred to as the gastro-intestinal loop and which occurs with greater or lesser modifications in all vertebrates. Where the loop is in its lowest position, it leaves a space or saddle on the dorsal side, and this space is taken up by the rudiment of the pancreas. The posterior part of the alimentary canal also becomes twisted into folds and loops as the result of the elongation of the alimentary canal which exceeds the elongation of the body. The folding and twisting of the duodenum and intestine is especially prominent in the tadpoles of frogs which, in connection with their herbivorous diet, have a very long intestine. The pattern of the twisting varies somewhat in different frogs and need not be followed here. In urodeles and also in fishes, the intestine does not elongate to the same degree and its folding may be very limited, though the gastro-duodenal loop is always present.

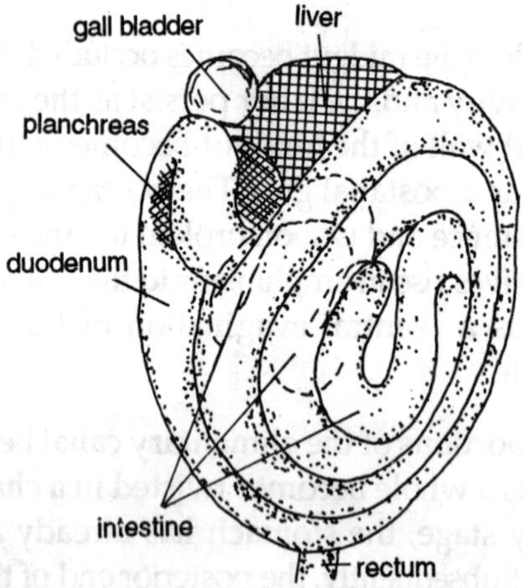

Fig. 6.4. Coiling of the intestine in a tadpole of the frog Xenopus laevis.

The Relation in Higher Vertebrates

In vertebrates having a meroblastic type of cleavage, the development of the alimentary canal presents very different problems, and the processes leading to the formation of the definitive cavity of the alimentary canal are quite peculiar. In the following we will describe this process as it occurs in birds and mammals.

It must be noted, in the first instance, that in birds the archenteric cavity is often lacking altogether, and if it is present as a canal leading forward from Hensen's node, it is very small and its walls are not endodermal. The starting point for the development of the alimentary canal proper is a sheet of endoderm lying flat under the ectodermal and mesodermal parts of the embryonic region of the blastodisc. The sheet of endoderm lies flat on the yolk of the yolk sac or forms the roof of the mammalian yolk sac, which is a space filled with fluid. In both cases, the cavity of the alimentary canal is separated from the cavity of the yolk sac by a process of infolding.

During this infolding, the median strip of the endoderm lying immediately under the notochord and somites remains in this position, while the immediately adjoining strips on the right and left become inflected downward, and the crests of the folds converge toward the middle and eventually fuse. The inner surfaces of the folds contribute to the formation of the floor of the gut, while the outer surfaces of the folds are continuous with the endodermal lining of the yolk sac. Concurrently with the movement of the endodermal layer in a transverse plane, a complicated shifting has been found to occur longitudinally: the median strip which forms the dorsal wall of the gut slides forward, while in the folds closing the gut laterally and ventrally endodermal material moves obliquely backward. During their movement downward, the lateral strips of endoderm are accompanied by the visceral layer of the lateral plate mesoderm, which closely adheres to the endoderm throughout this whole series of formative movements. It is probable tl at the dynamic force of the movement is due to the endoderm and mesoderm jointly.

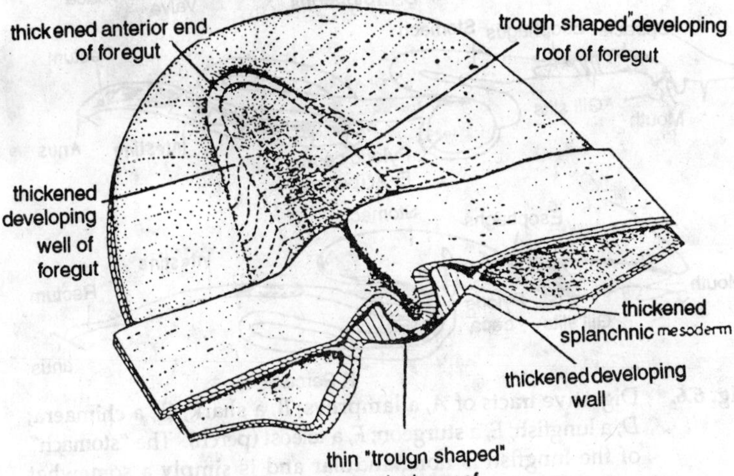

Fig. 6.5. Stage in the development of the foregut in a chick embryo. Ectoderm with neural folds and the parietal mesoderm are shown as being cut away at the anterior lend of the embryo.

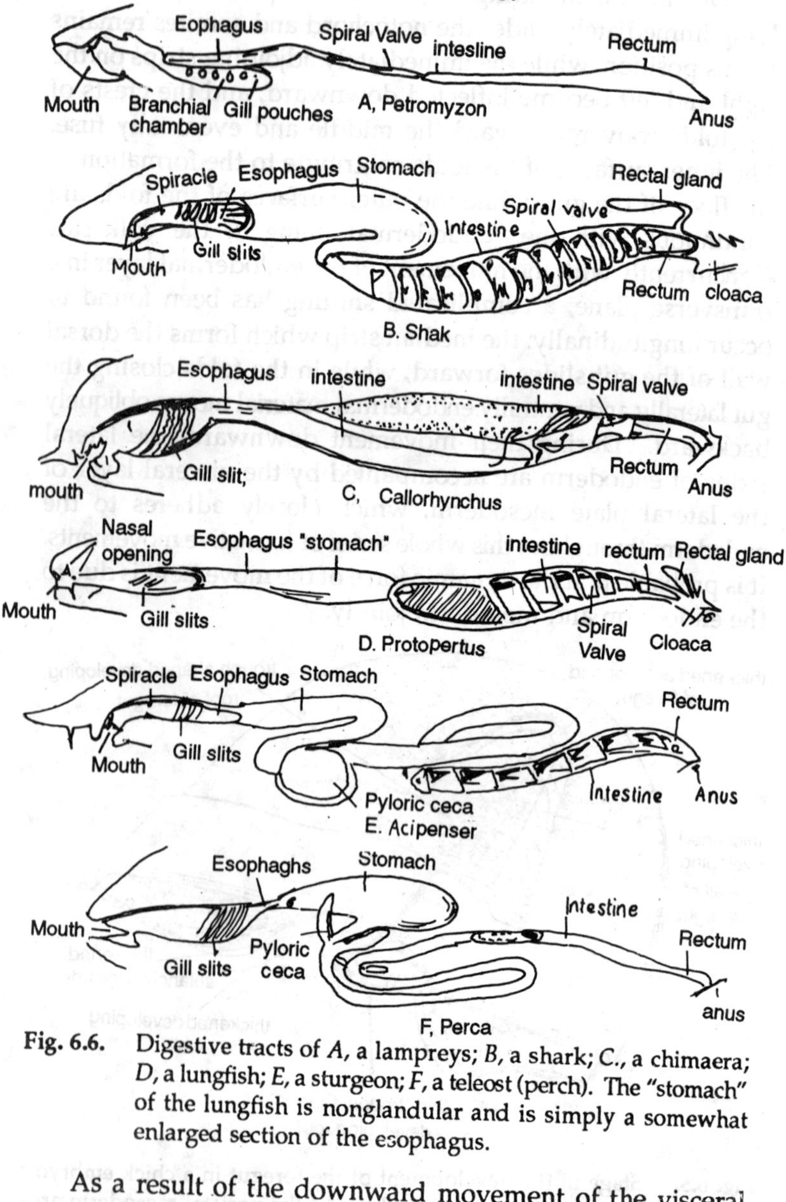

Fig. 6.6. Digestive tracts of *A*, a lampreys; *B*, a shark; *C.*, a chimaera; *D*, a lungfish; *E*, a sturgeon; *F*, a teleost (perch). The "stomach" of the lungfish is nonglandular and is simply a somewhat enlarged section of the esophagus.

As a result of the downward movement of the visceral layer of the lateral mesoderm, the coelomic cavity becomes considerably expanded locally. The coelom is reduced again, however, when the body folds undercut the embryo. The alimentary canal becomes separated from the yolk sac cavity first of all at the anterior end of the embryo. This part corresponds to the foregut of the amphibian embryo. Somewhat later, the posterior part of the endodermal groove closes into a canal, and this becomes the posterior part of the alimentary system which in the amniotes is called the *hindgut*, although most of it corresponds to the midgut of amphibian embryos. Between the foregut and the hindgut there remains a gap where the endoderm does not close to form a canal, or even a groove, and where it is in open communication with the yolk sac. This part, in amniotes, is referred to as the midgut.

The edges of the folds separating the foregut from the yolk sac where they meet in the middle form a ridge, known as the *anterior intestinal portal.* A similar edge at the anterior end of the hindgut is the *posterior intestinal portal.* The gap between the two is very large at first, but its relative size diminishes with the growth of the embryo, and eventually it is reduced to the opening of the *yolk stalk,* connecting the cavity of the gut to the cavity of the yolk sac.

From the beginning the foregut is much broader than the hindgut and is flattened in cross-section. As in the amphibians, the foregut gives rise to the endodermal lining of all anterior parts of the alimentary canal, including most of the duodenum, the liver and pancreas developing just in front of the anterior intestinal portal. The oral and especially the pharyngeal part of the foregut remain expanded in a transverse direction, and the pharynx becomes drawn out to form the *pharyngeal pouches,* but the posterior part of the foregut, corresponding to the esophageal, gastric and duodenal parts of the alimentary canal, eventually becomes round in cross-section. The esophagus in higher vertebrates is soon greatly elongated, in connection with the development of the neck. The hindgut is from the start narrower than the foregut and soon becomes round in cross-section too.

Although the foregut in higher vertebrates is quite different in shape from the foregut of amphibians, it develops the gastro-duodenal loop in much the same way. The anterior (cardiac) end of the stomach is displaced to the left side, the posterior (pyloric) end is turned downward and toward the middle, while the anterior part of the duodenum assumes a transverse direction, thus bringing about the familiar position of the stomach in adult mammals and birds. The intestine becomes convoluted in a pattern that varies greatly not only as between different classes but even in one class, as in various mammals. A peculiarity occurring in mammals is that sections of the intestine adjoining the yolk stalk, both anteriorly and posteriorly, sink down into the umbilical cord and lie for a time practically outside the body of the embryo proper as a sort of umbilical hernia. The convolutions of the intestine begin forming inside the umbilical cord, but well before birth the definitive intestine is withdrawn into the body and only the yolk stalk remains in the cord (this occurs in the human embryo during the third month of pregnancy).

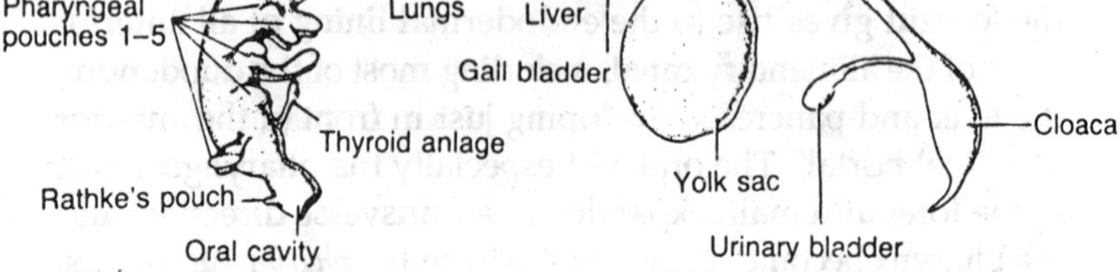

Fig. 6.7. A diagram of the digestive tract and its outgrowths in an amniotes embryo similar to that, but with the structures concerned shown as solid objects rather than in section. The first pharyngeal pouch here is the spiracle; the successive pouches are those numbered 1 through 4 elsewhere.

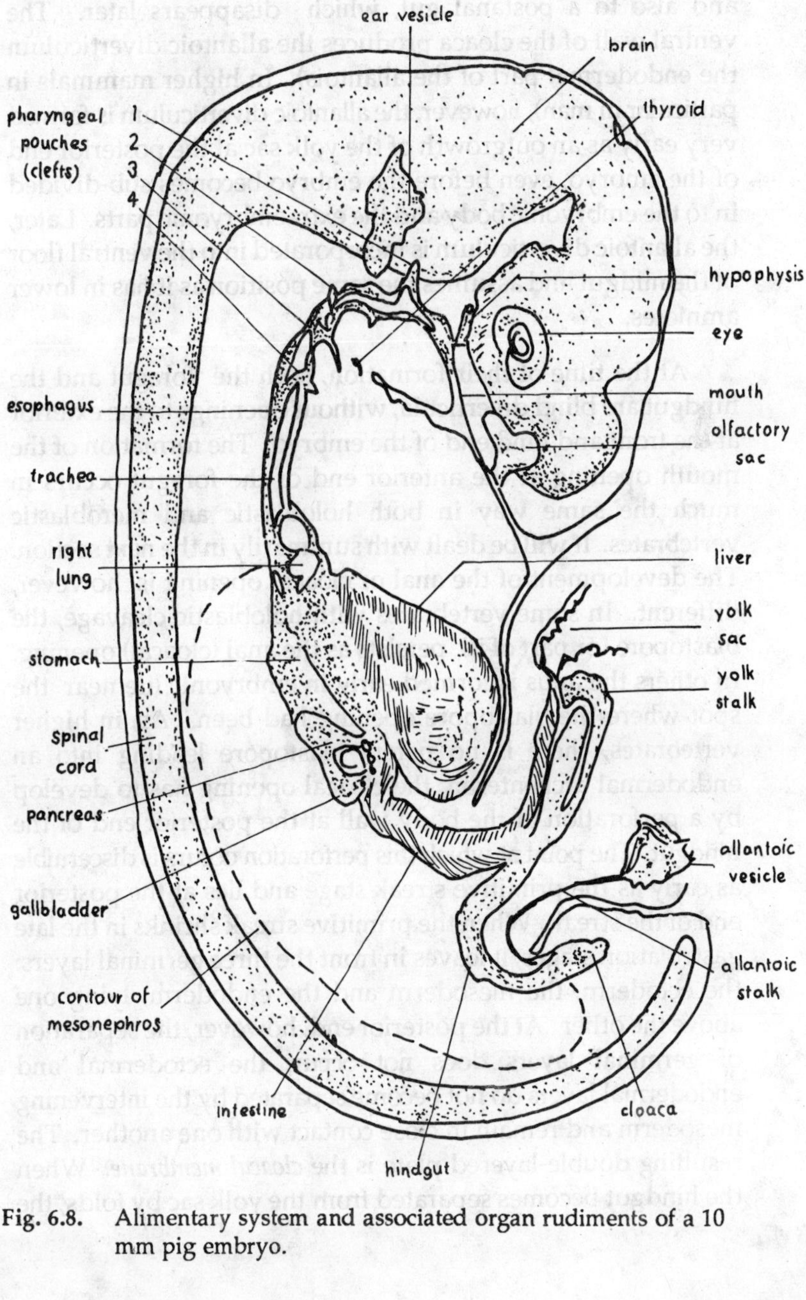

Fig. 6.8. Alimentary system and associated organ rudiments of a 10 mm pig embryo.

The posterior end of the hindgut gives rise to the cloaca and also to a postanal gut, which disappears later. The ventral wall of the cloaca produces the allantoic diverticulum the endodermal part of the allantois). In higher mammals in particular in man), however, the allantoic diverticulum is formed very early as an outgrowth of the yolk sac at the posterior end of the embryo, even before the embryo becomes sub-divided in to the embryonic body and the extraembryonic parts. Later, the allantoic diverticulum is incorporated into the ventral floor of the midgut and assumes the same position as it has in lower amniotes.

At the time of their formation, both the foregut and the hindgut are blind diverticula, without openings to the exterior at the front and hind end of the embryo. The formation of the mouth opening at the anterior end of the foregut occurs in much the same way in both holoblastic and meroblastic vertebrates. It will be dealt with summarily in the next section. The development of the anal or cloacal opening is, however, different. In some vertebrates with holoblastic cleavage, the blastopore (or part of it) persists as the anal (cloacal) opening. In others the anus is formed early in embryonic life near the spot where the blastopore opening had been. As, in higher vertebrates, there is no patent blastopore leading into an endodermal archenteron, the cloacal opening has to develop by a perforation of the body wall at the posterior end of the hindgut. The point at which this perforation occurs is discernible as early as the primitive streak stage and lies at the posterior end of the streak. When the primitive streak shrinks in the late gastrulation stages, it leaves in front the three germinal layers: the ectoderm, the mesoderm and the endoderm, lying one above the other. At the posterior end, however, the separation of germinal layers does not occur; the ectodermal and endodermal layers do not become separated by the intervening mesoderm and remain in close contact with one another. The resulting double-layered plate is the *cloacal membrane*. When the hindgut becomes separated from the yolk sac by folds, the cloacal membrane is incorporated into the wall of the gut. The ectodermal side of the cloacal membrane is originally dorsal, but this position is inverted by the development of the tailbud, occurring just anterior to the cloacal membrane. The tail rudiment protrudes backward, and the hindgut above the cloacal membrane develops a diverticulum entering the tail rudiment—the postanal gut. As a result, the cloacal membrane now comes to lie at the root of the tail with the ectodermal side facing downward. The part of the hindgut adjoining the cloacal membrane is somewhat dilated and becomes the rudiment of the cloaca into which open the mesonephric ducts. The ectoderm is slightly depressed in the region of the cloacal membrane, forming of the cloaca from the cavity of the proctodeum till late in embryonic development, but it is eventually ruptured and thus a free passage from the alimentary canal to the exterior is allowed.

Development of the Mouth

In all vertebrates, the mouth opening appears rather late in the embryonic life after all the primary organ rudiments have already been formed. The mouth opening breaks through where the anterior end of the endodermal part of the alimentary, canal touches the ectodermal epidermis beneath the front end of the neural tube. The ectodermal epidermis here sinks in to form a pocketlike depression, the ectodermal mouth invagination or the *stomodeum*.

The ectodermal epithelium and the endodermal epithelium fuse here and become an *oropharyngeal membrane*. The membrane becomes very thin, and eventually disappears. Forthwith, the boundary between the ectodermal and endodermal epithelia becomes indistinguishable, and it is not an easy matter to determine later which part of the oral cavity is derived from the stomodeum and which from the endodermal gut. There are, however, indications that the stomodeal epithelium reaches (in mammals) to about the middle of the tongue ventrally and to the beginning of the pharynx on the dorsal side.

The determination of the stomodeum begins at the end of gastrulation, and the development of the ectodermal mouth invagination is induced by the endoderm when the anterior end of the archenteron comes in contact with the ectoderm of the presumptive mouth region. No mesoderm ever penetrates here between the ectoderm and endoderm, and the endoderm remains in contact with the ectoderm throughout the subsequent stages until the rupture of the pharyngeal membrane. After the endoderm and ectoderm come into contact in the oral region, the oral ectoderm acquires, to some extent, and ability to differentiate as stomodeum. The anterior part of the archenteron may now be removed, and the remaining ectoderm produces a mouth invagination, although there are no endodermal oral and branchial parts behind it. The mouth invaginations thus developed are small, and their shape is not normal. If, however, the operation is performed a bit later, in the neurula stage, the stomodeum may develop in an almost completely normal way, with the sole reservation that it does not lead into the endodermal alimentary canal. As the oral ectoderm acquires the ability for self-differentiation, all the rest of the epidermis loses the competence for development into stomodeum. If the region where the mouth is to develop is covered with a flap of epidermis taken from a different part of the body, the development of the stomodeum is suppressed, in spite of the presence of oral endoderm.

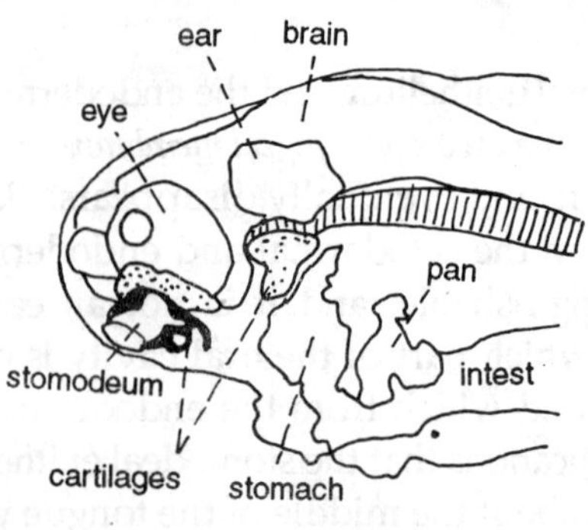

Fig. 6.9. Development of the mouth and stomodeal invagination in a newt embryo after removal of the anterior endoderm in the neurula stage.

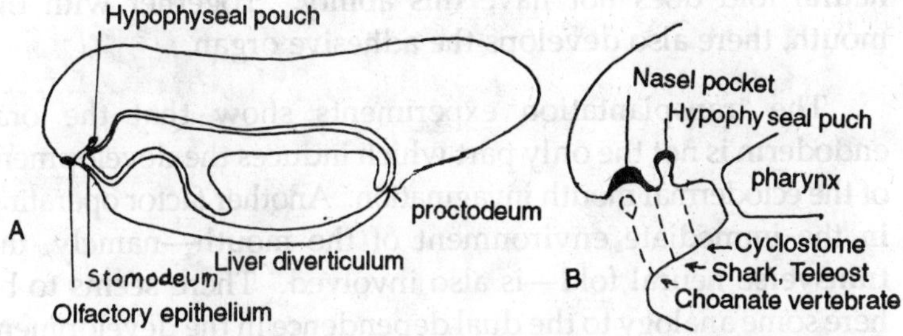

Fig. 6.10. A. Diagram of a larval amphibian, in longitudinal section, to show the extent of the endoderm (stippled) and its relation to structures of the oral region. B, Diagram to show the comparative position of the oral margins in various types of vertebrates.

Presumptive epidermis of the stomodeum may develop a mouth invagination in an abnormal position, but this it can do only under certain conditions. In newts, stomodeal epidermis of a neurula, when transplanted alone, will not develop a mouth, but when it is transplanted together with some adjoining endoderm, and if the site of transplantation is not too far from the oral region of the host, an additional stomodeal invagination is formed by the graft, and this invagination may break through into the endodermal cavity of the host.

In frogs, the determination of the mouth ectoderm proceeds more rapidly than in urodeles, and the capacity for differentiation in the determined direction is higher. Already, in the early neurula, the removal of oral endoderm does not interfere with mouth development, and transplants of presumptive mouth ectoderm may develop in a typical way, producing the characteristic horny jaws and horny teeth not only in the vicinity of the normal mouth but also in other locations, such as on the belly. It was found, however, that for the mouth to be developed in abnormal regions it is necessary to transplant a piece of the transverse neural fold, together with the presumptive mouth ectoderm. Only a narrow median piece of the neural fold can support mouth development; lateral neural fold does not have this ability. Together with the mouth, there also develops the adhesive organ.

The transplantation experiments show that the oral endoderm is not the only part which induces the development of the ectodermal mouth invagination. Another factor operating in the immediate environment of the mouth—namely, the transverse neural fold—is also involved. There seems to be here some analogy to the dual dependence in the development of the lens: the dependence on the head mesoderm, on the one hand, and dependence on the eye cup on the other hand. In the case of mouth differentiation, as in the case of the inductor of the lens, a certain variation in modes of development in different animals may be attributed to the relative strength of one or the other inducing factor.

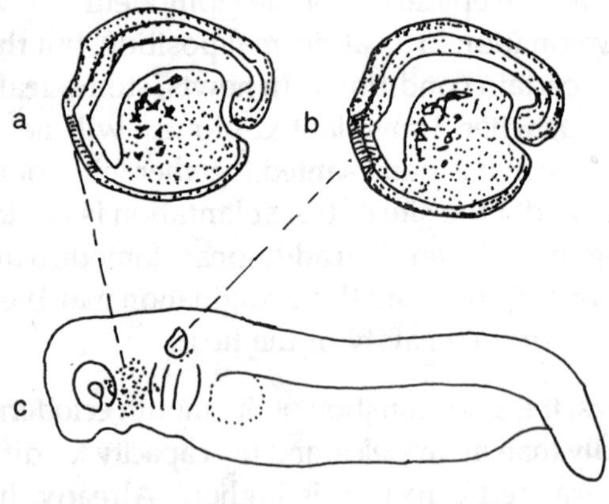

Fig. 6.11. Transplantation of the oral ectoderm alone (a) and with the underlying endoderm (b); diagram of operation. c, Position of the graft on the host embryo several days after operation. The position indicated is favorable for development of a stomodeum from grafted tissue.

Besides giving rise to a part of the oral epithelium, the stomodeal invagination furnishes the cells which become the rudiment of the anterior lobe of the *hypophysis*. The rudiment is formed as a solid but, or a small pocket *(Rathke's pocket)*, on the dorsal side of the stomodeal invagination, just in front of

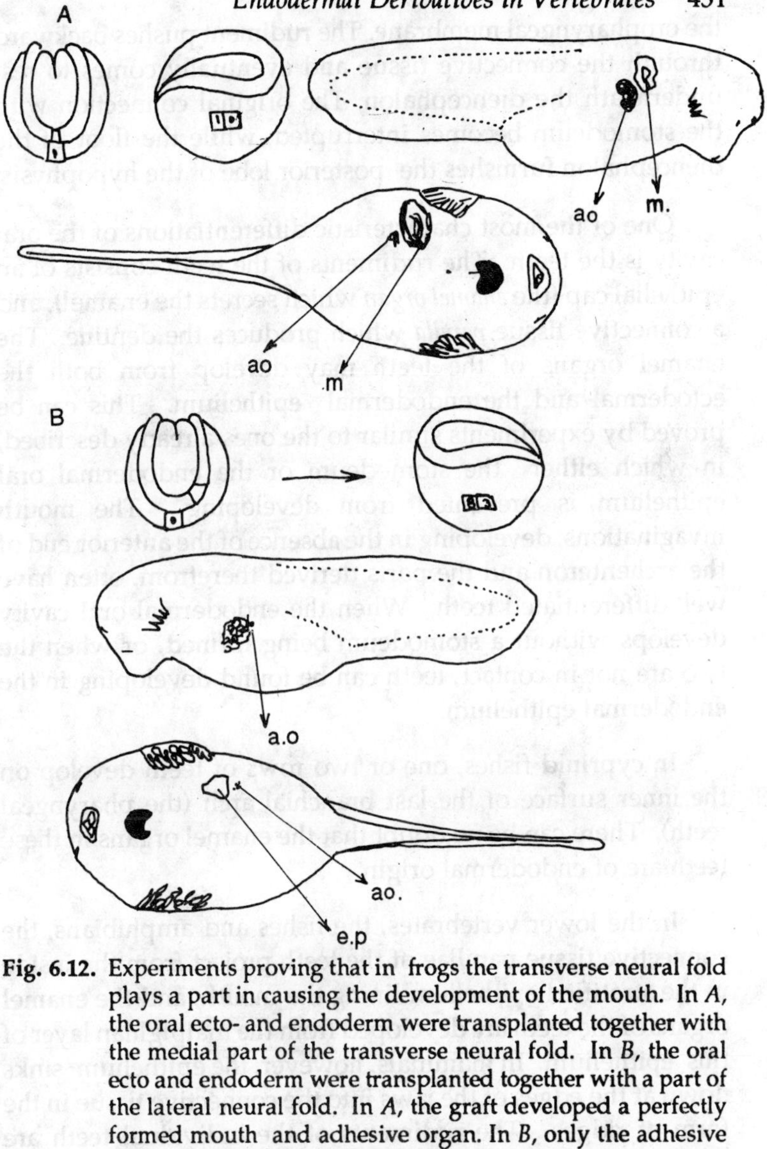

Fig. 6.12. Experiments proving that in frogs the transverse neural fold plays a part in causing the development of the mouth. In *A*, the oral ecto- and endoderm were transplanted together with the medial part of the transverse neural fold. In *B*, the oral ecto and endoderm were transplanted together with a part of the lateral neural fold. In *A*, the graft developed a perfectly formed mouth and adhesive organ. In *B*, only the adhesive organ developed a well as some skin epidermis; no mouth was formed. In both *A* and *B*, drawings above show diagrams of the operation, and below are the host embryos after 4 and 6 days of development respectively. *a.o.*, Adhesive organ; b, presumptive buccal ectoderm; *e.p.*, transplanted epidermis; *m*, mouth; 1,3 parts of the neural fold.

the oropharyngeal membrane. The rudiment pushes backward through the connective tissue and eventually comes to rest underneath the diencephalon. The original connection with the stomodeum becomes interrupted, while the floor of the diencephalon furnishes the posterior lobe of the hypophysis.

One of the most characteristic differentiations of the oral cavity is the teeth. The rudiments of the teeth consists of an epithelial cap (the *enamel organ* which secrets the enamel), and a connective tissue *papilla* which produces the dentine. The enamel organs of the teeth may develop from both the ectodermal and the endodermal epithelium. This can be proved by experiments similar to the ones already described, in which either the stomodeum or the endodermal oral epithelium is prevented from developing. The mouth invaginations, developing in the absence of the anterior end of the archenteron and the parts derived therefrom, often have well-differentiated teeth. When the endodermal oral cavity develops without a stomodeum being formed, or when the two are not in contact, teeth can be found developing in the endodermal epithelium.

In cyprinid fishes, one or two rows of teeth develop on the inner surface of the last branchial arch (the pharyngeal teeth). There can be no doubt that the enamel organs in these teeth are of endodermal origin.

In the lower vertebrates, the fishes and amphibians, the connective tissue papillae of the teeth project from the inside in the stratified epithelium lining the mouth, and the enamel organs of the teeth are developed from the malpighian layer of this epithelium. In mammals, however, the epithelium sinks down at the edges of the jaws into the connective tissue in the form of ridges. The rudiments of the individual teeth are formed at the edge of the dental ridges, at the expense of a connective tissue papilla and of the layer of innermost cells of the dental ridge which adjoin the papilla and form the enamel organ. The teeth are thus formed and begin to grow deep in the tissue of the jaw, and they erupt to the surface only when they have almost reached their full development.

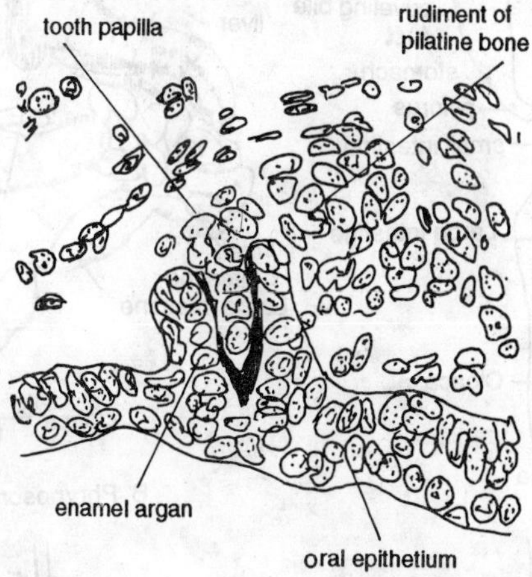

Fig. 6.13. Rudiment of tooth in a salamander embryo.

While the cells of the enamel organ may be ectodermal or endodermal, the connective tissue cells of the dental papillae are derived from the neural crest. The neural crest is also the source of skeletogenous cells for the development of the mandibular arch : the quadrate and the Meckel's cartilage. The development of the mouth as a whole depends therefore on the harmonious co-operation of the cells coming from different primary rudiments — the foregut, the epidermis and the neural crest.

The development of the mandibular arch is, in part at least, directly dependent on the influence exercised by the ectodermal mouth invagination: if the ectodermal mouth invagination fails to be formed, the ventral part of the mandibular arch is not formed. An additional Meckel's cartilage can develop in connection with a mouth invagination resulting from the transplantation of the presumptive stomodeal ectoderm. The quadrate, on the other hand, may develop even in the complete absence of any sort of oral cavity; it is thus not dependent on the latter.

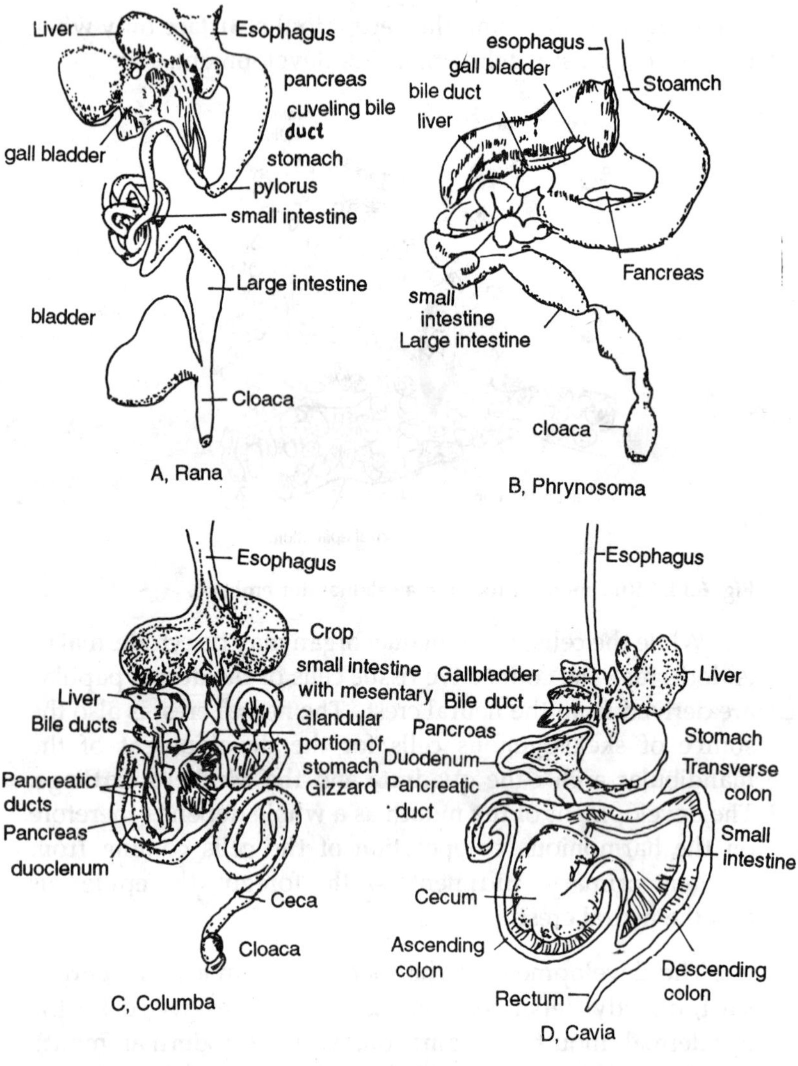

Fig. 6.14. Diagrams of digestive tract and appendages, seen in ventral view, in *A*, a frog; *B*, a lizard; *C*, a bird (pigeon); *D*, a mammal (guinea pig).

The development of structures surrounding the edges of the mouth contributes very much to the formation of the face in man and of corresponding parts in higher mammals. The structures in question are a number of swellings, consisting of actively growing mesenchyme and covered by ectoderm, which are formed around the stomodeal invagination. At the dorsal edge of the stomodeum medially there is a slight swelling—the *medial frontal process.* Lateral to the frontal process on each side there develops a U-shaped swelling encircling the nasal pit. The free ends of the U are directed downward. The inner branch lies just alongside the frontal process and impinges on the edge of the mouth as the *medial nasal process.* The outer branch, lying lateral to the nasal pit, is the *lateral nasal process,* and it does not quite reach to the edge of the mouth. Around the angle of the mouth on each side, another U-shaped swelling develops, with the upper branch ending on the edge of the mouth as the *maxillary process.* The lower branch spreads along the lower edge of the mouth as the *mandibular process.*

The two medial nasal processes grow downward and toward the midline until they fuse and exclude the frontal process from participating in the formation of the edge of the mouth. The maxillary processes grow forward and eventually fuse with the lateral edges of the medial nasal processes, thus completing the upper edge of the mouth. The opening of the nasal pit remains just above the line of fusion of the maxillary and medial nasal processes. The upper edge of the maxillary process also fuses with the lateral nasal processes. The furrow lying between the maxillary process and the lateral nasal process, leads from the angle of the eye to the nasal pit. The infolded epidermis lining this furrow gives rise to a ridge of epithelial cells, which later becomes hollowed out and establishes a communication between the space underneath the eyelids and the nasal cavity. This is the *nasolacrimal duct.*

The lower edge of the mouth acquires its final shape after the median fusion of the two mandibular processes.

The nasal maxillary and mandibular processes, as has been stated, are essentially proliferating masses of mesenchyme covered externally by ectodermal epithelium. The mesenchyme later ossifies and gives rise to some of the most important parts of the facial skeleton. The mesenchyme in the lower portions of the medial nasal processes ossifies as the pair of premaxillary bones. The maxillary bones are produced by the ossification of the mesenchyme of the maxillary processes. The mesenchyme of the mandibular processes gives rise to the mandibular bones. The upper parts of the medial nasal processes together with the medial frontal process develop into the back of the nose.

DEVELOPMENT OF THE BRANCHIAL REGION

The development of the branchial region is no less dependent on the co-operation of parts of different origin than is the mouth region. The leading part in this development belongs to the endoderm. The endodermal cavity in this region, from its beginning, is distended in a transverse direction. In the amphibians it is derived from the inflated part of the foregut. In the stage immediately following the closure of the neural tube, the lateral walls of the pharyngeal cavity bulge out and produce a series of outward directed pockets on each side. These pockets are the *pharyngeal pouches* or *branchial pouches.* The pharyngeal pouches are developed one after another, beginning with the first pair, that is, the one lying just posterior to the mandibular arch. As the endodermal pharyngeal pouches reach the epidermis, having pushed aside the intervening mesoderm, the epidermis becomes folded inward, to meet the pharyngeal pouches. A series of branchial grooves is thus developed on the surface of the embryo, each groove corresponding to an endodermal pouch. The outer wall of the endodermal pouch and the inner wall of the epidermal groove fuse into a *branchial membrane,* similar to the oropharyngeal membrane. *A gill cleft* is formed when the branchial membrane becomes perforated, so that an open communication is established between the pharyngeal cavity and the other medium. In aquatic vertebrates gill filaments are then developed on the walls of the gill clefts; these are the internal gills.

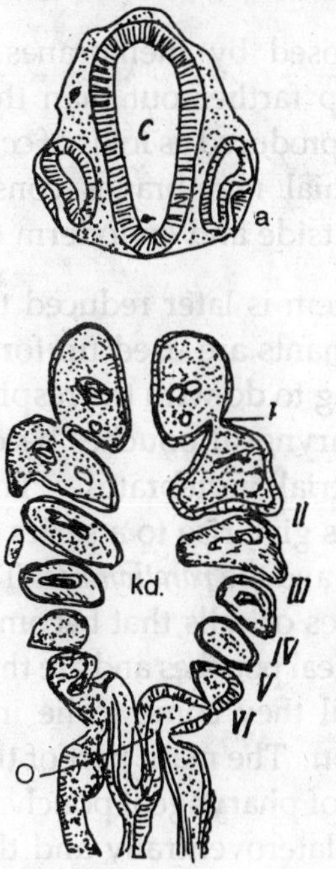

Fig. 6.15. Frontal section of an embryo of a skate, *Raja* showing pharyngeal pouches and gill clefts, which are indicated by Roman numerals, *a*, Eye; *c*, cavity of brain; *Kd*, cavity of pharynx; *o*, esophagus.

The development does not always reach the final stage — the formation of the gill filaments serving for respiration. Even in aquatic vertebrates, some of the pharyngeal pouches may not develop into gill clefts; thus in bony fishes and the amphibians, the first pair of pharyngeal pouches (the one lying behind the mandibular arch) does not reach the epidermis and remains a blind diverticulum of the pharyngeal wall. In the amniotes, the gills never function as respiratory organs, but four pairs of pharyngeal pouches nevertheless develop. Branchial grooves and branchial membranes are formed, and in the first three pairs of pharyngeal pouches the branchial membranes are broken so that gill clefts are formed. The fourth pouch does not open to the outside. The existence of open gill clefts is, however, of very short duration, and the clefts are again closed by membranes. The new closing membranes develop farther out than the original branchial membranes and are produced as folds of ectoderm only, whereas the original branchial membranes consisted of a layer of ectoderm on the outside and endoderm on the inside.

The whole system is later reduced to a greater or lesser extent, and its remnants are used up for the development of parts having nothing to do with the respiratory function; thus the first pair of pharyngeal pouches becomes the Eustachian tube in the terrestrial vertebrates. The third and fourth pharyngeal pouches give rise to a series of glands of internal secretion: the *thymus* and the *parathyroids*. These organ rudiments develop from masses of cells that become detached from the walls of the pharyngeal pouches and are then shifted downward and backward until they come to lie in the neck or in the anterior trunk region. The main part of the thymus is derived from the third pair of pharyngeal pouches; the pouch actually becomes extended lateroventrally and this extension is later used in the formation of the thymus rudiment. The development is of great importance, as it is now known that the thymus in mammals is the exclusive source of the cells which produce antibodies and which are thus responsible for the immune response in these animals and in man.

It has been generally believed that the cells giving rise to the parathyroid glands and the thymus are derived from the endodermal part of the pharyngeal pouches. There are indications, however, that at least a part of the thymus rudiment is derived from the ectodermal cells of the branchial grooves.

The important of these glands for the well-being of the animal gives a possible explanation for the persistence of the branchial pouches in the embryos of terrestrial animals, although they have long since lost their original functional significance.

A further important gland of internal secretion associated with the *pharynx* in its development is the thyroid gland. The thyroid gland develops in vertebrates as a ventral pocket in the floor of the pharynx. Subsequently the pocket becomes closed and separated from the pharyngeal wall. The thyroid rudiment is then displaced in a caudal direction, until, in terrestrial vertebrates, it comes to lie ventral to the trachea.

It has already been stated that the endodermal pharyngeal pouches are the initiators of all the developments in the branchial region. The epidermal branchial grooves are induced by the endodermal pouches when they touch the epidermis; without the endodermal pharyngeal pouches, the epidermal grooves do not develop. Neither do they develop if the endodermal pouch, though present, does not reach the epidermis (as in the case of the first pouch in amphibians and bony fishes).

The relations between the endoderm and the external gills have been discussed already. For the development of the external gills in the urodeles, it is also necessary that the endodermal pharyngeal pouch reach the epidermis; otherwise the gills fail to appear. The external gills in the urodeles are later supplemented by internal gills—gill filaments developing on the branchial arches. Both external and internal gills function simultaneously, until both are reduced during metamorphosis. In the anurans, the external gills function only temporarily, during a short period after the hatching of the larvae. Soon after the little tadpoles begin to swim, a fold of skin appears anterior to the external gills—the *opercular fold*. The opercular fold spreads backward over the gill region, covering both the external gills and the gill slits. Both are thus included in a branchial cavity. The posterior edge of the opercular fold becomes attached to the skin behind the branchial region, so that only a narrow opening—the *branchial aperture*—leads from the branchial cavity to the exterior. At the same time, the external gills are reduced in size, and the internal gills develop on the branchial arches beneath the external gills and function throughout the whole period of larval development.

The visceral skeleton is an important integral part of the branchial region. As has been stated, the visceral skeleton develops from cells of the neural crest. In their downward migration, the neural crest cells are split by the pharyngeal pouches into several streams, moving in between the adjacent pharyngeal pouches. Later, the branchial arches are formed by the chondrification of the neural crest mesenchyme in about the same position as the masses of migrating cells were to be found. The dependence of the visceral arch development on the gill clefts is, however, a more intimate one than would follow from this description. In the absence of the pharyngeal pouches, the neural crest mesenchyme does not chondrify, and no branchial arches are formed. If the number of pharyngeal pouches is reduced (after an operation in which part of the endoderm of the branchial region has been removed), the number of visceral arches is similarly reduced; one skeletal arch is developed on each side of the remaining pharyngeal pouches. Thus the number of arches is one more than the number of pouches present.

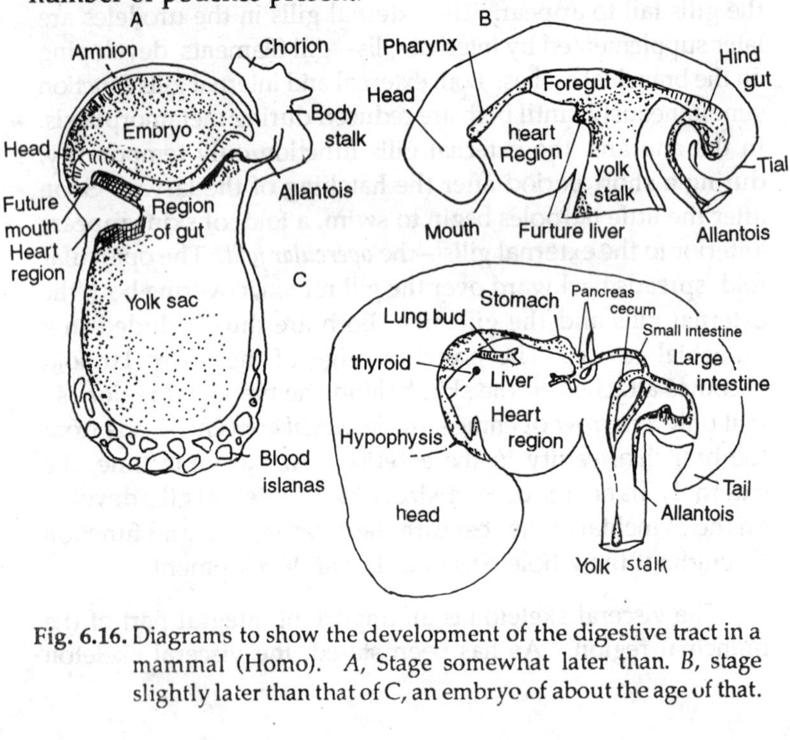

Fig. 6.16. Diagrams to show the development of the digestive tract in a mammal (Homo). *A*, Stage somewhat later than. *B*, stage slightly later than that of C, an embryo of about the age of that.

The hyoid arch shows the same kind of dependence on the pharyngeal pouches as the branchial arches proper.

For the development of the visceral arches, it is not necessary for the endodermal pharyngeal pouches to establish a connection with the epidermis and for the gill cleft to break through; the presence of the endodermal pouches, even if they are represented by blind pockets, is sufficient to induce the development of the skeletal arches.

We can now review the dependence of the development of the various parts of the visceral skeleton on the adjoining parts of the alimentary canal.

1. The upper part of the mandibular arch (the quadrate and the region of articulation with the lower jaw) develops independently of parts of the alimentary canal.

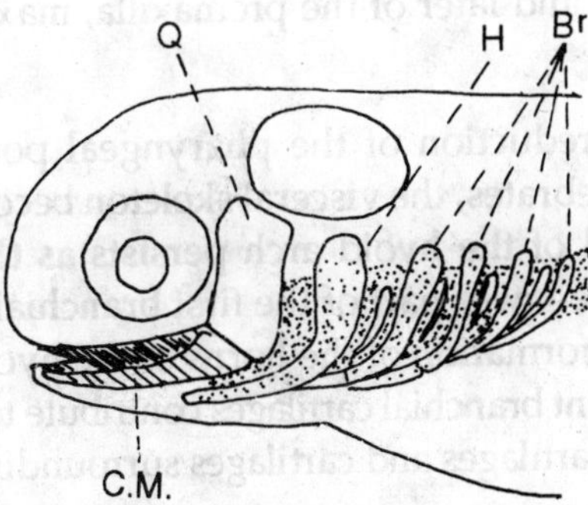

Fig. 6.17. Diagram of the head of a urodele amphibian, showing the dependence of parts of the visceral skeleton on different sections of the alimentary canal. The ectodermal stomodeum and that part of Meckel's cartilage (C.M.) that is dependent on it are hatched; the endodermal foregut and the visceral arches dependent on the endoderm are stippled. Q, quadrate; H, hyoid arch; Br, branchial arches.

2. The lower part of the mandibular arch is dependent on the ectodermal mouth invagination.

3. The hyoid arch and the branchial arches are dependent on the endodermal pharyngeal pouches.

A very peculiar in the development of the amphibian visceral skeleton is presented by the second basibranchial. The skeletal element is developed from mesodermal mesenchyme, not from the neural crest mesenchyme, and it is also independent of the endodermal pharyngeal pouches; it is formed even if the whole of the endoderm of the branchial region has been removed.

The visceral arches, which are dependent in their development on the epithelial parts of the alimentary canal, appear to exercise some influence on the development of the teeth. The tooth rudiments, consisting of the ectodermal or endodermal enamel organ and the papilla, derived from neural crest mesenchyme, are formed in connection with certain skeletal elements: Meckel's cartilage, the rudiments of the vomer and palatine bone, and later of the premaxilla, maxilla and dental bone.

With the reduction of the pharyngeal pouches in adult terrestrial vertebrates, the visceral skeleton becomes modified. The lower end of the hyoid arch persists as the body of the hyoid bone; the lower part of the first branchial arch becomes utilized in the formation of the horns of the hyoid bone. Parts of the subsequent branchial cartilages contribute to the formation of the thyroid cartilages and cartilages surrounding the trachea.

Development of the Accessory Organs of the Alimentary Canal: Lungs, Liver, Pancreas, Bursa Fabricii

The lungs. The lungs develop from a rudiment which is a pocketlike evagination of the endodermal epithelium on the ventral side of the alimentary canal, just posterior to the branchial region. The pocket at first projects straight downward. At its tip it bifucrates, and the two branches grow out to be the sides and backward. The unpaired medial part of the rudiment becomes the trachea; the two branches give rise to the two bronchi and to the lungs themselves.

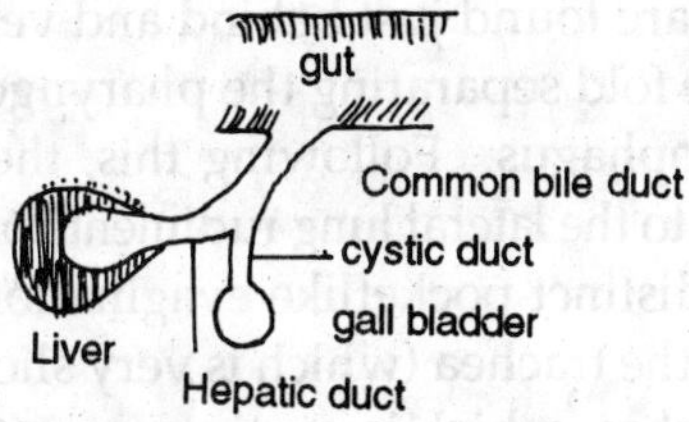

Fig. 6.18. Diagram of an early stage of liver developing to show the relationship of the ducts.

In the lower vertebrates, the lungs are developed as saclike expansions at the end of the bronchi, the walls of which become folded to various degrees. In the warm-blooded vertebrates, birds and mammals, in which the lung respiration attains the highest efficiency, the greater degree of differentiation of the air spaces in the lungs already becomes manifest in the earlier stages of lung development. In mammals, the distal ends of the bronchi as they grow out become branched in a more or less dichotomous fashion, the branches representing the secondary, tertiary, etc., bronchi and the bronchioles. The alveoli are eventually developed on the terminal branches of this system.

The unpaired part, the rudiment of the trachea, may elongate greatly by subsequent growth. Although the first visible rudiment of the lungs is ventral and unpaired, there is good reason to believe that the lungs are derived from originally paired and lateral rudiments. By local vital staining in amphibians, it has been ascertained that the presumptive epithelium of the lungs lies, in the neurula stage, in the lateral walls of the foregut, just posterior to the presumptive endoderm of the pharyngeal pouches. The presumptive lung endoderm later shifts downward the midline.

In the clawed toad *Xenopus* (and possibly also in other frogs), the lung rudiments first become noticeable as two separate lateral pockets in a stage when the parts of the alimentary canal are not yet clearly separated from one another. When the gastric part of the gut becomes inflected downward, the lung rudiments are found just behind and ventral to the crest of the transverse fold separating the pharyngeal section of the gut from the esophagus. Following this, the part of the gut cavity connected to the lateral lung rudiments protrudes forward and becomes a distinct pocketlike evagination. This cavity is the rudiment of the trachea (which is very short in frogs). The cavity of the trachea, which is continuous with the cavities of the lung rudiments, becomes temporarily separated from the esophagus and later opens into the pharyngeal cavity. This new opening is the glottis—it does not coincide with the mouth of the original invagination which gave rise to the trachea.

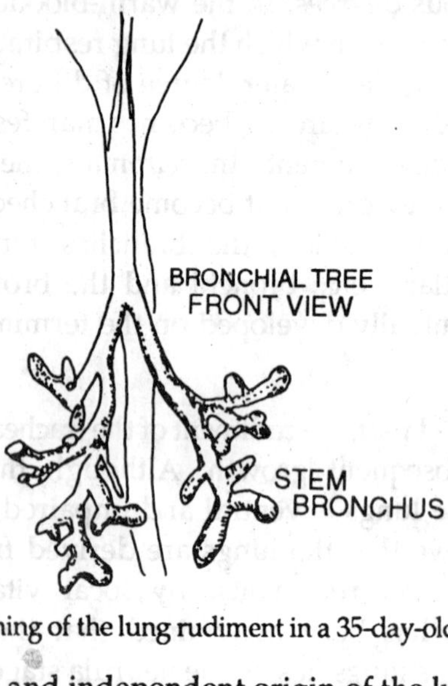

Fig. 6.19. Branching of the lung rudiment in a 35-day-old human embryo.

The lateral and independent origin of the lung rudiments in amphibians makes it probable that in the early history of the terrestrial vertebrates the lungs developed from the last pair of pharyngeal pouches, which failed to break through to the exterior and became adapted to the retention of air gulped in through the mouth; thus they became an organ in which the oxygen could diffuse into the blood vessels supplying the organ.

The swim bladder of fishes is similarly a pocket, growing out from the endodermal wall of the alimentary canal posterior to the branchial region. In many fishes the cavity of the swim bladder remains permanently in communication with the esophagus, and air can be taken into the swim bladder through this canal. It is fairly obvious that the unpaired lung of the lungfish *Neoceratodus* is the same organ as a swim bladder but adapted to respiration. Whether the rudiment of the swim bladder may be compared to a pharyngeal pouch has not been investigated.

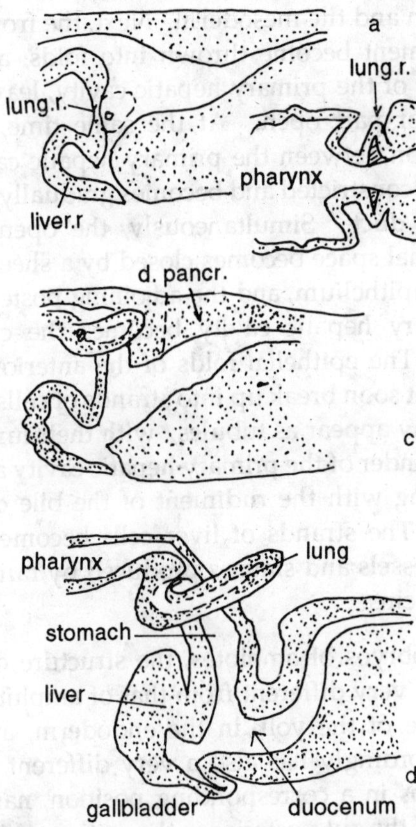

Fig. 6.20. Development of the lungs in a frog. *Xenopus laevis* (semi-diagrammatic). *a, c, d,* Projection on the median plane; b, frontal section.

The Liver. The liver in all vertebrates develops from the endodermal epithelium on the ventral side of the duodenum.

In amphibians, the site of liver development is the anterior wall of the liver diverticulum. At the stage when the main parts of the alimentary canal begin to take shape, the anterior wall of the liver diverticulum bulges forward, so that the slitlike cavity of the diverticulum enlarges locally. This pocketlike enlargement of the gut cavity is the *primary hepatic cavity.* At the time when this occurs, in frogs, the liver diverticulum is still in communication with the space between the endoderm and the mesoderm. Next, the front wall of the hepatic rudiment become thrown into folds, and the folds occlude most of the primary hepatic cavity, leaving only the most posterior part open. At the same time, the original communication between the primary hepatic cavity and the duodenum is constricted and becomes gradually transformed into the bile duct. Simultaneously, the opening into the submesodermal space becomes closed by a sheet of cuboidal endodermal epithelium, and the adjoining posterior remnant of the primary hepatic cavity becomes the cavity of the gallbladder. The epithelial folds of the anterior wall of the liver rudiment soon break up into strands of cells, which for a short time may appear as tubules, with their lumen opening into the remainder of the primary hepatic cavity and thus also communicating with the rudiment of the bile duct and the gallbladder. The strands of liver cells become interwoven with blood vessels and sinuses produced by ramifications of the vitelline veins.

In the embryos of amniotes, the structure of the gut in early stages is very different from that of amphibians, owing to the absence of the yolk in the endoderm, and the liver rudiment accordingly has also a very different appearance, but it develops in a corresponding position, namely, in the ventral wall of the gut posterior to the section which gives rise to the stomach. The first visible rudiment of the liver can be found as a pocketlike evagination on the anterior intestinal portal at a stage when the opening from the yolk sac into the definitive gut is still quite wide. As the floor of the gut continues to close in an anteroposterior direction, the liver rudiment is later found some distance in front of the anterior intestinal portal, well within the foregut. The endodermal cells then begin migrating forward from the original pocketlike evagination, in the form of solid strands or cords of cells. The strands of liver cells form a meshwork and enclose the blood vessels (the vitelline veins and their ramifications) lying in the region posterior to the heart. From an interaction of the cords of liver cells and the blood vessels, there emerges eventually the complicated structure of the adult liver. A fine lumen may be seen sometimes running along the length of the strands of epithelial cells, so that they may appear as thin tubules.

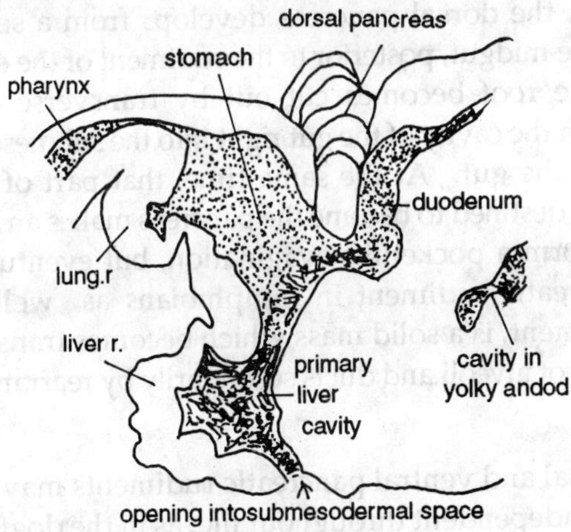

Fig. 6.21. Reconstruction of the lung, liver and pancreas rudiments of the frog. *Xenopus laevis.*

The gallbladder is formed as a secondary hollow outgrowth at the posterior edge of the original hepatic rudiment.

The liver increases in size very rapidly and soon becomes a large and massive organ, although the part of the duodenal wall from which it develops is relatively a very small one, and, even of that, a large proportion is used up for the formation of the gallbladder and cystic duct.

The Pancreas. The pancreas develops from two rudiments—a ventral rudiment (or two ventral rudiments) and a dorsal one. The ventral rudiment develops from a part of the ventral wall of the duodenum just posterior to and in close association with the rudiment of the liver. It may appear as a pocketlike evagination and very soon becomes subdivided distally into a system of proliferating epithelial tubules. The original evagination becomes the duct of the ventral pancreas.

The dorsal pancreatic rudiment in amniotes is also a pocketlike evagination of the duoderum, but it appears on its dorsal side slightly in front of the liver rudiment.

In frogs, the dorsal pancreas develops from a section of the roof of the midgut, posterior to the rudiment of the stomach. a part of the roof becomes cut out by transverse crevices reaching from the cavity of the gut right into the submesodermal space above the gut. At the same time, that part of the gut wall which is destined to become the pancreas makes an abortive attempt to form a pocketlike evagination, but eventually the dorsal pancreatic rudiment in amphibians as well as the ventral rudiment, is a solid mass which becomes transformed into a system of alveoli and ducts secondarily by rearrangement of cells.

The dorsal and ventral pancreatic rudiments may remain completely independent throughout life, as in the dogfish, but in amphibians and amniotes the two rudiments approach each other and fuse completely. The system of pancreas, ducts becomes reorganized in later life and is very variable in different vertebrates.

Bursa Fabricii. In birds a dorsal diverticulum of the cloaca becomes the bursa fabricii which is the sourse of antibody-producing cells. This structure in birds thus plays the same role as the thymus in mammals.

Determination of the Endodermal Organs

Experimental investigation of the endoderm in early stages of development shows that the endodermal organ rudiments, like those derived from other germinal layers, are not determined right from the start, but that the endodermal cells destined to participate in the formation of the various organ rudiments are no more determined for their respective fates than are the cells of the other germinal layers. In the earlier stages of development, their fate is a function of the position that each cell or group of cells occupies in the embryo as a whole. This can be proved by isolating parts of the presumptive endoderm and cultivating them apart from the rest of the embryo or by transplanting them into an abnormal position.

In an extensive series of experiments, pieces of presumptive endoderm of young gastrulae were cultivated in the "Holtfreter solution." Various tissues were observed to differentiate from such isolated pieces; some conformed to the normal destiny of the isolated parts, and some did not. The range of differentiations included not only oro-branchial epithelium, stomach epithelium, liver, pancreas and intestine, but also notochord and muscle, which should not have developed from the presumptive endoderm if it were to keep to its prospective significance. Thus the fate of the endoderm is not laid down finally in the early gastrula stage. Much greater deviations from the prospective significance of the various endodermal parts could be observed when these parts were placed in surroundings which, unlike the saline solution, could actively influence the differentiation of these parts.

In the early neurula stage, it is possible to separate the whole of the endoderm of a newt embryo from the ectoderm and mesoderm. The endoderm is removed as a whole through a slit on the ventral side of the embryo, leaving the ectoderm and mesoderm as an empty shell. The isolated endoderm can then be inserted again into the ecto-mesodermal shell of the same embryo, or of another embryo of the same species, or even into the ecto-mesodermal shell of an embryo of another species; the endoderm of the small *Triturus taeniatus* has been successfully implanted into the ecto-mesodermal shell of the larger *Triturus alpestris*.

The implantation may be carried out so that the orientation of the endoderm is in harmony with the orientation of the ecto-mesoderm, or the endoderm may be implanted in an inverted position. In the first case, a completely normal larva has been observed to develop. A normal embryo also develops if the endoderm is implanted with its dorsoventral orientation reversed. This shows that the determination of the dorsal and ventral parts in the endoderm is not fixed in the endoderm itself but is imposed on the endoderm by the surrounding ecto-mesoderm. However, if the anteroposterior axis of the endoderm was inverted with respect to the axis of the ecto-mesodermal shell, the development was highly abnormal, thus showing that the differentiation of the endoderm along the anteroposterior axis cannot be dominated by the ecto-mesoderm. Here we may recall that in the experiments with the transplantation of the dorsal lip or the blastopore (primary organizer) the endoderm is often observed to develop a secondary lumen of the midgut, just underneath the notochord developed from the transplanted organizer. This secondary lumen is of course part of the dorsal differentiation of the endoderm.

A similar result emerged in experiments in which small pieces of endoderm taken from a late gastrula or early neurula stage were implanted in various positions into another embryo. When the pieces of endoderm were taken from embryos in the gastrula stage, the grafts were often smoothly incorporated into the endoderm of the host. The use of heteroplastic transplantation made it possible to distinguish the grafted cells from the host cells (differences in cell size in grafts between *Triturus taeniatus* and *Ambystoma nuricanum*) and thus to make sure that the graft was not destroyed but had fitted into the construction of local tissues. Thus presumptive oro-branchial endoderm was found to be able to develop into intestinal epithelium and *vice versa.* Stomach epithelium was developed from endoderm having a different prospective significance. Occasionally, however, grafts differentiated out of harmony with their surroundings, and the later the stage of the embryo from which the graft was taken, the oftener this occurred. After the end of the neurulasion the grafts differentiated, in the main, according to their prospective significance.

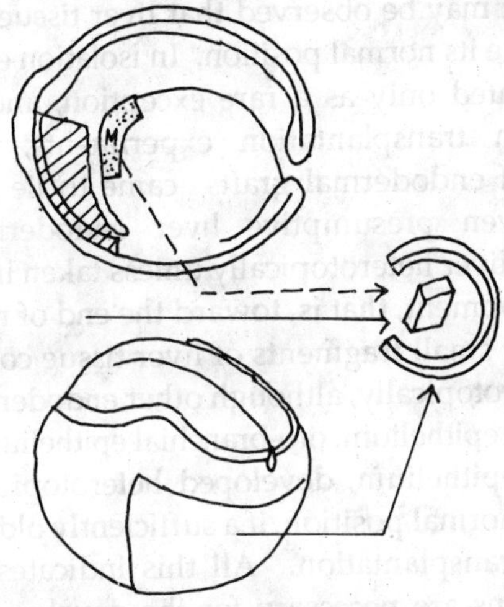

Fig. 6.22. Testing the determination of parts of the neurula endoderm in the newt *Triturus pyrrhogaster* (diagram of operation). *A,* Anterior endoderm *M,* middle endoderm.

In another experiment, different parts of the neurula endoderm were transplanted into parts of the ecto-mesodermal shell, either from the anterior part of the neurula or from the posterior part. The endoderm taken for this experiment was either a part of the foregut endoderm, mainly destined to become pharynx, or endoderm from the midgut normally differentiating as stomach and intestine. It was found that midgut endoderm grafted into the anterior ecto-mesoderm produced pharynx (in addition to other parts). Foregut endoderm surrounded by posterior ecto-mesoderm was in part differentiated as intestine. In both cases, endoderm produced parts which were not in accord with the prospective significance of the endodermal cells, and it seems plausible that these differentiations were induced by the adjoining mesoderm. Again we find that the endoderm, as well as the ectoderm, is dependent on the mesoderm in its differentiation.

In all experiments with explanation and transplantation of endoderm, it may be observed that liver tissue very rarely develops outside its normal position. In isolation experiments, liver differentiated only as a rare exception, and then only atypically. In transplantation experiments, liver often differentiated if endodermal grafts came to lie in the liver region, but even presumptive liver endoderm did not differentiate as liver heterotopically unless taken in rather late stages of development, that is, toward the end of neurulation. Even then, only small fragments of liver tissue could be seen to develop heterotopically, although other endodermal tissues, such as stomach epithelium, oro-branchial epithelium, pancreas and intestinal epithelium, developed heterotopically just as well as in their normal position, if a sufficiently old graft were taken for the transplantation. All this indicates that quite special conditions are necessary for the development of the liver rudiment, and that these special conditions are found only in the region where the liver develops normally. This is another instance (besides the dorsoventral differentiation of the endoderm) where the development of the endodermal rudiments is definitely shown to be dependent on the influence of the surroundings.

There is as yet very little information concerning the earliest determination of endodermal organs in higher vertebrates. In birds, some information has been derived from experiments in which parts of the chick blastoderm were grafted to the chorio-allantoic membrane of another chick embryo. Various endodermal organs were seen to differentiate from the grafts, namely, pharyngeal epithelium, thyroid, lung, liver, large and small intestine. However, these tissues developed without a very definite relationship to the origin of the grafts, if the grafts were taken from early blastoderms up to the head-process stage. In other words, the endodermal differentiations in grafts were by no means strictly in accord with the prospective significance of the areas transplanted. The relation between the prospective significance of parts of the blastoderm, in so far as the endodermal organs are concerned, and the differentiation obtained by explanting these same parts. The lack of correspondence is obvious. The interpretation of these experiments is complicated even more by the fact that the endoderm, taken alone, differentiates extremely poorly. Information on the differentiation of endodermal organs is obtained by grafting sections of the blastoderm containing all the germinal layers.

It has been concluded that in itself the endoderm has a very low power of differentiation, and that it needs interaction with mesoderm for its development, at least until the completion of gastrulation. It was noted that liver and thyroid usually differentiate in explants which also show the presence of the heart, and that intestine is accompanied by mesoderm forming coelomic spaces. This would be broadly in correspondence with what has been found in the amphibian embryo. The role of the upper layers of the blastoderm may possibly go even beyond that. It has been observed that the epiblast alone, without the hypoblast, when cultivated on the chorio-allantois, will produce various endodermal tissues such as thyroid, liver, pancreas and intestine, almost in the same way as a whole blastoderm consisting of both epiblast and hypoblast. This may mean that the actual formative cells of the endodermal organs are derived from the epiblast in conformity with what has been said about migration of cells from the primitive streak into the hypoblast.

Only when the endodermal gut becomes separated from the yolk sac during the second day of incubation does the differentiation of endodermal explants correspond to their prospective significance.

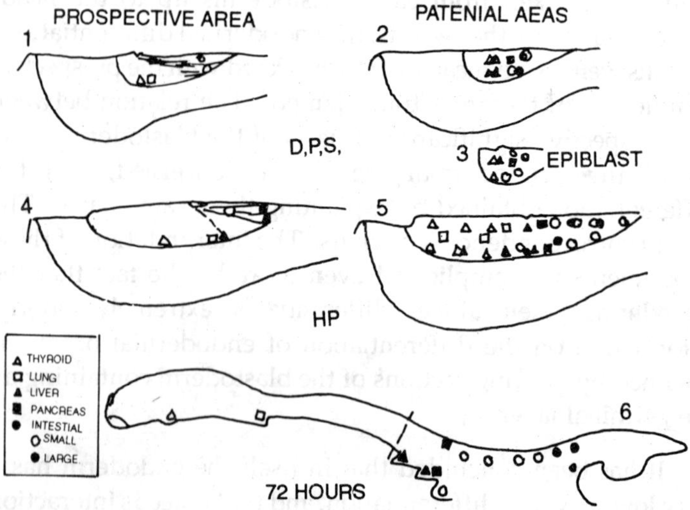

Fig. 6.23. Diagrams showing the development of prospective as compared with potential areas of the blastoderm: 1, Prospective areas for thyroid, lung, liver, pancreas (dorsal) and intestine in the definitive streak stage. The blastoderm is represented as folded longitudinally so that median structures appear in their future dorsal position. Data from marking experiments 2, Potential areas for the above organs in the definitive primitive streak stage, from results of differentiation in chorio-allantoic grafts. All three germ layers included in transplants. 3, Potential areas as in Diagram 2 when hypoblast was not included in the grafts. 4, Prospective areas as in Diagram 1, for the late head-process stage. 5, Potential areas in the head-process stage. 6, Diagram of the nearly closed gut tube of the three-day chick. At this stage most organ rudiments are visible, and potencies correspond with prospective significance.

Experiments have been reported concerning explanation of parts of early mammalian embryos to the chick chorio-allantois to omentum or under the connective tissue capsule of the kidney of adult rats. Only grafts from fairly advanced embryos (ten-day embryos of rats, eight-to nine-day embryos of rabbits) produced endodermal tissue. The earliest embryos whose endodermal tissues showed signs of differentiations in grafts were rabbit embryos already in the eight-somite stage. In mouse and rat embryos the determination of the pancreas was studied by the method of explantation in a culture medium. Normally the (dorsal) pancreas appears as an evagination on the ninth day of gestation when the gut is already closed into a tube and has become segregated from the yolk sac. A piece of gut, including the presumptive material of the pancreas, if explanted on the eighth day, will differentiate, producing pancreating tissue, provided that the adjoining mesodermal tissue is explanted together with the endodermal epithelium. Pieces of gut from younger embryos, in their seventh day of gestation, cannot develop pancreatic tissue although they produce stomach-like and liver-like tissue. Presumptive pancreatic endoderm explanted without mesoderm fails to differentiate, however, even if taken from 11-day embryos in which pancreas had already attained the stage of a pouch-like evagination. The mesoderm was thus necessary to promote the differentiation of pancreatic tissue.

7

Metamorphosis

Organ formation in animals is accomplished by the interactions of numerous cell types. In the preceding three chapters, we have seen how developmental interactions can be mediated by adjacent cell populations. In this chapter, we shall discuss the regulation of development by diffusible molecules that travel long distances from one cell type to another. Diffusible regulators of development can be divided into two general categories: *Hormones,* which are secreted by one cell type and which cause changes in the *differentiation* and *morphogenesis* of other tissues; and *Chalones*, which are secreted by an organ and which regulate the *growth* of that *same* organ.

The Hormonal Reactivation of Development

Because minute quantities of hormones can accomplish their action, it is exceedingly difficult to isolate them from embryos. The most thorough analysis of the hormonal control of development has therefore centered upon the reativation of development known as *metamorphosis.*

In many species of animals, embryonic development leads to a larval stage with characteristics very different from those of the adult organism. Very often, the larval form are specialized for some function, such as growth or dispersal. The pluteus larva of the sea urchin, for instance, can travel on ocean currents, whereas the adult urchin leads a sedentary existence. The caterpillar larva of butterflies and moths are specialized for feeding, whereas their adult forms are specialized for flight and reproduction and often lack the mouth parts necessary for eating. The division of functions between larva and adult is often remarkably distinct. Mayflies hatch from eggs and develop for several months. All this development enables them to spend one day as fully developed winged insects, mating quickly before they die.

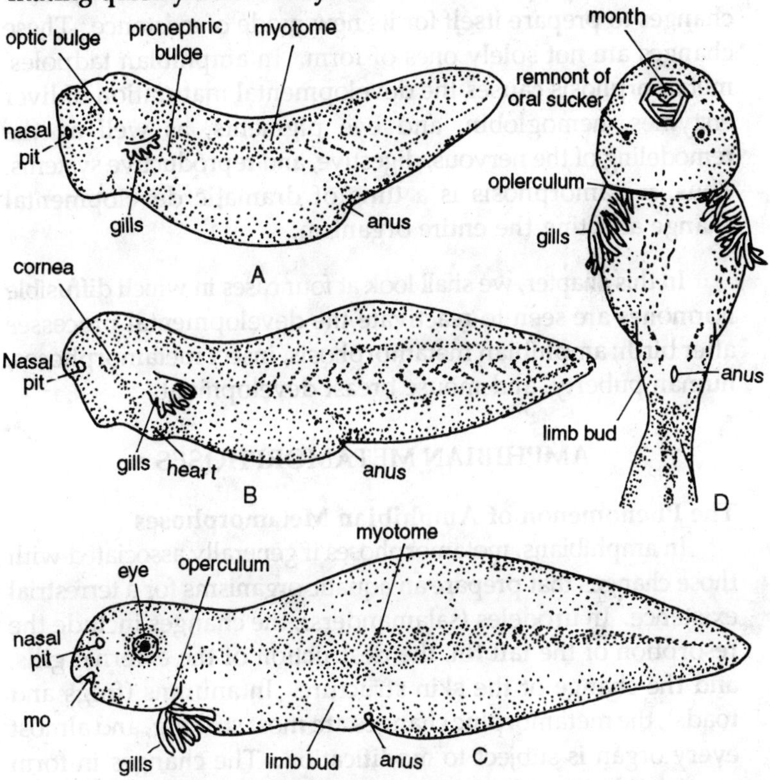

Fig. 7.1. Formation of external gills, tail and other larval structures in tadpol of frog: A—Larva at hatching, B—8 mm long tadpole, C—11 mm long tadpole, D—Ventral view of 11 mm long larva.

As expected the larval form and the adult form often live in different environments. Thus, natural selection works differently on the larval and adult organisms. The voracious gypsy moth caterpillar comes in many different colours, all of which metamorphose into the same adult. Conversely, the swallowtail butterfly *Papilio polytes* has many adult varieties but only one form of caterpillar.

The transition from the larva to the adult is called metamorphosis. Here, the developmental processes are reactivated by specific hormones, and the entire organism changes to prepare itself for its new mode of existence. These changes are not solely ones of form. In amphibian tadpoles, metamorphosis causes the developmental maturation of liver enzymes, hemoglobin, and eye pigments, as well as the remodeling of the nervous, digestive, and reproductive systems. Thus, metamorphosis is a time of dramatic developmental change affecting the entire organism.

In this chapter, we shall look at four cases in which diffusible hormones are seen to reactivate the developmental processes after birth; amphibian metamorphosis, insect metamorphoses, human puberty, and mouse breast development.

AMPHIBIAN METAMORPHOSES

The Phenomenon of Amphibian Metamorphoses

In amphibians, metamorphoses if generally associated with those changes that prepare an aquatic organisms for a terrestrial existence. In urodeles (salamanders), the changes include the resorption of the tail fin, the destruction of the external gills, and the change of the skin structure. In anurans (frogs and toads), the metamorphic changes are most striking, and almost every organ is subject to modification. The changes in form are very obvious, Regressive changes include the destruction of the tadpole tail and internal gills, as well as the loss of the tadpole's horny teeth. At the same time, constructive processes such as limb development and dermoid gland construction are also evident. For locomotion, the paddle tail recedes while the hindlimbs and forelimbs differentiate. The horny teeth constructed for tearing pond plants disappear as the mouth and jaw take a new shape and the tongue muscle develops. Meanwhile the large intestine characteristic of herbivores shortens to suit the more carnivorous diet of the adult frog. The gills regress and the gill arches degenerate. The lungs enlarge, and muscles and cartilage develop for pumping air in and out of the lungs. The sensory apparatus changes, too, as the lateral line system of the tadpole degenerates and the eye and ear undergo further differentiation. In the ear, the middle ear develops, as does the tympanic membrane so characteristic of frogs and toads. In the eye, both nictitating membrane and eyelids emerge. Moreover, the eye pigment changes. In tadpoles, as in freshwater fishes, the major retinal photopigment is *porphyropsin,* a complex between the protein opsin and the aldehyde of vitamin A_2. In adult frogs, the pigment changes to *Rhodopsin,* the characteristic photopigment of terrestrial and marine vertebrates. Rhodopsin consists of opsin conjugated to the aldehyde of vitamin A_1.

TABLE 7.1

Summary of Some Metamorphic Changes in Anurans

System	*Larva*	*Adults*
Locomotory	Aquatic, tail fins	Terrestrial; tailless tetrapod
Respiratory	Gills, skin, lungs; larval hemoglobins	Skin, lungs: adult hemoglobins
Circulatory	Aortic arches; aorta; anterior, posterior, and common cardinal veins	Carotid: systematic arch; jugular veins
Nutritional	Vegetarian: Long spiral gut—intestinal symbionts; small mouth—horny jaws, labial teeth	Carnivorous: Short gut—proteases; large mouth—long tongue
Nervous	Lack of nictitating membrane, porphyropsin, lateral line system—Mauthner's neurons	Development of ocular muscles, nictitating membrane, rhodopsin, loss of lateral line system—degeneration of Mauthner's neurons; tympanic membrane
Excretory	Mesonephros: Largely ammonia, some urea (ammonotelic)	Mesonephros: Largely urea, high activity of enzymes of ornithine-urea cycle (ureotelic)

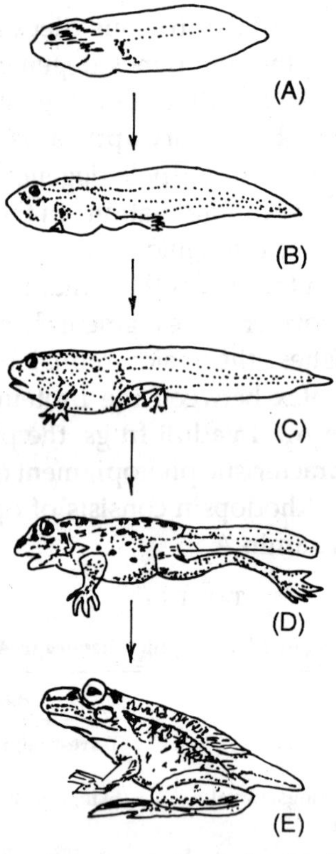

Fig. 7.2. Sequence of metamorphosis in the frog *Rana pipiens.* *(A)* Premetamorphic tadpole; *(B)* prometamorphic tadpole showing hindlimb growth; *(C)* onset of metamorphic climax as forelimbs emerge; *(D, E)* climax stages.

Other biochemical events are also associated with metamorphosis. Tadpole hemoglobin binds oxygen faster and releases it more slowly than adult hemoglobin. Moreover, Riggs (1951) showed that the binding of oxygen by tadpole hemoglobin is independent of pH, whereas frog hemoglobin (as in most other vertebrate hemoglobins) shows increased oxygen binding as the pH rises (Bohr effect). Probably the most striking biochemical change in metamorphosis is the induction of those enzymes necessary for the production of urea. Tadpoles, like most freshwater fishes, are *Ammonotelic;*

that is, they excrete ammonia. Adult frogs, however, like most and vertebrates, are *Ureotelic,* excreting urea. During metamorphosis, the liver develops those enzymes necessary to create urea from carbon dioxide and ammonia. These enzymes constitute the *urea cycle,* and each of them is seen to arise during metamorphosis.

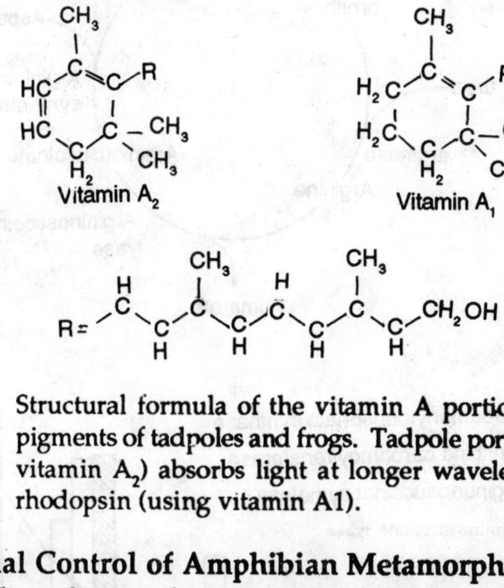

Fig. 7.3. Structural formula of the vitamin A portions of the visual pigments of tadpoles and frogs. Tadpole porphyropsin (using vitamin A_2) absorbs light at longer wavelengths than frog rhodopsin (using vitamin A1).

Hormonal Control of Amphibian Metamorphosis

All these diverse changes are brought about by the diffusion of the hormones thyroxine (T_4) and Triiodothyronine (T_3) from the thyroid during metamorphosis. It is now believed that T_3 is the active hormone, as it will cause the metamorphic changes in thyroidectomized tadpoles in much lower concentrations than will T_4. In both urodeles and anurans, the thyroid produces small amounts of T_3 and T_4 in the larva, but they are overbalanced by the diffusion of the hormone prolactin from the anterior pituitary. Prolactin acts as a larva growth hormone and inhibits metamorphosis. During metamorphosis the thyroid hormones T_3 and T_4 increase in concentration, thereby causing the tadpoles to become frogs and stimulating larval newts to become land-dwelling efts. In salamanders, thyroxine will become overpowered by prolactin later in development, thereby causing them to return to the water to spawn as adults. Thus, in some

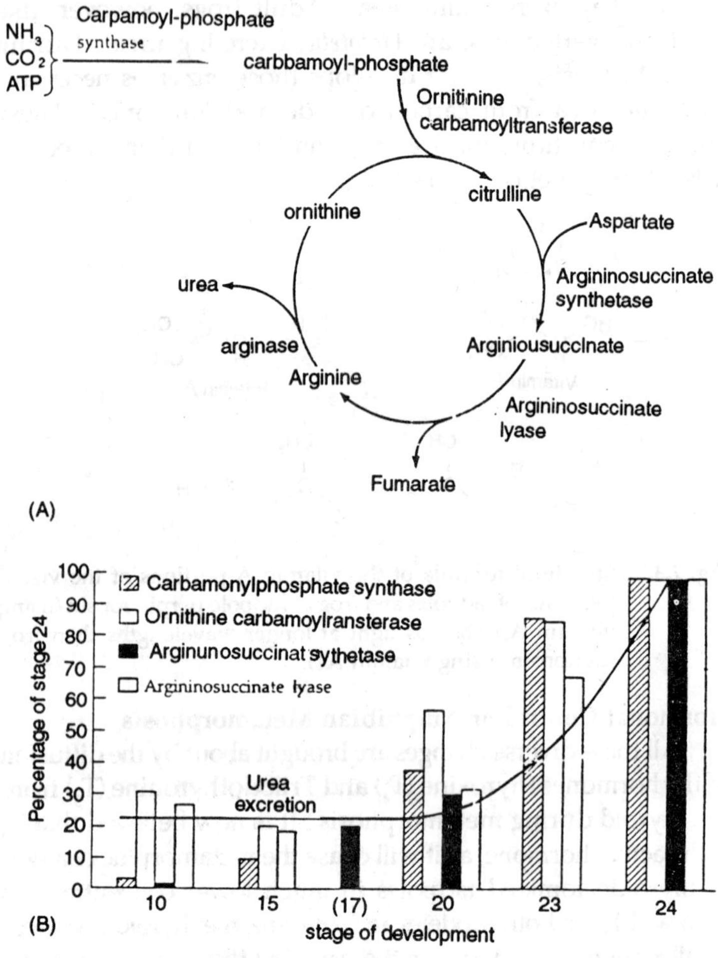

Fig. 7.4. Development of the urea cycle during anuran metamorphosis. (A) The major features of the urea cycle, by which nitrogenous wastes can be detoxified and excreted. (B) Emergence of urea cycle enzyme activities correlated with metamorphic changes in the frog *Rana Catesbeiana*.

salamanders, there are two metamorphoses. The first stimulated by thyroxine, the second induced by prolactin. Moreover, even after the second metamorphosis, an injection of thyroxine will cause the adult newt to return to land as a dry-skinned eft.

The control of metamorphosis by thyroid hormones was shown earlier in this century when Gudernatsch (1912) found that tadpoles metamorphosed prematurely when fed powered sheep thyroid gland. Allen (1916) and Hoskins and Hoskins (1917) found that when they removed the thyroid rudiment from early tadpoles; the larva never metamorphosed, becoming instead giant tadpoles.

The release of T_3 is under the control of the hypothalamic portion of the brain. During the period of larval growth, this portion of the brain is underdeveloped, and so the brain does not exert much control over the anterior pituitary. In the absence of hypothalamic regulation, the pituitary secretes high levels of prolactin but little or no Thyroid Stimulating Hormone (TSH). Thus, T_3 levels are low and prolactin levels are high. As the hypothalamus develops, its production of *thyroid stimulating hormone releasing factor* (TSH-RF) causes a rise in the level of TSH. This TSH causes the thyroid to increase T_3 and thyroxine synthesis. Thus, T_3 is released from the thyroid and its concentration gradually increases until the first changes of metamorphosis (Prometamorphosis) appear. During the time, the hindlimbs start to enlarge. The rising T_3 levels also stimulate the further development of the median eminence of the pituitary, which mediates the flow of TSH-RF to reach the anterior pituitary. There is, then, a *positive feedback* wherein increased T_3 production causes more T_3 production. In addition, the hypothalamus begins secreting a substance—probably dopamine that inhibits the pituitary synthesis of prolactin. Thus, the ratio of T_3 to prolactin increases enormously. This leads to the *metamorphic climax* in which most of the development associated with metamorphosis takes place. One of the effects of metamorphosis is the partial degeneration of the thyroid, and it is also possible that high levels of T_3 have an inhibitory effect on TSH or TSH-RF production. Thus, a new balance of hormones is acquired.

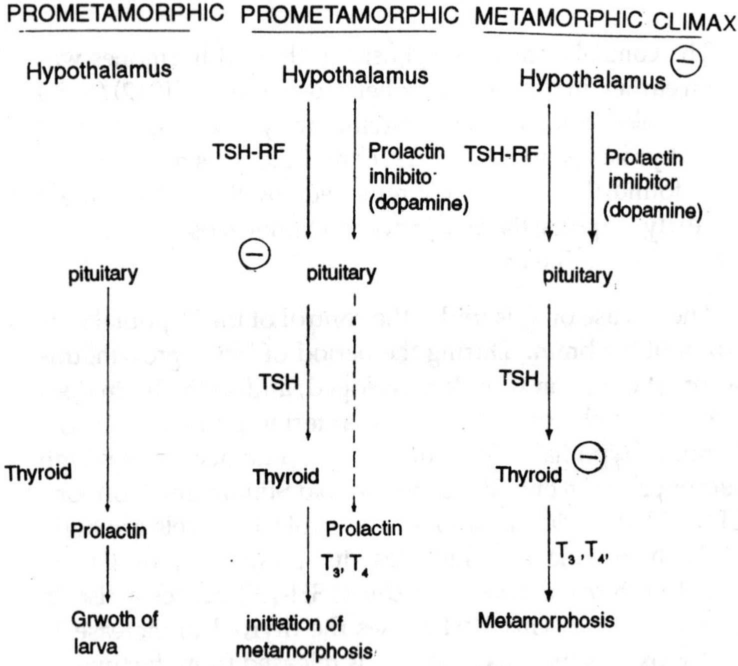

Fig. 7.5. The hypothalamus-pituitary-thyroid axis during different stages of anuran metamorphosis. As the hypothalamus develops, it stimulates the pituitary to instruct thyroid hormone secretion and inhibit prolactin secretion, T_3, Triiodothyronine, T_4 thyroxine, TSH, thyroid stimulating hormones TSH-RF, thyroid stimulating hormone releasing factor.

The different organs of the body respond differently to hormonal stimulation. The same stimulus will cause certain tissues to degenerate while causing others to develop and differentiate. For instance, tail degeneration is clearly associated with the increasing levels of thyroid hormones. This can be shown in vitro where isolated tail pieces can be placed in agar dishes and subjected to chemical treatments. Those tails grown in untreated medium remain healthy, whereas those placed into medium containing thyroid hormones undergo characteristic regression. Moreover, prolactin will inhibit the degeneration of the tail induced by thyroid hormones. The regression of the tail is thought to occur in three states. First, protein synthesis decreases in the striated muscle cells of the tail. Then, there is an increase in the lysosomal enzymes: cathepsin D (a protease), RNase, DNase, collagenase, phosphatase, and glycosidases in the epidermis, notochord, and nerve cord cells. Cell death is probably caused by the release of these enzymes into the cytoplasm. After this death, macrophages collect in the tail region, digesting the debris with their own proteolytic enzymes. The result is that the tail becomes a large sack of proteolytic enzymes.

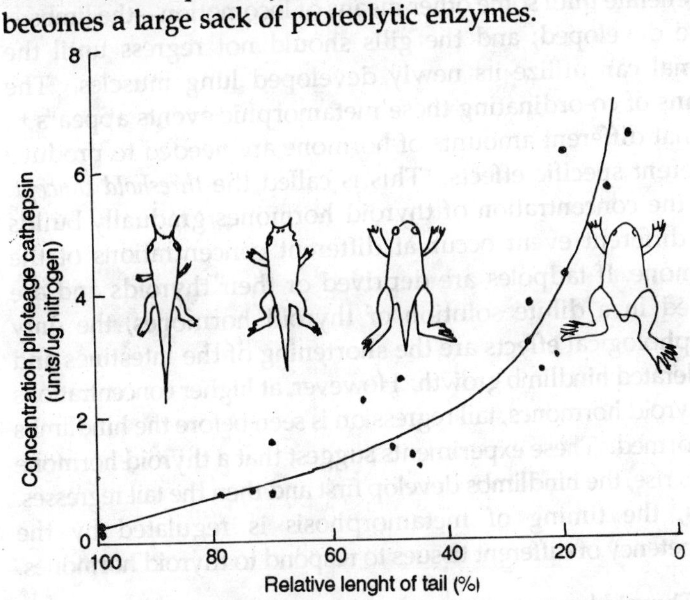

Fig. 7.6. **Increase of lysosomal protease cathepsin during tail regression in Xenopus laevis. The lysosomal enzymes are thought to be responsible for digesting the tail cells.**

The response to thyroid hormones is intrinsic to the organ itself and is not dependent on other factors. This can be seen when tail tips are transplanted to the trunk region or when eye cups are placed in the tail. The extra tail tip placed into the trunk is not protected from degenerating, and the eye retains its integrity despite its being within the degenerating tail. Thus, the degeneration of the tail represents a programmed cell death. Only specific tissues die when a signal is given. Such programmed cell deaths are important in molding the body. In humans, such degeneration occurs in the tissues between our finger and toes, and the degeneration of the human tail during week 4 of development resembles the regression of the tadpole tail.

One of the major problems of metamorphosis is the co-ordination of developmental events. The tail should not degenerate until some other means of locomotion—the limbs—have developed; and the gills should not regress until the animal can utilize its newly developed lung muscles. The means of co-ordinating these metamorphic events appears to be that different amounts of hormone are needed to produce different specific effects. This is called the *threshold concept.* As the concentration of thyroid hormones gradually builds up, different event occur at different concentrations of the hormone. If tadpoles are deprived or their thyroids and are placed in a dilute solution of thyroid hormones, the only morphological effects are the shortening of the intestines and accelerated hindlimb growth. However, at higher concentrations of thyroid hormones, tail regression is seen before the hindlimbs are formed. These experiments suggest that a thyroid hormone levels rise, the hindlimbs develop first and then the tail regresses. Thus, the timing of metamorphosis is regulated by the competency of different tissues to respond to thyroid hormones.

Thyroid hormones can cause existing tissues to break down or can remold the tissues to their adult function. The cells of the tadpole liver, for instance, are not destroyed and replaced during metamorphosis. Rather the structure of the existing liver cells is changed. This change is accompanied by dramatic increases in ribosomal and messenger RNA synthesis, and the rate of protein synthesis increases nearly 100-fold within 4 hours of thyroid hormone stimulation. Many of these new mRNAs are those coding for the new functions of the adult liver.

Fig. 7.7. Organ specificity during frog metamorphosis. (A) Tail tips regress even when transplanted to the trunk whereas eye cups (B) remain intact even when transplanted into the regressing tail.

As shown, in Fig. 7.8, carbamoylphosphate synthase (a urea cycle enzyme) is synthesized after the burst of RNA synthesis. Mori and co-workers (1979) have shown that the level of in vitro translatable mRNA for this enzyme increases shortly after thyroid hormones are given to bullfrog tadpoles. Thus, metamorphosis is to some extent controlled at the transcriptional level. Other evidence also shows the importance of transcriptional regulation. Weber (1967) demonstrated that the injection of actinomycin D into normal prometamorphic tadpoles inhibited tail regression and head remolding. Because T_3 is known to bind to nucleosomes in mammalian liver chromatin, it is probable that T_3 can reprogramme the genome to transcribe new types of messages for the adult organism. This does not mean that transcription is the only level of gene regulation working during metamorphosis, but it is obviously an important one.

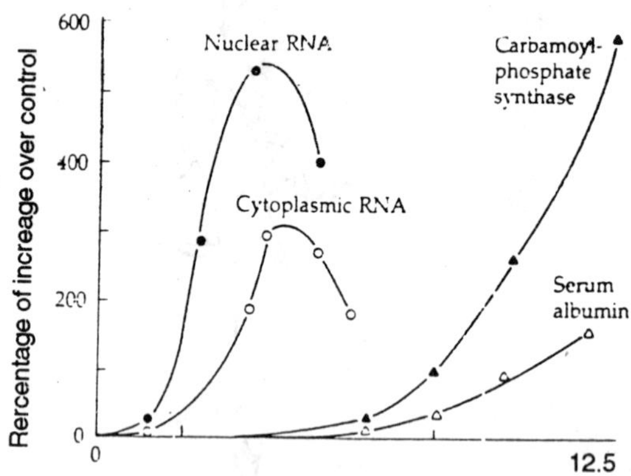

Fig. 7.8. Molecular synthesis in Rana catesbeiana liver cells after treatment of tadpoles with triiodothyronine. Increases first in nuclear and then in cytoplasmic RNA are observed before increases in liver-specific proteins, especially those of the urea cycle.

Neoteny

Many salamanders are able to retain their larval form throughout their lives, becoming sexually mature without undergoing metamorphosis. Such retention of juvenile characteristics in sexually mature individuals is called *neoteny*. The degree to which metamorphosis occurs differs from species

Fig. 7.9. Metamorphosis induced in the axolotl. (A) normal condition of the axolotl. (B) Specimen treated with thyroxine to induce metamorphosis.

to species. The Mexican axolotl, *Ambystoma mexicanum*, does not undergo metamorphosis in nature because it does not release an active TSH to stimulate its thyroid glands. Thus, when investigators gave *Ambystoma mexicanum* either thyroid hormones or TSH, they found that the salamander will metamorphose into an adult not seen in nature. Other species, such as *Ambystoma tigrinum*, metamorphose only if given cues from the environment. Otherwise, they become neotenic, successfully mating as larvae. In part of its range, *A. tigrinum* is a neotenic salamander, paddling its way through the cold ponds of the Rocky Mountains. However, in the warmer portion of its range, the larval form of *A. tigrinum* is transitory, leading to the land-dwelling tiger salamander. Neotenic populations from the Rockies can be induced to undergo metamorphosis simply by placing them in water at higher temperatures. It appears that the hypothalamus of this species cannot produce TSH-RF at the low temperatures.

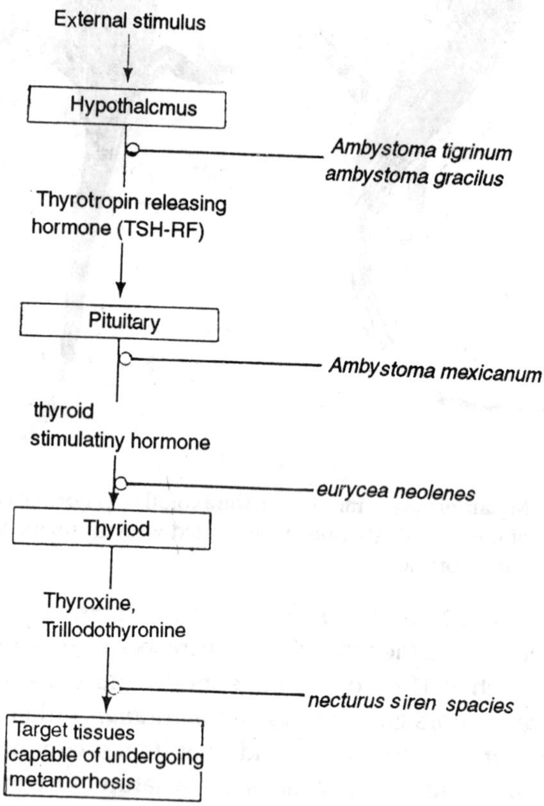

Fig. 7.10 Metamorphosis induced in the axolotl. (A) normal condition of the axolotl. (B) Specimen treated with thyroxine to induce metamorphosis.

Some salamanders, however, are permanently neotenic, even in the laboratory. Whereas thyroxin was able to produce the long-lost adult form of *A. mexicanum*, the neotenic species of *Necturus* and *Siren* remained refractile to thyroid hormones. It appears that the target tissues have lost their capacity to respond to thyroid hormones; thus their neoteny is permanent. The genetic lesions responsible for Neoteny in several species as shown.

Gould (1977) has speculated that Neoteny is a major factor in the evolution of more complex taxa. By retarding the development of somatic tissues, natural selection is given a flexible substrate. Neoteny would "provide an escape from specialization. Animals can slough off their highly specialized adult forms, return to the lability of youth, and prepare themselves for new evolutionary directions."

Metamorphosis in Insects

Whereas amphibian metamorphosis is characterized by the remolding of existing tissues, insect metamorphosis often involves the destruction of larval tissues and its replacement by an entirely different population of cells. Like amphibian metamorphosis, insect metamorphosis involves the balance between a hormone necessary for continued larval growth and hormone capable of stimulating the new developmental changes.

There are three major patterns of insect development. A few insects, such as springtails, have no larval stage and are said to undergo *direct development*. Other insects, notably grasshoppers and bugs, undergo a gradual, *hemimetabolous* metamorphosis. Here, adult organs are formed without any profound discontinuity. The rudiments of the wing, genital organs, and other adult structures are there at hatching, and they become more mature with each molt. At the last molt, the emerging insect is a winged and sexually mature adult. The larval form of a hemimetabolous insect is a nymph.

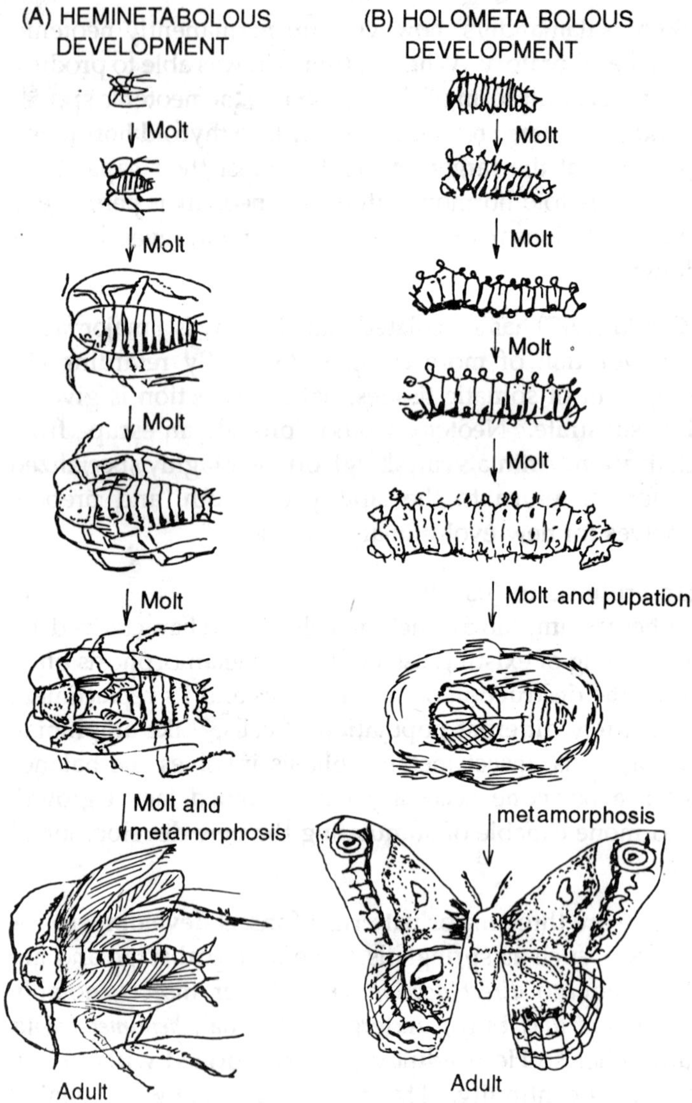

Fig. 7.11. (A) Hemimetabolous (incomplete) metamorphosis.
(B) Holometabolous (complete) metamorphosis.

In the *Holometabolous* insects (flies, beetles, moths, and butter-flies), there is a dramatic and sudden transformation between the larval and adult stages. The juvenile larva undergoes a series of molts as it becomes larger. The newly hatched insect larva is covered by a hard *cuticle.* The insect must shed its old cuticle and then replace it with a larger one in order to grow. Thus, the postembryonic development of these insects consists of a succession of molts. The number of molts before becoming an adult is characteristic for the species, although environmental factors can lengthen or shorten the number. The stages between these molts are called *instars.* Here, growth of the larva is the major characteristic. After its last instar stage, the larvae undergo a *metamorphic molt to* become a pupa. The pupa does not feed, and its energy must come from those foods ingested by the larva.

It is within this pupa that the transformation of juvenile into adult occurs. The old body of the larva is systematically destroyed as new adult organs develop from undifferentiated nests of cells, the *histioblasts* and the *imaginal discs.* When the adult organism *(imago)* is developed, the *imaginal molt* sheds the pupa, and the mature insect emerges. In holometabolous larvae, then, there are two cell populations: the larval cells, which are used for the functions of the juvenile, and the imaginal cells, which lie in clusters awaiting the signal to differentiate.

We have already encountered imaginal discs in our previous discussion of transdetermination and differential gene transcription. In *Drosophila,* there are 10 major pairs of imaginal discs, which reconstruct the entire adult integument (except for the abdomen), and a genital disc, which forms the reproductive structures. The abdominal epidermis forms from a small group of imaginal cells called histioblasts, which lie in the region of the larval gut, whereas nests of imaginal cells located throughout the larva form the internal organs of the adult. The imaginal discs can be seen in the newly hatched larva as local thickenings of the epidermis such as that shown for a second thoracic leg disc. In *Drosophila,* these newly hatched eye-antenna, wing, haltere, leg, and genital discs contain

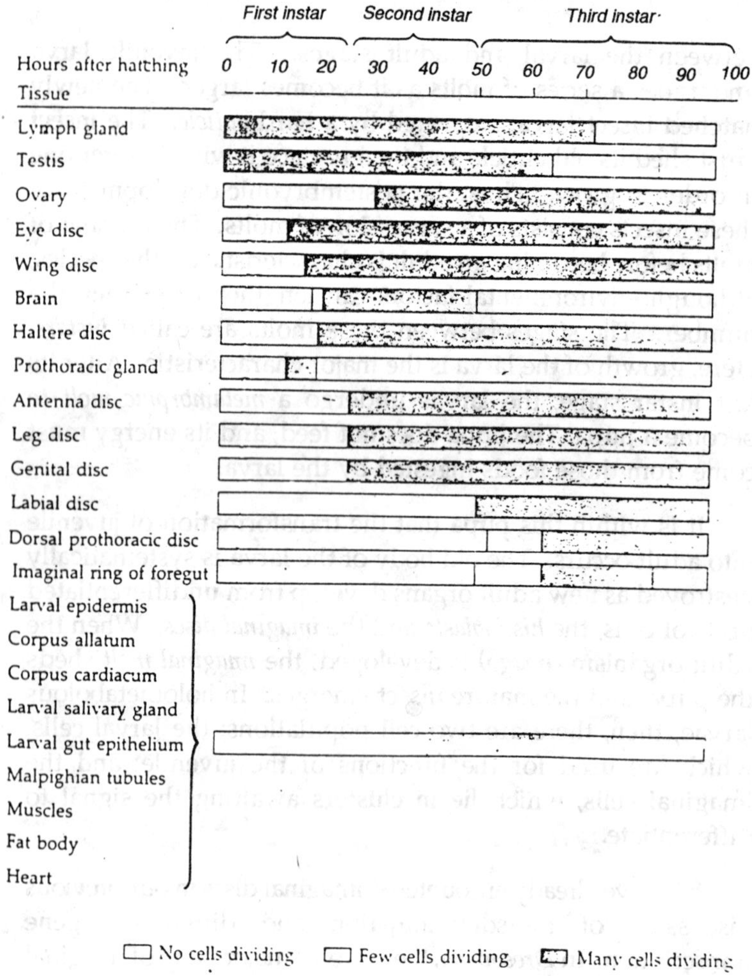

Fig. 7.12. **Patterns of cell division for several larval and imaginal tissues during the development of *Drosophila* larvae.**

70, 38, 20, 36-45 and 64 cells, respectively. Whereas most of the larval cells have a very limited mitotic capacity, the imaginal discs divide rapidly at specific characteristic times. As the cells proliferate, they form a tubular cpithelium that folds in upon itself in a compact spiral. The largest disc, that of the wing, contains some 60,000 cells, whereas the leg and haltere discs contain around 10,000. At metamorphosis, these cells differentiate and *evert*. The fate map and eversion sequence of the leg disc is shown in figure 7.14. The cells at the center of the disc telescope out to become the most distal portions of the leg—the claws and the tarsus—and the outside cells become the proximal structures—the coxa and the adjoining epidermis. After differentiating, the cells of the appendages and epidermis secrete a cuticle appropriate for the specific region. Although the discs are primarily composed of epidermal cells, a small number of *adepithelial* cells migrate into the disc early in development. During the pupal period, these cells give rise to the muscle and nerves that serve that structure.

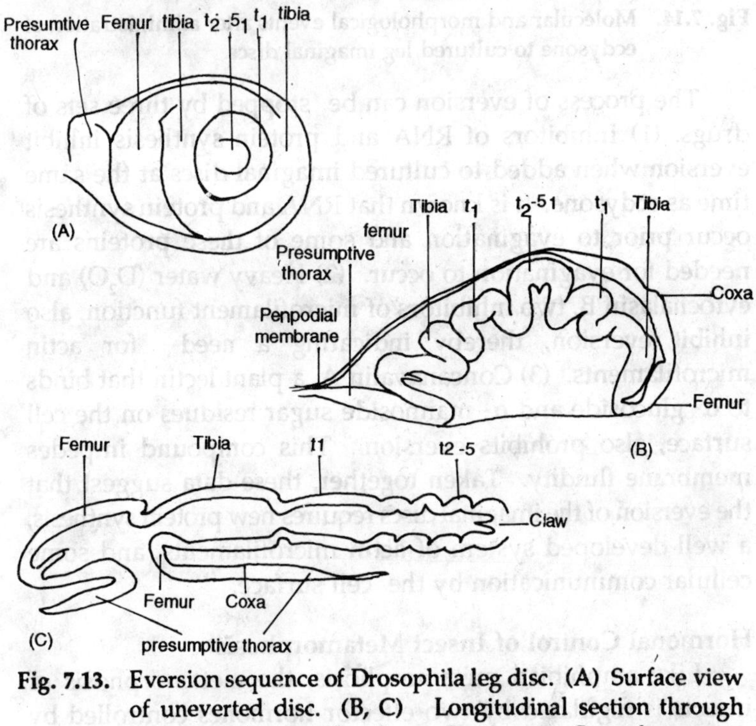

Fig. 7.13. **Eversion sequence of Drosophila leg disc. (A) Surface view of uneverted disc. (B, C) Longitudinal section through everting and everted leg disc. t_1, Basitarsus; t_2-5 tarsal segments 2-5.**

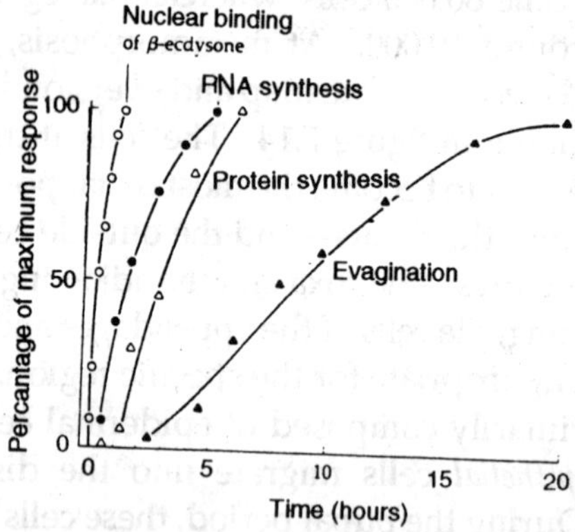

Fig. 7.14. Molecular and morphological events after administration of ecdysone to cultured leg imaginal discs.

The process of eversion can be stopped by three sets of drugs: (1) Inhibitors of RNA and protein synthesis inhibit eversion when added to cultured imaginal discs at the same time as ecdysone. It is known that RNA and protein synthesis occur prior to evagination and some of these proteins are needed for evagination to occur. (2) Heavy water (D_2O) and cytochalasin B, two inhibitors of microfilament function, also inhibit eversion, thereby indicating a need for actin microfilaments. (3) Concanavalin A, a plant lectin that binds to α- glucoside and α- mannoside sugar residues on the cell surface, also prohibits eversion. This compound impedes membrane fluidity. Taken together, these data suggest that the eversion of the imaginal discs requires new protein synthesis, a well-developed system of actin microfilaments, and some cellular communication by the cell surface.

Hormonal Control of Insect Metamorphosis

Like amphibian metamorphosis, the metamorphosis of insects is regulated by two effector hormones controlled by neurosecretory peptide hormones in the brain. Although the detailed mechanisms of metamorphosis differ between species, a general scheme has been elucidated for several organisms. The molting process is initiated in the brain, where neurosecretory cells release *prothoracicotropic hormone* (PITH) in response to neural, hormonal, or environmental factor. Prothoracicotropic hormone is a peptide hormone with a molecular weight of approximately 40,000 and it stimulates the production of *ecdysone* by the *prothoracic gland*. Ecdysone, however, is not an active hormone; rather it is a prohormone that must be converted into an active form. The conversion is accomplished by a heme-containing oxidase in the mitochondria of peripheral tissues such as the fat body. Here, the ecdysone is changed to the active hormone *20-hydroxyecdysone*.

Each molt is occasioned by two pulses of ecdysone. The first pulse produces a small rise in the ecdysone concentration in the larval hemolymph (blood) and is probably important only during the metamorphic molt. The second, large pulse in ecdysone initiates the events associated with molting, independent of the hormonal conditions of the larva during the first ecdysone pulse. The 20-hydroxyecdysone produced by this pulse stimulates the epidermal cells to synthesize enzymes that digest and recycle the components of the cuticle. In some cases, environmental conditions can control molting, as in the case of the *Cecropia* moth. Here, PTTH secretion ceases after the pupa has formed. The pupa remains in this suspended state, called diapause, throughout the winter. If not exposed to cold weather, *diapause* lasts indefinitely. Once exposed to 2 weeks of cold, however, the pupa can molt when returned to a warmer environment.

The second major effector hormone in insect development is juvenile hormone (JH). The structure of a common juvenile hormone active in *Cecropia* caterpillars. Juvenile hormone is secreted by the *corpora allata*. The secretory cells of the corpora allata are active during larval molts but are inactive during the metamorphic *molt*. This hormone is responsible for preventing metamorphosis. As long as juvenile hormone is present, the ecdysone-stimulated molts result in a new larval instar stage.

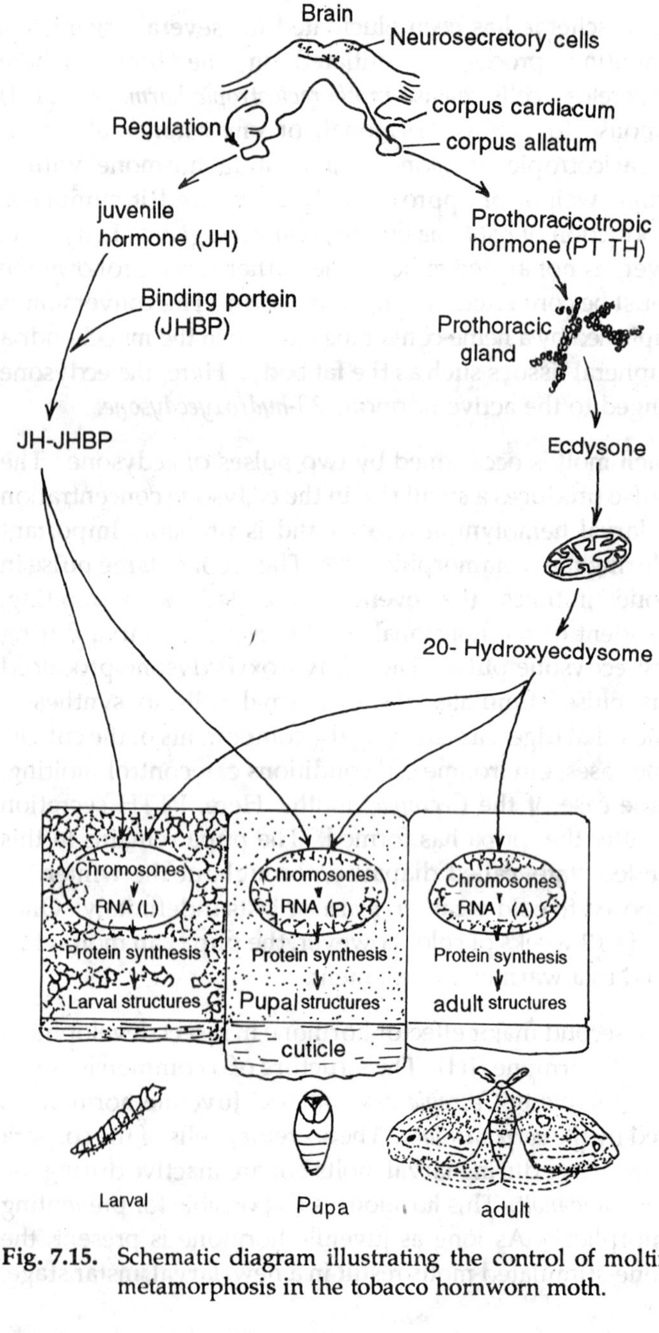

Fig. 7.15. Schematic diagram illustrating the control of molting and metamorphosis in the tobacco hornworn moth.

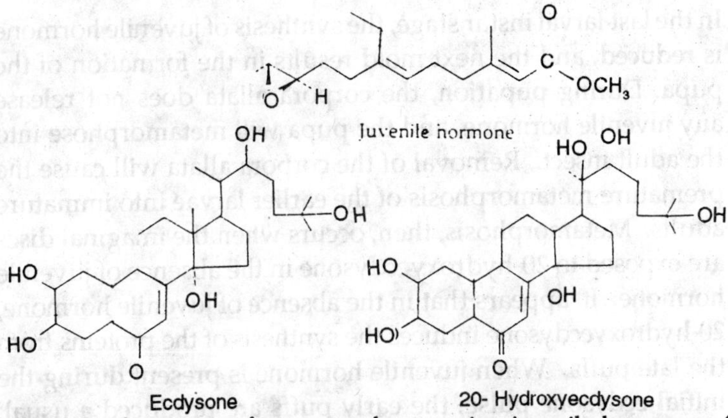

Fig. 7.16. **Structures of a commonly occurring juvenile hormone, of ecdysone, and of the active steroid 20-hydroxyecdysone.**

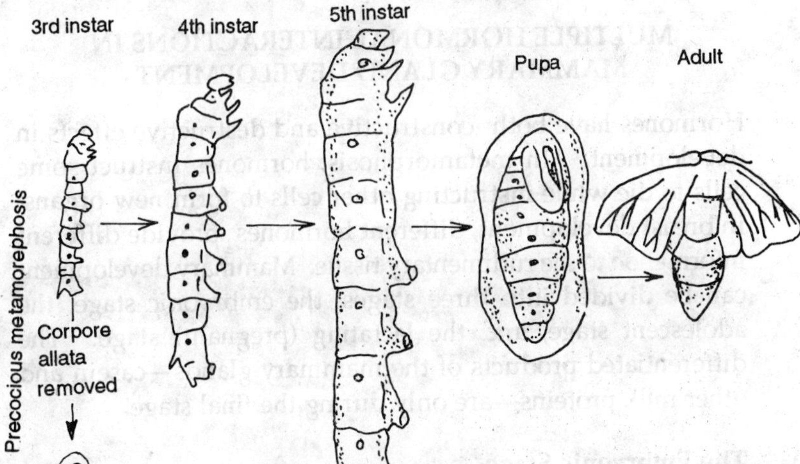

Fig. 7.17. Precocious metamorphosis of silkworm moths caused by the removal of the corpora allata during the third instar stage. Rather than continue molting through the fifth instar stage, the larva initiates pupation directly.

In the last larval instar stage, the synthesis of juvenile hormone is reduced, and the next mold results in the formation of the pupa. During pupation, the corpora allata does not release any juvenile hormone, and the pupa will metamorphose into the adult insect. Removal of the corpora allata will cause the premature metamorphosis of the earlier larvae into immature adults. Metamorphosis, then, occurs when the imaginal discs are exposed to 20-hydroxyecdysone in the absence of juvenile hormone. It appears that in the absence of juvenile hormone, 20-hydroxyecdysone induces the synthesis of the proteins from the late puffs. When juvenile hormone is present during the initial ecdysone pulse, the early puffs are produced a usual, but the later puffs are repressed.

MULTIPLE HORMONAL INTERACTIONS IN MAMMARY GLAND DEVELOPMENT

Hormones have both constructive and destructive effects in development. In metamorphosis, hormones instruct some cells to die while instructing other cells to form new organs. In breast development, different hormones provide different information to the rudimentary tissue. Mammary development can be divided into three stages: the embryonic stage, the adolescent stage, and the lactating (pregnant) stage. The differentiated products of the mammary glands—casein and other milk proteins—are only during the final stage.

The Embryonic Stage

In the normal development of the female mouse, two bands of raised epidermal tissue appear on both sides of the ventral midline on day 11. This tissue is called *mammary ridge.* Within this ridge, cells collect at centers of concentration and remain there, forming the *mammary buds*. In the mouse, there are five of these buds on each side; in humans, only one per side. In the days immediately prior to birth, the epithelial cells at these places proliferate rapidly, giving rise to the *mammary cord.* This cord opens at the skin at one end, forming the nipple, while it other end begins branching into ducts. Here development ceases until puberty.

The development of mammary tissue in male mice is identical to that of females until days 13-15 of gestation. At this time, the mesenchyme condenses around the center of the mammary bud, and the cells of the cord die. Thus, a small cord of epithelial cells is detached from the skin and the mammary gland does not extend to the surface. No further development occurs.

This cell death in the mammary cord of males has been studied by culturing the mouse mammary buds in vitro. Such buds from female mice normally develop lobes connected to the surface. However, when testosterone is added to the culture medium, the buds degenerate. Mammary buds from male mice will also produce lobes, provided they are cultured in the absence of testosterone. Thus, the hormone testosterone prevents mammary development in the male. Testosterone causes this specific cell death by instructing the mesenchymal cells to destroy the epithelial cord. This was shown by a series of recombination experiments. There exists in mice (and in humans as well) a mutation called *Androgen Insensitivity syndrome,* in which male (XY) individuals do not make a functional testosterone receptor. Thus, even though these individuals have tests that are actively secreting testosterone, they are unable to respond to it. One of the results is that these individuals have female breast development. Kratochwil and Schwartz (1975) isolated mesenchyme and epithelial cells from normal and mutant mammary buds and cultured them in various combinations. Some cultures were given testosterone and some were not. When both mesenchyme and epithelium were wild type, the rudiment developed into breast tissue. When testosterone was added, the mesenchyme condensed around the bud and the cord was served. When normal epithelium was cultured with mutant mesenchyme (which could not respond to testosterone), normal breast development occurred in the presence of testosterone. However, when the

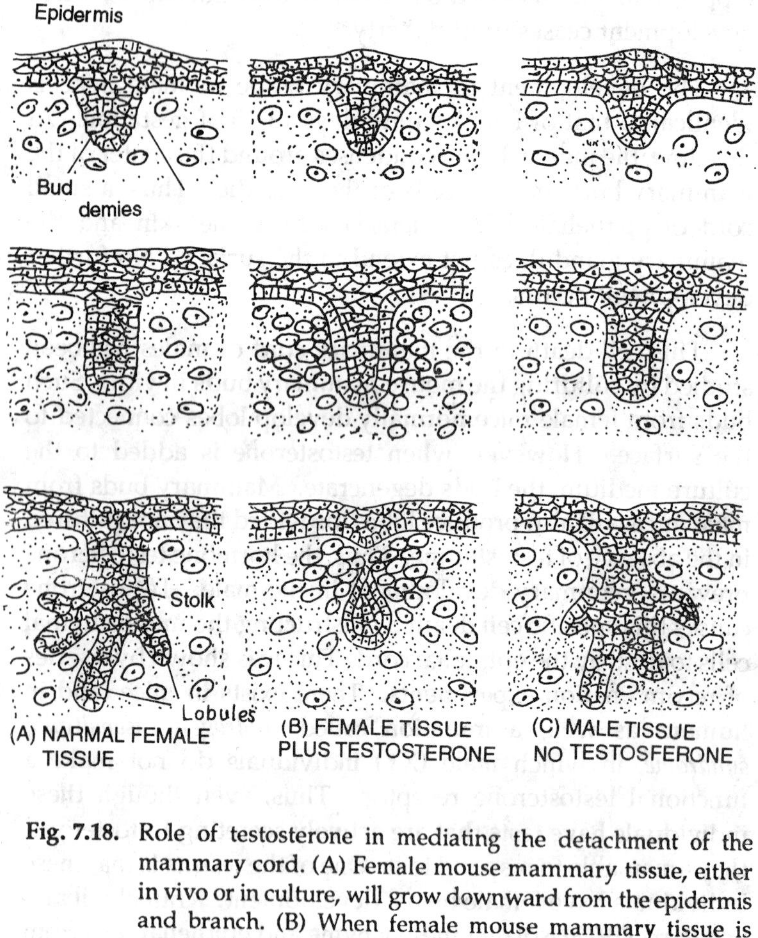

Fig. 7.18. Role of testosterone in mediating the detachment of the mammary cord. (A) Female mouse mammary tissue, either in vivo or in culture, will grow downward from the epidermis and branch. (B) When female mouse mammary tissue is cultured in the presence of testosterone, the bud elongates, but mesenchymal cells aggregate around the stalk and the lower portion is cut off, just as in normal male development. (C) When male mouse mammary tissue is cultured in the absence of testosterone, it develops as it would in the female mouse.

mesenchyme was normal and the epithelium was mutant, testosterone was able to cause the degeneration of the mammary cord. Thus, the target of testosterone is the mesenchyme, not the epithelium. The mesenchyme must be responsive to testosterone for its action to occur. In males the testosterone induces the mammary mesenchyme to destroy its adjacent epithelium. The effect is specific for the organ in that no other mesenchyme will kill the mammary epithelium and no other epithelium can be destroyed by mammary mesenchyme.

Adolescence

During adolescence the duct system of the mammary gland proliferates extensively. The extensive cell division is under the control of estrogen and growth hormones. The milk-secreting alveolar cells at the tips of the ducts have not differentiated yet, and no milk is produced.

Pregnancy

Between adolescence and pregnancy, the mouse breast cells are mitotically dormant and undifferentiated. This is changed during the second half of pregnancy. Under the influence of the hormones estrogen and *progesterone* (the latter from the placenta), new ducts are formed and the distal cells of the ducts begin to develop the characteristics of secretory tissue.

When such midpregnancy mammary glands are cultured in vitro, most of the cells have little rough endoplasmic reticulum and Golgi apparatus, and no casein granules. When insulin or other promoters of DNA synthesis is added to these cultures, the cells become responsive to other hormones. Glucocorticoids can induce the formation of the rough endoplasmic reticulum, and prolactin can then stabilize the casein mRNA to ensure appropriate milk protein synthesis.

The development of the mammary gland, then, involves a complex interplay of several hormones at three different stages of life: embryonic, adolescent and pregnant. The mammary gland never develops in normal males and does not become a fully differentiated organ in females until the middle of pregnancy in the adult organism.

PUBERTY AS A VARIATION ON THE THEME OF METAMORPHOSIS

In mammals, one of the most striking displays of hormonal control of differentiation occurs during human puberty. If metamorphosis is understood as the processes of dramatic change whereby a juvenile reaches maturity, both sexually and biochemically, then human puberty may be considered a variation on the metamorphic theme. Recent research suggests that the processes of metamorphosis and puberty may be quite similar. A good summary of puberty is given by Styne and Grumbach (1978):

> Puberty is the period of transition between the juvenile state and adulthood; during this stage of development, secondary sex characteristics appear and mature, the adolescent growth spurt occurs, fertility is achieved, and profound psychologic effects are observed. These changes result directly or indirectly from maturation of the hypothalamic-pituitary gonadotropin unit, stimulation of the sex organs, and the secretion of sex steroids.

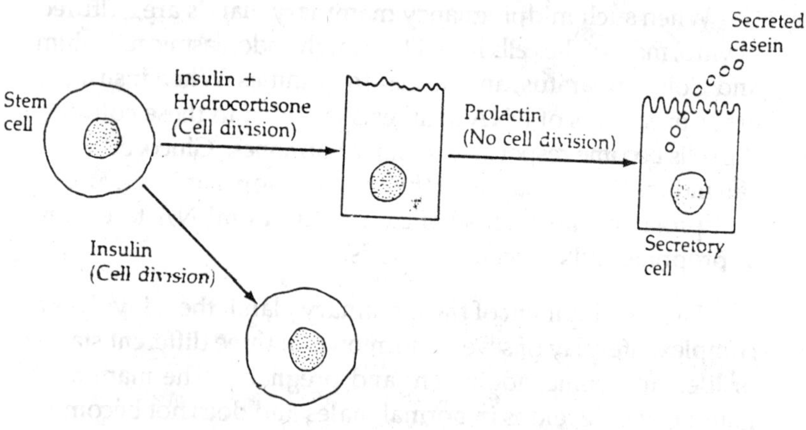

Fig. 7.19. Schematic diagram of the hormone-dependent development of the mammary gland in vitro.

Certainly, striking changes in body form occur during this time. At the beginning of puberty, boys and girls have the same proportion of muscle mass, skeletal mass, and body fat. By the end of puberty, men have 1.5 times the skeletal and muscle mass as women, whereas women develop twice as much body fat as men. Secondary sex characteristics also develop at this time, marking the change from the juvenile form to the sexually mature adult. In women, the development of breasts is controlled by a surge of estrogen secreted by the ovaries, and in men, the maturation of the penis and testes is controlled by the testosterone released from the testes. In both sexes, public and auxiliary (underarm) hair development is regulated by testosterone, which is secreted by the testes in males and by the adrenal gland in females. Men also undergo testosterone dependent enlargement of the larynx and its associated muscles and cartilage, leading to a deepening of the voice. In both sexes, puberty is the time when major sexual development takes place. In males, fertility is achieved when meiosis begins in the male germ cells and the spermatic ducts hollow out to form a channel for the sperm cells to pass from the testes into the urethra. In females, menarche, the first menstrual period, represents the new integration of the hormonal cycles releasing the developing egg from the ovary.

The hormonal basis of puberty is thought to be very similar to that of metamorphosis. The metamorphosis of amphibians and insects were both seen to be regulated by hormonal changes that were initiated by the neurohormones of the brain (TSH-RF) and prothoracicotropic hormone, respectively). The changes of human puberty—in both sexes— are initiated by *Luteinizing Hormone Releasing Factor* (LRF) from the hypothalamus of the brain. Like TSH-RF, this factor is released from the hypothalamic neurons to the *Medial Eminence* (infundibulum) of the pituitary gland. Similarly, the LRF is then transported by the blood vessels of the pituitary to the anterior lobes of that gland. Once in the anterior pituitary, the releasing factor causes the release of a tropic hormone. In human puberty, LRF releases

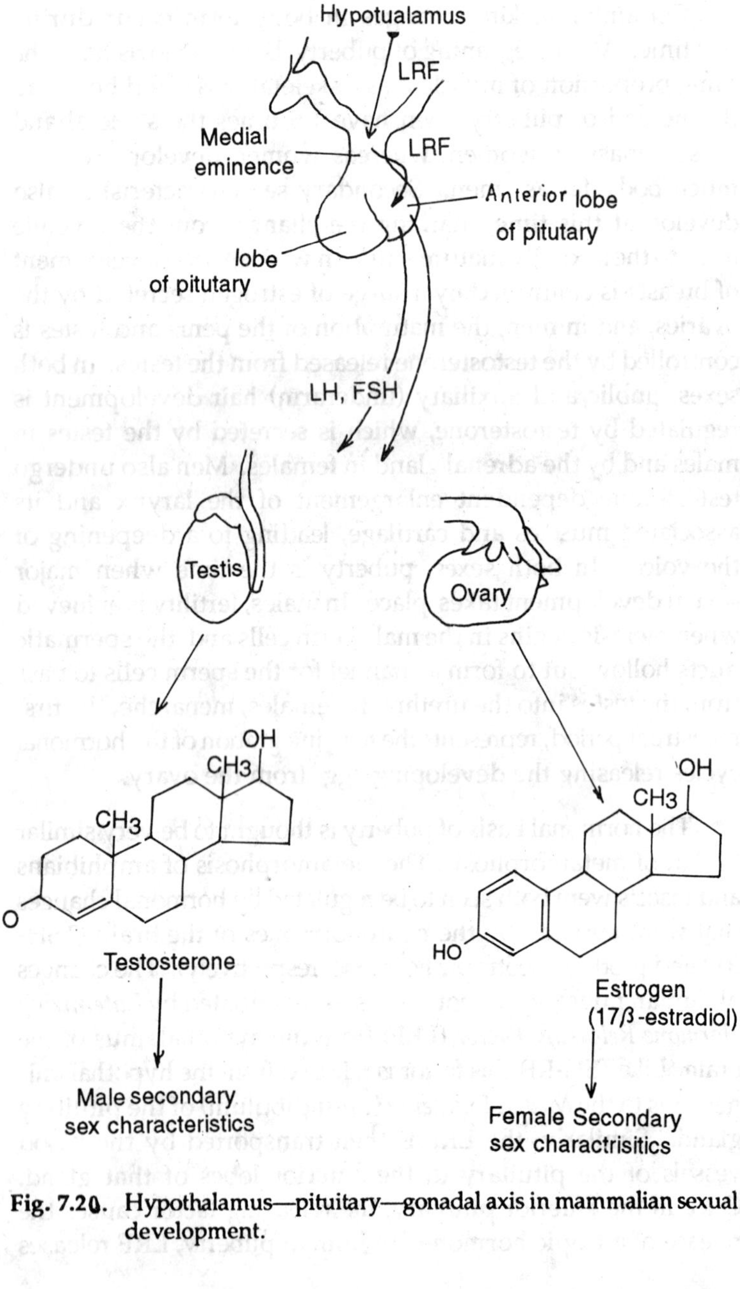

Fig. 7.20. Hypothalamus—pituitary—gonadal axis in mammalian sexual development.

luteinizing hormone (LH) and *Follicle Stimulating Hormone* (FSH). *Collectively*, these two hormones are called *gonadotropins* because they stimulate the development of testes in males and ovaries in females.

As a result of this stimulation, the gonads secrete the sex hormones: testosterone from the testes and estrogen from the ovaries. The various morphological and behavioural changes of puberty are due to the actions of these hormones on the various target tissues. As in metamorphosis, there appears to be a maturation-inhibiting hormone whose activity decreases to permit the reactivation of development. In humans, this hormone is probably melatonin, whose serum concentration decreases as that of LH rises.

Although the details concerning the initiation of puberty are not known, Grumbach and co-workers (1974) have proposed the following mechanism. Before puberty, the child secrets a small amount of LRF, so that the amounts of circulating LH and FSH are very small. Therefore, the gonad remains immature and secretes little estrogen or testosterone. Moreover, the hypothalamus is highly sensitive to these sex hormones and "turns off" LRF production when the circulating sex hormones get anywhere above their very low levels. The onset of puberty is thought to involve the maturation of the hypothalamus. At this time, the hypothalamus becomes less responsive to the negative feedback of testosterone and estrogen. Thus, if takes more or these hormones to turn off LRF secretion. Thus, the hypothalamus makes more LRF, ultimately causing the further differentiation of the gonads and the release or more testosterone or estrogen. This, in turn, causes the development of those secondary sexual characteristics.

It can readily be seen, the, that puberty is normally controlled reactivation of development leading to sexual maturity and changes in bodily form and physiology. As such, it has many parallels to the metamorphic changes seen elsewhere in the animal kingdom.

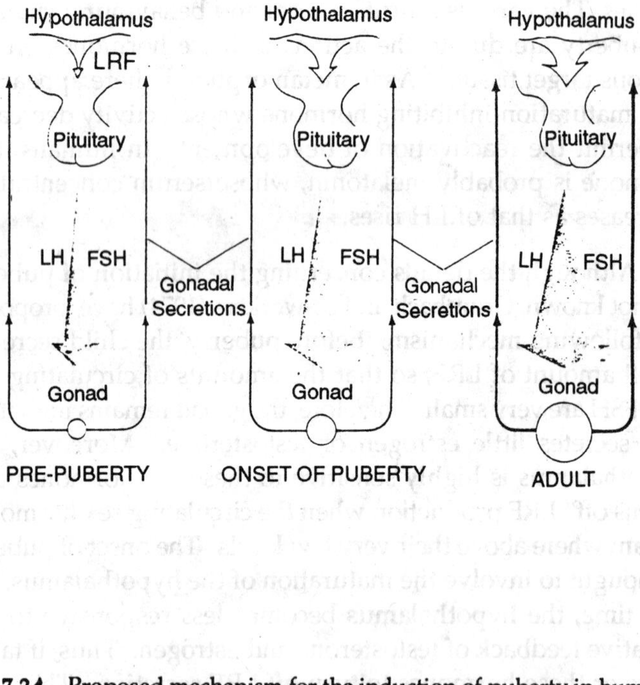

Fig. 7.21. Proposed mechanism for the induction of puberty in humans. Before puberty, the hypothalamus is sensitive to very small concentrations of sex steroids and stops LRF production, thereby halting further steroid synthesis. Sex steroids are kept at low levels. At the onset of puberty, the hypothalamus becomes progressively less sensitive to the sex steroids, thereby allowing more of them to be synthesized, until finally the adult level of sex steroids is achieved. Relative widths of arrows indicate levels of hormone production.

CONTROL OF CELL PROLIFERATION: CHALONES

So far, we have focused our attention on hormones, substances that circulate throughout the body and instruct other tissues to activate specific genes. There may be other diffusible substances, called *chalones,* that act to regulate growth. When the outer layer of epidermal cells is peeled off the skin of vertebrates, the cells responsible for forming the epidermis undergo extra rounds of cell division to restore the missing layer. Once this layer is made, the precursor cells revert back to their normal rate of cell division. The simplest explanation is that some substance is made by the epidermal cells to inhibit the division of their precursors. When the epidermal cells are removed, the precursors are no longer under this inhibition, so they multiply faster. Eventually, the new epidermal cells make their own chalones to inhibit further divisions.

Chalone may also be responsible in the growth of one of a pair of organs after the removal of the other. This is called Compensatory Hyperplasia. For instance, if one kidney is removed from an animal, the other kidney will grow larger, as if to make up for this loss. Similarly; if part of a liver is removed, the cells of the remaining lobes will proliferate until the original mass is restored. It is thought that these tissue-specific chalones are proteins and that, like hormones, they travel by means of the bloodstream.

It is difficult to demonstrate that a given protein is a natural inhibitor and not merely a cytotoxic chemical. The in vitro assays for chalones are complicated by the fact that they often work in combination with other substances. Epidermal chalones, for example, will inhibit the mitosis of epidermal precursor cells in vitro only if epinephrine is also present. The investigation as to whether or not chalones actually exist has taken on new importance as certain diseases, such as psoriasis, may be the result of overproliferation of stem cells. In this disease, the epithelial precursors divide at a very rapid rate (as if they could not respond to chalone or as if chalone were not being made), thereby causing the epidermis to thicken and the cells to be shed before they have a normal time to differentiate.

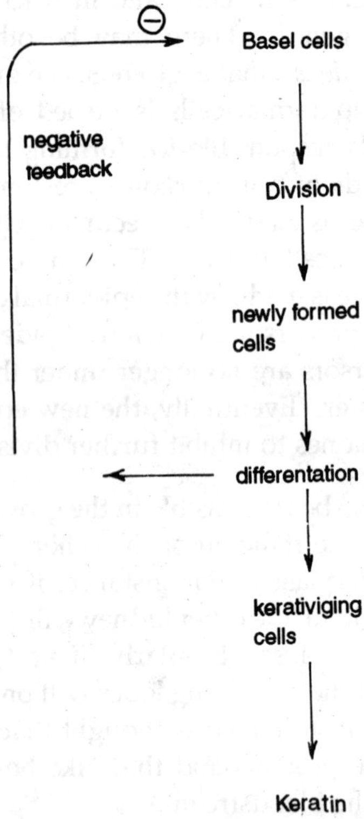

Fig. 7.22. Schematic diagram of the proposed regulation of epidermal cell division. One of the products of the differentiating epidermal cells would be a chalone that would signal the basal stem cells to cease dividing.

We have seen that diffusible regulation of cell-cell interactions is also important in regulating development. In the next chapter we shall study the proximate and hormonal tissue interactions occurring in the development of sexual phenotype.

8

Differentiation of Sexes

In vertebrates the sexes are usually separate with without sexual dimorphism. The identification of the sexes is made on the basis of secondary sexual characters. During development the organism takes place in the same way in both the sexes but as the development proceeds further the formation of ovary and testes occurs in different ways. Thus, the differentiation of the sexes is brought about by several genetical hormonal as well as ecological factors. The mechanism of sexual differentiation can be studied under the following heads :

Genic and Chromosomal Differentiation

Xenopus laevis is a distinctly gonochoristic species. Occasionally in sections of testes one sees aberrant ovocytes within seminal tubules, but so far only one true hermaphrodite has been found with a separate testicular segment in otherwise ovarian gonads. The sex ratio in laboratory-bred control stock populations of 3,620 was 1000 females to 104.4 males. The female being the heterozygous sex, either the W chromosome may be more often carried into the polar body than Z, or the chromosomal ZW condition may have a lower survival capacity than ZZ during early developmental stages.

The practicability of sex conversion makes it possible to keep laboratory breeding stocks of ZZ, ZW and WW genotypes of either sex.

The conformity of obtained and expected results in all breeding combinations removes every doubt about the reality of a chromosomal ZW-ZZ mechanism even though morphologic studies have failed to identify the sex chromosomes(4).

Considerations similar to those now found applicable to mammalian work distinctly suggest a feminizing potency for the *W* chromosome. The Z-chromosome and/or the autosomal complexes therefore must be expected to have some masculinizing value. However, the sex chromosomes of *Xenopus* exert barely more than a kind of direction-giving influence. They have not a prerogative or any specific importance for the development of primary and secondary sex characteristics. ZZ females have been bred with ZZ males continuously over five generations. Despite the constant absence of *W* chromosomes they have developed and retained all the physical and behavioural characteristics exhibited by normal "wild type" Z*W* females. So far also three consecutive generations of WW males have been bred with *WW* females. Likewise the absence of Z chromosomes throughout their lineage has not diminished the masculine instincts nor affected the development of male secondary sex characters and spermatogenesis. Therefore, the assumption is unavoidable that complete sets of sex factors are also located in autosomes and/or in non-sex-determining portions of the Z and *W* chromosomes.

The Postgenetic Mechanics of Sex Differentiation

In a discussion on how in the course of embryonic development sexual differentiation comes about, attention must be focused not only on the *activation of genes but also on the establishment of target organs*. Their relationship is not of a one-two nature, but that of reciprocal interactions. Sometimes the target organ is an essential factor in shaping the environment of the gene and may assume the role of an activator. This indeed is an essential tenet of the corticomedullary induction theory. However in this case, induction is not to be considered as a sole prime mover but as a summary expression for a multiplicity of morphologic and biochemical processes. The inductor tissues are already the products of several genetically-controlled evolvements.

Development of the Gonads

The major target organ, the *gonad*, encompasses in fact a multitude of reactive tissues with varied responses. Nieuwkoop and Faber have given a condensed outline of gonad morphogenesis in *Xenopus*. At standard stage 24a, about one to two dozen primordial germ cells still heavily laden with yolk platelets lay assembled at the dorsal root of the mesentery. Presently they separate into two strands that move laterally. By the end of the first week (St. 25a) a pair of *primitive gonad ridges* have formed which protrude into the body cavity. The germ cells are only thinly covered by squamous epithelial cells. During the second week, mesonephric blastema cells migrate through the hilum, accompanied by mesenchyme and pigment cells. As a result, by the end of the second week (St. 25c), *indifferent gonads* have become established. Each has acquired a cortex and 14 to 18 nodular medulla cords. Fine blood vessels enter through the hilum into the sparse mesençhyme of the primary gonad cavity.

Differentiation of *testes* and *ovaries* proceeds throughout the third week (St. 25d and 26). In the females the medullary cords remain slender and the germ cells, now much smaller after dissolution of the yolk platelets, remain in the cortex, while in the males they take places deeper in the thickening medullary cords. Morphologic differences become more distinct later. At St. 28 the ovarian medulla cords develop central cavities which soon widen to spacious ovarian sacs (St. 29). In males medullary cells start arranging in epitheloid fashion, forming primitive testicular tubules. By the mode of their development ovarial sacs and testicular tubules (rete and seminal tubules) prove to be homologous with mesonephric tubules. Meiosis starts in female larvae at St. 30 shortly before the break-through of the arms (St. 31). At the end of metamorphosis, the ovaries through accumulation and growth of ovocytes and inflation of the ovarian sacs become much larger than the testes. First spermatocytes are usually found only 2 or 3 months after metamorphosis.

Conversion of Genic Males by Estrogenic Hormones

Hormonal feminization is possible during the week or 10 days in which physiologic sex differentiation ordinarily takes place. Earlier and later estrogen administrations produce no conversions. If exposures are of short duration and cover only a part of the sensitive period, hermaphrodite gonads will develop.

It is not only of theoretical but also of practical interest that after treatment in the short sensitive period, physiologic sex conversion is complete and permanent. This makes it possible to use *conversion females* (ZZ) as breeders, exactly like true females. Actual records show persistent egg production in some cases over 10 years. If mated with ZZ males, conversion females produce 100 percent male offspring.

Only relatively few estrogens have been tested for capacity to convert male (ZZ) larvae to females. It can be assumed that besides estradiol (25 to 50 μg/1), equilin (1 to 100 μg/1) and equilenin (10 to 100 μg/1), some further phenolic C_{15} steroids would produce the same effect. Equilenin and equilin are slightly more effective than estradiol. Also diethyl stilbestrol is competent, though only within a narrowly restricted dosage (100 μg/1. This is noteworthy since in *Rana-sylvatica* stilbestrol is entirely ineffective, while equilenin again is superior to estradiol) (1).

The advantages of working throughout with homozygous breeders are obvious. Every offspring is of known (here of ZZ) constitution (case 2). Evaluation of experimental results no longer relies on statistic certification. Any deviation from the male sex would be an experimental effect. To the contrary, in the "all female" estrogenized offspring from a ZW parent (case 1), the desired homozygous (ZZ) females must be sorted out by the time-consuming breeding tests.

Conversion of Genic Females by Testicular Antagonist

As mentioned in the introduction, the neutral C_{19} steroids—

testosterone and androsterone — support only the development of secondary sex characteristic including sex behaviour but do not interfere with ovarian differentiation in genic females. Rather to the contrary, they may have a partially feminizing effect on testis development.

While the impact of estrogens is restricted to the short period of physiologic sex differentiation affecting indifferent gonad segments at the moment when they enter the differentiation stage, the testicular antagonist depresses cortical development at any larval and postmetamorphic stage. However, actual conversion and sex reversal may possible also be restricted to early larval stages. Further studies are to resolve this. Two procedures have been used to expose genically female gonads (*ZW* or *WW*) to testicular antagonist : ***parabiosis*** and *testis implantation.* Parabiosis may be performed at early embryonic state or after metamorphosis (unpublished). Implantation is the preferred method for production of converts. A small testis of a newly metamorphosed *ZZ* male is pushed into the body cavity of an anesthetized larva, of female genetic sex (*ZW* or *WW*). So far it has not been attempted to secure the graft in some desired position. At laparotomy, at the age of 5 month or later, it is most often found attached to the body wall or to a fat body. Considerable depression of cortical development follows in every instance of successful attachment, but compensatory growth and differentiation of medullary cords into testicular nodules seems to eventuate only after early grafting (St. 27 or younger larvae). In some cases, the gonads are still ovaries but less than 0.5 per cent the normal size. The cortex contains scattered ovogonia: ovocytes degenerate at early synaptic stages. A few auxocytes persist, indicating that some active ovarian development had occurred before establishment of the graft.

The production of *WW* males involves a series of breeding and grafting work, extending over several years. To sort out *ZZ*, *WW* and *ZW*, males from heterogeneous implantation groups, it is necessary to testbreed animals that acquire nuptial pads with conversion females. In a first step, *ZW* conversion males must be separated from ZZ males by breeding individuals from mixed groups with ZZ females.

Then *ZW* males are bred with *ZW* or *WW* females and their entire offspring is testis-implanted. Finally, there follows the task of recognizing *WW* conversion males. Table 8.2 deals with the simpler second alternative. Breeding of samples of 20 or even as few as 10 offspring is adequate to decide with high probability whether the paternal constitution is *ZW* or *WW*.

It is to be mentioned here that no indication of transfer germ cells from the ZZ implant to the host gonad has ever been found. If it would occur, exceptional males should appear among the offspring of *WW* fathers bred to ZZ mothers, which so far has never been observed.

As mentioned above, male larvae that were treated with estradiol only during part of the sensitive period (3rd week) develop hermaphrodite gonads. At the beginning of these experiments, the possibility had been envisaged that here the antagonism might not be very pronounced, since the testicular and ovarium segments are of the same animal. More recent extensive observations over longer periods show that antagonism gains expression in all hermaphrodites. In fact, in most animals having been treated over a short two-day period, the ovarian segments become entirely eliminated within the first year. Only the abnormal shape and small size of the testes will then testify to the earlier hermaphrodite phase of development. The case is one of the few in which some eggs have developed to the pigmented stage. A complete study of a large amount of material of this type is in progress. It should reveal interesting details about the working of the antagonist.

Having gained the possibility of procuring purely male and purely female cultures at almost any desired time, it was tempting to take up investigations on the biogenesis of steroid hormones in amphibians sex development.

TABLE 8.1

Testbreeding of Possible ZZ or ZW Males with ZZ Temales

Parento	*Sex of F_1*	*Conclusion*
1	63o	P o1 is a normal ZZ- type
2	98o	P o2 is a normal ZZ- type
3	57o	P o3 is a normal ZZ- type
4	29o + 30o	P o4 is a conversion ZW- type
5	44o + 40o	P o5 is a conversion ZW- type
6	28o + 37o	P o6 is a conversion ZW- type

TABLE 8.2

Testbreeding of Possible ZW or WW Males with ZZ Females

Parent	*Sex of F*	*Conclusions*
1	26o + 8o	P o1 is a conversion ZW- type
2	17o + 14o	P o2 is a conversion ZW- type
3	19o + 17o	P o3 is a conversion ZW- type
4	20o	P o4 is a conversion WW- type
5	24o	P o5 is a conversion WW- type
6	89o	P o6 is a conversion WW- type

The most characteristic indicators of androgens and of estrogens in circulation are the *nuptial pads* (on forelimbs) and the oviducts. The former make their appearance in males 6 to 10 months old, i.e., about the time when complete spermatogenesis has become established. Funnels and upper portions of the *oviducts* begin to develop in both sexes during climactic metamorphosis (St. 30 to 33). When the tails of the little frogs have become fully resorbed, the tips of the oviducts reach the upper level of the gonads. Within the following 2 months they continue down-growth along the lateral edge of the mesonephric bodies until they reach the cloaca, with ends fused, at the age of 4 months. In males they persist only if castration is performed not later than at 7 months. If this operation is delayed another 3 months, the ducts have regressed up to about midway and only the upper parts are permanently retained. Castration at the end of the first year meets with the condition of complete regression which is characteristic of normal male development at this age.

Less obvious secondary sex characteristics are the large *larynx* of the male and the protruding *cloacal labia* of the female. Both structures start sex-specific enlargement following maturity of the gonads at the age of 6 to 8 months and later.

There is only little evidence of spontaneous production of ovarian hormones before the sixth to the eighth month, which is the time when yolk platelets appear in the largest ovocytes. The oviducts of females, which after the third month have only slightly larger diameters than those of males, then lengthen again become convoluted and their walls begin to mature histologically. Males and females that were castrated at 7 months or earlier remain practically identical throughout adult life. They have nearly straight filamentous oviducts, barely protruding cloacal lips, and lack nuptial pads. The larynges are not enlarged. In summary, well recognizable effects of gonadal hormones appear in both sexes only at about the sixth or seventh month of age. They identified traces of estrogenic steroids in alchol extracts of gonads of 5-month-old female *Xenopi,* and appreica-larger amounts at 8 and 11 months.

Hormone responsiveness of the primordia of secondary sex characteristics is, however, established long before the normal start of sex-dimorphic development. Injected androgens cause differentiation of the *pad forming arms of arins and fingers* immediately after termination of metamorphosis in both sexes. Size and shape of the responding skin areas are not determined nor altered by hormonal factors. The larynges enlarge precociously. Administered androgens also activate the copulation instincts, the sexes clasping indiscriminately.

The formation of primordial funnels and upper ducts at stages 31 to 33 is not hormone determined. However, even the earliest rudiments respond to exogenous estrogen stimulating. By such treatment, they become hyperplastic; the cells of walls and budding tips become cytologically adultized and therefore are rendered incapable of proliferating further.

After castration at the juvenile phase (7 months or earlier), the secondary sex characteristics retain their responsiveness to hormonal stimulation. In both sexes the oviducts assume functional condition following estrogen administration. Likewise, nuptial pads develop with androgen injections. The reactions are independent of genic sex constitution and of sex type at the time of castration.

Direct chromatography and indirect chemical methods have provided evidence of adrenal steroidogenesis starting at early larval stages, immediately following histologic organization of cortical islands (St. 25d). The entire larval production is corticoid; sex steroids have not been identified. The latter negative finding agrees wit the slow histologic development of the larval gonads and with the apparent unimportance of androgens and estrogens in premetamorphic development.

Catabolism in Sex Steroids

Steroid tests with larval amphibians have been conducted by addition of hormones to the aquarium water. For quantitative work it is essential to obviate the possibility of bacterial interference. To that end the animals are pretreated in baths of 0.1 per cent sodium sulfadiazine, and penicillin is added to the incubation fluids.

It was realized early that liver in vitro catabolizes testosterone and estradiol at rates nearly equal to those of intact larvae (St. 26 to 29). Lateral liver was found to be quantitatively six to seven times as efficient as adult liver in converting administered hormones to less active compounds, reflecting the well-known fact that metabolic rates in general are particularly high at larval stages. In experiments neither qualitative nor quantitative differences were found between the activities of female and male livers of same age.

In a recent experiment, male ZZ-larvae were incubated with tritiated estradiol. Treatment was give on the 17th day, i.e., at the stage when the hormone is inducing sex conversion. It was possible to identify estrone very early in the livers but not in the gonad region. Obviously the change of steroid structure occurs only in the liver and is not an essential part of the mechanics of sex conversion by estradiol.

In collaboration with the laboratory of Professor Breuer, the steroid metabolism of Xenopus was reinvestigated, making use of radiocarbon-marked hormones and some of the newer, more sophisticated methods of chromatography. Table 8.3 lists the [4-^{14}C$_{19}$ and C$_{18}$ steroids that were offered to larvae or incubated with adult liver slices and the subsequently identified metabolites. Some of the latter must have gone undetected since large parts of the residues (sometimes more than half) were not, or were only incompletely analyzed.

The animal material includes several stages from 8-days larvae (St. 25a) to adult males and females. Even though analyses were not equally complete in all cases, they indicate clearly enough that the metabolites from testosterone and from 17-estradiol are the same in both sexes and over the full range of larval and adult stages. Adult liver was not incubated with estradiol. An experiment of Lisboa with estrone yielded only estradiol and no estriol (unpublished). The importance of this finding remains unclear. It is difficult to bring it into accord with the reported estradiol metabolism of larvae.

Not yet resolved is the question of a possible *quantitative difference* between male and female steroid catabolisms. On the basis of early results we had opined that more unchanged testosterone is recoverable from male than from female groups, and *vice versa* that females catabolize estradiol more sparingly than males. But of 16 now available pairs of assays only 10 sustain that rule. Even though it would not be inconceivable that the genic sex differences might have an influence on rate of liver function, independent of gonad physiology, direct evidence is missing or at least insufficient. In an experiment with adult livers, the recovery values of testosterone were 53 per cent of ZZ males, 47 percent of *ZW* females, and 46 per cent for ZZ conversion females (Lisboa, unpublished). If the differences are at all meaningful, then physiologic rather than genic sex should seem to be decisive. Repeated experiments with larvae very considerably by substrate concentrations and by differences of age. Hence, they do not yet give a critical decision about this question. At any rate, a sex difference could be only one of minor degree. This material also is not quite suited for demonstration of *qualitative sex differences* in relative proportions of metabolites. Recovered quantities always are small, and few suggestive observations were confirmed by repetition.

TABLE 8.3

Metabolites after Incubation with C_{19} and C_{18} [4-^{14}C] Steroids

Recovery of Substrates 0 to 70 per cent in 34 Analyses

Substrates	*Biologic materials*	*Metabolites*
Dehydroepi-androsterone	Larvae stage 25c	Δ5-androstenediol, var. (no Δ4-3-ketosteroids)
Testosterone	Larvae Stages 25b to 31 Adult liver slices	Androstenedione Androsterone Androstanediol Androstanedione Hydroxy compounds var. polar fractions
17 β-Estradiol	Larvae Stages 25a to 25d	Estrone Estriol 6-Hydroxyestrone 15-Hydroxyestrone
Estrone	Adult liver slices	Estradiol

A discussion of the enzyme systems that are already present in the livers of 1-week old larvae, i.e., before the organization of gonads, will be given elsewhere. Of particular interest is the absence of a hydroxysteroid dehydrogenase in 13-day larvae (St. 25c), as must be concluded from the absence of androstenedione and testosterone among metabolites after incubation with dehydroepiandrosterone.

Biogenesis of Sex Hormones

Adopting techniques that have given interesting results in the mammalian field, the possible steroidogenic capacities of *Xenopus* larvae were tested by offering probable precursor substances marked by radiocarbon. The incubation procedures were those already used in the study of liver catabolism. However, interest was now directed toward the developing gonads and the biogenesis of sex hormones. Beginning with St. 29, testes and all ovaries were dissected out — free of adhering somatic tissues — for separate testing. At earlier stages the entire larvae were incubated and extracted.

In a first set of experiments 17-αhydroxyprogesterone was incubated with 10-day (St. 25b) and 17-day (St. 25d) larvae for various lengths of time, from 3 to 24 hours. These early stages demanded special attention since it is here that initiation of sex differentiation actually occurs. The substrate was continuously metabolized, mainly to corticoids that were not further determined. Since previous work has shown that cortisol is released normally from adrenals at the end of St. 25 and the beginning of St. 26, it is interesting that it appears now (St. 25d) as a metabolite of administered hydroxyprogesterone.

In the present series C-steroidogenesis starts at St. 29. A small fraction of radioactivity was recovered in the form of androstenedione. This is sufficient evidence of the presence, now of side-chain splitting desmolase (17-αhydroxyprogesterone C_{17}-C_{20}-lyase). The yield doubles in the later stages of testicular development (St. 31, 33, frogs of 3 months), but remains low in the ovaries. Neither testosterone nor estradiol were obtained at these metamorphic stages. Examining again the histologic state of ovarian and testicular development, one misses cellular elements of distinctly endocrine character. However, considering that quantitatively, steroid production is extremely low, it seems not improbable that the walls of the ovarian sacs and the sparse interstitium of the testicular medulla may be competent to exert such endocrine function. Doubts have been expressed by other authors. Pesonen and Rapola (24) applying the Wattenberg steroid-β3-ol-dehydrogenase test on adult females and males of *Xenopus* observed strongly positive reactions in the adrenals but only trace deposits of formazan granules in testes and none whatsoever in ovaries. They consider it possible that sex-steroid synthesis might be nonexistent in amphibian gonads. Such a hypothesis is untenable now in view of the above-mentioned effects of castration and gonad transplantation of *Xenopus,* and the evidence reported above of desmolase activity in gonads, resulting in the production of androstenedione, testosterone, and estradiol. Independently, Rao and Dale (unpublished) obtained evidence of sex-hormone biogenesis in testicular and ovarian tissues from adult animals.

Reference should also be made to the work by Dale and Dorfman on androstenedione and testosterone synthesis by bullfrog testes from progesterone substrate in vitro. Ozon in a very complete study of steroid metabolism in a newt *(Pleurodeles waltlii),* finds similar chains of conversions. In the testis; progesterone —> 17β-hydroxyprogesterone —> androstenedione —> testosterone. In the ovary this biosynthetic pathway is extended by aromatization of testosterone with resultant 17β-estradiol formation. It seems safe enough to assume that steroid biogenesis in the adult testes and ovaries of *Xenopus* follows essentially identical patterns.

DISCUSSION

Sex Chromosomes

Since males as well as females can develop and become functional with any of the three diploid sex chromosome combinations, one must conclude that Z and *W* carry similar arrays of male *(M)* and female *(F)* sex-determining genes. The latter kind obviously prevail in the *W* chromosomes and under non-experimental conditions, direct ovarian type differentiation. This is the case in *ZW* diploids and also in *WW* diploids and various polyploids. Haploids described by Gallien were all of Z type. Accordingly, they developed testes (sometimes partially feminized by parabiotic union with females). It is of particular interest that the entire functional importance of the seemingly very powerful genes is restricted to the period of less than 10 days during which initial sex differentiation occurs. The histologic appearance of the gonads at this time changes little. The major developments in secondary sex characteristics follow much later, i.e., after having become independent of either male or female genic constitution.

Steroid Organs

Even though steroid hormones may occasionally be produced in trace amounts by cells outside the gonad-adrenal complex, they are essentially secretions of tissue complexes which derive from the primitive intermediate mesoderm. Their morphologic separation is accompanied by functional differentiation. With respect to steroidogenesis, the *adrenal* is first to become productive, synthesizing cortisol as early as St. 26. The *gonadal medulla* gives indications of androstenedione manufacture in minimal amounts only at St. 29. No evidence of estrogen synthesis or activity was found before the end of metamorphosis. Nevertheless, the two gonad components — cortex and medulla — assume important other functions at early embryonic stages. The *primitive cortex,* first a narrow strip inserted in the celomic epithelium, guides the migration of the primordial germ cells from the crest of the endoderm to the mesial edges of the mesonephric bodies (St. 20 to 24). Apparently the cortical cells are releasing a chemical substance into the surrounding tissues which attracts germ cells until they reach the gonad site and become invested in the cortex itself. Such attraction substances, for which the designation *telopherons* is proposed, have been long recognized, particularly in studies on human (29) and avian embryos. Nothing is known about their chemical nature.

Activating the Sex Genes

Every cell of a normal tadpole carries identical genes, including the sex genes. However, the female-determiners (F) become activated only in the cortical-cell combine and the male-determiners *(M)* only in the medullary system of differentiating gonads. This activation occurs in the restricted time while larvae pass through St. 25c to 26. At temperatures of 2^0 to 22° C, this occurs during the third week after fertilization. Like other events in embryogenesis, activation arises "spontaneously" at a hereditarily fixed stage. Beyond doubt it is causally linked with specific physiochemical events, possibly with a definite phase in overall RNA synthesis. On the basis of the work by Alfrey and Mirsky we can assume that intranuclear enzymes start the process of acetylation of the specific histones which so far had held the *F* or the *M* genes in the dormant condition.

The possibilities of parallel and of consecutive cortical and medullary development—both leading to hermaphrodism—are realized in some other amphibian species. However, in *Xenopus* immediately following activation, the epistatic sex inhibits the development of the hypostatic gene complex. Thus, in a genic female *(ZW)* the activated *F* genes not only produce RNA sequences, enzymes, and inductors of ovarian differentiation, but also repressors (possibly ribonucleases) which cross over from the cortex to the medulla and prevent the activation and/or RNA production of the *M* genes. The same or similar antagonists prevent copious growth and male-type differentiation of the medullary cords. Instead, transformation into ovarian sac is induced.

In genic males *(ZZ)* activation of the *M* genes follows a similar course. However, antagnosim, besides inhibiting activation of the *F* genes, also leads to disaggregation of the cortex. Germ cells and somatic elements become detached from the surface epithelium and are transferred to the medulla. The complicated relationships were more fully discussed and diagrammatically illustrated in an earlier publication.

Mechanisms of Sex Reversal

Recent work on uterine growth stimulation in the rat by estradiol provides some information that also seems applicable to the case of sex conversion by the same hormone in ZZ larvae of *Xenopus*. Literature and pertinent facts are reviewed and their bearing on developmental biology discussed by Segal. Of prime importance is the discovery that cells of target organs withdraw the hormone from circulation and accumulate it at a particularly high rate. In the case of the gonad of *Xenopus*, estrogen receptors most likely are cortical cells in which activation of the *d* genes is set to start. Experiments with tritiated estradiol are intended to test this assumption. If the animal is of the ZZ constitution, ordinarily F-gene activation would be inhibited by antagonists from the medulla. However, as shown by Libby estradiol increases the rate of histone acetylation in nuclei of target cells. Apparently, in the case of differentiating *Xenopus* larvae such increase is sufficient to reverse the constitutional M>F power ratio and bring about conversion of sex development, i.e., feminization of genic males. The precision with which this estradiol process works is impressive. Withdrawal of the hormone before completion of the differentiation period is immediately followed by return to purely genic control, i.e., testicular development.

The female to male conversion by testis implants is less understood. Testes suppress ovarian development at any developmental stage, even in the adult. Whether the emitted *antagonist* is always the same agent as the one acting during normal male differentiation at larval stages is not clear. Certainly it is not a steroid hormone. Observations in parabiosis suggest rather a proteinic nature. Testicular development is permitted, but probably not stimulated, by the disappearance of the cortical inhibitor. Usually only one or a few medullary cords start compensatory testicular development. Their growth becomes secondarily stimulated by gonadotropins after removal of the implant testis.

9

Placentation in Mammals

It was only in the 17th century that some scientific facts were established. *Harvey* (1657) was the first to define a *placenta.* According to him, it is an organ elaborated from maternal blood. *Mathew* afterwards found out that the placenta functions as a *fetal-lung. John William Hunter* found out that the placenta is an organ, in which the embryonal blood circulates and becomes separated by lacunae or spaces, in which maternal blood is found. *Jenkinson* described in his book *"Vertebrate Embryology"* that the placenta is an organ, in which mucosa of uterina-wall becomes connected with the chorion. His further investigations led him to the conclusion that the placenta serves as a physiological exchange of the material between the embryo and the mother. *Ohogrosser,* with whom most of the students of vertebrate embryology agree, found out that, in the placenta, there is apposition of foetal tissue and embryonal tissue, but in no case there is any mixing of the blood of the embryo with that of the mother. *Mossman* (1937) defined placenta as an apposition or fusion of foetal membranes to the *uterine mucosa* for physiological interchange between the vascular system of mother and embryo.

Thus, he minute mammalian egg passes down the oviduct and lies on the inner surface of the maternal uterus. The small quantity of the yolk is sufficient only for first three days after copulation. But, due to the activity in the growth of the egg, embryo is formed, which is wrapped in an *amnion,* filled with the fluid. The *chorion* and *allantois* grow up, and make contact with the uterus, whereby embryonic blood-vessels on the allantois come close to the maternal blood-vessels in the uterine wall. Processes or *villi* of the chorion are embedded in the depressions of the uterine-surface and, thus, a cord-like structure is formed joining *embryonic-maternal* structure. However, there is no direct connection between maternal and embryonic circulations.

FORMATION OF VARIOUS TYPES OF PLACENTAE

A mammalian placenta is a composite structure based on double origin. it is produced by the development and apposition of extra-embryonic membranes with the endometrium of uterus (*decidua basalis*) for the purpose of physiological exchange. It therefore, follows that the placenta from the point of development (origin), consists of following components.

The *foetal part* of the placenta is furnished by the extra-embryonic membranes i.e., amnion, chorion, yolk-sac and allantois. The first of these, the amnion may be ruled out, as it makes no direct contribution to the placenta. Out of the remaining three, the chorion, being the outermost membrane makes immediate contact with the endometrium of uterus. But the chorion plays its role by way of a vascular supply, which it receives from the allantois. The *maternal parts* of the placenta is furnished by a single component, the uterine endometrium which includes both the compact and the spongy layers. In between these two parallel plates (the chorionic membrane and the endometrium) a huge blood sinus, the *intervillous space*, contains an enormous number of chorionic villi.

In mammalian placenta, there are two possible sources of chorionic vascularization—the *vitelline circulation* provided by the yolk sac and and the *allantoic circulation* provided by the allantois. On the basis of above vascularizaiton there exists two main types of mammalian placentae. (i) the *chorio-vitelline placenta* and (ii) the *chorio-allantoic placenta*.

(i) Chorio-vitelline placenta

The egg-laying monotreme mammals naturally does not develop a placenta. Some marsupials (*Didelphys, Macropus*), after a brief gestation period, give birth to immature young ones. Their allantois remains relatively small and never makes contact with the chorion; whereas the yolk-sac becomes very large and fuse broadly with the chorion, to complete a transitory *yolk-sac placenta* or *chorio-vitelline placenta*. In this form, the chorion gains its blood supply from the network of vitelline blood vessels of yolk-sac. Further, the chorion never advances beyond a smooth membrane in close apposition with the vascular uterine lining, the endometrium.

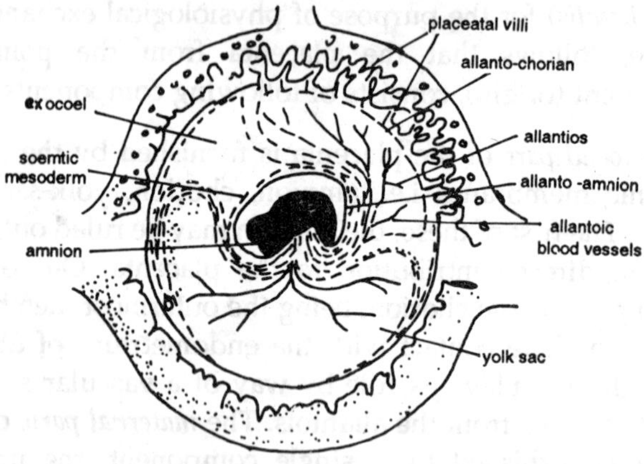

Fig. 9.1. Embryo of eutherian mammal showing allantochorionic villi.

(ii) Chorio-allantoic placenta

In some other marsupials (*Parameles, Dasyurus*) and all eutherian mammals, the yolk-sac remains rudimentary but the allantois becomes well-developed and vascularized; and it fuses with the chorion to produce a composite membrane. This membrane serves as component part of the so called *chorio-allantoic placenta*. The chorion does not remain smooth but grows out into root like vascular processes, the *chorionic villi* which engage the uterine mucosa in a more or less intimate functional relaiton persisting through out pregnancy.

When the allantois is insignificant of lacking, the organ is often designed as a *chorionic placenta.*

The development of chorio-allantoic placenta is as follows:

When the mammalian embryo enters the uterus, it remains bathed in the uterine fluid containing organic substances produced by the tubular glands of the uterine wall. The early embryo may absorb some of these substances through its epithelial cov ering as long as a closer connection with the uterine wall has not been established. For its further development, however, the embryo completely depends on substances supplied to it from the maternal tissue. However, the allantochorion placenta involves the following tissue : (i) an outer layer of chorionic trophoblast, (ii) a thick middle layer of fused mesoderm of allantois and chorion consisting numerous lacunae, and (iii) an inner layer of allantoic endoderm. These tissues constitute the *foetal part* of the placenta. The uterine wall, into which the villi of the allantochorion come to lie, consists of (i) the uterine mucosa, and (ii) uterine submucosa the two constitute the *maternal part* of the placenta. Nevertheless, it is found that the closeness of this connection between foetal and maternal tissues differs greatly withing eutherian mammals.

The *chorionic villi* are most important part of the allanto-chorionic placenta because they furnish the means by which all kinds of interchanges take place between the mother and foetus. The early villi are compact, bush-like tufts with a few small branches. Their main stems arise from the chorionic membrane and almost all the ends of anchoring villi attach the mingle with the eroded surface of compact endometrium (deciduabasalis). Side branching begins inthe early period of prgenancy and produces many free villi as well. During the middle and later months of pregnancy, the villi become much more tree-like, with prominent trunks and long and slender branches bearing innumerable small twigs. Many terminal branches fuse with the decidua-basalis of the endometrium at some distance from their ends, and then recurve into the intervillous space in the J-shaped manner.

All parts of the villous tree have the same structural plan. At the centre is a connective tissue core, which embeds blood vessels. These appears as arterioles and venules, but taper to prominent a capillaries which continue to have the villous tips where they complete continuous system of closed vessels. The connective-tissue core is covered with a double layer of trophoblast. Inside, next to the connective tissue, is a single layered *cytotrophoblast* or *cellular trophoblast* with its separate cuboidal cells; it is also known as the *layer of Langhans*. This cellular layer gives rise to syncytium, the *syntrophoblast*, which clothes the villi externally. During later pregnancy, the cytotrophoblast of the villi becomes progressively more interrupted and hence scarcer. This is because, it is almost completely used up in the formaiton of syncytium which forms the sole covering of the villi.

On the basis of intimacy of foetal and maternal tissues the placentation can be classified into following types :

1. Non-deciduate placenta

In most mammals, the plantaiton is superficial, i.e., the blastocyst embryo lies in the cavity of th uterus in contact with the uterine wall. The contact may be made intimate by the surface of the embryo with the formation of chorionic villi which are simple and minute, penetrate into the depressions in the wall of the uterus. These villi are initially formed by the trophoblast, later on the connective tissue and blood vessels invade the outgrowth. The chorionic villi and the endometrium lie apposing one another and do not fuse, such as placenta is also called *semiplacenta*. In this case, the nutrient and gases which reach the embryo diffuse across all the six tissues of placenta.

At the time of birth, when parturition (separation of the foetus and its membranes from the mother's body) occurs, the chorionic villi are simply drawn out from the depressions in the wall of the uterus, and thus maternal and foeta tissues are separated without any damage to the uterine wall, hence no bleeding occurs. This type of placenta, found in pigs, cattle and some other animals, is known as a *nondeciduous placenta.*

2. Deciduous placenta or placenta vera

In other mammals, however, the degree of intimacy between maternal and faetal tissues becomes further increase. The wall of the uterus becomes eroded to various degrees through the action of the trophoblast and the embryonic tissues penetrate into the uterine wall establishing a more intimate contact and faciliating the passage of substances fromthe mother to the faetus and from the faetus to the mother. Here because the chorionic villi fuse with the eroded uterine mucosa, such placenta is called *placenta vera* (true placenta). At the end of pregnacy the uterine wall is no longer intace and when the faetus with its membranes including the chorion is removed, more or less extensive haemorhage from the uterine wall ensues (i.e., at birgh, when such placenta is discharged, the uterine lining also tears away with some bleeding). Such a type of placenta found in higher eutherian mammals and is called *deciduate* or *deciduous placenta.*

The maternal tissues which are expelled at birth in the case of deciduate placenta are called *deciduae.* There are three regions : (i) The part which lies between the chorionic vesicle and the muscles of the uterus wall is the *decidua basalis.* (ii) The part which surrounds the chorionic sac and separates it fromt he cavity of the uterus is the *decidua capasularis.* (iii) The part which forms the inner lining of the rest of the uterus is *decidua parietalis.*

The haemorrhage at parturition is normally stopped by the same mechanism as serves for the expulsion of the netborh—the contration of the muscular wall of the uterus constricts the blood vessels and thus, slows down the flow of blood, untoil clothing of the blood stops the haemorrhage altogether.

3. Contradeciduate placenta

In *Perameles* and *Talpa* (mole), somewhat modified type of deciduate placenta occurs, which is called *contra-deciduate placenta*. In such case, not only thee is a loss of maternal tissue but also of the faetal portion of the placenta, both of which absorbed in situ by maternal leucocytes.

CLASSIFICAITON OF MAMMALIAN PLACENTA

According to different criteria, the non-decidous and decidous type chorio-allantois placentae of different eutherian mammals have been classified as follows :

(i) Central or superficial implantation

When implantation occurs in such a way that the embryo remains within the lumen of the uterus while the extra embryonic membranes make a superficial attachment to the uterine mucosa, it is known as *central or superficial implantation*. This type of implantation is common among lower vertebrates, among marsupial mammals it is present in Perameles and Dasyurus and among the eutherian mammals in the pig, cow, rabbit, sheep, dog, cat, etc.

(ii) Eccentric implantation

In mouse and rat the early blastocyst comes to lie between the uterine epithelial folds in an antimesometrial position. These folds soon enclose the blastocyst among completely. This type of implantaiton is called the *eccentric implantation*.

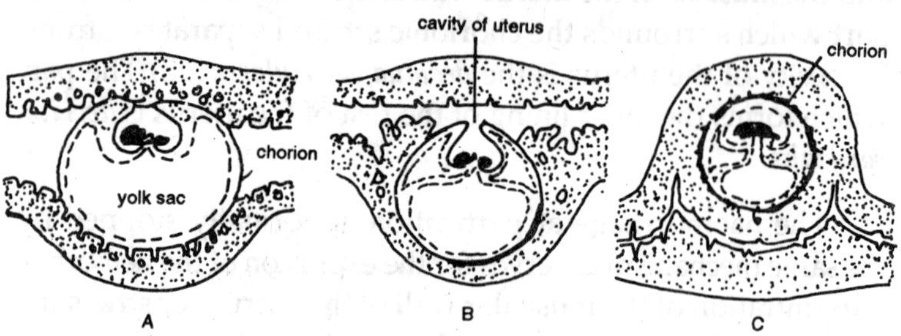

Fig. 9.2. Implantation (a) Central, (b) Eccentric, and (c) Industrial.

(iii) Interstitial implantation

In guinea pig, man, chimpanzee, the embryo burrows

into the uterine mucosa below the epithelium and in this way becomes surrounded completely by the endometrial tissues of the uterus. This type of implantaiton is called the *interstitial implantation.*

TYPES OF PLACENTAE ACCORDING TO THE ARRANGEMENT OF VILLI ON THE BLASTOCYST

In different mammals the placenta is further classified into six types according to the distribution and arrangement of villi on the surface of chorion.

1. Diffused placenta

The villi are uniformly distributed all over the surface of the chorion except the extreme ends. It is found in the ungulates, e.g., *pig, mare* etc.

2. Cotyledonary placenta

In *cattle, sheep* and *goat,* the villi are found in groups or patches called cotyledons while the rest of the chorion surface is smooth. The placenta of this type is known as the *cotyledonary placenta.*

3. Intermediate placenta

In this type the isolated villi are scattered between the "cotyledons". Thus the blastocyst is provided both with the villi and the cotyledons, e.g., *camel* and *giraffe.*

4. Zonary Placenta

In carnivores, e.g., *dog, cat, lion, tiger, fox, mongoose* etc., the villi are developed in the form of a belt around the middle of their blastocyst which is generally elliptical in shape.

5. Discoidal placenta

The villi are restricted to a small disc-shaped area of the blastocyst. The functional placenta therefore, has the shape of a disc and is known as the *discoida placenta.* Examples are *insectivores, bat, bear, rodents* (rat, guinea pig, porcupine) and *rabbit*. In the *monkeys*, the placenta consists of two discs and is, therefore, called a *bidiscoidal placenta*.

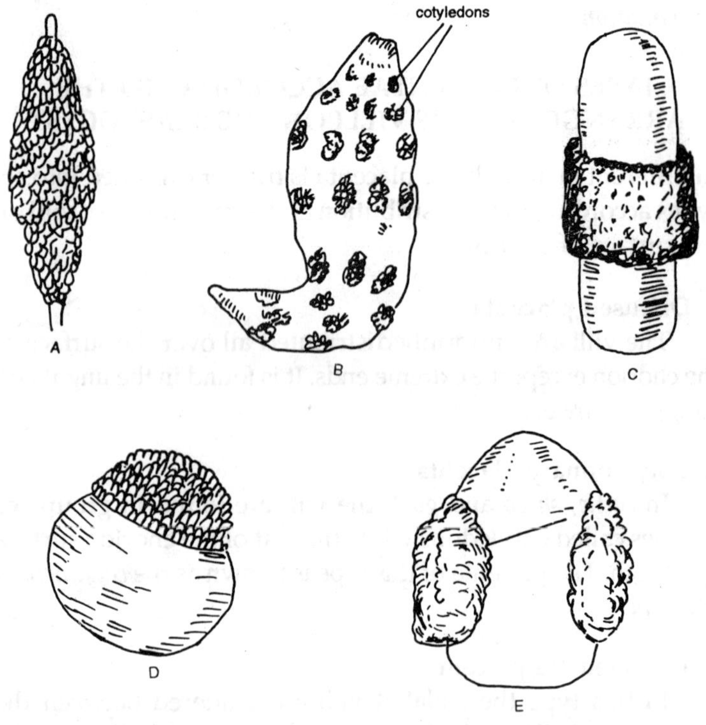

Fig. 9.3. Types of mammalian placentae on the basis of distribution of chorionic villi. A—Diffused placenta (pig); B—Cotyledonary placenta (calf); C—Zonary placenta (dog); D—Monodiscoidal placenta (man); E—Bidiscoidal placenta (monkey).

TYPES OF PLACENTAE ACCORDING TO THE HISTOLOGICAL RELATIONS BETWEEN THE MATERIAL AND FOETAL TISSUES

The *foetal placenta* has allantoic blood vessels (in case of chorio-allantoic placentae) or the vitelline blood vessels (in the case of chorio-vitelline placentae). These blood vessels run to and from the foetus whereas maternal vessels run to and from the *maternal placenta*, furnished by the uterine endometrium. By virtue of this arrangement, the two circulations are brought very close together. However, there is *no fusion of the foetal and maternal blood systems.* Foetal blood dos not circulate in the mother; nor maternal blood circualtes in the foetus. All food materials, oxygen, carbon dioxide and urea pass from one circulation to the other through tissue barriers lying between the blood of the mother and of the embryo.

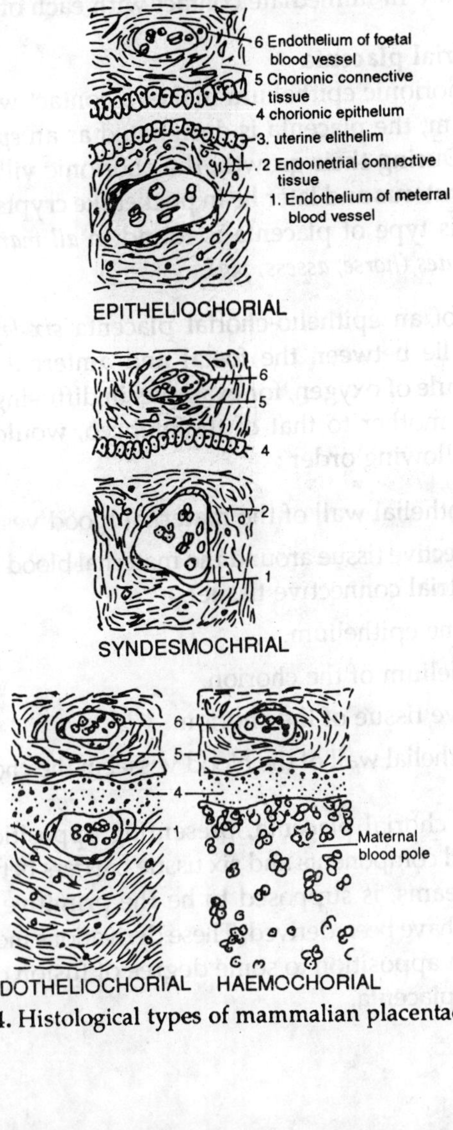

Fig. 9.4. Histological types of mammalian placentae.

The thickness of the partition between the foetal and maternal blood may be decreased by the removal of some of the layers of tissue or barriers lying between. Depending on which layers have disappeared, several types of placentae may be distinguished. The names given to the various histological types indicate the two tissue—one foetal, the other maternal which are in immediate contact with each other.

1. Epithelio-chorial placenta

When the chorionic epithelium makes a contact with the uterine epithelium, the placenta is designated as an *epithelio-chorial placenta*. During their growth, the chorionic villi push in the wall of the uterus and later lie in pocketlike crypts of the uterine wall. This type of placenta is found in *all marsupials* and in the *ungulates (horse, assess, pig)*.

In the case of an epithelio-chorial placenta *six layers of tissue or barriers* lie between the foetal and amternal blood streams. A molecule of oxygen, for example, in diffusing from the blood of the mother to that of the embryo, would pass through in the following order :

1. The endothelial wall of the maternal blood vessel.
2. The connective tissue around the maternal blood vessel (endometrial connective tissue).
3. The uterine epithelium.
4. The epithelium of the chorion.
5. Connective tissue of the chorion.
6. The endothelial wall of the blood vessel in the chorion.

The epithelio-chorial placenta, presenting apposition of foetal and maternal components and six tissue barriers between the two blood streams, is supposed to be the primitive type from which others have been derived. These derivations indicate a changeover from apposition to some degree of fusion of the two halves of the placenta.

2. Syndesmo-chorial placenta

In this type of placenta, the foetal and maternal components are fused so intimately as to result in a destruction of the uterine epithelium. As a result, chorion comes into contact with the endometrial connective tissues (desmos). Only *five barriers*, therefore, intervene between the two blood streams. Syndesmochorial placenta is found in the *ruminant* (cud-chewing), *ungulates, e.g., camel, giraffe, goat, sheep, buffalo, cow and reindeer.*

3. Endothelio-chorial placenta

If the destruction involves the uterine epithelium and the underlying connective tissue which is reduced, the chorionic epithelium comes into direct contact with the endothelial walls of amternal blood vessels. A placenta is then formed which is called an *endothelio-chorial placenta.* In this case the number of *barriers* between foetal and maternal blood streams is reduced to *four*. A placenta of this type is characterstic of the carnivores, *e.g., dog,* all "cats" including *house cat, lion,tiger,* etc., *fox, bear, mongoose, walrus.*

4. Haemo-chorial placenta

A reduction of the *barriers* to *three* is found in the primates (e.g., lemur, apes, women) many insectivores, bats and rodents. This type of placenta is called the *haemo-chorial placenta* in which the endothelium of the maternal blood vessels also disappears and the chorionic villi are bathed directly in maternal blood. actually, the chorionic villi are surrounded by sinuses into which maternal blood enters through the arteris of the uterus and from which the blood flows into the uterine veins.

5. Haemo-endothelial placenta

In this type of placenta all the three maternal tissues and two foetal tissues, i.e., chorionic epithelium and chorionic connective tissue are completely eroded so that the foetal blood vessels dip into blood lacunae of the uterus. The number of barriers between the maternal and foetal blood streams, therefore, is reduced to just one. This type of placenta is found in higher rodents, e.g., rat, guinea pig and in the rabbit.

PHYSIOLOGY OF PLACENTA

In all histological types of placentae, there exists a *placental barrier* which may include to two six kinds of tissues. Due to the placental barrier, the blood of faetus and mother is never blended. Physiologically, the placental barriers is like the *ultrafilter* or semi-permeable membrane. Accordingly, there is a relationship between placental transmission and molecular size, smaller molecules passing more readily through placental barrier than the larger ones. This means that water, oxygon, carbon dioxide, soluble inorganic materials as chloride and phosphates of sodium, potassium and magnessium and soluble organic substances as monosaccharides, hormones, vitamins and urea pass by *diffusion*. Macromolecules of polysac-charides, lipids and proteins may be absorbed by trophoblast cells by *pinacocytosis*.

The *antibodies*, which have developed in the blood of mother who has acquired immunity to certain diseases, such as *diphtheria, scarlet fever, small pox* and *measles*, are passed to the foetus, which thus, becomes passively immunized and unsusceptible to these illnesses in the first period after birth. Certain pathogenic organisms and virsus can penetrate through the placental barrier and can infect the faetus, if the mother is infected by those pathogens. This is known to happen with *syphilis, small pox, chicken pox, measles,* and *rubella or German measles*.

Many drugs used medically many penetrate the placental barrier and sometimes cause most adverse effects on the embryo. This it is believed that, the drug *thalidomide*, which was used as a sedative, when taken by women in early pregnancy (25 to 44 days), caused very extensive deficiencies in the development of libs, the alimentary canal (non-perforation of the anus) and the heart.

Besides, acting as to and fro passage for maternal and faetal bio-chemical molecules, the placenta stores materials such as fat, glycogen and iron; it participates in the metabolism of proteins, and is also acts like a endocrine gland. Like a endocrine gland it secretes many harmones such as *estradiol, progesterone, chorionic gonadotropin and placental lactogen* in human female most plcental mammals. In some animals, such as rabbit, the placenta is a significant source of *relaxin,* which is a protein hormone for the *relaxation* of the pelvic ligaments to facilitate the birth of the young.

EXTRA EMBRYONIC MEMBRANES OR FOETAL MEMBRANES

Taking foetal membranes into considerations, the vetebrates can be divided into two main groups; Anamniota and Amniota. *Anamniota* includes amphibians and fishes while reptiles, birds 'and mammals are included in *amniota*. As the name suggests, anamniota means without amnion while amniota means animals possessing amnion. Among anamniote vertebrates the eggs are laid under aquatic conditions while that of amniotes are laid under terrestrial conditions or retained inside the body of mother for a definite period. The terrestrial eggs, for protection, are wrapped with specific envelopes. These protective membranes two in number are the *amnion and chorion*). The latter is also called sometimes the *serosa or false amnion.* (The amnion, which is inner membrane), formed by somatopleure, is continuous, with the somatopleure of the embryo. The amnion (encloses with in a fluid—*the amniotic fluid*. The embryo being attached to the yolk sac by umbilical stalk, floats in the so-said fluid. The chorion which is outer membrane, is also formed by somatopleure but its cavity continuous with the coelom is called as *exocoel*). As alread mentioned, the two membranes-amnion and chorion, are developed from somatopleure but are united on the dorsal side of the embryo. Besides, these two are associated with another extra-embryonic structure the *allantios,* during development.

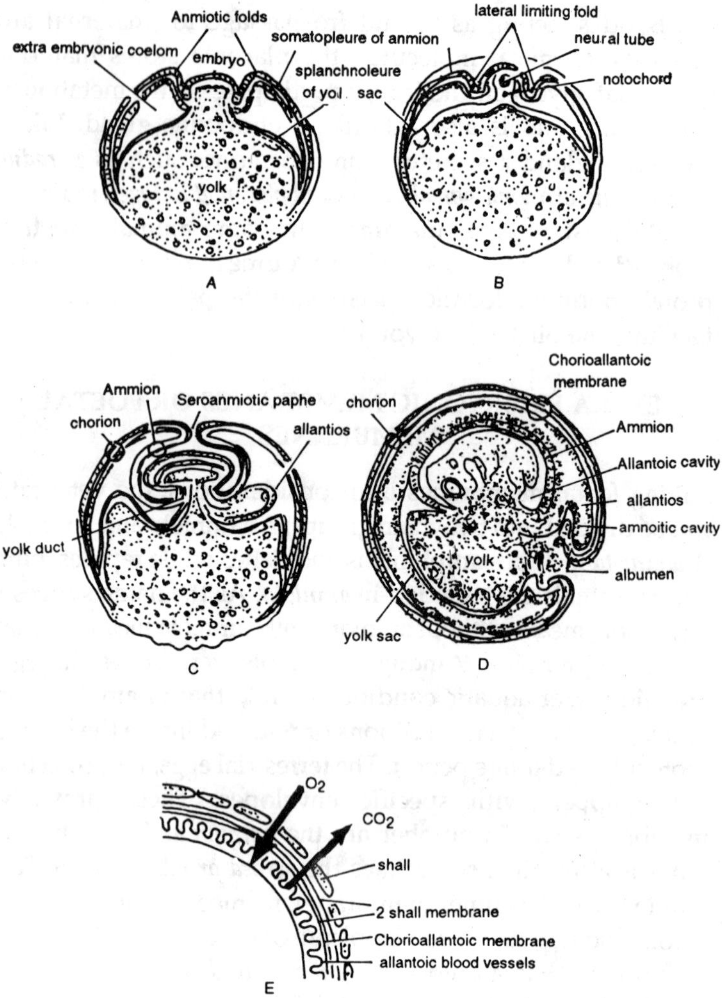

Fig. 9.5. Drawings illustrating the formation of extraembryonic membranes in the chick. A, Sagittal view showing developing yolk sac and amnion. B, Transverse section through A. C., Sagittal section showing beginning of allantois. D, Sagittal view showing chorion, allantois, amnion, and yolk sac. E, Diagrammatic section through chick membrranes showing layers through which the exchange of gases occur.

All these extra-embryonic membranous sacs are composite structures, and involve two germ layers. The amnion and chorion are composed of extra-embryonic ectoderm and somatic layer of mesoderm (*somatoplure*) while th eyolk sac and allantois are composed of extra-embryonic endoderm and splanchic layer of mesoderm (*splanchnopleure*).

The *placenta* is a distinctive membrane that has been elaborated by the higher mammals. To this compound organ the chorion and, to a greater or lesser extent, the allantois furnishes a part, while the uterine lining supplies the rest. The *umbilical cord* is vascular cable that serves as a communication between the foetus and its placenta.

All these membranes eventually will separate fromt he foetus by natural methods of amputation or withering, and are then discarded.

FORMATION OF EXTRAEMBRYONIC MEMBRANES IN REPTILES AND BIRDS

The embryo, perched on the top of a large mass of yolk, develops from the pellucida region of the blastoderm. The area opaca of the blastoderm continues to spread peripherally over the yolk mass. The three germ layers delineated in the region of the embryo continue to differentiate laterally in the area opaca. Here they form the extraembryonic *somatopleure* and *splanchnopleure*. The amnion and chorion form from the extraembryonic somatopleure; the yolk sac and allantoic sac develop from the extraembryonic splanchnopleure. The membranes of all amniotes form in essentially the same way and carry out the same functions.

Yolk sac

The yolk sac is the first membrane to form. The gun region of the early embryo has a dorsal roof and sides, but no floor. The ventrally open gun rests on the yolk mass. The extraembryonic mesoderm accompanied by extraembryonic endoderm closely applied to the yolk, extend peripherally, eventually surrounding the yolk.

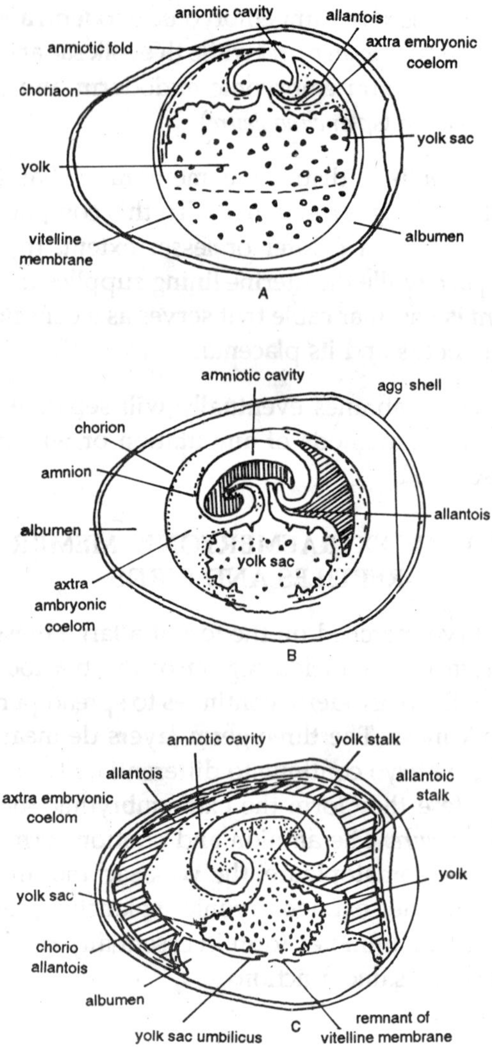

Fig. 9.6. Stages in the development of foetal membranes in chick.

As the body of the embryo grows, in the anterior region folds that roll under the head area appear, separating it from the underlying yolk and forming a subcephalic pocket. Eventually, similar folds develop on either side of the body, undercutting it laterally as shown in cross section in Figure. Posterior liming folds from about the third day of incubation and left the tail region from the surrounding yolk area. As a result, the subcephalic, lateral, and subcaudal folds define the body of the embryo and separate the embryonic from the extraembryonic regions. As these undercutting folds move towards one another, they establish a floor to the gut except for a small region of the midgut. These folds do not completely merge, but a small aperature (the yolk duct) remains, opening into the yolk sac. The yolk stalk connects the body of the embryo with the yolk sac and consists of an inner layer of endoderm and an outer layer of splanchnic mesoderm, the same combination of layers comprising the yolk sac.

The constriction of the yolk stalk region brings together the omphalomesenteric blood vessels (arteries and veins) located in the splanchnic mesoderm of the embryo. They continue to the wall of the yolk sac as vitelline vessels, with their arteries and veins composing the *vitelline arch.* The yolk sac with its richly enmeshed circulatory system suspends the yolk from the body. As the embryo ages, the ling of the yolk sac develops folds, considerably increasing its surface area. The endodermal cells lining the yolk sac secrete digestive enzymes that break the yolk down into soluble form. it is then absorbed into the vitelline veins and carried by their embryonic continuations, the omphalomesenteric veins, into the body proper. From here, the general circulatory system carries the soluble food material to all parts of the embryo for its metabolic use. As absorption of the yolk continues, the yolk sac shrivels and withdraws into the body and disappears completely by the time of hatching.

Amnion and chorion

The origins of these two membranes are considered together since they form simultaneously from the extraembryonic somatopieure. The *amnion* encloses the embryo in a sac, which fills with fluid, the *"ancestral pond"* referred to earlier. The amniotic fluid not only protects the embryo from desiccation but also acts as a watery cusion, shielding it from mechanical injury. The contraction of smooth muscle fibers in the wall of the amnion circulates the fluid that moves the parts of the embryo about and prevents them from growing together. The chorion, closely applied to the inner surface of the shell, and the allantois function in the respiratory exchange of the embryo.

Anterior to the chick's head region, the extraembryonic somatopleure forms a fold after about 30 hours of incubation. As viewed from the dorsal surface, this cresent-shaped thickening grows progressively backward over the head of the embryo. It is essentially a sheet of somatopleure folded back upon itself, seen best in a diagrammatic sagittal view.The wall of ectoderm and mesoderm next to the embryo becomes the amnion; the outer layer gives rise to the chorion. As the embryo develops, it pushes forward into the fold at the same time the fold itself grows posteriorly. The anterior amniotic fold fuses with the lateral ones, covering much of the anterior and middle regions of the body. On about the third day of incubation, a posterior tail fold forms, similar to the one in the anterior region. It grows anteriorly where it fuses with the anterior and lateral folds, completely encasing the emrbyo in a doublewalled pocket. The point of fusion leaves a scar, the *seroamniotic raphe*.

The fusion of the fold produces two sac like membranes and two cavities. The cavity formed between the embryo and the first fold, the amnion, becomes the *amniotic cavity*. Ectoderm lines its inner surface, the embryo side, and somatic mesoderm, its outer surface. The outer layer of the original fold gives rise to the chorion, with an inner surface of mesoderm and an outer surface of ectoderm. The cavity between the chorion and the amnion is the *seroamniotic cavity*, or extraembryonic coelom. Somatic mesoderm, derived from the inner wall of the chorion and the outer wall of the amnion, lines the coelom. As the chorion and amnion form, the somatopleure continues to grow peripherally and laterally, until it completely covers the embryo and all the otuerh membranes that are forming at this time.

Allantois

By about the sixtieth hour of incubation, an outpocketing of endoderm appears in the floor of the hindgut. The pocket, which is the beginning of the allantois, continues to grow into the extraembryonic coelom, eventually filling it. This endodermal evagination of the gut carries with it the closely associated splanchnic mesoderm. The combination of these two layers, splanchnopleure, as in the example of he yolk sac, forms the wall of the allantois. As the pocket enlarges and spreads outward, the splanchnic mesoderm comes in contact with the somatic mesodermal lining of the chorion. These two mesodermal layers fuse and from a single membrane, the *chorioallantoic membrane*, which becomes highly vascularized.

The growing membranes rupture the vitelline membrane (which still surrounds the egg) and continue to expand outward toward the shell membranes. They push the albumen to one side, where a pocket of the allantois envelops it. The blood vessels that cover the allantoic walls absorb the water and albumen. These vessels form in the splanchnic layer of the allantois and consist of two allantoic arteries and two allantoic veins. The highly vascular chorioallantoic membrane expands, eventually, assuming a position adjacent to the shell membrane. It functions as the primary respiratory organ until hatching. Oxygen passes through the porous shell and diffuses through the shell membranes, the chorioallantoic membrane, and the walls of the allantoic blood vessels. Carbon dioxide moves in the opposite direction. The allantoic sac is also an excretory organ. The relatively insoluble uric acid, deposited in the sac, remains with the shell at the time of hatching. This process represents another evolutionary modification that made it possible for animals to move permanently from an aquatic to a terrestrial environment.

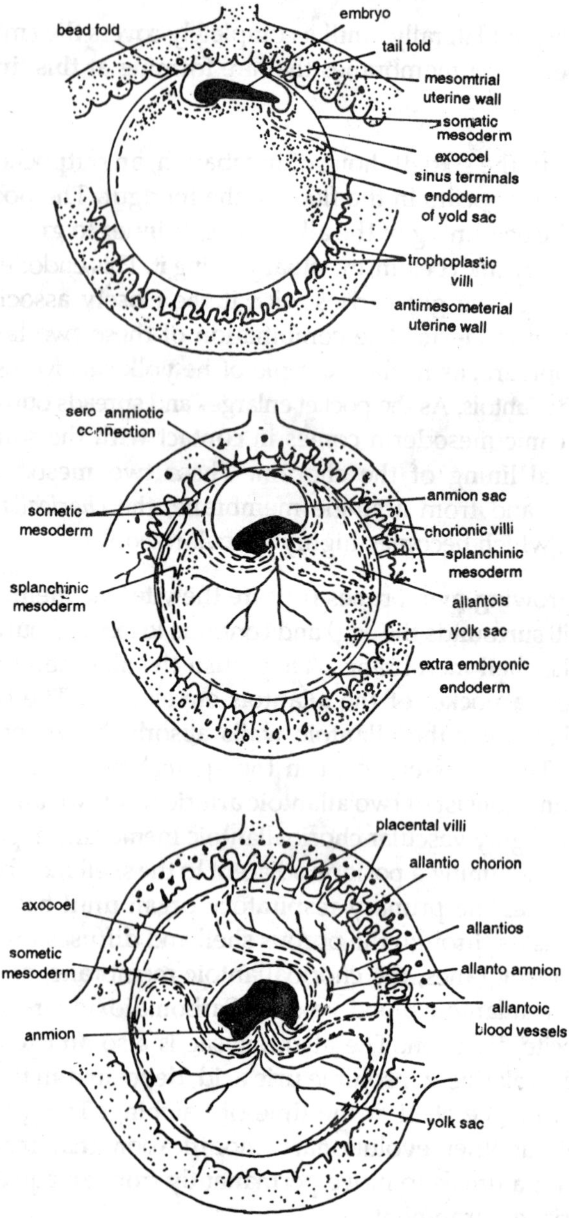

Figu. 9.7. Implanted blastocyst and development of foetal membranes in a mammal.

FORMATION OF EXTRAEMBRYONIC MEMBRANES IN MAMMALS

Amnion

The amnion is essentially the same in all mammals, although it arises by several different mechanisms. In some animals, such as the marsupial, the pig, and most insectivores, the amnion forms after gastrulation by folding in much the same way as in the chick and reptile. Many gradations exist between the fold method and that employed by the human embryo. In the monkey or man, the amniotic cavity develops precociously and a space appears by *cavitation* in the inner cell mass between the future embryonic disk and trophoblast cells. Except in the area of the future body stalk, amniogenic cells of the human embryo eventually separate from the inner layer of the chorion and from a distinct amniotic membrane. Apparently there is a correlation between the method of amnion formation and the time of implantation; embryos of those animals utilizing the somatopleure-fold method tend to implant later than those in which vacitation occurs.

Regardless of how the amnion forms, the mammalian embryo lies batched in the fluid of the amniotic cavity like the embryos of the reptiles and birds. At first the amniotic cells secrete the fluid of the cavity, but later the fetal urine also collects in it. Human fetuses shallow their amniotic fluid, which the cells of the gut then absorb. The fluid with its dissolved materials eventually enters the fetal circulation and passes into the mother's bloodstream via the placenta. Her kidney's eliminate both her metabolic wastes and those of the fetus.

Yolk sac

Evidently the mammals evolved early from the stem reptile since their first traces appear in the late Triassic rocks. At this time the mammalian egg, no doubt, resembled the reptilian egg and contained a large supply of yolk. However, sometime during the Triassic and Jurassic days some of he mammals became viviparous, and the mother supplied the necessary food to her young while she carried them in her uterus. Apparently, during the early part of the Mesozoic era, a reduction of the yolk occurred and the yolk finally disppeared in the advanced subclass of the mammals, the Eutheria, or placental mammals. Although no yolk remained in the egg, the yolk sac continued to form when the ectoderm and endoderm differentiated. Since it is an organ that was more fully developed and functional at an early stage of evolutionary development, the yolk sac is considered a *vestigial* structure.

The vitelline blood vessels develop int he walls of the vestigial secondary yolk sac in much the sam way that they did in the functional sacs of the reptiles and birds. The blood cells in the very young embryos arise in the walls of the yolk sac from mesenchyme cells probably derived from the splanchnic mesoderm. In many, during the second month of gestation intraembryonic organs, especially the liver and spleen, assume the function of blood cell formation.

Allantois

The allantois, which forms from the hindgut of all amniote embryos, probably evolved from the amphibian bladder. It is essentially an extra-embryonic respiratory and excretory organ in the reptile and bird. In the mammal, it also functions in absorption.

As in the reptile and bird, splanchnopleure composes the mammalian allantois. The amout of endoderm present in the splanchnopleure varies among the mammals. It ranges from a large endodermal sac in the dog and pig to a much reduced structure in the monkey and man. In some rodents it is entirely absent. However, the mesoderm component ramins very important in all forms. The mesodermal layer contributes to the formation of the umbilical blood vessels of the mammals. These vessels are homologous with the allantoic vessels of reptiles and the chick and are the only vessels found in the body stalk, or umbilicus, of the human embryo. The allantois is important since the allantoic mesoderm and its umbilical vessels fuse with the chorionic mesoderm and its trophoblast covering to produce the fetal portion of the mammalian placenta.

Trophoblast and chorion

The *trophoblast* is considered a primary fetal membrane of the mammal. Its origin traces back to the sueprficial layer of the morula. It functions in implantation. Later when mesoderm lines its cavity, it becomes the chorion and contributes to the formation of the placenta.

EMBRYONIC INDUCTION

The early embryo is a mass of similar cells that eventually become different. All are derived from the single fertilized egg; all contain identical genes. How do they become different? This is a major area of interest in developmental biology today. It has already been discussed that certain special types of cytoplasm or surface regions can be segregated to specific cells during the cleavages process. The interaction of cells with neighboring cells plays an important role in causing cells to differentiate. This was easily seen in neural tube differentiation, in which the prospective neural ectoderm must interact with the archenteron roof for neural differentiation to occur. This interacting system is a major topic of this chapter. The concept of embryonic induction originated in studies of this system. Embryonic induction simply means that one tissue (*inducing tissue*) interacts with another tissue (*responding tissue*), causing the responding tissue to differentiate. This sort of interaction was also seen in organogenesis. The tip of the optic vesicle induces prospective lens ectoderm to differentiate into lens; the tip of the growing ureter induces metanephrogenic mesoderm to differentiate into metanephric kidney tubules. In short, inductive interactions between cells play a major role in causing differentiation. These inductive interactions can involve a chemical inducer that passes from the inducing tissue to the responding tissue. This inducer may activate specific genes that code for the specific proteins required for cell differentiation. In this chapter we shall focus on the nature of the interactions involved in embryonic induction and the factors responsible for *"turning cells on"*, causing them to differentiate.

ORIGIN OF THE INDUCTION CONCEPT

Let us begin with a few of the experiments that led to the concept of embryonic induction. In one of the first of these, carried out almost one hundred years ago, Hans Driesch separated the sea urchin embryo at the two-cell stage and found that each blastomere gave rise to a complete, normal (smaller) larva. *E.B. Wilson* did the same experiment with the *Amphioxus* embryo, with the same results. Each cell could give rise to an entire embryo.

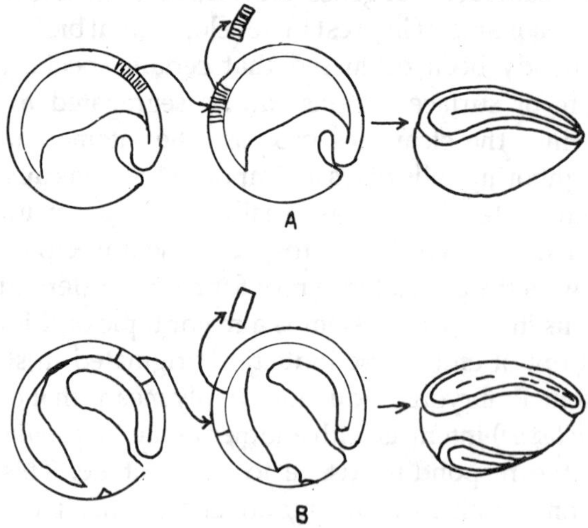

Fig. 9.8. Spemann's experiment demonstrating the determination of the presumptive neural plate area. Transplantation at an early gastrula stage (A) leans to normal development of the host whereas a similar experiment performed at late gastrula stage (B) results in the formation of an extra neural plate at the belly side of the donor.

G.A. Schmidst was the first, in 1933, to accomplish the more difficult task of separating the amphibian embryo at the two-cell stage. In baout half the cases, at least one of the two blastomeres formed a complete embryo. In many cases, both blastomeres did so. *Hans Spemann* showed that if the first cleavage plane passed through the *gray crecent* region of the zygote, each blastomere gave rise to a complete embryo. The gray crecent is a surface region of the zygote that is exposed soon after fertilization when the outer pigmented surface of the zygote moves toward the point of sperm entry. We shall discuss the gray crecent later in this chapter.

These results suggest that the gray crescent contains something essential for normal embryonic development. The gray crescent gives rise to the dorsal lip of the blastopore. During gastrualation the dorsal lip invaginates to form the roof of the archenteron (*prospective notochord*) that eventually lies under the prospective neural tube. Neural differentiation does not occur unless the archenteron roof underlies the prospective neural ectoderm. This can be demonstrated by removing the outer membranes from the amphibian egg and growing the embryo in a solution of lithium chloride or in a hypertonic solution. Under these conditions, the propsective archenteron roof grows outward rather than inward. The archenteron roof does not underlie the prospective neural ectoderm, and the prospective neural ectoderm fails to differentiate into the neural tube.

In 1924, a set of beautiful experiments by *Spemann* and *Mangold* showed that the interaction of the archenteron roof with the prospective neural tube is essential for neural differentiation. (These experiments earned *Spemann* and *Mangold* a Nobel Prize.) *Spemann* and *Mangold* transplanted the dorsal lip of the blastopore of an early gastrula of one species of newt into the blastocoel of an early gastrula of another species of newt. Because the species differed in pigmentation, the origin of specific structures forming in the host embryo could be determined. The host embryo developed two neural systems, and in some experiments two entire embryos. In another experiment, the dorsal lip of one embryo was transplanted near the lateral lip of the blastopore of another embryo. The transplant invaginated and induced a second neural tube, as in the previous experiment. The differences in pigmentaiton between the host and dorsal lip donor embryos showed that both neural tubes in the host embryo were derived almost entirely from host tissue. Thus the transplanted dorsal lip did not give rise to a neural tube; instead, it induced the host embryo to form a neural tube. In cases where whole second embryos fomed, the neural tubes were of host origin, but the donor dorsal lip differentiaed into a variety of tissues that normally form from this region. Spemann and Mangold called the dorsal lip the *organizer* because it could cause a complete second embryo to form.

THE INFLUENCE OF ONE GROUP OF CELLS ON ANOTHER

Embryonic induction is an *interaction of two or more groups of embryonic cells that alters the development of at least one of the groups*. We can see many instances of induction in embryonic development—for **example, in the** formation of parts of the vetebrate eye and in the formation of the nervous system in vertebrate embryos, which is influenced by the mesoderm. The characterstics of the different cases of induction vary greatly; for instance, some are specific and indispensable for the formation of a particula rkind of tissue, whereas others are more general in nature and affect the timing of an event. At the first extreme are cases in which unique and essential *instructive information* passes between the partners. A fine example here is the induction of the neural tube. At the other extreme, even the sepaated partners can develop into tissues they would normally form; the rate of development of each may be only slightly accelerated by the presence of the other or by the presence of a wide variety of other kinds of cells. In these latter instances, one particular partner creates *premissive conditions* for the other to complete differentiation, but the ultimate fate of the responding cell may depend on the interaction in only a small way.

Some embryologists apply the term *embryonic induction* to all interactions in which the developmental behavior of a particular cell is affected by neighboring cells or by the extracellular products of neighboring cells. By this definition, the usbject of induction would include all intercellular aspects of development. Other embryologists restrict the term *induction to early, large-scale interactions between major groups of embryonic cells*. We shall use this more restrictive definition in this chapter. It may well be, however, that one day embryonic induction will be understood as a special case of the cell-cell interactions occurring not just in embryos but in adults as well.

Induction Involves the Transient Passage of a Stimulus

There are several general features of the embryonic induction. First, when the two participating cell groups interact, one produces a stimulus and the other receives it; that is, one is *inductively active* and releases an *inductor*, whereas the other is *receptive* to the inductor and presumably possesses the appropriate receptors. More correctly, the receptive partner is said to be *competent* to respond to the inductor in a specific way. Because the responding cell is often able to respond differently to different inductors or to different levels of the same inductor, its competence is defined as the *complete set of possible developmental responses that the cell can make, give all inductive situations.* A second characterstic of inductive interactions is that they last only briefly. Thus, by the time the next developmental stage begins, the inductive tissue may no longer produce the inductor and the receptive tissue may no longer have the same competence. As a result, if the progress of either partner is delayed, the interaction will fail.

TISSUE, STAGE, AND REGIONAL SPECIFICITY IN NEURAL INDUCTION

A question that now comes to mind is, will *any* tissue respond to the archenteron roof and differentiate into neural structures? The answer is no. If a piece of dorsal lip is transplated onto endoderm, no neural differentiation occurs. Only ectoderm will form neural structures as a result of interaction with the dorsal lip. We say that only the ectoderm is *competent* to respond to induction by the dorsal lip. Competence for neural induction is tissue specific.

What about the developmental stage? Can ectoderm from any embryonic stage respond to the inducer by differentiating into neural structures? Again, the answer is no. Only gastrula-stage ectoderm is competent to respond to induction. A dorsal lip will not induce neural differentiation if granted into a neurula. If a dorssal lip is transplated into a blastula-stage embryo, the ectoderm differentiates into neural structures only after gastrulation is complete, when the normal host neural system develops.

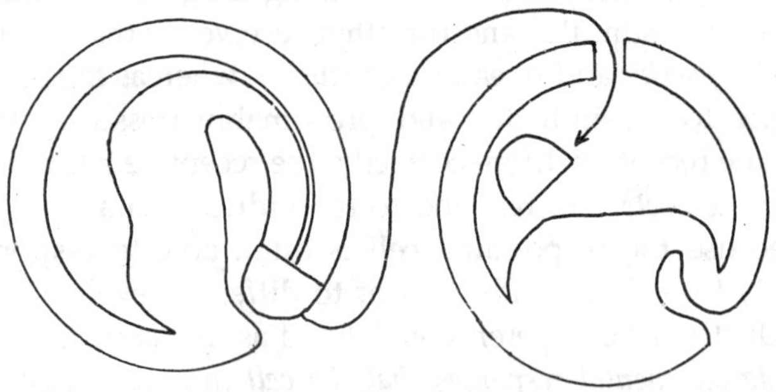

Fig. 9.9. *Regional inducing specificity of the dorsal lip.* (a) Early dorsal lip (early gastrula from amphibian) induces a secondary head in recipient embryo. (b) Late dorsal lip (last gastrula from amphibian) induces a secondary trunk and tail in recipient embryo.

Does a dorsal lip taken from an early gastrula have the same inducing capacity as a dorsal lip taken from a later gastrula? Normally, the early dorsal lip invaginates and comes to lie beneath the anterior part of the prospective neural tube. A later dorsal lip comes to lie beneath the posterior part of the prospective neural tube. An early dorsal lip, or the anterior portion of the archenteron roof, can indeed be successfully transplated into another embryo. There, it induces mostly head neural structures in the ectoderm (forebrain, eyes, nose rudiments, hindbrain, and ear vesicles). A late dorsal lip, or the posterior portion of the archenteron roof, induces mostly posterior structures, such as the spinal cord and associated trunk and tail organs. Thus the dorsal lip or archenteron roof has *regional inducing specificity*.

CELLULAR CONTACT IN EMBRYONIC INDUCTION

What physical conditions are needed for induction to occur? Is cell-to-cell contact between the archenteron roof and the prospective neural tube required for normal neural induction? The answers to these questions remain an enigma.

One approach to investigating the need for cell-to-cell contact is to separate these two tissues with different artificial barriers. If induction is caused by a molecule passed from the archenteron roof to the overlying ectoderm, then a filter between them that allows molecular passage but not cell passage should not inhibit induction. Material that does not allow even small molecules to pass between the inducing and responding tissues should prevent induction.

In early experiments, membrane barriers were placed between the archenteron roof and the ectoderm. Cellophane, which does not allow the passage of molecules, did indeed prevent induction. When permeable membrane filters were used instead, induction still occurred. These results suggested that cell contact was not needed for induction, and that the key was the passage of molecules between the inducing and responding tissue. Later, however, when the permeable filters were examined with an electron microscope, thin cell processes were sometimes observed in the filter pores. Thus the issue of cell contact was not really resolved.

Toivonen, and *Saxen* reinvesigated this problem using filters with tiny straight pores (0.1. μm) produced by shooting neutron beams through the filters. These filters were inserted between

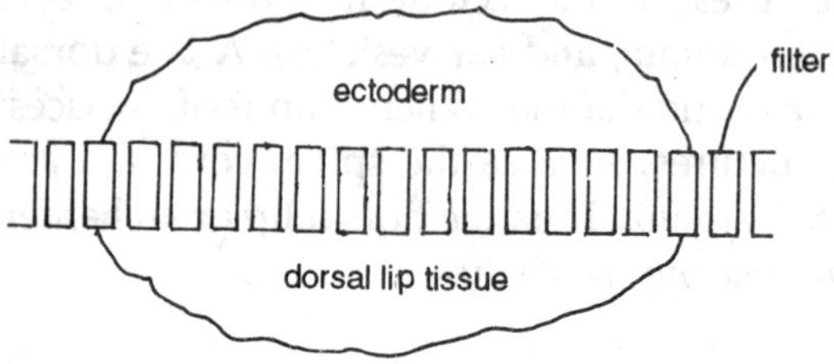

Fig. 9.10. Separation of dorsal lip and responding ectoderm by porous filter permits induction to occur.

the inducing the responding tissues for a very brief time. The tissues were then separated, and the ectoderm was cultured separately. Neural differentiation did indeed occur in the ectoderm. When the filter pores were analyzed under the electron microscope, no cell processes seemed to have entered the filter. This is perhaps the best evidence to date that neural induction does not require contact between the cells of the archenteron roof and the prospective neural tube. This induction may occur in response to the diffusion of molecules from the archenteron roof to the responding ectoderm.

MECHANISM OF INDUCER ACTION

We have not, as yet, definitively identified the chemical nature of the neural inducer. However, this is not the only important aspect of induction. Other key questions include; What is the nature of the response by the ectoderm to the inducer? What physical conditions are needed for induction to occur? Is contact between the archenteron roof and the prospective neural tube needed for induction to occur? Let us examine these issues.

What is the nature of the response to the inducer? The response is controlled by genes in the responding ectoderm. Active genetic informaiton fromt he dorsal lip is probably irrlevant to the response. In other words, the inducer provides permissible rather than instructive information to the responding ectoderm. Permissive information *"tuns on"* the responding tissue's genes, instructive information provides actual genetic information provides actual genetic information to the responding ectoderm. These condlusions are supported by the following experiments involving several species of salamanders.

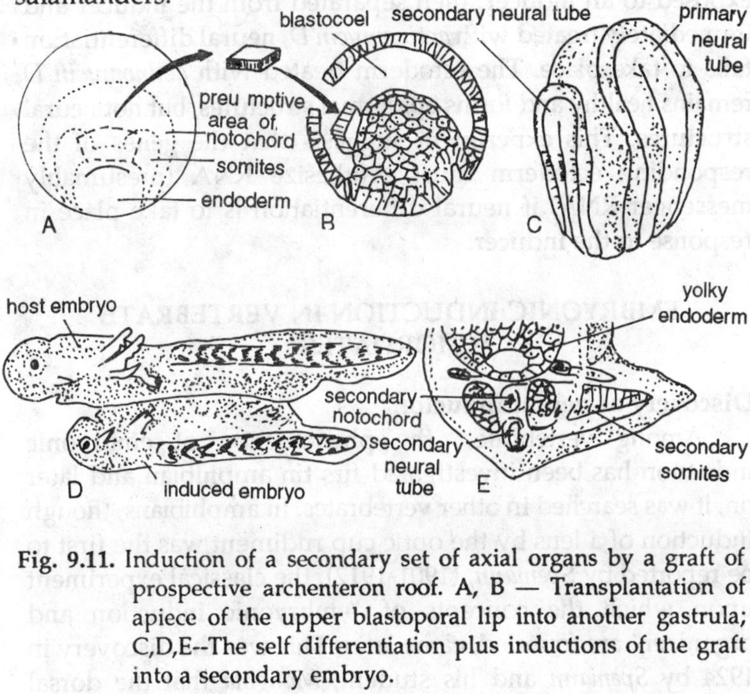

Fig. 9.11. Induction of a secondary set of axial organs by a graft of prospective archenteron roof. A, B — Transplantation of apiece of the upper blastoporal lip into another gastrula; C,D,E--The self differentiation plus inductions of the graft into a secondary embryo.

Only some salamander larvae have balancers, which can be used as markers in experiments with salamanders. Balancers are induced in the epidermis by the anterior portion of the archenteron roof. Experiments have shown that a balancer can be induced in a species that does not possess a balancer by replacing a piece of ectoderm of the host with ectoderm from the species possessing a balancer. Thus the archenteron roof of the species without a balancer can induce a balancer in ectoderm transplanted from a species competent to form a balancer. The reverse experiment showed that a balancer cannot be induced in ectoderm that is not competent to form a balancer.

The ectoderm must possess the genetic code to be activated by the inducer. Thus the inducer dos not alter the genetic response of the responding tissue.

Another experiment along the same lines uded the RNA synthesis inhibitor *actinomycin D*. If newt gastrula ectoderm is exposed to an inducer, then separated from the inducer and immediately treated with *actinomycin D*, neural differentiation fails to take place. The ectoderm treated with *actinomycin D* remains healthy and forms epidermal structures, but not neural structures. This experiment suggests that the genes of the responding ectoderm must synthesize RNA (presumably messenger RNA) if neural differentiation is to take place in response to the inducer.

EMBRYONIC INDUCTION IN VERTEBRATE CHORDATES

Discovery of neural inductor

Among vertebrates, the phenomenon of embryonic induction has been investigated firs tin amphibian and later on, it was searched in other vertebrates. In amphibians, though induction of a lens by the optic cup rudiment was the first to be reported by *Spemann*, (1901-1912), the classical experiment upon which the concepts of "embryonic induction and organizer" are founded, came later. This was the discovery in 1924 by *Spemann* and his student, *Mangold* that the dorsal blastoporal lip from a salamander (of newt) early gastrula, when implanted into a ventral or lateral position of another gastrula, turns inward, develops into a notochord and somites, and induces the host ectoderm to form a neural tube, i.e., it induced the formation of the axial organs of a more or less complete secondary embryo. This secondary embryo accordingly may have a notochord, somites, kidney tubules, each of which is composed partly of grafted, partly of the induced host material. In this experiment, thus, the graft becomes "self-differentiated" (viz., it forms the very structures it would have given rise in its normal setting-notochord, mesoderm, etc.), but it induces the adjoining host tissue to form spinal cord and other structures, including somites and kidney tubules.

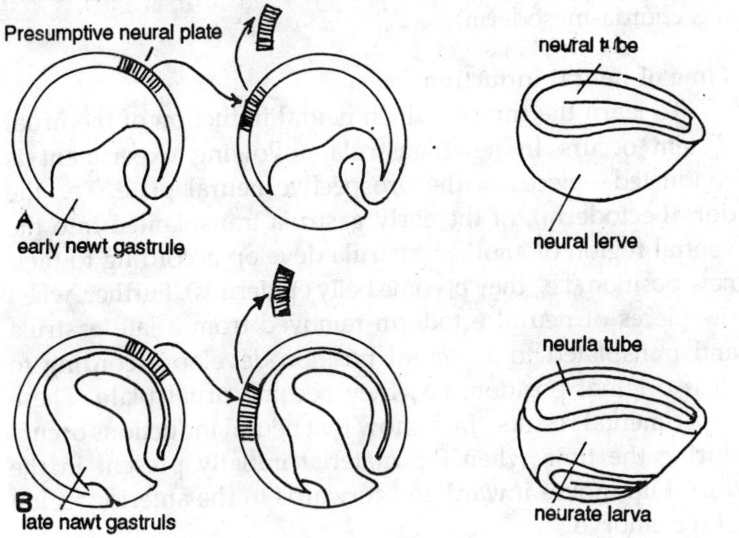

Fig. 9.12. A—Piece of presumptive neural plate removed from an early newt gastrula, and grafted onto the ventral side of another of the same age. No induction occurs, but the graft develops in accordance with its new surroundings; B—A corresponding piece removed from a late gastrula develops in the new surroundings acording ot its original fate.

Visualizing these results, *Spemann* and *Mangold* spoke of the dorsal lip as the *"organizer"* or *"primary organizer"* because it caused the development of a complete secondayr embryo. It is `primary' in the sense of being the first in sequenmce of inductions. However, it is not an "organizer" in the sense attributed to it by its discoverer (*Elbert and Sussex*, 1970). Actually, the organization of the secondary embryo results from a series of both inductive interactions and self-differentiative changes in the host and donor tissues. Today, therefore the more general term *embryonic induction or inductive interaction* are preferred. Accordingly, since the primary organizer of *Spemann and Mangold*, induces the ectoderm to become differentiated into neural tube, it is called *neural inductor or primary inductor*. Then neural inductor of newt (*Triturus* sp.) is found to be composed of presumptive material for notochor and dorsal mesoderm (i.e. chorda-mesoderm).

Time of neural induction

To learn the time at which neural induction of the nrual system occurs in newt gastrula, following experiment is conducted—pieces of the prospective neural plate (i.e., the dorsal ectoderm), of the early gastrula transplanted into the ventral region of another gastrula develop according to their new position (i.e., they become belly epidermis). Further, when the pieces of neural ectoderm removed from a late gastrula and transplated to a ventral position develop according to their original position, i.e., they become nrual plate. These experimental results, thus, show that neural inductions occurs during the time when the material initially present in the dorsal lip moves inward and forwards to the anterior region of the embryo.

Morphology of neural inductor

Vital-staining experiments of *Vogy* with newt eggs have shown that the material successively forming the dorsal blastoporal lip moves forward as the archenteric roof. Transplants taken form this region are also able to induce a secondary embryo or the belly of a new host, i.e., the archenteron roof acts as a primary inductor in essentially the same way as does the dorsal lip tissue proper.

(i) Regional differentiation of neural inductor

The inductions of neural inductor is found to be regionally specific and the regional specificity is imposed on the induced organs by the inductor. In other words, the inductive capacity of the blastoporal lip varies both regionally and temporally. Most of the dorsal and dorsolateral blastoporal material is necessary for a graft to induce a more or less complete secondary embryo. If only prospective endomesodermal material of dorsal lip is grafted into another gastrula, it induces a partial embryo containing ectodermal head structures such as brain, sense organs, mouth structures, etc. Moreover, explants or grafts of the median portion of the dorsal lip tend to form eyes and olfactory pits, lateral parts tend to form posterior cephalic structures and still more lateral parts of the marginal zone tend to induce spinal cord and mesenchyme of a trunk-tail.

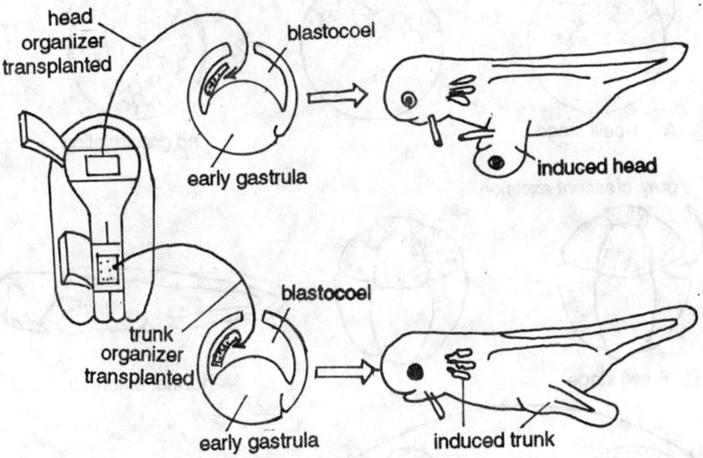

Fig. 9.13. The separation of the neural inductor (primary organizer) induced head and trunk organizers.

As invagination continues, however, and the dorsal lip no longer consists of prospective head endomesoderm but progressively becomes prospective trunk mesoderm, it acts as a trunk-tail inductor. The most caudal-region of the archenteron roof, in fact, specifically induces tail somites and probably other mesodermal tissues. The archenteron roof (containing chordamesoderm and some endodermal cells), thus, induces entirely different classes of tissues : various neural and mesoectodermal tissues by its anterior region and various mesodermal tissues by its most posterior region. From this fact, *Lehmann* (1945) concluded that differences in specific induction capacities exist between head and trunk level of the archenteron roof, and are related to the regional differentiation of the neural tissue into archencephalic (including fore-brain, eye, nasal pit), *deuterencephalic* (including hindbrain ear-vesicle) and *spinocaudal* components. Accordingly, the archenteron roof can be said to consist of an anterior head inductor including an *archencephalic inductor* and a *deuterencephalic inductor* and a *trunk or spinocaudal inductor.*

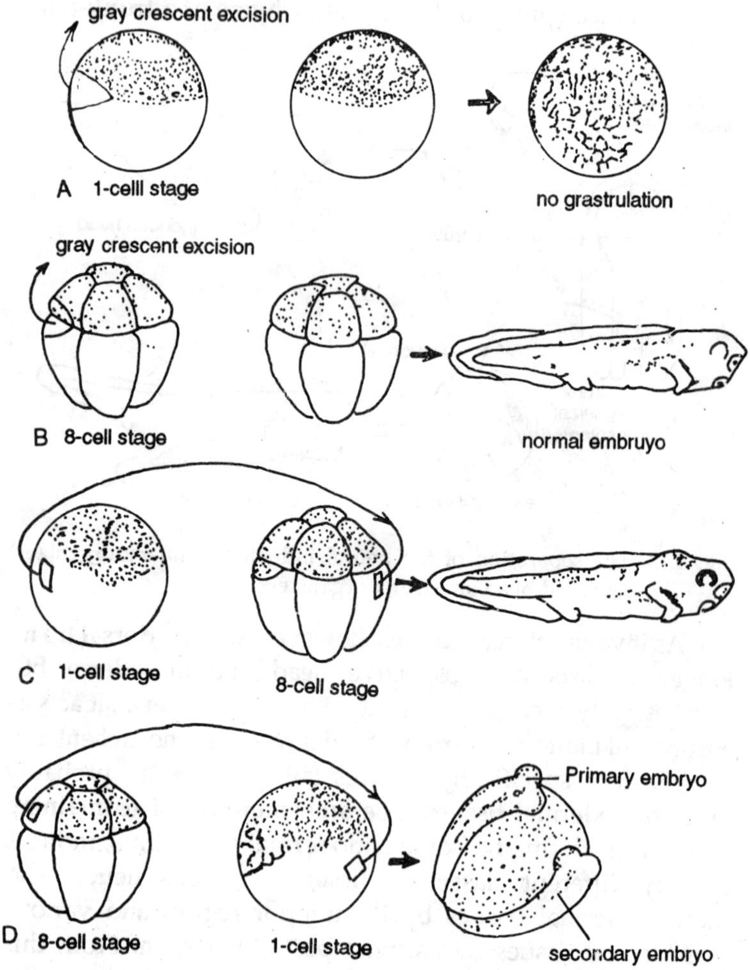

Fig. 9.14. Experiments of A.S.G. Curtis of *Xenopus*. A—Excision of the cortical gray-crescent area at the one cetls stage : no gastruation; B—Same experiment at the eight cell stage : normal embryo; C—Graft of the gray crescent cortex of one cells stage to the

ventral part of the eight cells stage does not result in the induction of a secondary embryonic axis; D—Graft of the graycrescent cortex from the eight cell stage to the ventral margin of the one cell stage induces a secondary embryonic axis (after)

(ii) Primary (neural) inductor and gray crescent

The dorsal lip region of the blastopore at the onset of gastrulation can be traced back to the gray-crescent of the undivided fertilized amphibian egg. It was conceived by certain developmental biologists that the crescent material of egg cortex initiated gastrulaiton and had the capacity of neural induction. To test these hypotheses, *A.S.G. Curtis* (1960, 1963) performed a series of experiments of transplating parts of the cortex of the fertilized egg of the clawed-toad, *Xenopus laevis* at the beginning of cleavage. In one of the experiment, the gray-crescent cortex was excised from the fertilized egg, then it was observed that the cell division through proceeded undisturbed, the gastrulation failed to take place. In other experiment, the gray-crescent cortex of a uncleaved fertilized egg was excised and transplanted into a ventral position of a second egg, so that the egg receiving the graft had two gray crescent on oppositie sides. As a result, egg cleaved to form a blastula which underwent two separate gastrulation movements to produce two separate primary nervous systems, notochords and associated somites (i.e., a primary and a secondary embryo were formed). Similar experiments conducted on the eight-cell stage, however, showed that something had happened during the short interval represented by the first three cleavages. Gray-crescent cortex of the eight-cell stage still retained its inductive capacity when grafted to younger stages, but the eightcell stage itself no longer reacted to grafts of the crescent material. Also, excision of the gray crescent at this stage no longer inhibited subsequent gastrulaiton and normal development, the missing crescent properties being replaced from adjacent cortical regions. Accordingly *Curtic* concluded that a change in cortical organization spreads across the surface of the egg during the second and third cleavages, starting from the gray crescent, when this change is completed, interaction, probably of a biophysical nature, can take place among various parts of the cortex.

EPITHELIO-MESENCHYMAL INTERACTIONS GOVERN ORGANOGENESIS

Most of the recent work related to the specificity problem has been performed on model systems for epithelio-mesenchymal interactions, i.e., morphogenetic interactions between differentiating epithelium and mesenchymal stroma. Although only a few of the many interactive sitnations explored can be described here, it can be stated that all the experiments show that the diff4erentiation of epithelium and mesenchyme depends on interactions between them. But, as will be seen, such interactions are not necessarily identical in mechanism and the influence of the mesenchyme may vary from determinative (*'instructive'*) actions to entirely non-specific, stablizing effects.

Dermal mesenchyme determines epidermal differentiation

Differentiation of the vetrebrate epidermis and its derivatives is known to be directed by the underlying mesenchyme, and here we find further examples of an actual 'instructive' influence. Mesenchyme from various sources, when combined experimentally with embryonic chick epidermis, alters epidermal differentiation over the whole range from the keratinizing, squamous type to ciliated, cuboidal epithelium. In such experiments, the regional differences can be demonstrated even within the same embryonic dermis. When dermis from the feeather-bearing back is combined with epidermis of the normally featherless ventral skin, feathers from, whereas the reciprocal combination does not support development of feathers. Juding from various transplantation experiments, the feather-forming bias of the epidermis is determined very early in development, but for its expression requires specific, local mesenchymal induction.

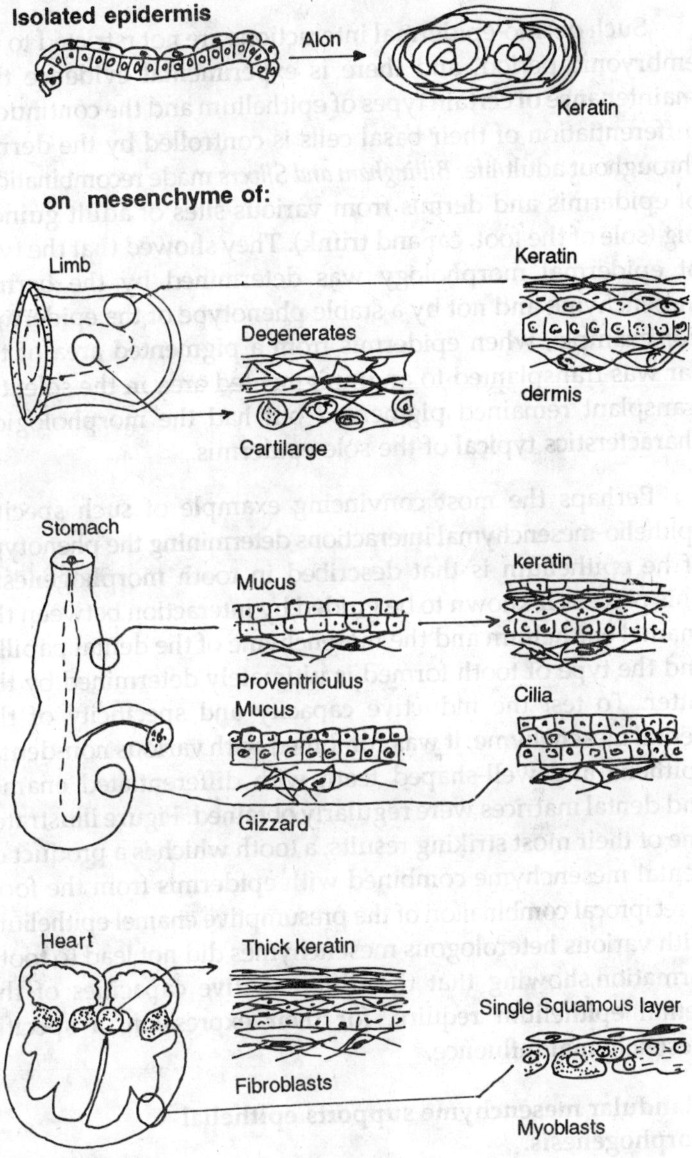

Fig. 9.15. Differentiation of embryonic chick epidermis in combination with various heterologous mesenchymes.

Such dermo-epidermal interactions are not rstricted to the embryonic period, and there is experimental evidence that maintenance of certain types of epithelium and the continuous differentiation of their basal cells is controlled by the dermis throughout adult life. *Billingham and Silvers* made recombinations of epidermis and dermis from various sites of adult guinea-pig (sole of the foot, ear and trunk). They showed that the type of epidermal morphology was determined by the bermal mesenchyme and not by a stable phenotype of the epidermis. For example, when epidermis from a pigmented area of the ear was transplanted to an unpigmented area in the sole, the transplant remained pigmented but had the morphological characterstics typical of the sole epidermis.

Perhaps the most convincing example of such specific epithelio-mesenchymal interactions determining the phenotype of he epithelium is that described in tooth morphogenesis. This process if known to be guided by interaction between the enamel epithelium and the mesenchyme of the dental papilla, and the type of tooth formed is ultimately determined by the latter. To test the inductive capacity and specificity of the dental mesenchyme, it was combined with various non-dental epithelia and well-shaped teeth with differentiated enamel and dental matrices were regularly obtained. Figure illustrates one of their most striking results, a tooth which is a product of dental mesenchyme combined with epidermis from the foot! A reciprocal combinaiton of the presumptive enamel epithelium with various heterologous mesenchymes did not lead to tooth formation,showing that the differentiative capacities of the dental epithelium require, for their expression, a specific mesenchymal influence.

Glandular mesenchyme supports epithelial morphogenesis.

Glandular epithelium regularly interacts with its meschymal ceunterpart and this system has been used in a great variety of experiments. Pancreatic, salivary, mammary, hepatic and many other glandular epithelia have been shown to be morphogenetically dependent on their mesenchymal stroma. Unlike the readily modified epidermis discussed above, the glandular epithelium, in most instances studied, seems to be strictly limited in its developmental alternatives. A good example is the pancreatic epithelijm, which even at a stage when the gland rudiment is not more than a small stalk does not depend on the presence of normal pancreas mesenchyme and can continue its development if cultured with mesenchyme from almost anywhere in the embryo. Morphogenesis and enzyme synthesis proceed after combination of this isolated epithelium with a varieyt of heterologous mesenchymes. Salivary epithelium at a variety of heterologous mesenchymes. Salivary epithelium at a comparable stage is similarly committed but much more selective in its mesenchymal requirement, as very few of the mesenchymes tested so far can support its normal branching and adenomere formation. The basis for such variations in the specificity of the mesenchymal requirements of these and other glandular epithelia is not know, but most of them respond to the mesenchyme in an all-or-none fashion. Alteration of the morphogenesis of these epithelia by heterologous mesenchyme is rare. The most striking example is the salivary-like branching pattern of mammary epithelium in combination with embryonic salivary mesenchyme.

Parallel with the restriction of their differentiative potentialities, these epithelia gradually acquire independence of specific mesenchymal influences during embryogenesis. This is shown in the following series of experiments. At an early stage an explant of the presumptive pancreatic region of the gut surrounded by its homologous mesenchyme will form pancreatic acini *in vitro,* but fails to do so if the mesenchyme is removed or replaced by a heterologous mesenchyme. Somewhat later stages of the epithelial anlage still requires mesenchymal support, but this becomes less specific, and finally the epithelium reaches a stage at which the mesenchymal influence can be simulated by high concentrations of embryo extract in the culture medium in the total absenc e of mesenchymal cells.

Comparable phases of interactions of varying specificity have been demonstrated in the development of the liver epitheliu. At the 5-somite stage of the chick embryo the liver entoderm is determined by the liver-heart mesoderm, which cannot be replaced by any heterologous mesenchyme. Later, this determined entoderm still requires mesenchymal support to become organized into typical liver cords, adn now the effect can be simulated by several heterologous mesenchymes.

If such experiments on the varying specificity of the inductive interactions are tentatively generalized and simplified, the scheme might be as follows. During embryogenesis, extrinsic influences guide the differentiation of cells and cell populations through various phases. Initially pluripotent cells become determined and their developmental alternatives are restricted (*neuralization, epithelialization,* etc.). Subsequently, their organization into polarized, synchronized structures is governed by increasingly less specific, permissiue influences (*organogenetic tissue interactions*). Finally, stages of increasing stability of the organ are reached ans tablizing influences are replaced by maintenance effects which are supporting and probably act rhough nonspecific channels. Obviously, there are differences in this simplified scheme between different tissues and organs. For example, the constantly renewing epidermis may represent a target cell population which never reaches the stable stage of differentiation, and instead of maintenance effects, requires constant guidance by the underlying dermis. This scheme also emphasizes the great multitude of interactive processes usually lumped under the title 'embryonic induction' and makes it obvious that the various types of interactive mechanisms should be considered separately, as will be done below.

Induction of the vertebrate lens

In vertebrates of all types the development of the eye is essentially the same. It is now well proved that neurulation produces and ectodermal neural tube with a wide lumen anteriorly in the presumptive brain region. Constrictions appear which divide the brain into fore-, mid- and hind sections, and from the forebrain a bulge (the *optic vesicle*) grows out from each side and comes into contact with the ectoderm at the surface of the head. The optic vesicle then becomes indented, forming the double-walled optic cup; subsequently the inner wall forms the optic retina and the outer wall, which becomes much thinner, the layer of retinal pigment cells (the only melanocytes which are not derived from the neural crest). The rim of the ucp gives rise to the epithelial rudiment of the iris and the circular opening to the pupil of the eye. The stem of the cup is the optic stalk, and when the retinal cells differentiate as sensory neurones (*retinal ganglion cells*) their axons grow back along it to the brain, so that it is transformed into the optic nerve. The region of surface ectoderm is contact with the optic cup becomes thickened to form a placode, which sinks in to form a hollow vesicle which rounds off and become detatched to give the lens rudiment. Later the inner vesicle wall becomes very much thicker than the outer; its cells elongate greatly, filling the lumen, then begin to synthesize crystallins—the characterstic luxury molecules of differentiated lens cells—to give the lens. Over the lens vesicle the ectoderm which remains at the surface becomes the outer layer of the transparent corneum, whose completion through immigration of neural-crest mesenchyme cells betweenthe ectodermal epithelium and a thin mesodermal endothelium.

Spemann showed that, in many amphibians, the lens is formed in the ectoderm only as a response to an inductive interaction resulting from contact with the optic cup. If the optic cup is removed before induction has occurred, the ectoderm in the eye region differentiates as epidermis, but if (within a certain period within which the ectoderm is said to be competent) presumptive flank ectoderm is grafted into this region, or if an optic cup is transplanted under the ectoderm in the flank region, this ectoderm, which would have become epidermis now produces a lens. More recent work by *Jacobson* in 1958 has shown that this classic induction process is in fact only the culminaiton of a sequence of inductive interactions (leading to the formation of the lens in the correct location at the right time), which begins with a reaciton to the anterior archenteron roof in early gastrulation, followed by one to the presumptive heart mesoderm in the late gastrula, and concludes only with the response to contact with the optic cup. The influence of the three is similar and additive; the order of their activity may be changed experimentally, and one or other may be omitted so long as contact with the other is prolonged.

Instructive and permissive interactions

The optic cup/lens system illustrates several general points. Firstly, induction is followed by a phase of organogenesis, whcih may involve any one or several of the cellular activities we have discussed in connection with it, succeeded by a phase of cytodifferentiation in which specific proteins are synthesized and the cells assume their characterstic differentiated form. Secondaly, induction can be recognized only by experimental intereference with the normal course of development, and it is the response of tissues to the presence or absence of other tissues which enables us to characterisize the process. If the inducter is (1) present, the tissue it lies against will develop according to its presumptive fate but, if the inducer is (2) absence, it will not; but furthermore, (3) this inducer may be capable of producting the same development in cells whose presumptive fate is different—in this case, lens from presumptive flank epidermis. Where this third point can be shown experimentally, the induction is said to be *instructive* but, where only points 1 and 3 can be demonstrated, the induction may merely be *permissive*.

An instructive induction implies that the responding cells have been directed into a pathway of development which they would not otherwise have followed, i.e. that induction here has affected their state of deteremination; a permissive induction affects only the extent to which an already determined cell can express that state through differentiation in its particular pathway, and several permissive interactions will always succeed an instructive one.

Figure 9.16. Lens induction in amphibian embryos.

The necessity for permissive interactions has especially complicated the interpretation of studies on induction done *in vitro*, where suboptimal conditions may inhibit differentiation which would take place if the culture environment were improved; thus *Ellison* and *Lash* showed in 1971 that chondrogenesis could occur in somitic mesenchyme in the absence of the neural inducer which had been thought, until then, to be required if the right culture conditions were provided. Again, cells proliferation is generally required for differentiation, and isolated epithelia in culture will not mitose; an underlying mesenchymal layer is required for this, but almost any mesenchyme will serve. In 1966 studies by *Hauschka and Konigsberg* stimulated a great interest in the effect of collagen in cell cultures; *Dodson* showed in 1967 that embryonic chick epithelium survived and keratinized on a collagen gel, and suggested that the epithelium is reacting to it in the same way as to the extracellular products of its normal dermis. *Bernfield* and colleagues followed up this lead in 1972 in the development of salivary gland epithelium, and found that not the collagen in the extracellular matrix but acid mucopolysaccharides in the basal lamina, whose synthesis depended in some specific way ont he underlying mesenchyme, provided the crucial factor.

Besides the optic cup/lens system, many interactions between epidermis and underlying dermis are clearly instructive, e.g., where chick dermis from feather-forming skin regions which is implanted beneath the corneal epithelium of the eye induces feather formation in that tissue. But *"instructive"* in this sense does not necessarily, and almost certainly does not, involve a transference of a complex message; each cell at any point in this development will have only a restricted number of pathways open to it, perhaps even only two, and an instructive induction merely switches development into one or other of those pathways by activating a particular subroutine in the genetic programme. Thus though the response may be complex, the signal may be rather simple.

Examples of inductive interactions are found in development of the neural tube and somites (*primayr induction),* which being more complicated will be considered last; in limb development; in the development of the skin and its appendages; and in the development of a wide variety of mesodermal tissues, e.g., in the differentiation of cartilage (*chondrogenesis*), development of the pancreas, salivary and mammary glands, and of kidney tubules. Epithelio-mesenchymal interactions such as the last four have bene intensively studied (see the reports in the book edited by *Fleischmajer and Billingham* in 1968) on account of their suitability for *in vitro* culture and bio-chemical analysis, notable the pancreas by *Rutter* and *Wessells* in the 1960s, and *Rutter* and *Ronzio* in 1973. Most of the interactions between an epithelial and a mesenchymal component appear to be permissive in that, though a wide range of inducers may lead to the normal development of a particular structure, only one specific responding tissue will develop into a particular structure; the responding tissues appear to have been determined, but their differentiation may be nonspecifically stimulated by a variety of mesenchyme, or, as in the case of kidney tubules, where neural tube will have the same effect, by other tissues. There is one striking exception, reported by *Kratochwill* in 1969, where mammary gland epithelium will branch in the manner of salivary gland epithelium if it is combined with salivayr mesoderm, but even here its original biochemical properties persist, and in the appropriate hormonal situation it will secrete milk.

Induction of metanephric kidney tubules in vertebrates

The kidneys are formed in a strip of intermediate mesoderm on each side of the embryo which lies between the somites and the laternal plate. From this a series of tubules is formed, divided into three groups produced in sequence : the pronephros, mesonephros and — in birds and mammals, in which it is the only one which survives into postembryonic life—the

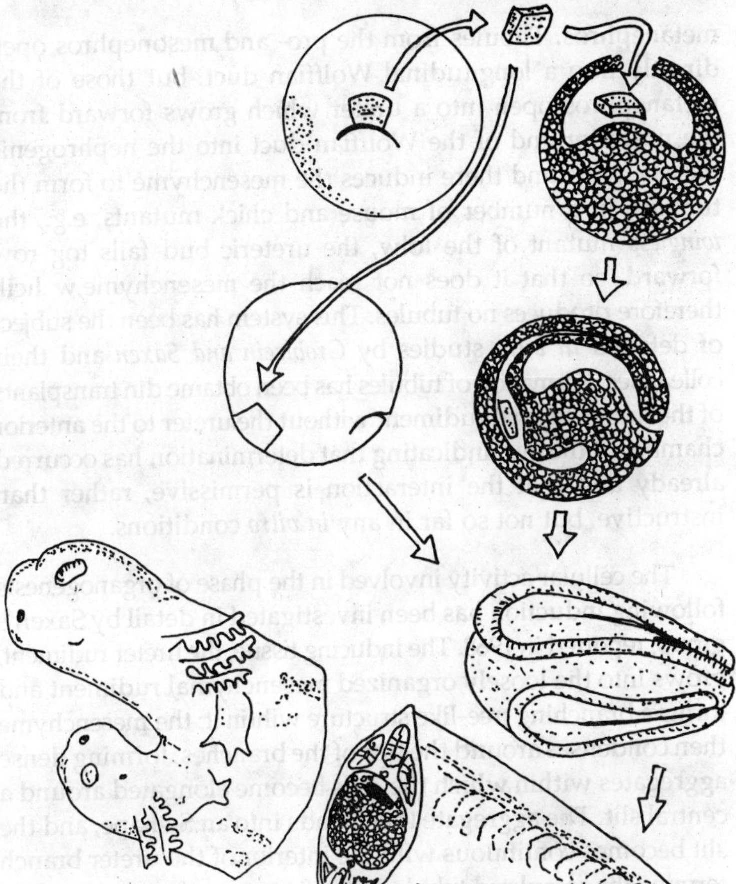

Fig. 9.16. Primary embryonic induction in amphibians.

A piece of dorsal blastopore lip material, taken from one newt early gastrula and implanted into the ventral region of another, causes the overlying host tissue to develop a second embryonic axis, producing a secondayr embryo attached to the original one. An alternative method is to insert the blastopore material into the blastocoele of a late blastula, where the gastrulation movements will carry it into a ventral position adjacent to the host tissues.

metanephros. Tubules from the pro- and mesonephros open directly into a longitudinal Wolffian duct, but those of the metanephros open into a ureter which grows forward from the posterior end of the Wolffian duct into the nephrogenic mesenchyme and there induces the mesenchyme to form the tubules. In a number of mouse and chick mutants, e.g., the *wingless* mutant of the folw, the ureteric bud fails tog row forward, so that it does not reach the mesenchyme,w hcih therefore produces no tubules. This system has been the subject of detailed *in vitro* studies by *Grobisein and Saxen* and their colleagues. Formation of tubules has been obtaine din transplants of the mesenchymal rudiment without the ureter to the anterior chamber of the eye, indicating that determination has occurred already and that the interaction is permissive, rather than instructive, but not so far in any *in vitro* conditions.

The cellular activity involved in the phase of organogenesis following induction has been investigated in detail by Saxen's group, reported in 1968. The inducing tissue, the ureter rudiment, grows into the loosely organized mesenchymal rudiment and forms a branching tree-like structure wihtin it; the mesenchyme then condenses around the tips of the branches, forming dense aggregates within which the cells become elongated around a central slit. The aggregate then bends into an S-shape, and the slit becomes continuous with the interior of the ureter branch forming the completed tubule. In the first stage, the mesenchymal cells mvoe at random, then become trapped within the condensations, within which they gradually lose their mobility. The condensations become first spherical, with radially oriented wedge-shaped cells, which suggests that one of the first results of induction is increased cell-to-cell adhesion. That there is some cell surface change is further supported by acquisition, during the aggregaiton phase, of a resistance against polyoma-virus infection. More details have beem added by the studies of *Gossens and Unsworth* in 1972.

Primary embryonic induction

In amphibians, determination of the neural ectoderm occurs

between late gastrulation and early neurulation; if presumptive neural ectoderm from a gastrula is implanted into presumptive ventral ectoderm of another gastrula at the time of dorsal-lip invagination, it develops into skin in the same way as the surrounding epidermis; but if the same operation is performed on an early neurula, a neural plate is formed in that position as well as the normal one. Grafts of dorsal-lip material and the chorda mesoderm invaginated there will induce a new neural plate wherever they are implanted beneath competent ectoderm; a piece of the dorsal lip introduced into ventral ecotderm will form a secondary blastopore and move in as it would do at its original site, or it can be simply pushed into the blastocoele, when the morphogenetic movements will force it up against the ectoderm. All the events of neurulation and somitogenesis follow,a nd then organogenesis, to give a complete secondary embryo. Most work done since the beginning of the 1950s, e.g. by *Nieuwkoop* and colleagues, and by *Toivonen* and *Saxen*, has been done not on whole embryos but using uclture techniques in which competent ectoderm is wrapped around the inducter (or combination of inducers) and—since they can be obtained in large quantities—often using as artificial inducers material from quite different sources.

Neurulation does not produce a simple tube, but a tube with regional differences, and this appears to be produced by corrdsponding regional differences in inductive activity; pieces of chordamesoderm taken from the anterior end of the archenteron roof will produce secondary embryos with good head but poor tail development, and *vice versa* for pieces from the posterior end. Unnatural inducers will mimic these effects : guinea-pig liver cells or heat-inactivated Hela cells will produce neurulation elading to forebrain structures, and guinea pig bone marrow will produce mosodermalization, i.e. it will cause ectodermal cells to produce mesodermal structures such as somites and notochord. A mixture of the two will inducte hindbrain structures and spinal cord. This led *Saxen and Toivonen,* in 1961 to propose a double-gradient model for primary induction, suggesting hat two such factos exist in the chordamesoderm, and that the type of structure induced in a particular region depends upon the balance between the two at that level. The two factors probably act sequentially, neurulization occurring first, uniformly, by a factor produced by the invaginating dorsal-lip cells, whcih is then modified in a regional way by a mesodermal factor produced by the invaginated mesoderm of the archenteron roof. Where gradients are observed in inductive interactions, the distinction between induction and positional information models becomes blurred, and may disappear with more precise knowledge of a process at the molecular level.

The nature of the inductive signal

In all inductions the interacting tissues are in close proximity, but it is not easy to establish what the precise degree of contact is in particular cases. *Grobstein,* developing the *in vitro* metanephric mesenchyme/ureter (or neural tube) system in 1960s, showed that induction could occur across a Millipore filter at least 30 um thick and with pores of 0.45 um diameter, in which no cytoplasmic process could be detected, suggesting that some transmissible factor was involved; but work by Wartiovaara and others in Saxen's laboratory showed that if Nucleopore filters (which have straight cylindrical pores of more uniform size) are used, long cytoplasmic processes can be seen extruding into them, suggesting that actual contact between the plasma membranes of inducers and responding cells can be made, and that previous transfilter results should be reinterpreted. welss suggested "contacted orientation" between molecules of adjacent cells as a mechanism of induction, but proposed an idea that was further developed by *Grobstein*—that the two surfaces would not require to be in direct apposition, but that this type of interaction could also occur if they were separated by a narrow space filled with intercellular matrix secreted by the cells and containing a population of molecules whose free mobility was restrained. Investigating the nature of this extracellular matrix and its role in developmental

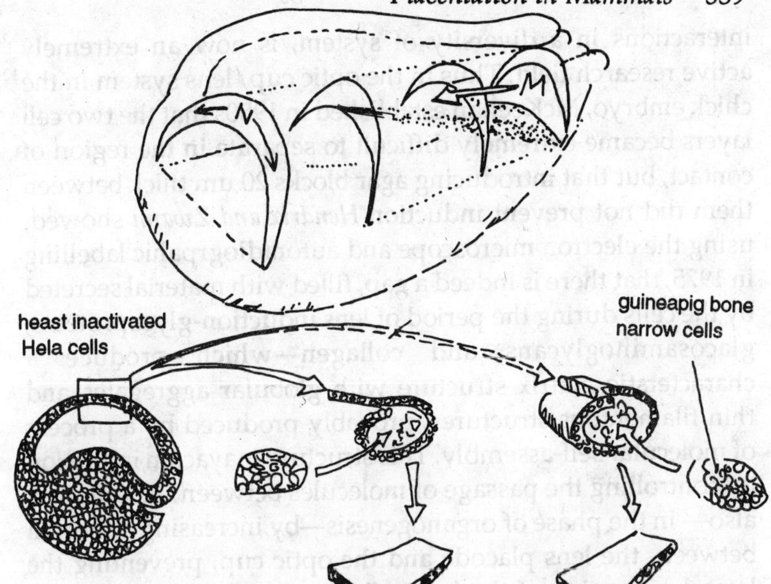

Fig. 9.17. Induction of regional differences ın the amphibian neural tube.

A. Saxen and Toivonen's double-gradient model of neural induction, superimposed on an amphibian neurula.

B. Some experimental evidence using heat-inactivated Hela cells as neural factor, and guineapig bone marrow cells as mesodermal factor, enclosing them in rolls of presumptive ectoderm from an early gastrula. The first induces fore-brain structures, the second, spinal cord.

interactions in a diversity of system, is now an extremely active research field. Thus in the optic cup/lens system in the chick embryo, McKeehan established in 1950s that the two cell layers became extremely difficult to separate in the region of contact, but that introducing agar blocks 20 um thick between them did not prevent induction *Hendrix and Zwann* showed, using the electron microscope and autoradiogrpahic labelling in 1975, that there is indeed a gap, filled with material secreted by the cells during the period of lens induction-glycoproteins, glacosaminoglycans and collagen—which produce a characterstic matrix structure with globular aggregates and thin filamentous structures, probably produced by a process of molecular self-assembly. This structure mayact in induction by controlling the passage of molecules between the cells and also— in the phase of organogenesis—by increasing adhesion between the lens placode and the optic cup, preventing the lateral expansion of the lens cell population and causing by simple cell crowding many of the characterstic cell shape changes which produce the lens vesicle.

Since work in the 1930s showed that in amphibians pieces of heat-killed mesoderm induced neurulation, and therefore that some chemical agent was involved, attempts to isolate and characterize diffusible factors of induction have never ceased. One possibility is that large informaiton-carrying macromolecules might be transmitted fromc ell to cell; in the 1940s *Brachet* produced some evidence which suggested that RNA-rich particles were so transmitted in neural induction, and proposed that the process and comparable to virus intection. Proteins have also been proposed as agents : *Yamada and Takata* in 1961 extracted a mesoderm-inducing protein fraction from the guinea-pig bone marrow used as an artificial inducer int he experiments described above, and in 1968 *Tiedemann* isolated a protein with similar properties from 9-day chick embryos. But all of these results are open to alternative interpretaitons and, in the light of current concepts of determination, and of the nature of cell communication in early embryos, it is considered more likely that no very complex informational transfer takes place. There is much interest in the investigations by *Barth and Barch*, e.g., in 1972, which indicate that, on the contrary, simple organic cations (Na+, Ca2+ and Li+) induce differentiation of neurones from presumptive ectoderm in anurans, and at a later stage of ectodermal competence the same ions induce development of these neurones into pigment cells. These ions are clearly acting by operating switch mechanisms within the responding cells, and it may well be that they function in this way in normal development, either directly or indirectly, In 1974, *McMahon* proposed a hypothesis for induction which incorporates this concept and links inducers with other chemical messengers, both in the embryo and the adult. This hypothesis suggests that molecules which function as neurotransmitters (e.g. serotonin, acetycholine) in physiological activity, are secreted by the induces cells and act at the surface of responding cells by controlling their intracellular concentration os cyclonucleotides cyclic AMP (adenosine monophosphate) and cyclic GMP (guanosine monophosphate) (the so-called second-messengers of nonsteroid hormones) whose effects and metabolism are intricately interwoven with the ionic conditions within and outside of the cell, which will act upon the gene regulatory mechanism to lead development into particular developmental pathways. The role of neurotransmitters aas information carriers in eukaryotes may therefore be much wider than previously supposed, beginnin in evolution as intracellular messengers, then serving as intercellular messengers for the relatively slow communication of developmental information, and finally evolving as the agent of rapid intercellular neural communication in adults.

Ancient evolutionary limb patterns, which have led to more complex patterns through the superimposition of positional information systems as a sort of fine-tuning. In the *talpid* mutant, and in limbs produced from reaggregated mesenchyme cells, the fine-tuning appears to have broken donw, resulting in a limb pattern which is more like that of the crossopterygian fish fins from which the tetrapod limb is thought to have evolved. How far this fine-tuning is carried, and how far development depends upon each cell or small group of cells being assigned a particular state in the course of determination, on which its future developmental activities depend, is an important problem which arises in its most striking form in the development of the nervous system.